# Infrared and Raman Spectroscopy of Biological Materials

# PRACTICAL SPECTROSCOPY
## A SERIES

*ADDITIONAL VOLUMES IN PREPARATION*

# Infrared and Raman Spectroscopy of Biological Materials

edited by

## Hans-Ulrich Gremlich

**Novartis Pharma AG**
**Basel, Switzerland**

## Bing Yan

**ChemRx Advanced Technologies, Inc.**
**South San Francisco, California**

MARCEL DEKKER, INC.                    NEW YORK · BASEL

**ISBN: 0-8247-0409-6**

This book is printed on acid-free paper.

**Headquarters**
Marcel Dekker, Inc.
270 Madison Avenue, New York, NY 10016
tel: 212-696-9000; fax: 212-685-4540

**Eastern Hemisphere Distribution**
Marcel Dekker AG
Hutgasse 4, Postfach 812, CH-4001 Basel, Switzerland
tel: 41-61-261-8482; fax: 41-61-261-8896

**World Wide Web**
http://www.dekker.com

The publisher offers discounts on this book when ordered in bulk quantities. For more information, write to Special Sales/Professional Marketing at the headquarters address above.

Current printing (last digit):
10  9  8  7  6  5  4  3  2  1

**PRINTED IN THE UNITED STATES OF AMERICA**

# Preface

Rapid advances in the biological sciences and medicine have led to an increase in the demand for chemical and structural information on biological materials. Due to its unique fingerprinting capability, vibrational spectroscopy plays a significant role in providing structural and mechanistic information.

Vibrational spectroscopy is currently undergoing a renaissance triggered by many new developments in infrared and Raman instrumentation, such as lasers, array detectors, step-scan, imaging, vibrational circular dichroism, Raman optical activity, time-resolved and surface-enhanced techniques. In addition, as with many other disciplines, the rapid development of vibrational spectroscopy has benefited from the astonishing development in computer and computational methodologies, e.g., artificial neural networks.

These developments are the basis for novel applications of vibrational spectroscopy in biology and medicine, making accessible the specific molecular investigations of complex molecules under difficult experimental conditions. Therefore, an increasing number of researchers in biophysics, biochemistry, biomedicine, and biomaterials are using vibrational spectroscopy to address key questions at the molecular level.

This book provides a thorough introduction to the exciting new methodologies as well as to their biological applications for future use in both academic and industrial environments. On the other hand, this volume will serve as an up-to-date reference source for those who are already familiar with this field.

The success of this volume is a testament to the contributors, who, writing about techniques and developments in their own laboratories and specific research areas, produced chapters of very high quality. This volume is conceived as a comprehensive overview of vibrational spectroscopy for molecular spectroscopists interested in biological applications as well as for researchers in biological areas interested in the use of molecular spectroscopy for life-sciences problems.

Hans-Ulrich Gremlich
Bing Yan

# Contents

# Contributors

**Dieter Baurecht**  Institute of Physical Chemistry, University of Vienna, Vienna, Austria

**Robert Callender**  Department of Biochemistry, Albert Einstein College of Medicine, Bronx, New York

**Elizabeth A. Carter**  School of Chemistry, University of Sydney, Camperdown, Australia

**Hua Deng**  Department of Chemistry, Albert Einstein College of Medicine, Bronx, New York

**Howell G. M. Edwards**  Department of Chemical and Forensic Sciences, University of Bradford, Bradford, West Yorkshire, England

**Teresa B. Freedman**  Department of Chemistry, Center for Science and Technology, Syracuse University, Syracuse, New York

**Urs P. Fringeli**  Institute of Physical Chemistry, University of Vienna, Vienna, Austria

**Klaus Gerwert**   Institute of Biophysics, Ruhr University Bochum, Bochum, Germany

**Hans H. Günthard**   Institute of Physical Chemistry, Swiss Federal Institute of Technology, Zurich, Switzerland

**H. Michael Heise**   Institut für Spektrochemie und Angewandte Spektroskopie an der Universitat Dortmund, Dortmund, Germany

**Timothy A. Keiderling**   Department of Chemistry, University of Illinois at Chicago, Chicago, Illinois

**Ira W. Levin**   Division of Intramural Research, Section on Molecular Biophysics, Laboratory of Chemical Physics, National Institute of Diabetes and Digestive and Kidney Diseases, National Institutes of Health, Bethesda, Maryland

**Steven M. LeVine**   Department of Molecular Integrative Physiology, University of Kansas Medical Center, Kansas City, Kansas

**E. Neil Lewis**   Laboratory of Chemical Physics, National Institute of Diabetes and Digestive and Kidney Diseases, National Institutes of Health, Bethesda, Maryland

**Henry H. Mantsch**   Institute for Biodiagnostics, Division of Spectroscopy, National Research Council of Canada, Winnipeg, Canada

**Koichi Murayama**   Faculty of Agriculture, Kobe University, Kobe, Japan

**Laurence A. Nafie**   Department of Chemistry, Center for Science and Technology, Syracuse University, Syracuse, New York

**Dieter Naumann**   Biophysical Structure Analysis Laboratory, Robert Koch Institute, Berlin, Germany

**Yukihiro Ozaki**   School of Science, Kwansei Gakuin University, Nishinomiya, Japan

**Michael D. Schaeberle**   Laboratory of Chemical Physics, National Institute of Diabetes and Digestive and Kidney Diseases, National Institutes of Health, Bethesda, Maryland

**Jüergen Schmitt**   Spectroscopy Group, FB VI, Department of Hydrology, University of Trier, Trier Germany

**Thomas Udelhoven**   Digital Imaging, FB VI, Department of Remote Sensing, University of Trier, Trier Germany

**David L. Wetzel**   Microbeam Molecular Spectroscopy Laboratory, Kansas State University, Manhattan, Kansas

# Infrared and Raman Spectroscopy of Biological Materials

# 1

# Historical Survey of Infrared and Raman Spectroscopy of Biological Materials

**Henry H. Mantsch**
*National Research Council of Canada, Winnipeg, Canada*

## 1 PREAMBLE

With the information overload we experience today, it is imperative to pause occasionally in order to take stock of the progress in one's chosen field of research and also to identify new trends and emerging areas. Yet, in order to look ahead, it is often useful to scrutinize the past. Clearly, those of us active in the natural sciences are much less aware of the historical value of our professional records than are our colleagues in the arts and humanities, where the preservation of personal records has always been recognized as necessary. It is true that most of our scientific contributions are documented explicitly in journal publications; however, these reveal little about the circumstances under which the work was carried out or about the personal motivations and the historical background that led to the final product as we know it today.

Moreover, the historical perspective also introduces a human element into the austere science. Long gone are the days when scientists like Justus Liebig began their scientific reports with "Today is a lovely, sunny April day." Editorial pressures for conciseness, along with stiff publication charges, have forced most scientific papers into a clipped telegraphic style. If one is curious about the human interests behind the research, one has to rely on personal contacts with the people themselves (and, although the Internet provides almost unlimited access to a wealth of scientific data, it is these personal contacts that most often enable one's scientific and personal development). As one involved in biospectroscopy for the last 40 years, I shall attempt to take such a retrospective look at the marriage between vibrational spectroscopy and biological molecules from a more personal perspective.

## 2 WHY BIOLOGICAL MOLECULES?

To start with, we have to ask just what biological molecules are. This question was elegantly answered by Albert Lehninger in his textbook on biochemistry (1). Indeed, we have to go back some 20 billion years, when the universe arose with a cataclysmic explosion that hurled hot, energy-rich subatomic particles into space. Gradually, as the universe cooled, the hundred or so chemical elements appeared. Every single atom in the universe, including those of all living organisms, was thus born out of the "big bang," and all living things are literally made of stardust. However, the molecules from which living organisms are constructed are highly specific and occur on earth only as products of biological activity. These building blocks, called *biomolecules*, were selected during the course of biological evolution for their fitness to perform specific functions. It is therefore quite legitimate to ask what the purpose or the specific function of a given biomolecule in a living organism may be. When these biomolecules are isolated and examined individually, we see that they conform to all the physical and chemical laws that describe the behavior of inanimate matter. Yet living things possess unique properties not shown by collections of inanimate molecules. It is therefore understandable that these biomolecules present a distinct challenge to the biospectroscopist.

## 3 WHY VIBRATIONAL SPECTROSCOPY?

In view of their diversity and complexity, it is not surprising that an array of physical techniques has been used to study biological molecules, including many types of spectroscopies. Still, vibrational (infrared and Raman) spectroscopy offers certain advantages over other spectroscopic techniques. First, the experimental accessibility to a large number of infrared (IR) and Raman active transitions that originate from specific functional groups/moieties provides information germane to spatially localized regions within these biomolecules. Second, vibrational spectroscopic data are obtained in a noninvasive manner from intrinsic "molecular probes"; that is, neither infrared nor Raman spectroscopy requires extrinsic probe molecules such as spin labels or fluorescent probes. Third, there is no limit to the size, allowing the study of high-molecular-weight biopolymers such as DNA. Fourth, the molecular events that are monitored by the vibrational spectroscopic experiment are atomic motions on the picosecond time scale, and thus provide an instantaneous "snapshot" of all molecular conformations. The line broadening that occurs, for instance, in magnetic resonance spectra, due to relaxation phenomena leading to time-averaged structures, is absent in the vibrational spectra. Without denying or overlooking the merits of other spectroscopic modalities, it is clear that by now vibrational spectroscopy has gained a firm place within the spectroscopic arsenal used to investigate biological materials.

## 4  HOW IT ALL BEGAN: LAYING THE FOUNDATIONS

The roots of infrared spectroscopy go back almost 200 years, if we consider that William Herschel discovered the infrared region of the electromagnetic spectrum in 1800. Yet, except for some early developments, notably those by William Coblentz at the beginning of this century (2), infrared spectroscopy lay dormant for a long time. Only after Gerhard Herzberg published the molecular spectroscopy bible, "Infrared and Raman Spectra of Polyatomic Molecules," in 1945 (3) did vibrational spectroscopy become an autonomous research tool. Applied infrared spectroscopy gathered momentum during World War II, particularly in the United Kingdom and the United States, but under conditions of wartime restrictions on publication. Wartime applications were particularly concerned with petroleum fractionation and the fingerprinting of hydrocarbon blends in aviation gasoline. The knowledge of the infrared spectra of various types of straight, branched, and cyclic hydrocarbons helped the Allied Forces trace the origin of the gasoline used by the German Luftwaffe. After the war, when commercial infrared spectrometers became more readily available, these instruments were continuously at work accumulating a vast library of infrared spectra, primarily of organic compounds. A careful scrutiny of all these spectra led to the creation of the very useful concept of *group frequency* and culminated in the publication of early reference works such as Bellamy's "Infrared Spectra of Complex Molecules" in 1954 (4) and Norman Jones' and Camille Sandorfy's "The Application of Infrared and Raman Spectrometry to the Elucidation of Molecular Structure" in 1956 (5). Werner Brügel published the first IR reference work in German in 1954 (6). Early infrared spectra of biological materials were published in the 1950s, however, many of these studies were actually concerned with steroids (7), natural products (8), or complex heterocyclic structures (9), which at that time were still considered the domain of organic chemistry.

In terms of history, Raman spectroscopy is certainly junior to infrared spectroscopy. Although the basic theory of the Raman effect was developed earlier, the birth of Raman spectroscopy is generally associated with a series of publications by Raman in 1928, that is, 128 years after the discovery of infrared. Only a few years later, in 1936 (10), John Edsall recognized the potential of Raman spectroscopy for the study of aqueous solutions of biological molecules. Yet the experimental difficulties encountered with the use of mercury arc excitation, long photographic exposure time, and the requirement for large amounts of optically pure compounds rendered such applications impracticable. Indeed, few Raman spectrometers were to be found in analytical laboratories prior to the introduction of laser sources in the mid-1960s. That changed rapidly as instrument manufacturers were prompt to produce commercial spectrometers with laser sources. The early models used weak but stable helium-neon lasers with one excitation frequency in the red. These were soon replaced by more powerful krypton and argon ion lasers, which provided a range of stable frequencies throughout the visible and could be extended into the ultraviolet by frequency-doubling devices. Both organic chemists and biochemists were quick to apply this new technique to study a range of biological molecules. A first roadmap to the strategy and tactics in the Raman spectroscopy of biomolecules was provided by Dick Lord in his Lippincott Medal Address at the 1976 FACSS Meeting in Philadelphia (11). Comprehensive reviews on Raman spectroscopy, including the momentum-gaining resonance Raman spectroscopy of biological systems, were written by Paul Carey in 1980 (12) and 1982 (13). A major advantage of Raman over infrared was the fact that water, omnipresent in biological systems, is less of a problem in Raman or resonance Raman. However, Raman spectroscopy could not bask in the sun for too long, for the introduction of laser Raman spectroscopy was soon followed by the arrival of Fourier transform infrared (FT-IR) in the late 1970s, which also was able to address some of the problems that dispersion infrared failed to cope with.

## 5 THE COURTSHIP YEARS: THE RECENT PAST

So far, organic chemistry had benefited most from the emergence of infrared and Raman spectroscopy as new analytical tools. Only after molecular biologists took charge of biochemistry did it become clear that we are dealing in fact with four primary classes of biomolecules: (a) the proteins, which include the enzymes and biologically active peptides, (b) the nucleic acids, (c) the lipids or fats, which include the biomembranes, and (d) the carbohydrates or sugars. Each of these four groupings has a distinct vibrational spectroscopic history.

The *proteins* (from *proteos*, ''first'') are not only the most numerous but also the most versatile of all biomolecules. The vibrational spectra of proteins provide information at three different levels of structural organization: (a) assessment of protein secondary structure from an analysis of the amide bands, (b) mechanistic information about protein function/structure and protein-ligand interactions at the level of individual chemical groups, and (c) information regarding the accessibility of protein domains to hydrogen–deuterium exchange.

The use of infrared spectroscopy to study the conformational structure of proteins and polypeptides dates back to the early work of Elliot and Ambrose in 1950 (14). These pioneers were the first to suggest that a correlation exists between the position of certain bands in the infrared spectra of proteins and the conformation or secondary structure of the polypeptide backbone. This was later confirmed by detailed theoretical calculations of the amide normal modes by

Sam Krimm and coworkers (for a review, see Ref. 15). For the next few decades, however, the applications of IR spectroscopy remained rather limited in scope. Susi et al. (16) further pursued this application in the mid-1960s; at that time, however, there were still serious obstacles that prevented a more widespread use of infrared spectroscopy in the study of biological systems: (a) the poor sensitivity of traditional dispersive instruments, (b) problems arising from the very strong infrared absorption of water, and (c) the intrinsic broadness of individual infrared bands with instrumentally unresolvable multicomponent band contours. A new era of biological infrared spectroscopy began in the 1970s, when Fourier transform infrared instrumentation became widely available, offering not only increased accuracy and reproducibility, but also a dramatic improvement in the achievable signal-to-noise ratios and the ability to perform measurements with strongly absorbing (low-throughput) samples. The last was particularly important for aqueous solutions of proteins, because it permitted the detection of weak protein bands superimposed on the very strong absorption band of water.

An equally important consideration in the evolution of biological IR and Raman spectroscopy was the electronic revolution, a factor that cannot be understated. This is not to suggest that the advent of computers was an innovation in spectroscopic research; however, it made life much easier for molecular spectroscopists. Computers were now used not only for data acquisition and data storage or retrieval, but also for data analysis or interpretation. Yet in the 1960s, computers still had limited storage capacities, they were cumbersome, and programming was an agonizing operation. When the program aborted (which happened often), one had to trace it back through the algorithm step by step to locate the source of trouble. A first system, based on digital recording of infrared spectra, using punched paper tape, was developed by Norman Jones at the NRC in Ottawa in the mid-1960s (17). Needless to say, the fragile paper tapes jammed and ripped easily. Throughout the 1970s, large-sized digital computers with huge hard disks and racks of magnetic tapes became common sights in places where infrared and Raman spectroscopy was practiced. From there on, progress occurred at a frightening pace, and today data acquisition and storage have become trivial tasks and computers are used primarily for data analysis. In 1982, at a Fourier transform spectroscopy workshop in San Jovite, Quebec, Heino Susi first learned about Fourier self-deconvolution and other band-narrowing procedures (18). As soon as he returned home, he, along with Michael Byler, started applying this methodology to the analysis of the amide bands in proteins (19), which triggered an avalanche of articles on this subject (for reviews, see Refs. 20–22).

In the 1970s, a number of spectroscopists also started to use infrared as a method to measure hydrogen–deuterium exchange to probe the tightness of protein folding and to study protein conformational dynamics. Both the protein backbone amide II mode at 1550 cm$^{-1}$ (23) and the amide A mode at 3400 cm$^{-1}$ (24) were used (for a review, see Ref. 25). Another early use of infrared spectroscopy was to analyze local molecular states in oxygen-binding proteins. Infrared studies of heme complexes in native proteins were first reported by Jim Alben and Winslow Caughey in 1968 (26). In a very imaginative way they used carbon monoxide as an exchangeable ligand, since it is competitive with dioxygen for binding of Cu(I) or Fe(II) and thus serves as a good structural probe of local molecular interactions.

The original use of infrared spectroscopy to study *nucleic acids* is linked to the name of Masamichi Tsuboi from the University of Tokyo. He published a first paper in 1957 (27) and then a comprehensive review of the subject in 1969 (28). Initial band assignments were made by correlations between IR spectra of simple bases, nucleosides, nucleotides, and their oligomers or polymers. Many of these original assignments, which were aided by isotopic substitutions using $^{15}N$, $^{18}O$, and hydrogen–deuterium exchanges, withstood the test of time. Later they were refined and extended to complex nucleic acids by younger biospectroscopists like Eliane Taillandier and others (29). In North America, Dick Lord had established a center for bioanalytical spectroscopy at MIT in the early 1960s. His studies with Mike Falk (30) of the controlled hydration of DNA and the first vibrational spectroscopic studies of viruses with George Thomas (31) are classical early biological IR work. Later Lord reminisced about the history and personal relationships among spectroscopists in an interview article published by European Spectroscopy News that provides an illuminating window on the state of vibrational biospectroscopy at that time (32). The study of metal–DNA interactions and the effect of metal ions in general were championed by Theo Theophanides in the late 1970s (33) and have since become an active area of research.

The application of infrared spectroscopy to the analysis of *lipids* was pioneered by the Molecular Biophysics Group around Dennis Chapman at Unilever in Great Britain in the mid-1960s. Interestingly, this happened in an industrial lab and not in an academic laboratory. Early infrared spectra of phospholipids were published in the inaugural volume of *Chemistry and Physics of Lipids* in 1967 (34). For the first time, IR was used side by side with other techniques—thermal analysis, x-ray diffraction, and nuclear magnetic resonance (NMR)—for investigating the physical properties of lipids, in particular their thermotropic and lyotropic mesomorphism. The first work on Raman spectroscopy of lipids came in 1971 from the group of Warner Peticolas in Eugene, Oregon (35); it dealt with the effect of cholesterol on the conformational changes in lecithin multilayers. Soon other Raman publications followed, both on pure lipids and on lipid model membranes (36–39). In fact, the main impetus for many of the vibrational spectroscopic studies of lipids that started in the late 1970s came from the desire to understand the structure and function of biological membranes. During this time, the group around Ira Levin at the National Institutes of Health in Bethesda was extremely active (for reviews, see Refs. 40 and 41), as was our Molecular Spectroscopy Group at the National Research Council of Canada in Ottawa (for a review, see Ref. 42). An early review on the application of laser Raman and infrared spectroscopy to the analysis of membrane lipids was published in the Membrane section of Biochimica Biophysica Acta by Wallach et al. in 1979 (43); it covered the literature up to 1977 (123 references). All IR studies described in this review were still performed with conventional, scanning (i.e., dispersive) instruments. It was not until 1978 (44) that the first publication appeared in the literature that used FT-IR instrumentation in a study of the gel-to–liquid crystal phase transition of 1,2-dipalmitoyl-sn-glycero-3-phosphocholine (DPPC), the workhorse of so many lipid spectroscopists. Raman studies, on the other hand, were already performed with laser Raman instruments, though the spectra were still plagued by fluorescence. A milestone in the 1980s was the introduction of near-infrared excitation in combination with interferometric recording of near-IR FT-Raman spectra, which finally took care of the perennial fluorescence problem associated with impurities found in most biological samples.

Much of the present knowledge on membrane structure has come from infrared and Raman studies of synthetic or naturally occurring lipids. In the early 1960s, Bob Snyder at the U.S. Shell Research Station at Emeryville, California, performed the first detailed vibrational analyses of n-alkanes (45). These seminal publications laid the foundation for all the spectroscopists who later used his experimental data to interpret the vibrational spectra of acyl and alkyl chains in lipid bilayers and membranes. Interestingly, later, after he quit Shell and had moved to Berkeley, Bob also became interested in biological systems.

An important milestone in the history of the analysis of lipid spectra was the elaboration of novel data-processing techniques in the early 1980s, in particular the development of the band-narrowing procedure of Fourier self-deconvolution. When Jyrki Kauppinen, a high-resolution spectroscopist from Finland, joined the Molecular Spectroscopy Group at the National Research Council in Ottawa in 1980, he was dismayed by the broad, unattractively looking infrared bands he was faced with when he had to analyze the infrared spectra of aqueous lipid suspensions. We told him that if he did not like them, he should do something about it, and so he did (46). We had many discussions before settling on the term *Fourier self-deconvolution*. Indeed, today everybody uses this methodology as though it was always part of the software that resides in the spectrometer computer.

The class of biomolecules that is least well investigated by vibrational spectroscopy is that of *carbohydrates*. This has to do with the fact that unlike the nucleic acids, the proteins, or the lipids, these biomolecules lack prominent polar, infrared-active functional groups with heteroatoms and multiple bonds. The predominance of C—C and C—O bonds in carbohydrates and the similar mechanical properties of these bonds give rise to broad, unresolved infrared absorption bands. The specificity of carbohydrates arises from the geometry of the many O—H groups and the configuration of the C—O, C—C, and C—H bonds in the skeletal base configuration, which makes Raman a more useful tool (see Chapter 8 in Ref. 13). While crystalline mono- or oligosaccharides give nice infrared and Raman spectra (Ref. 47; see also Chapter 8 in Ref. 25), this is not the case for the vibrational spectra of complex polysaccharides, where one often needs pattern recognition algorithms to differentiate among individual spectra. Using such pattern recognition algorithms, Dieter Naumann introduced a new modality for the characterization and identification of pathogenic bacteria, based on the vibrational spectra of the cell walls (48). Vibrational spectroscopy has also been successful when applied to the analysis of complex glycolipids (49).

## 6 FROM THE PAST TO THE PRESENT

It is now necessary to define the boundary between past and present. In the context of this presentation, I shall take the liberty of placing this in the mid-1980s. I do this because by then the marriage between vibrational spectroscopy and biological molecules was consummated and had turned into a mature relationship that was blooming. The widespread utilization of interferometric FT-IR instrumentation and the explosive expansion of laser Raman and resonance Raman spectroscopy opened the vast domain of the aqueous world of biomolecules.

As more and more vibrational spectroscopists had turned their attention to this world and started working with complicated biological systems, the need became apparent to find a common emporium and to create a forum where those who had embarked upon this road could share their experience and their frustrations. Perhaps the first opportunity to achieve this goal was provided by a NATO summer school, organized by two Canadians, Camille Sandorfy and Theo Theophanides, in the summer of 1983 in southern Italy. These popular summer schools were sponsored by the NATO Science Committee and aimed at the dissemination of advanced scientific and technological knowledge. As was the tradition, the proceedings were promptly published (50). Even during the height of the Cold War, these summer schools were often attended by scientists from Eastern Europe. The meetings were always held in affable surroundings, and the 1983 Advanced Study Institute on the Spectroscopy of Biological Molecules, in Acquafredda di Maratea, was no exception. Many of the early and present-day vibrational spectroscopists engaged in the investigation of biological molecules were there (and had a good time!). There the idea was born to meet again in two years. With no clear sponsor in sight, one of the participants, Lucien Bernard, the powerful rector of the University of Reims, offered to provide a venue for the next meeting. Along with his students Alain Alix and Michel Manfait, he organized a meeting in Reims in September of 1985. The famous cathedral in Reims and the free-flowing champagne contributed to a very successful meeting, which became known as the First Conference on the Spectroscopy of Biological Molecules (51). This meeting was followed by regular biennial conferences in

Freiburg, Germany, 1987 (52); Rimini, Italy, 1989 (53); York, England, 1991 (54); Loutraki, Greece, 1993 (55); Lille, France, 1995 (56); and El Escorial, Spain, 1997 (57), with the 8th meeting being planned for 1999 in Enschede, Netherlands. Though the name *European* stuck, it really is an international conference.

The published proceedings of these meetings bear witness to the evolution of the field and to the progress made over the last 15 years in vibrational biospectroscopy. A cursory inspection of these and other documents reveals that today at least 50% of all vibrational spectroscopic studies of biological molecules deal with proteins. These studies cover the whole range from simple peptides to highly complex enzymes. About 20% each, vibrational biospectroscopy is dedicated to nucleic acids and lipids. Again, these studies range from simple nucleotides to complex viruses and from basic phospholipids to complicated membrane structures. Furthermore, as in nature, the four types of biomolecules often interact with each other; a number of vibrational spectroscopic studies deal with such complex structures as lipoproteins, nucleoproteins, and glycoproteins. Less than 10% of all vibrational biospectroscopic studies are dedicated to carbohydrates and other minor biological molecules, such as the vitamins and the steroids.

It is interesting to compare the history of vibrational biospectroscopy with that of the much younger bio-NMR spectroscopy. In 1953, at a meeting of the American Chemical Society, papers presented by Herb Gutowsky signaled the advent of a new instrumental technique for chemical analysis—nuclear magnetic resonance (NMR) spectroscopy (58). Some time elapsed before commercial NMR spectrometers became available in sufficient numbers to have an impact on the analytical laboratory, but a stream of papers on proton NMR on organic structure analysis appeared in the 1960s, and these snowballed in the 1970s, particularly with the advent of multinuclear FT-NMR. This gave chemists and biochemists the "chemical shift" as a measure of molecular structure, just as vibrational spectroscopy had given them the "group frequency" about 20 years earlier. Although both NMR and vibrational spectra provide information at the submolecular level, by the mid-1980s more papers were published that used MR spectroscopy to investigate biological molecules than IR and Raman combined (meanwhile, the N in NMR has been dropped because the term nuclear is compromised). In spite of the fact that MR instruments are quite expensive compared to IR or Raman spectrometers, the technique seems to be more glamorous and certainly exerts a greater attraction on graduate students and their teachers. Actually, as each technique reveals its particular advantages and disadvantages (see Sec. 3), MR and vibrational spectrometry complement each other in the investigation of biological materials. An excellent and succinct account of the early history of magnetic resonance spectroscopy was published by Ted Becker in 1996 (59). About 10 years earlier, in 1985, Norman Jones similarly produced an excellent account of the early history of vibrational spectroscopy (60).

## 7  THE LEAP FROM BIOLOGY TO MEDICINE

One of the more recent developments in vibrational biospectroscopy is that of
medical applications. While the study of isolated biomolecules is interesting in
itself, the mystery of the human body has always exerted a special attraction to
many biospectroscopists. As far as infrared spectroscopy is concerned, its history
goes back to the early attempts by Elkan Blout and Robert Mellors in 1949 (61)
and Donald Woernley in 1952 (62), who first reported that infrared spectra of
human and animal tissues provide information concerning the molecular structure
of the tissue. Unfortunately, the instrumentation at that time did not allow a meaningful study of such complex systems, and very little was done during the next
few decades. Furthermore, all the early attempts to measure Raman spectra of
tissues were severely limited by the highly fluorescent nature of biological samples, as well as the long integration time and the required high-power density.
Meanwhile, the availability of inexpensive diode lasers and charge-coupled device (CCD) cameras sensitive to the near infrared has made it possible to attain
high-quality Raman spectra of tissue.
      Another attraction of vibrational biospectroscopy is to noninvasively deter-
mine metabolites in tissues and biological fluids such as blood. The Holy Grail
is to monitor glucose levels noninvasively, but this goal still remains rather elu-
sive. Along these lines, Nils Kaiser registered a first German patent in 1958 (63),
yet progress with tissues and biological fluids was slow, and these early pioneers
(or dreamers) were a bit ahead of their time. However, in the late 1980s and
1990s many new developments occurred that profoundly altered the arena of
biomedical infrared and Raman spectroscopy (for recent reviews, see Refs. 64
and 65). As yet, however, it is difficult to evaluate their impact on a long-term
basis with the objectivity that hindsight brings to earlier historic events; they are
simply too close in time to be assessed from a historical perspective.

## 8  WHERE TO FIND IT?

A dilemma that many vibrational biospectroscopists face today is that of where
to publish, that is, finding the best avenue to communicate their results. Moreover,
as more vibrational spectroscopists now work at the interface between spectroscopy and biology or medicine, the question also arises as to where to find publications on a given subject. Some of the major publishing houses have started to
dedicate multiauthor books to this interdisciplinary subject, for instance, volumes
13 (66), 20, 21 (67,68), and 25 (69) within Wiley's series on Advances in Spectroscopy, edited by Clark and Hester, or the stand-alone book *Infrared Spectroscopy of Biomolecules*, edited by Mantsch and Chapman (70). A critique of all
of these books is that individual accounts reflect the area of expertise of the
contributing authors, because they address the specialist rather than nonspecial-

ists. On the other hand, the present volume, although written by specialists, is intended as a review of vibrational spectroscopic applications to biological materials for all interested in biological molecules.

Journals dedicated to spectroscopic studies of biological molecules include the *Biophysical Journal, Biochemistry, Biophysical Chemistry*, and *Biospectroscopy*. Recently, journals such as *Spectrochimica Acta* and the *Journal of Physical Chemistry* also have set aside special sections for biospectroscopy. However, many aspects of vibrational biospectroscopy are now published in journals that specialize in a particular topic, such as photochemistry, time-resolved spectroscopy, biopolymers, biomembranes, and protein science, or in various medical journals. Fortunately, powerful search engines that scan the literature make it possible to dig out publications on a particular topic or subject that are spread over dozens of journals. Finally, it should be mentioned that a number of interesting contributions to vibrational biospectroscopy are being reported at various meetings, but many excellent posters or oral presentations linger only in the minds of those present at the meeting and are lost for posterity.

## 9   WHERE DO WE GO FROM HERE?

Clearly, infrared and Raman spectrometry now provide biospectroscopists with a powerful tool to investigate the aqueous world of biomolecules. A tendency has developed for extreme specialization that affects all of analytical biospectroscopy. These specialized areas include spectroscopy under extreme conditions, such as very high pressure, low temperature, high salt concentrations, or incubation at temperatures around zero, to probe the metastability of lipid assemblies. In addition, the use of isotope-edited spectra keeps gaining prominence, though some of the earlier studies made good use of deuterium and carbon-13 labeling. As we look at the future, my task becomes increasingly more difficult. Historians are supposed to look backwards, not forwards; but if I have to speculate about the future, I would refer to Alan Kay's statement that ''the best way to predict the future is to invent it.'' As the founding father of Silicon Valley in California, he has some credibility. My very personal dream about the future is to see Raman and infrared imaging (IRI) play catch up to magnetic resonance imaging (MRI). While IRI could/should not compete with MRI, it may create its own niche in functional imaging, for instance, of skin, and thus provide an in vivo window into metabolism (71).

## 10   APOLOGIA

In a short article such as this, it is not possible to acknowledge all the many biospectroscopists who have made important contributions to our subject. I have had to be selective, and no doubt this selectivity reflects my own bias and preoccupation with certain aspects of the field. I apologize to my colleagues in the vibrational spectroscopic community for my sins and omissions.

## REFERENCES

1.  AL Lehninger. Biochemistry. New York: Worth, 1975.
2.  WW Coblentz. From the Life of a Researcher. New York: Philosophical Library, 1951.
3.  G Herzberg. Infrared and Raman Spectra of Polyatomic Molecules. New York: Van Nostrand Reinhold, 1945.
4.  LJ Bellamy. The Infrared Spectra of Complex Molecules. London: Methuen, 1954.
5.  RN Jones, C Sandorfy. The application of infrared and Raman spectrometry to the elucidation of molecular structure. In: W West, ed. Chemical Applications of Spectroscopy. New York: Interscience, 1956.
6.  W Brügel. Einführung in die Ultrarotspektroskopie. Darmstadt, Germany: Steinkopff Verlag, 1954.
7.  K Dobriner, ER Katzenellenbogen, RN Jones. Infrared Spectra of Steroids: An Atlas. Vols. 1 & 2 (760 spectra). New York: Interscience, 1953.
8.  ARH Cole. Infrared spectra of natural products. In: L Zechmeister, ed. Fortschritte der Chemie organischer Naturstoffe. Vienna: Springer-Verlag, 1956, pp. 1–69.
9.  AR Katritzky, P Ambler. Infrared spectra. In: AR Katritzky ed. Physical Methods in Heterocyclic Chemistry. New York: Academic Press, 1963, pp. 161–360.
10. JT Edsall. J Chem Phys 4:1–8, 1936.
11. RC Lord. Appl Spectrosc 31:187–194, 1977.
12. RP Carey, VR Salares. Raman and resonance Raman studies of biological systems. In: RJH Clark, RE Hester, eds. Advances in Infrared and Raman Spectroscopy, vol. 7. Chichester, England: Wiley 1980, pp. 1–58.
13. PR Carey. Biochemical Applications of Raman and Resonance Raman Spectroscopies. New York: Academic Press, 1982.
14. A Elliot, E Ambrose. Nature 165:921–922, 1950.
15. S Krimm, J Bandekar. Adv Protein Chem 38:181–364, 1986.
16. H Susi, SN Timasheff, L Stevens. J Biol Chem 242:5460–5473, 1967.
17. RN Jones. Pure Appl Chem 18:303–321, 1969.
18. T Theophanides, ed. FT-IR Spectroscopy: Industrial Chemical and Biochemical Applications. Dordrecht, The Netherlands: Reidel, 1984.
19. H Susi, DM Byler. Biophys Biochem Res Com 115:391–397, 1983.
20. DM Byler, H Susi. Biopolymers 25:469–487, 1986.
21. WK Surewicz, HH Mantsch. Biochim Biophys Acta 952:115–130, 1988.
22. M Jackson, HH Mantsch. Crit Rev Biochem Mol Biol 30:95–120, 1995.
23. L Pershina, A Hvidt. Eur J Biochem 48:339–344, 1974.
24. EV Brashnikov, YN Chirgadze. J Mol Biol 122:127–135, 1978.
25. FS Parker, ed. Applications of Infrared, Raman and Resonance Raman Spectroscopy in Biochemistry. New York: Plenum Press, 1983.
26. JO Alben, WS Caughey. Biochemistry 7:175–183, 1968.
27. M Tsuboi. J Am Chem Soc 79:1351–1354, 1957.

28. M Tsuboi. Appl Spectrosc Rev 3:45–90, 1969.
29. E Taillandier, J Liquier, JA Taboury. Infrared spectral studies on DNA conformations. In: RJH Clark, RE Hester, eds. Advances in Infrared and Raman Spectroscopy. vol. 12. New York: Wiley, 1985, pp. 65–114.
30. M Falk, KA Hartman, RC Lord. J Am Chem Soc 84:3843–3846, 1962.
31. RC Lord, GJ Thomas Jr. Biochim Biophys Acta 142:1–11, 1967.
32. RC Lord. Europ Spectrosc News 56:10–15, 1984.
33. T Theophanides. Infrared and Raman Spectroscopy of Biological Molecules. Dordrecht, The Netherlands: Reidel, 1979.
34. D Chapman, RM Williams, BD Ladbrooke. Chem Phys Lipids 1:445–475, 1967.
35. JI Lippert, WL Peticolas. Proc Nat Acad Sci USA 68:1572–1576, 1971.
36. BJ Bulkin, N Krishnamachari. J Am Chem Soc 94:1109–1112, 1972.
37. K Larsson. Chem Phys Lipids 10:165–176, 1973.
38. RC Spiker, IW Lewin. Biochim Biophys Acta 388:361–373, 1975.
39. H Akutsu, Y Kyogoku. Chem Phys Lipids 14:113–122, 1975.
40. RC Lord, R Mendelsohn. Raman spectroscopy of membrane constituents and related molecules. In: E Grell, ed. Membrane Spectroscopy. Berlin: Springer Verlag, 1981, pp. 377–436.
41. IW Levin. Vibrational spectroscopy of membrane assemblies. In: RJH Clark, RE Hester, eds. Advances in Infrared and Raman Spectroscopy. vol. 11. Chichester, England: Wiley Heyden, 1984, pp. 1–48.
42. HL Casal, HH Mantsch. Biochim Biophys Acta 779:381–401, 1984.
43. DFH Wallach, SP Verma, J Fookson. Biochim Biophys Acta 559:153–208, 1977.
44. DG Cameron, HH Mantsch. Biochem Biophys Res Commun 83:886–892, 1978.
45. RG Snyder, JH Schachschneider. Spectrochimica Acta 19:85–168, 1963.
46. JK Kauppinen, DG Moffatt, HH Mantsch, DG Cameron. Appl Spectrosc 35:271–277, 1981.
47. HH Mantsch, RN McElhaney. Chem Phys Lipids 57:213–226, 1991.
48. D Naumann, D Helm, H Labischinski, P Griesbrecht. The characterization of microorganisms by FT-IR spectroscopy. In: WH Nelson, ed. Modern Techniques for Rapid Microbiological Analysis. New York: VCH, 1991, pp. 43–96.
49. U Seidel, K Brandenburg. Supramolecular structure of lipopolysaccharide and lipid A. In: DC Morrison, JL Ryan, eds. Bacterial Endotoxic Lipopolysaccharides. vol. 1. Boca Raton, FL: CRC Press, 1992, pp. 225–250.
50. C Sandorfy, Th Theophanides, eds. Spectroscopy of Biological Molecules. Dordrecht, The Netherlands: Reidel, 1983.
51. AJP Alix, L Bernard, M Manfait, eds. Spectroscopy of biological molecules. Proceedings of the First European Conference on the Spectroscopy of Biological Molecules, Reims, France, 1985. Chichester, England: Wiley, 1985.
52. ED Schmidt, FW Schneider, F Siebert, eds. Spectroscopy of biological molecules: new Advances. Proceedings of the Second European Conference on the Spectroscopy of Biological Molecules, Freiburg, Germany, 1987. Chichester, England: Wiley, 1988.
53. A Bertoluzza, D Fagnano, P Monti, eds. Spectroscopy of biological molecules. Proceedings of the Third European Conference on the Spectroscopy of Biological Molecules, Bologna, Italy, 1989. Bologna: Societa Editrice Esculapio, 1989.

54. RE Hester, RB Girling, eds. Spectroscopy of biological molecules. Proceedings of the Fourth European Conference on the Spectroscopy of Biological Molecules, York, England, 1991. Cambridge, England: Royal Society of Chemistry, 1991.

55. Th Theophanides, J Anastassopoulou, N Fotopoulos, eds. Fifth International Conference on the Spectroscopy of Biological Molecules, Loutraki, Greece, 1993. Dordrecht, The Netherlands: Kluwer Academic, 1993.

56. JC Merlin, S Turrell, JP Huvenne, eds. Spectroscopy of biological molecules. Proceedings of the Sixth European Conference on the Spectroscopy of Biological Molecules, Lille, France, 1995. Chichester, England: Wiley, 1995.

57. P Carmona, R Navarro, A Hernanz, eds. Spectroscopy of biological molecules: modern trends. Proceedings of the Seventh European Conference on the Spectroscopy of Biological Molecules, Madrid, Spain, 1997. Dordrecht, The Netherlands: Kluwer Academic, 1997.

58. DM Grant, RK Harris. eds. Encyclopedia of Nuclear Magnetic Resonance. vol. 1. Chichester, England: Wiley, 1996.

59. ED Becker. Appl Spectrosc 50:16A–28A, 1996.

60. RN Jones. Analytical applications of vibrational spectroscopy: a historical review. In: JR Durig, ed. Chemical, Biological and Industrial Applications of Infrared Spectroscopy. London: Wiley, 1985, pp. 1–50.

61. EK Blout, RC Mellors. Science 110:137–138, 1949.

62. DL Woernley. Cancer Res 12:516–523, 1952.

63. N Kaiser. German patent DBP-K 36 308 IX/42 1, 1958.

64. M Jackson, MG Sowa, HH Mantsch. Biophys Chem 68:109–125, 1997.

65. HH Mantsch, M Jackson. Proceedings of Infrared Spectroscopy: New Tool in Medicine, 28–30 January 1998, San Jose, CA.

66. RJH Clark, RE Hester, eds. Spectroscopy of Biological Systems. Chichester, England: Wiley, 1986.

67. RJH Clark, RE Hester, eds. Biomolecular Spectroscopy Part A. Chichester, England: Wiley, 1993.

68. RJH Clark, RE Hester, eds. Biomolecular Spectroscopy Part B. Chichester, England: Wiley, 1993.

69. RJH Clark, RE Hester, eds. Biomedical Applications of Spectroscopy. Chichester, England, Wiley, 1996.

70. HH Mantsch, D Chapman, eds. Infrared Spectroscopy of Biomolecules. New York: Wiley-Liss, 1996.

71. JR Mansfield, MG Sowa, JP Payette, B Abdulrauf, MF Stranc, HH Mantsch. IEEE Transactions Medical Imaging 17:1011–1018, 1998.

# 2

# Biological and Pharmaceutical Applications of Vibrational Optical Activity

**Laurence A. Nafie and Teresa B. Freedman**
*Syracuse University, Syracuse, New York*

## 1 INTRODUCTION

Vibrational optical activity (VOA) is a relatively new area of molecular spectroscopy (1–7). Discovered experimentally approximately 25 years ago (8–11), VOA consists of an infrared form, called *vibrational circular dichroism* (VCD), and a Raman form, called *Raman optical* activity (RŎA). In both areas, one measures the differential response of a chiral molecule to left versus right circularly polarized radiation that induces a vibrational transition in the molecule. Over the past quarter century, VOA has evolved in sophistication, both experimentally and theoretically, so that today this spectroscopic technique is making unique contributions to our understanding of the structure and dynamics of chiral molecules.

Chiral molecules differ from higher-symmetry molecules in that they lack spatial constraints that increase our knowledge of their conformational possibilities. On the other hand, chiral molecules, without such constraints, have greater potential to influence the course of stereochemical reactions in particular ways. Fortunately, nature provides an additional spectroscopic probe for chiral molecules, one not active for molecules with mirror symmetry, called *optical activity*. This phenomenon arises from the existence of two orthogonal states of radiation polarization that are related by mirror symmetry, namely, left- and right-circular polarization states. These two states of polarization can interact diastereomerically with chiral molecules, much in the same way that left- and right-handed gloves interact differentially with your left and right hands.

Vibrational circular dichroism is defined as the difference in the absorbance of a molecule for left versus right circularly polarized infrared radiation for a vibrational transition. An energy-level diagram illustrating this definition is provided in Fig. 1, where the vibrational transition is between the g0 and g1 vibrational sublevels of the ground electronic state. The definition of ROA is analogous, but owing to the presence of both incident and scattered light, there are more variations in the energy-level diagrams for polarization modulation. In particular, as given in Fig. 1, for a Raman scattering transition between the same initial and final vibrational states as used in the definition of VCD, there are four basic forms of ROA. The first corresponds to the original form of ROA, called *incident circular polarization* (ICP) ROA. Here, only the polarization of the incident laser beam is modulated between left- and right-circular states. The next form of ROA, first measured in 1988, is called *scattered circular polarization* (SCP) ROA. The last two forms are in-phase and out-of-phase dual-circular polarization ($DCP_I$ and $DCP_{II}$) ROA, in which both the incident and scattered beams are synchronously modulated in and out of phase with respect to one another. Because Raman scattering is a coherent two-photon process, the DCP forms of ROA are distinct in information content from ICP and SCP ROA.

The combination of optical activity, with its inherent stereochemical sensitivity, and vibrational spectroscopy, with its extreme sensitivity to basic molecular structure, was sought for many years before the discovery of VOA. Finally, in the late 1960s and early 1970s, technological advances in electronics, infrared detectors, lasers, and polarization modulators paved the way for the independent discoveries of VCD and ROA a few years later. And VCD and ROA have followed different courses to their present status as mature techniques. It proved easier to measure VCD reliably, and VCD was boosted by the advent of Fourier transform methodology in the early 1980s (12–14). Also, VCD was calculated first using ab initio quantum mechanical methods (15,16). Because VCD crossed key experimental and theoretical hurdles at an earlier stage, there is a more extensive literature of VCD covering measurement techniques, theoretical calculations, and applications to molecules of biological and pharmaceutical interest. Today VCD enjoys the advantage of a fully dedicated, commercially available instrument (17) as well as commercially available software for the calculation of VCD intensities from first principles (18).

All of this is not to say that ROA lags terribly far behind or is not worthy of continued attention, for ROA has many intrinsic advantages, in much the same way that Raman spectroscopy has unique advantages relative to infrared (IR) spectroscopy. Within the past decade there have been significant advances in ROA spectroscopy that have included the discovery of dual-circular polarization methods (19,20), the adaptation of holographic optical components, near-quantum limited CCD detectors, and fiber optics, which promise soon to close the gap between VCD and ROA technologies (21,22). It is likely that in the

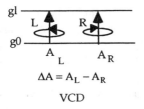

$$\Delta A = A_L - A_R$$

VCD

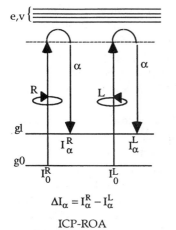

$$\Delta I_\alpha = I_\alpha^R - I_\alpha^L$$

ICP-ROA

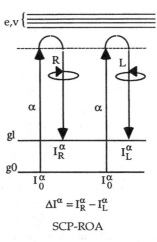

$$\Delta I^\alpha = I_R^\alpha - I_L^\alpha$$

SCP-ROA

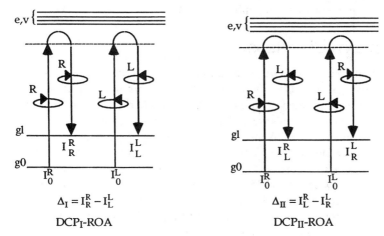

$$\Delta_I = I_R^R - I_L^L$$

DCP$_I$-ROA

$$\Delta_{II} = I_L^R - I_R^L$$

DCP$_{II}$-ROA

**Fig. 1** Energy-level diagram illustrating the polarization modulation schemes of the four basic forms of vibrational optical activity.

coming years, commercial ROA instruments will make their appearance, giving those in molecular stereochemistry a choice of VOA techniques, VCD, ROA, or both. In addition, a few years after the first ab initio VCD calculations, ROA was calculated from first principles (23,24), although this software is not yet commercially available.

Concomitant with these advances in VOA has been growing interest among chemists in the synthesis and control of chiral molecules. Biochemists have long had a deep interest in the structure and action of chiral molecules, since most molecules of stereochemical significance in biology are chiral, including amino acids, peptides, proteins, sugars, carbohydrates, nucleic acids, and natural products. Interest in chiral molecules has also increased dramatically in the pharmaceutical industry, where the importance of enantiomerically pure drugs has gained wider recognition for a variety of reasons, not the least of which has been safety concern for the consumer.

In short, VOA is a new spectroscopic tool for diagnostic and research applications. This review is aimed at the reader with practical interests in the areas of biomolecular structure and pharmaceuticals. The material that follows includes brief descriptions of the latest technological advances in VOA from a user's perspective. If further technical details are desired, the reader is referred to the list of review articles and papers included in the references list. By contrast, the application molecules of biological and pharmaceutical interest are given greater emphasis. It is hoped that these applications will encourage nonspecialists to apply VOA to their research and application needs.

## 2  INSTRUMENTAL METHODS

The early measurements of VOA were carried out with dispersive grating spectrometers. In the case of ROA, the requirements of stray light rejection using conventional gratings dictated the use of double or triple monochromators. As a result, these early measurements were extremely inefficient by today's standards. Dramatic improvements in VOA measurements were achieved with the successful application of multiplex techniques. For VCD, this was the application of Fourier transform (FT) methods (12,13), and for ROA it was the introduction of multichannel-array detectors (25–27). Collecting VOA data simultaneously across the entire spectrum not only improved the overall efficiency and speed of the VOA measurements, but it also eliminated the problem of changes in sampling conditions having an effect on one part of the spectrum and not the other. With modern instrumentation, quality VCD (28) and ROA (21,29,30) spectra for favorable samples can be obtained as quickly as 10 seconds (21,28), whereas 20 years ago, hours were required to scan a few hundred wavenumbers for such samples. This represents an improvement of approximately four orders of magnitude. The basic optical layout for VCD and ROA measurements is illustrated in

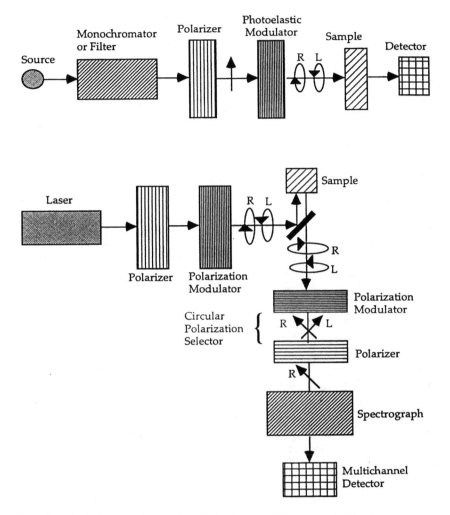

**Fig. 2** Block diagrams illustrating the basic optical layout needed for the measurement of VCD and ROA.

Fig. 2. Here, in the simplest terms, the elements of VCD and ROA measurements can be understood and compared.

## 2.1 Vibrational Circular Dichroism Instrumentation

For VCD, the basic instrument consists of a thermal source, such as a silicon carbide glower, and an FT-IR spectrometer followed by the polarization optics needed for VCD measurements (28). Alternatively, a combination light chopper and monochromator may be substituted for the FT-IR spectrometer, but the monochromator design is older and is used today only for more specialized applications where one is concerned with a limited region of the spectrum. The VCD optics can be built into the body of the FT-IR spectrometer, as in the design of the commercially available Chiral*ir* from Bomem/BioTools, or they may be added as an accessory bench at the exit port of the FT-IR spectrometer. In either case, the VCD optics consist of a polarizer followed by a photoelastic modulator (PEM), which in turn is followed by the sample. The PEM modulates the polarization of the IR light beam between left- and right-circular polarization states in a sinewave fashion in the tens of kilohertz frequency range. The beam is then imaged by a lens to a semiconductor detector, such as HgCdTe or InSb, which is cooled by liquid nitrogen. There are two distinct signals at the detector. One is the interferogram associated with the ordinary IR spectrum of the sample, and the other is the interferogram associated with the VCD spectrum of the sample. A lock-in tuned to the polarization modulation frequency of the PEM is used to demodulate the VCD interferogram. After Fourier transformation of these interferograms, with proper phase corrections and some further normalization and intensity calibration steps, the final spectra can be presented, as shown in Fig. 3 for the IR and VCD of neat (−)-*cis*-pinane. Each band in the VCD spectrum is either positive or negative, depending on whether left- or right-circular polarized IR light respectively, is absorbed more strongly, in accordance with the sign convention of VCD given in Fig. 1. Each VCD band can be associated with an IR band in the spectrum above it in Fig. 3. There is no particular relationship between the intensity of an IR band and the corresponding intensity of its daughter VCD band. It is also apparent that the VCD intensity scale is approximately four orders of magnitude smaller than the corresponding IR scale. This is the reason that VCD went undetected for several decades before the needed technology became available.

## 2.2 Raman Optical Activity Instrumentation

In the case of ROA, one starts with a laser, an argon ion laser or a frequency-doubled solid-state CW laser such as $Nd:YVO_4$ (21,29,30). Before reaching the sample, the incident laser beam passes through optics that modulate the polarization between left- and right-circular states in a square-wave cycle in the frequency range of fractions of a hertz to tens of hertz, depending on the Raman counting electronics. The light scattered from the sample may be analyzed for its circular polarization content before being imaged on the entrance slit of a single-stage spectrograph equipped with a holographic grating. The spectrometer images a band of frequency-dispersed Raman light onto a charge-coupled device (CCD) detector array, where it is further processed electronically. The Raman counts

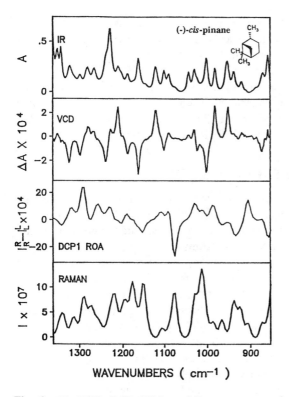

**Fig. 3**   IR, VCD, DCP$_I$-ROA, and Raman spectra of a neat liquid sample of (−)-*cis*-pinane.

for right- and left-circular polarization states are accumulated separately. The sum of the two registers is the Raman spectrum representing the total of all the Raman counts collected, and the difference, right minus left, is the ROA spectrum. The lower two spectra in Fig. 3 are the DCP$_I$ ROA and Raman spectra of (−)-*cis*-pinane for the same frequency range as the VCD and IR spectra above them. The ROA intensity scale is approximately three orders of magnitude smaller than the Raman scale, rather than four orders of magnitude smaller as is the case for VCD and IR spectra. This is because Raman scattering occurs in the visible region of the spectrum, where the wavelength of light is an order of magnitude smaller than in the IR region and hence is an order of magnitude more sensitive to molecular chirality. Not only is it possible to compare the ROA and Raman spectra in the same way that VCD and IR spectra were compared earlier, but it is also possible to compare all four spectra to one another. These spectra represent four different vibrational spectra of the same molecule spanning the same set of normal modes and band frequencies. It is clear that there is no particular correlation between strong and weak bands in any of the spectra, and there is no apparent sign correlation between the VCD and ROA spectra. The reason for the uniqueness of VCD and ROA spectra, along with IR and Raman spectra, is because they arise from different interactions between radiation and molecules. The lack of correlation of sign patterns between VCD and ROA also signifies that molecular vibrations do not possess an inherent chirality that is independent in sign of the way they interact with IR or Raman radiation.

## 3  SPECTRAL INTERPRETATION

The VCD and ROA spectra can be interpreted on different levels of sophistication. At the simplest level, they can be correlated empirically with the structures of the molecules being compared. And VOA can also be used to assess the optical purity of the chiral sample. If the VOA spectrum of a sample with known optical purity is measured with good intensity accuracy, it is then possible to determine the optical purity of any sample of the same molecule. Another use of VOA spectra is the determination of absolute configuration. This can be accomplished empirically with the use of marker bands. Since VOA spectra contain a large number of vibrational bands representing widely differing structural regions of a chiral molecule, it is not difficult to find bands that are conserved in sign when structural modifications are made to the molecule without changing its center or centers of chirality. All of these applications involve essentially making empirical or statistical correlations between closely related VOA spectra.

An intrinsically more powerful application of VOA is its use for the determination of absolute configuration in the absence of any previous VOA studies of structurally related molecules. This can be accomplished by comparing the experimental VOA spectrum to the corresponding VOA spectrum calculated using ab initio quantum mechanical calculations. This has now been accomplished with impressive certainty dozens of times in recent years as quantum mechanical methods using density functional theory have come to be more widely used. In addition, programs using these sophisticated methods for the calculation of VCD spectra have become commercially available in the past year in the quantum chemistry program *Gaussian 98* (Gaussian, Inc., Pittsburgh, PA) (18), and this has made possible the accurate simulation of VCD spectra for anyone interested in carrying out such calculations.

A final level of application of VOA spectra is their use to determine the solution-state structures of chiral molecules. While in some cases this can be achieved through careful empirical studies, the most direct way to accomplish this is through the same kind of quantum mechanical calculation used for the determination of absolute configuration discussed earlier. Several different formalisms have been developed for the calculation of VCD intensities. The most widely used ab initio formalisms to date are magnetic field perturbation (MFP) theory (15,31), which as already noted is commercially available (18), vibronic coupling sum-over states theory (VCT) (32–34), and the approximate locally distributed origin (LDO) gauge model (35,36), which has been shown to be accurate for larger systems in the hydrogen stretching region. An ab initio implementation (37) of the localized molecular orbital model (38) has also been employed. Also of use recently for theoretical analysis of coupled vibrational modes in large biological molecules is the extended coupled oscillator (ECO) model of VCD advanced by Diem and coworkers (39). The theoretical analysis of VCD spectra can become complex when more than one conformation is present in solution. The percentage population of these conformers can be determined from VCD by first calculating the VCD spectra of the most stable conformers and then comparing these conformer spectra to the measured VCD spectra. The accuracy of this approach is still limited to some degree by the accuracy of calculating the effects of solvent on the conformer structures. This in turn leads to conformer distributions that deviate substantially from the prediction of their Boltzmann population based on the calculated energies. Nevertheless, good agreement between measured and simulated VCD spectra lends encouragement to the notion that these efforts are on the right track.

We now discuss further the background of two practical applications of the analysis of VOA. In subsequent sections we then provide a series of more detailed examples of applications of VOA to particular classes of biological and pharmaceutical molecules.

### 3.1 Determination of Enantiomeric Excess

The percent enantiomeric excess (%ee) is defined as the excess of the number of moles of one enantiomer over that of the opposite enantiomer as a percentage of the total number of moles of both enantiomers. For a sample consisting of a mixture of enantiomers, the VCD is directly proportional to the %ee. In the case of a sample of only one enantiomer, the VCD is full strength and the %ee is 100. For a racemic mixture, there is no excess, and hence there is 0%ee and zero VCD intensity. In Fig. 4, we present three VCD spectra in the mid-infrared region for (+)-α-pinene for %ee values of 100, 95 and 90 (17). It is clear that the magnitude of the VCD is decreasing across the spectrum as the %ee decreases. The optimum way to analyze these spectra for the %ee associated with each of these spectra is to employ statistical methods, such as partial least squares. One first obtains the best fits for training spectra, such as these in Fig. 4, and then uses the resulting database to make predictions of unknowns. Recent analyses of optical purity in a number of cases has led to the conclusion that in favorable cases, that is, for spectra with good signal quality, an accuracy of prediction at less than 1%ee can be achieved (3,17). This compares favorably with chiral chromatography, which

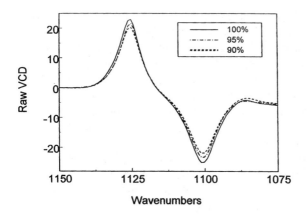

**Fig. 4**   VCD spectra in the mid-infrared region between 1150 and 1075 cm$^{-1}$ for neat (+)-α-pinene for three samples with enantiomeric excess of 90%, 95% and 100%.

can reach an accuracy of 0.1% when a complete separation of enantiomers is available. An advantage of VCD is that %ee determinations can be carried out without the need for physical separation. Another advantage of VCD is the multiplicity of chromophores available for analysis. In comparison to optical rotation, which has been used extensively for %ee determinations, VCD usually has a number of strong VCD bands in the spectrum from which to choose. On the other hand, values of optical rotation can often be small in the absence of chromophores absorbing in the mid- to near-UV region. Another advantage of VCD is its absence of strong temperature dependence; optical rotation is very sensitive to the density of the sample for the long pathlengths used, and this can pose a problem in obtaining an accurate determination of %ee.

And ROA can also be employed for the determination of optical purity. Here again, the magnitude of the ROA varies directly with the %ee, becoming zero at 0%ee and reaching its maximum value at 100%ee. Two recent studies (40,41) have indicated sensitivity of ROA to %ee at the level of 1% or better. At present, with the availability of commercial VCD instrumentation, VCD is the more accessible VOA approach to the measurement of optical purity.

## 3.2   Determination of Absolute Configuration

Until recently, there has been no established spectroscopic method for unambiguously determining the absolute configuration of a chiral molecule from a single solution sample. Now VOA provides such method, and that method is comparison of VOA spectra determined from ab initio theory and from experiment. Presently, there are several limitations, but these should gradually become less significant as further progress is achieved both experimentally and theoretically. An obvious limitation is the size of the chiral molecule in question. A practical limit to accurate quantum calculations is approximately 20 heavy atoms, such as carbon, nitrogen, and oxygen. Another complication is the conformational flexibility of the chiral molecule. The easiest case is when a single predominant solution-state conformation is present. Predicting absolute configuration for the case of multiple conformations of up to three or four important conformations is feasible; but if the molecule possesses many conformations, the level complexity may rise too high. On the other hand, there is generally some structural rigidity near the chiral center, or centers, and those vibrational modes near chiral centers are generally less perturbed by conformational variation, and they also tend to have large VCD intensities. Some other limitations include the solubility of the molecule in a suitable solvent and overall weak VCD intensity. Experience gained over the years has revealed that most organic or biomolecules of small to medium size show VCD intensities large enough to measure with good signal quality. Large molecules, such as proteins and nucleic acids, also have strong VCD, but at present, they are beyond the scope of ab initio calculations. In addition, for such molecules comprised of L-amino acid or D-sugar subunits, the issue of absolute configuration is usually not a concern. The largest need of information about absolute configuration arises in studies of chiral synthesis and the design and development of new chiral pharmaceutical products.

In Fig. 5, we provide an example of a comparison of experimental and theoretical VCD spectra that illustrates the determination of absolute configuration using VCD. We show the IR absorbance and VCD of (+)-fenchone in the mid-IR region (42). The experimental FT-IR and FT-VCD spectra, a and c, were measured approximately 10 years ago in our laboratory at Syracuse University (43), while the theoretical spectra, b and d, were calculated recently in the laboratory of Philip Stephens at the University of Southern California using ab initio quantum calculations. These calculations were carried out using density functional theory (DFT) using hybrid functionals, and the VCD intensities were determined using the magnetic field perturbation (MFP) method with a gauge-invariant atomic orbital (GIAO) basis set. The modes in all four spectra are numbered to facilitate comparisons between spectra. It is clear from this example that the level of agreement between virtually all the bands in the IR and VCD spectra is very high. There is no doubt about the chemical identity and absolute configuration of this molecule based solely on the ab initio calculations.

Also, ROA can be used for the determination of absolute configuration. A particularly striking example is the recent determination of the absolute configuration of bromochlorofluoromethane (44). In this case, the use of optical rotation was not accurate enough due to the small rotation angles, whereas the ROA spectrum was found to be strong and distinctive. The corresponding ab initio calculation agreed in sign with all the observed ROA bands. Others determina-

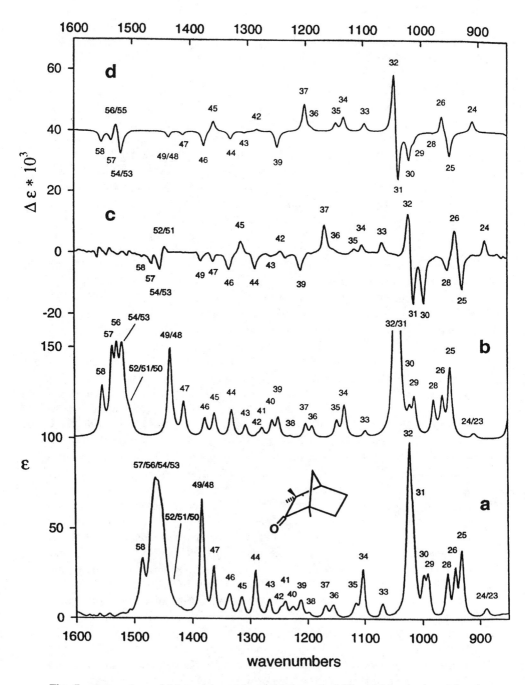

**Fig. 5** Comparison of (a) experimental and (b) theoretical IR and (c) experimental and (d) theoretical VCD for neat (+)-fenchone in the mid-infrared region.

tions of the absolute configuration of chiral molecules using a combination of ROA experiment and theory have been reported (45,46), but in these cases the context was the simulation of an ROA spectrum for a molecule with previously known absolute configuration.

## 4 VIBRATIONAL CIRCULAR DICHROISM APPLICATIONS

Applications of VCD to biological and pharmaceutical molecules before 1996 have been included in earlier reviews of the field (2,6,47–55). We summarize here some of the important early applications and provide more extensive examples from recent publications for several research groups.

### 4.1 Simple Biological Molecules

#### 4.1.1 Amino Acids and Peptides

Some of the earliest measurements and applications of VCD were for simple amino acids (56–62), amino acid metal complexes (63–65), and simple peptides (66–68). These studies initially focused on the hydrogen-stretching regions. In particular, the intense methine stretch in chiral amino acids (56–62), amino acid transition metal complexes (63–65), and simple peptides (66,67) was identified as a marker band for absolute configuration, with a positive methine-stretching band observed for L-amino acids and L-amino acid residues at central and C-terminal positions. For amino acids this signal is greatly reduced at low pH, but is maintained at high pH, whereas for C-terminal peptide residues the methine-stretching VCD intensity is intense and fairly independent of pH (62,67). For N-terminal residues, the methine stretching VCD is weak, independent of pH (67). A characteristic VCD couplet is also observed for the methine deformations in amino acids (69); in simple peptides an overall couplet pattern that identifies the absolute configuration of the C-terminal residue is observed, even though the methine deformations couple with the amide III mode (69–72). A study of protected $N$-urethanyl amino acids (73) revealed amide I VCD features assigned to a 7-membered $\gamma$-turn intramolecularly hydrogen-bonded ring. A more recent application of VCD to a set of protected peptides R′—CO—AA—NH—R″ (R′ = Me and tertBu; AA = Ala, Leu, Val, and Phe; R″ = Me, isoBu and neoPe) associates NH-stretching VCD features with monomers in a $C_5$-conformation and $C_5$-$C_5$ dimers (74).

Recent density functional calculations of alanine used explicit water molecules to stabilize the zwitterion. The geometries and calculated VCD spectra were sensitive to the positions of the water molecules (75). The VCD (and ROA) spectra of $N$-acetyl-L-alanine $N'$-methylamide have also been calculated (76), again considering explicit water molecules, which stabilized structures not stable for the isolated peptide.

### 4.1.2  Carbohydrates

Several early studies also focused on carbohydrates, including CH-stretching and mid-IR studies of sugars and sugar derivatives (77–80). These studies provided an opportunity to identify marker bands for absolute configuration and investigate the effects of adjacent stereogenic centers in a wide variety of sugars. A band was identified in the mid-IR region at $\sim 1150$ cm$^{-1}$ that correlates with the overall orientation of the hydroxy groups around the ring (79,80).

### 4.1.3  Terpenes

Naturally occurring terpenes have been extensively investigated by VCD, due to their availability, variety, and often rigid structures (31,42,81–86). In the CH-stretching VCD for a number of terpenes with six-membered rings in the chair conformation, a $(+ - +)$ pattern, which arises from the inherently dissymmetric chromophore $CH_2$—$CH_2$—$C^*H$, was identified in an early study as a VCD marker for absolute configuration and conformation (84). A later study of the 900–1500 cm$^{-1}$ region compared the VCD and ROA spectra of four terpenes that illustrated the lack of correspondence and thus complementary nature of the two types of measurement (85). The recent studies have used terpene VCD spectra to test advances in computational methodology (86) and demonstrate the close agreement between calculation and experiment when modern ab initio density functional theory methods are employed with the magnetic field perturbation approach (calculations on camphor, fenchone, and α-pinene) (31,42). The vibronic coupling theory approach has been used in a study of camphor and 2-vinyl-exo-borneol (87) and of the solution conformations of (1S,2R,5S)-(+)-menthol (34). A recent, new excitation scheme methodology (88) has also been used to calculate terpene VCD spectra, including a study of camphor and α-pinene. The VCD spectra of camphor, α-pinene, and borneol were recently measured and compared to MFP and VCT calculations, which were found to underestimate the absolute rotational strengths (89).

### 4.1.4  Methyl Lactate Derivatives

We recently completed a study of methyl lactate and related molecules to identify abundant solution conformations and establish correlations between OH- and methine-stretching VCD intensity and the chiral environments of these oscillators (90). By using specific deuteration to eliminate overlapping absorptions from achiral portions of the molecules, the anisotropy ratios ($g = \Delta A/A$) for the methine stretches could be accurately measured. For the methine stretching VCD of the molecules with both α-oxy and α-C=O substituents, (S)-methyl-d$_3$ lactate, (S)-methyl-d$_3$ 2-(methoxy-d$_3$)-propionate, di(methyl-d$_3$) D-tartrate, (S)-methyl-d$_3$ mandelate, (S)-methyl-d$_3$ O-(acetyl-d$_3$)-mandelate, and (S)-benzoin, anisotropy ratios between $+2.1 \times 10^{-4}$ and $+2.8 \times 10^{-4}$ were measured. Calculations of low-energy conformations and VCD intensities, utilizing the vibronic coupling theory (VCT) methodology, demonstrated that an approximately cis planar arrangement of O=C—C*—O in these molecules correlates with the large methine-stretching VCD, which serves as a marker for both absolute configuration and solution conformation. In Fig. 6 we compare the mid-IR VCD for (S)-methyl lactate with the calculated spectra for the most abundant OH—O=C hydrogen-bonded conformer, carried out at the ab initio DFT (B3LYP-6-31G(d)) level with both the MFP and VCT computational methodologies. The agreement between the two calculations and with experiment clearly demonstrates that these formally equivalent computational methods yield comparable results, and again illustrates that VCD spectra combined with high-level calculations can be used to unambiguously identify absolute configuration and major solution conformations.

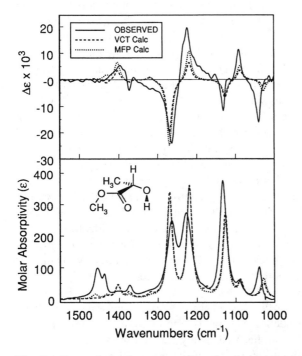

**Fig. 6** Comparison of experimental molar absorptivity ($\varepsilon$) and VCD ($\Delta\varepsilon$) spectra of (S)-methyl lactate (0.2 M in CCl$_4$ solution, 100-$\mu$m pathlength, 4.0 cm$^{-1}$ resolution) with calculated spectra [ab initio DFT, B3LYP/6-31G(d)] utilizing the magnetic field perturbation (MFP) and vibronic coupling theory (VCT) methodologies.

## 4.1.5   Cyclic Peptides

For larger biologically relevant molecules, simplified models for VCD still can be employed to interpret spectra. An example is a study of cyclic peptides by Diem and coworkers (91) for *cyclo*-(-Gly-Pro-Gly-D-Ala-Pro-), which adopts a type II β turn in the crystal, and a VCD study of *cyclo*(Pro-Gly)$_3$ and its interaction with cations (92). These studies employed an extended coupled oscillator calculation for the amide I (C=O stretch) vibrations, which provided reasonable agreement with experiment. Further studies of β turns in cyclic peptides (93,94) demonstrated the sensitivity of the VCD spectra to changes in solvent polarity and hydrogen-bonding ability, and identified VCD patterns for type I and II β turns in cyclic peptides.

## 4.2   Pharmaceutical Molecules

Most VCD studies of pharmaceuticals have focused on identification of absolute configuration and solution conformation. One approach to interpreting the VCD spectra of fairly large molecules is to carry out VCD intensity calculations on a suitable chiral molecular fragment, since achiral portions of the molecule do not make significant contributions to the VCD spectra, particularly in the hydrogen-stretching regions. A recent study utilizing this approach included the antiarrhythmic and anti-inflammatory drugs shown in Fig. 7 (36). Quinidine (I), RAC 109 (II), and flecainide (IV) are class I antiarrhythmic drugs that function as sodium channel blockers. Members of this class of drugs contain an aromatic portion for lipid solubility and an amino group for water solubility; the enantiomeric forms can differ in activity and binding constant to cardiac sodium channels. The VCD spectra of the hydrogen-stretching modes for these three drugs were interpreted by using ab initio calculations of low-energy conformations and VCD spectra of suitable chiral fragments of the drugs (the latter utilizing the approximate LDO model for VCD). For RAC 109, the VCD spectra were used to identify both the absolute configuration and the most abundant solution conformer. Quinidine is a fairly rigid molecule, with a single dominant solution conformation; VCD marker bands for configuration and conformation were identified in the OH- and CH-stretching regions. For flecainide, the VCD spectra and calculations (Fig. 8) are consistent with the presence of two low-energy solution conformers stabilized by 7-membered hydrogen-bonded rings (Fig. 9). The solution structures of quinidine and flecainide deduced from this VCD study can be used to understand the contrasting stereospecific binding properties of these two drugs. For (+)-quinidine and its diastereomer (−)-quinine, the ratio [IC$_{50}$(+)]/[IC$_{50}$(−)] for binding to cardiac sodium channels is 0.29, whereas the corresponding ratio for the two flecainide enantiomers is 1.03. The relative positions of the amino nitrogen and

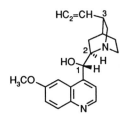

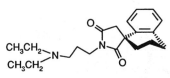

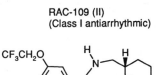

Quinidine (I)
(Class Ia antiarrhythmic)

RAC-109 (II)
(Class I antiarrhythmic)

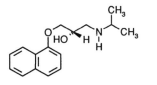

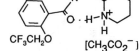

(S)-Propranolol (III)
(Class II antiarrhythmic)

(S)-Flecainide acetate (IV)
(Class Ic antiarrhythmic)

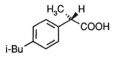

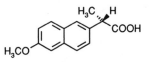

(S)-Ibuprofen (V)
(anti-inflammatory)

(S)-Naproxen (VI)
(anti-inflammatory)

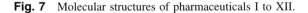

(1S,2R)-norephedrine, (1S,2S)-norpseudoephedrine
[R=H, R'=H]

(1S,2R)-ephedrine, (1S,2S)-pseudoephedrine
[R=H, R'=CH₃]

Ephedra drugs
(VII-XII)

(1S,2R)-N-methylephedrine,
(1S,2S)-N-methylpseudoephedrine
[R=CH₃, R'=CH₃]

**Fig. 7**  Molecular structures of pharmaceuticals I to XII.

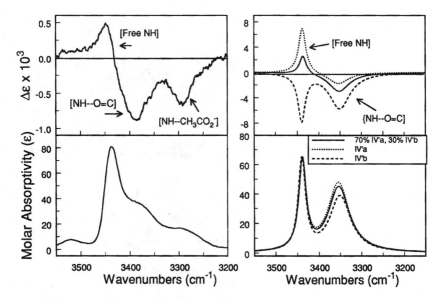

**Fig. 8** Comparison of experimental molar absorptivity ($\varepsilon$) and VCD ($\Delta\varepsilon$) spectra of (S)-flecainide acetate (0.033 M in $CDCl_3$ solution, 0.2-cm pathlength, 20 $cm^{-1}$ resolution) with calculated spectra (LDO model, 3-21G basis) for (S)-flecainide fragment IV′ in conformations IV′a and IV′b, with composite sum spectra for 70% IV′a and 30% IV′b.

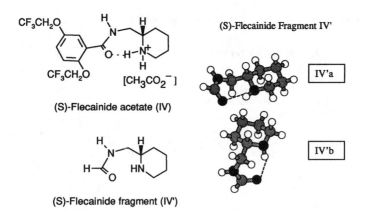

**Fig. 9** Structures of flecainide acetate, flecainide fragment used in calculations, and calculated lowest-energy fragment conformations.

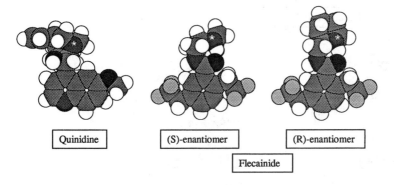

Quinidine  (S)-enantiomer  (R)-enantiomer

Flecainide

**Fig. 10**  Class I antiarrhythmic drugs: proposed active site conformations for binding of quinidine and (S)- and (R)-flecainide to cardiac sodium channels, in overlapping sphere representations, with an asterisk showing the location of amino nitrogen.

aromatic portion of the rigid quinidine structure, shown in Fig. 10, differ from those of the approximately mirror-image quinine structure (opposite chirality at carbons 1 and 2 in Fig. 7), and these two diastereomers interact differently with the cardiac sodium channel binding site. In contrast, (S)-flecainide in solution conformation corresponding to IV′b and (R)-flecainide in conformation IV′a assume a very similar disposition of the amino nitrogen and aromatic moieties, close to that of quinidine (Fig. 10); this flexibility of flecainide results in similar binding constants for the two enantiomers.

For propranolol (III, Fig. 7), a class II antiarrhythmic drug that functions via β-adrenergic blockade, several low-energy hydrogen-bonded conformations were identified by calculations on fragment III′ (Fig. 11). The calculated OH-/NH-stretching VCD spectra for the fragment reveal distinct VCD signatures that can be used to identify the presence of these conformers in solution from the corresponding bands in the experimental VCD spectrum of propranolol (Fig. 12). Although these conformations interconvert on an NMR time scale, on the much faster time scale for vibrational transitions, bands arising from various types of OH—N, OH—O, and NH—O hydrogen bonded conformations are uniquely identified.

For the analgesics ibuprofen and naproxen (V and VI), the positive VCD intensity in the hydrogen-stretching region is associated with the S-configuration of the drugs (36). In this case, the VCD signal can be used to identify absolute configuration even in the presence of strong background IR intensity arising from the presence of hydrogen stretches from achiral portions of the molecule and from the OH stretch of carboxylic acid dimers formed in nonaqueous solution.

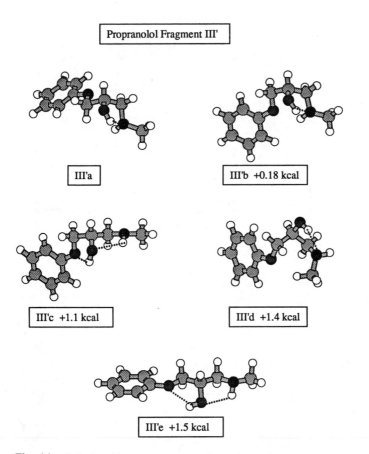

Propranolol Fragment III'

III'a

III'b  +0.18 kcal

III'c  +1.1 kcal

III'd  +1.4 kcal

III'e  +1.5 kcal

**Fig. 11**  Calculated lowest-energy conformations of propranolol molecular fragment III'
used for VCD calculations, indicating relative energies and hydrogen bonding.

An earlier study (95) focused on pharmaceuticals in the ephedra class (VII-
XII, Fig. 7), whose functions range from bronchodilators and decongestants to
appetite suppressants (depending on the configuration at centers 1 and 2 and on
the amino substituents). In this study, the OH-/NH-stretching VCD spectra for
dilute $C_2Cl_4$ solutions of the ephedra drugs were interpreted in terms of contributions from intramolecularly hydrogen-bonded OH—N and NH—O conformers
and non-hydrogen-bonded conformations.

The cyclosporins are a class of immunosuppressive drugs with the cyclic
undecapeptide structures shown in Fig. 13. In the crystal and in nonaqueous solution, the structure of cyclosporin A is stabilized by four intramolecular hydrogen
bonds; this structure is altered when the configuration at residue 11 is changed

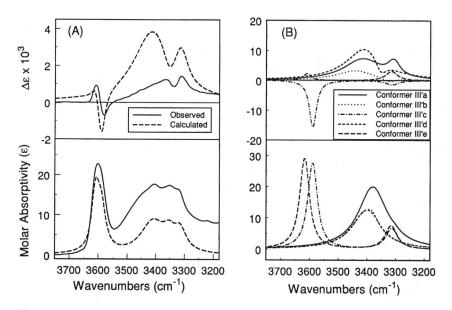

**Fig. 12** Comparison of observed molar absorptivity ($\varepsilon$) and VCD ($\Delta\varepsilon$) spectra of (S)-propranolol (III) in the OH-NH-stretching region (0.1 M in CDCl$_3$ solution, 2.8-mm path-length, 16 cm$^{-1}$ resolution, 6 hr for each enantiomer) with calculations on fragment III′. Calculated spectra in (A) (---) are composite weighted sums of calculated spectra for propranolol fragment III′ conformers (LDO model, 3-21G basis set), consisting of III′a (35%), III′b (20%), III′c (15%), III′d (10%), and III′e (20%). Calculated spectra of individual conformers and shown in (B).

from L to D and in the presence of magnesium ion. The sensitivity of the VCD spectra to these changes is apparent in the comparison of the VCD spectra of cyclosporins A and H in CDCl$_3$ solution and in the presence of Mg$^{2+}$ in CD$_3$CN solution (Fig. 14) (96,97).

The VCD of two pharmaceutical molecules used as inhalation anesthetics, isoflurane [(CF$_3$)CHFO(CF$_2$H)] (98) and desflurane [(CF$_3$)CHClO(CF$_2$H)] (99), have been investigated by Polavarapu and coworkers. There is a twofold difference in effectiveness of the two enantiomers of isoflurane (100). The VCD studies focused on identifying the absolute configuration from the VCD spectra in the ~1000–1450 cm$^{-1}$ region. Two stable, room-temperature conformers with low energy were identified for each molecule from ab initio geometry calculations. Assuming equal populations of the two conformers, VCD spectra calculated with the localized molecular orbital model provided an unambiguous assignment of both (+)-isoflurane and (+)-desflurane to the (S)-configuration (98,99).

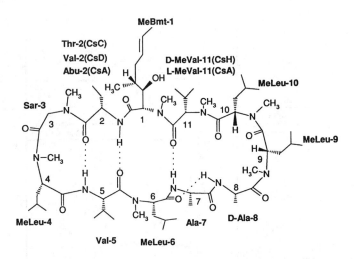

**Fig. 13** Structures of cyclosporins, showing intramolecular hydrogen bonds for cyclosporin A in nonaqueous solution.

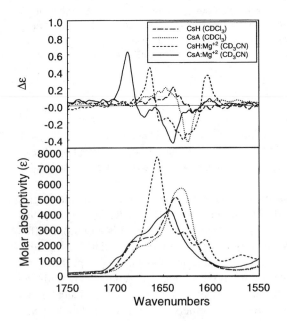

**Fig. 14** Comparison of molar absorptivity (ε) and VCD (Δε) spectra of cyclosporins A and H in the amide I region, in the presence and absence of Mg²⁺ ion.

## 4.3  Pheromones

To highlight a recent application of VCD to biomolecules from the laboratories of Prof. H. Wieser at the University of Calgary and Prof. P. Stephens at the University of Southern California, we show in Fig. 15 a comparison of the observed and calculated VCD spectra of the chiral pheromone frontalin (1,5-dimethyl-6.8-dioxabicyclo[3.2.1]octane) (101). An ab initio calculation utilizing density functional theory and the magnetic field perturbation (MFP) methodology for computing VCD intensities was carried out for conformers a and b. Comparison of observed and calculated VCD spectra firmly establish that the (1R,5S)-

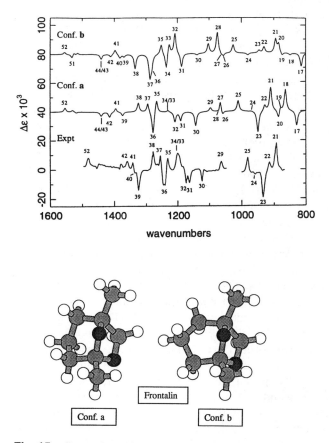

**Fig. 15**  Comparison of experimental VCD ($\Delta\varepsilon$) spectra of (1R,5S)-(+)-frontalin in CCl$_4$ solution with calculated spectra (MFP formalism, B3LYP functional, GIAO/6-31G(d) basis set, Lorentzian bandshape with 4 cm$^{-1}$ half-width at half-height) for conformers a and b. (From Ref. 101, with permission of Elsevier Science, Ltd.)

($+$) enantiomer adopts the lower-energy conformation a, with a chair conformation for the six-membered ring and a boat conformation of the seven-membered ring. The (1S,5R)-($-$) enantiomer has been identified as the active configuration for this pheromone of the Southern pine beetle.

## 4.4   Large Biological Molecules

### 4.4.1   Polypeptides

The VCD of polypeptides has been a focus of numerous studies and review articles (49,52,55). The first measurements (102–105) revealed characteristic VCD patterns in the amide A, amide I, and amide II regions of $\alpha$-helical oligopeptides. Studies of poly-L-lysine (106,107) in $D_2O$ solutions of different pH and salt concentration revealed characteristic amide I′ VCD patterns for $\alpha$-helix, $\beta$-sheet, coil, and disordered forms, and patterns for $3_{10}$-helices (108,109) and the poly(L-Pro)$_n$ II extended helix have also been identified (49,55,110). The "random coil" conformation in fact has regions of local order consistent with a left-handed helical arrangement (107,111). Characteristic features for helix, $\beta$-sheet, and coil have also been observed in the amide II region (112). A more recent study of blocked oligopeptides containing proline, focusing on the amide A (NH-stretching) region, assigned features to the $C_7$ $\gamma$ turn, the $C_{10}$ type II $\beta$ turn, a $C_5C_7$ conformation, and intermolecular hydrogen-bonded forms at high concentration (113). A study combining VCD, FT-IR, and electronic CD spectra of peptides with alternating L- and D-proline residues indicate an alternate cis-trans structure for the amide bonds. The decreasing VCD intensity with increases in chain length for these peptides suggests an equilibrium between helices of opposite sense in the longer polypeptides, where the chiral end groups determine which will be in excess (114). Electronic and vibrational CD have been used to characterize $\beta$-bend ribbon spirals (a subtype of $3_{10}$ helices), which were shown to require five to six residues in a sequence (L-Pro-Aib)$_n$ for full development of the structure (115). Based on the sense of amide I couplets observed in VCD spectra in $CDCl_3$ solution, homo-oligopeptides with $C^\alpha$-methylated residues were found to adopt a right-handed helical conformation with D-($\alpha$-Me)Phe ($C^\alpha$-methyl phenylalanine) peptides and the opposite, left-handed helical sense with a D-isovaline pentapeptide (116). Electronic and CD spectra have been combined to understand the conformations of other polypeptides. A transition from coil to $\beta$ sheet was found to be induced by both increase in concentration and increase in salt concentration in polypeptides with mixed Lys and Leu residues (117), and a stabilization of $3_{10}$ helices with aggregation was found for blocked homo-oligopeptides with $C^\alpha$-methyl valine residues (118). Also, VCD and FT-IR spectra have been used with factor analysis to study thermal unfolding in $D_2O$ solution for alanine-rich peptides (119), and a combined FT-IR, electronic, and vibrational CD study (120) of the mutarotation of poly-L-proline I to poly-L-proline II in $D_2O$ has demonstrated the presence of an intermediate structure.

A VCD study of chiral tripodal peptides that are precursors for model siderophores identified propeller-type conformations in solution, which are stabilized by interchain hydrogen bonds (121). The sense of the bisignate VCD couplets observed for the C=O stretches in the tripodal peptides correlates with the handedness of the propeller, which corresponds to that of the natural siderophore.

## 4.4.2 Proteins

Protein VCD studies, a major focus of research of Prof. T. A. Keiderling at the University of Illinois, Chicago, have been reviewed separately (6,54,55,122–125). These studies build on the oligopeptide studies to use VCD, in combination with other methods, to understand protein secondary structure and conformational changes. A major focus of this work is to use statistical analysis of electronic and vibrational circular dichroism and FT-IR spectra of proteins, in $H_2O$ solution (126), to predict protein secondary structure (122,125,127–131). In one study, IR and VCD spectra of the amide I and II bands and electronic CD spectra of 23 proteins were utilized in conjunction with principle component factor analysis methods to predict $\alpha$-helix and $\beta$-sheet fractions in proteins not included in the set. Recent applications of VCD to proteins include a comparison of electronic and vibrational CD conformational studies of milk proteins (132), correlation of bandshape with protein secondary structure utilizing VCD spectra in the amide III region (133), and an analysis of relationships between protein folds and optical spectra (134).

An interesting application of VCD in proteins is the study of ligands bound to iron in heme and nonheme proteins (135–140). The anisotropy ratio ($\Delta A/A$) for the antisymmetric azide stretch for azide covalently bound to low-spin iron in hemes is between $-6.5 \times 10^{-3}$ and $-1 \times 10^{-3}$ for heme proteins such as myoglobin and hemoglobin from several species (horse, human, insect, elephant, carp) (137,138) and $-1 \times 10^{-4}$ for hemerythrin (138). The anisotropy ratio for $CN^-$ is approximately $+2.4 \times 10^{-3}$, opposite in sign to that of the azide ligand (136,138). Little or no VCD is observed for the azide bound ionically to high-spin iron or bound to $Mn^{3+}$-exchanged heme (high spin), when distal E-11 Val is replaced by Asp, or when distal E-7 His is replaced by Gly. Exchange of F-8 proximal His with Gly reversed the sign of the azide-stretch VCD. The anomalously large anisotropy ratio for the azide and cyanide stretches when bound to heme has been attributed to interactions with low-lying magnetic-dipole-allowed electronic transitions, either charge-transfer states (138) or low-lying d→d $Fe^{3+}$ transitions (139). A recent study of azide bound to a $C_2$-chiral strapped iron porphyrin (140) found a large anisotropy ratio ($2 \times 10^{-3}$) with no apo protein and opposite signs for the VCD of azide bound on the strapped and unstrapped sides of the heme. The peripheral substituents of the porphyrin are asymmetrically arranged and present enantiotopic faces. These VCD studies demonstrate that the strap and the distal chiral environment in the heme proteins produce distereotopic faces with a preferred side for azide binding.

### 4.4.3 Nucleotides

The first observation of VCD for nucleotides, for synthetic poly(ribonucleic acids) (141), demonstrated VCD couplets arising from base C=O- and C=C-stretching modes, with intensities increasing with degree of order. The first measurements for model deoxyoligonucleotides, including observation of the B→Z phase transition (39) also showed VCD multiplet patterns that are sensitive to the handedness of the polymer helix. Subsequent studies (142–144) showed distinct VCD spectra for the B and Z forms (opposite sign patterns and frequency shifts). For both RNA and DNA, an exciton coupling model, either a coupled oscillator or extended coupled oscillator model for the base ring stretches (contributions from C=O, C=C, and/or C=N), reproduced the pattern and sense of the VCD features (39,142–145). A significant but smaller change in the VCD spectrum is observed for the B→A transition (146); this study also correlates similarity in VCD spectra with the similar structures for A-form DNA and tRNA. The sense of DNA helicity is also indicated by the VCD of the symmetric $PO_2^-$ stretch (147). The VCD spectra in both the base C=O-stretching and $PO_2^-$-stretching regions were used to study the solution conformations of poly(dI-dC)poly(dI-dC) (148) and the thermal denaturation of poly(rA)poly(rU) (149). Triple helix nucleic acids were found to exhibit VCD spectra quite different from the duplexes (150). For small DNA fragments, self-complementary tetranucleotides exhibited VCD at low ionic strength characteristic of right-handed polymers (151). The dinucleotides 5′(CG)3′ and 5′(GC)3′ at low ionic strength and 5°C were found to exhibit $(-,+,-)$ and $(-,+)$ VCD patterns near 1160 cm$^{-1}$ (low to high frequency), which were reproduced by coupled oscillator calculations, and attributed to duplexes formed in solution. Two recent reports utilize a DeVoe polarizability theory approach to analyze the VCD, absorption, and linear dichroism spectra in the base-stretching region of double- and triple-stranded polyribonucleotides, showing good agreement with previously measured experimental data (152,153). This method includes all C=O, C=N, and C=C vibrations of the bases (with frequencies, mode compositions, and dipole oscillator directions as assigned from IR spectra of 5′-AMP, 5′-UMP, 5′-GMP, and 5′-CMP) coupled to all orders, and shows improved agreement with experiment compared to the earlier coupled oscillator calculations. Three recent studies of nucleic acids have been reported from Prof. H. Wieser's laboratory at the University of Calgary. A study of six octadeoxynucleotides identified $PO_2^-$ and furanose modes independent of the base sequences, as well as C=O and skeletal vibrations that were characteristic of the sequence (154), and a study of other octadeoxynucleotides has also been initiated (155). Absorption and VCD spectra of deoxyoctaneucleotides complexed with daunorubicin (156) indicate preferential insertion of the drug between the terminal CG base pairs.

## 5 RAMAN OPTICAL ACTIVITY APPLICATIONS

In many respects, applications of ROA parallel those of VCD. Both techniques address questions of biomolecular structure and dynamics from the perspective of small biological molecules, including molecules of pharmaceutical interest, and large biological molecules, including proteins and nucleic acids. As mentioned earlier, both VCD and ROA spectra are measured with multiplex spectroscopic instrumentation, and in both areas ab initio quantum mechanical calculations have been carried out that give very close correspondence between measured and simulated spectra. Along with these similarities are differences that will be addressed in the following sections. The focus of these sections will be applications of ROA since roughly 1995. Earlier work has been reviewed extensively in a number of reviews that have appeared in recent years (1,2,4,5,49,157–161). Of these reviews, several review progress in both ROA and VCD (1,2,49,161) and offer comparisons between these two techniques.

### 5.1 Comparative Applications

There is considerable interest in the relative sensitivities of VCD and ROA in the same chiral molecules. In one study (85), a survey of the VCD and ROA spectra of four terpene or terpene-related molecules was carried out in the region 835–1345 cm$^{-1}$. Aside from the occurrence of several modes exhibiting strong VOA peaks in the corresponding VCD and ROA spectra and several modes exhibiting weak VOA intensities in each of these spectra, little other correlation could be found between the ROA and VCD. This reinforces the point that VCD and ROA are complementary techniques reflecting different ways in which a chiral molecule interacts with radiation during IR and Raman transitions. There was no statistical evidence to support the idea that a vibrational mode carries an intrinsic chirality that gives rise to the same sign for its VCD and ROA band. In another study (162), VCD and ROA were compared for the molecules 1-amino-2-propanol and 2-amino-1-propanol having the same absolute configuration at each of the two chiral centers. Since the ROA of these two molecules was rather similar, but the VCD strikingly dissimilar, the conclusion was reached that VCD is more sensitive to the molecule as a whole through longer-range coupling and ROA is more sensitive to local stereochemistry. In another comparative study (163), the VCD and ROA spectra of the cyclic dilactone, (3S,6S)-3,6-dimethyl-1,4-dioxane-2,5-one, were measured and calculated with ab initio density functional methods. Quantitative agreement was obtained between the measured and calculated VCD and IR spectra, whereas only qualitative agreement was obtained, using the same geometry and vibrational force field, for the ROA and Raman spectra. This illustrates the point that the theoretical formalism for calculating VCD spectra is more reliable than that currently available for ROA.

## 5.2 Small Biological Molecules

### 5.2.1 Amino Acids and Peptides

Several studies of the ROA of alanine and alanyl peptides have appeared recently. In the first, the experimental $DCP_I$-ROA of the alanyl peptides, L-alanylglycine (L-Ala-Gly), glycyl-L-alanine (Gly-L-Ala), L-alanyl-L-alanine [$(L-Ala)_2$], and L-alanyl-L-alanyl-L-alanine [$(L-Ala)_3$], were compared from an empirical perspective (164). It was found that the sum of the ROA spectra of L-Ala-Gly and Gly-L-Ala was nearly the same as the ROA of $(L-Ala)_2$. This indicates that the ROA associated with the N-terminus and C-terminus L-alanyl residues is essentially conserved for all three molecules and provides strong evidence that ROA is sensitive to the local stereochemistry of each residue and not very sensitive to its neighbor. Earlier studies of the VCD in the hydrogen-stretching region showed larger differences between these molecules, indicating a longer-range sensitivity to stereochemical environment for VCD. The ROA spectra of L-alanine and its deuterated isotopomers have been measured and calculated using ab initio quantum mechanical methods (45). For these molecules good agreement was obtained between theory and experiment, as illustrated for L-alanine in $H_2O$, shown in Fig. 16. Finally, a comparison of experimental to ab initio theoretical ROA for the molecule N-acetyl-N'-methyl-L-alaninamide has been carried out, with good success (46). The ROA spectra were measured in three different solvents, and the ROA calculations were able to predict the likely conformation present in each solvent.

### 5.2.2 Terpenes

Terpenes have been studied extensively by ROA and VCD. These molecules exhibit strong VOA due to their conformational rigidity, and they are widely soluble in organic solvents. Many can be measured as neat liquids, which further increases their accessibility. In addition to the VCD-ROA comparative study discussed earlier, the $DCP_I$-ROA for a series of terpenes was measured and analyzed empirically (165). Many of the compounds investigated had been studied over the years using right-angle ICP-ROA, and the new results were in close agreement with these earlier studies. A number of stereochemical marker bands were identified that were conserved in sign between molecules bearing similar structural fragments. This result is even further evidence for the local sensitivity of ROA.

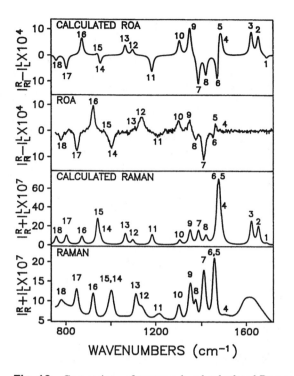

**Fig. 16** Comparison of measured and calculated Raman and ROA for L-alanine in $H_2O$ solution. (From Ref. 45 with permission of the American Chemical Society.)

### 5.2.3 Pharmaceutical Molecules

Molecules of pharmaceutical interest have on occasion been the focus of ROA studies. An early empirical study with a new $DCP_I$-ROA spectrometer contrasted the ROA of several ephedra molecules and demonstrated ROA marker bands sensitive to each of the two chiral centers in these molecules (166). More recently, the ROA spectra of naproxen and ibuprofen have been recorded, where the ROA of the former molecule is seen in resonance with a single excited electronic state (22). In general, ROA should be quite useful in the investigation of chiral molecules of pharmaceutical interest, and it is expected that further applications will appear in the future.

### 5.2.4 Saccharides and Nucleosides

The sugar ring is a source of rich ROA spectra. Over the past decade, ROA spectra have been measured for various forms of saccharides (167–169), and more recently the common pyrimidine nucleosides (170). From these empirical studies, one can deduce the nature of the expected ROA in particular regions of molecular structure of these molecules. In the case of the pyrimidine nucleosides, one has the additional source of molecular chirality arising from the linkage of the base to the sugar ring. The bases, in the monomeric nucleosides, are not a source of chirality, and their vibrational modes do not possess ROA spectra. Detailed vibrational assignments are not attempted, owing to the complexity of the underlying vibrational mode structure and the various conformations of these molecules that exist in solution. Nevertheless, these studies provide an empirical basis for studying structures of even more complex biological molecules, to be taken up in the following sections.

## 5.3  Large Biological Molecules

### 5.3.1  Proteins

Recent studies of various proteins using ROA have revealed a subtle sensitivity to the structure and dynamics of proteins that is not available with more conventional forms of spectroscopy. Unlike x-ray diffraction and NMR, ROA resolves structures in proteins that persist on the picosecond timescale. Further, its hypersensitivity to stereochemistry provides a structural enhancement beyond that obtained by ordinary Raman scattering. It complements VCD through its relative ease of sampling in aqueous solutions as well as its more local sensitivity to molecular environment, as discussed earlier for small molecules.

   One of the more interesting aspects of recent protein-ROA studies is the sensitivity of ROA to so-called tertiary fold or loop regions of proteins. Like VCD, ROA possesses characteristic sensitivity to the common secondary structural features of proteins, $\alpha$ helix, $\beta$ sheet and turns, and random coil; but beyond that, ROA is sensitive to those regions that connect these secondary structural motifs and can even be used to determine the degree of rigidity of these loop regions. Examples of early studies of loop structure were carried out for the proteins lysozyme (171) and alpha-lactalbumin (172). Subsequently, additional features of loop regions were investigated. It was found that in the ROA temperature profiles of lysozyme, there is evidence for a secondary, cooperative, entropic phase transition at 11°C, where the signature band for loop structure at $\sim$1340 cm$^{-1}$ gains intensity by approximately 40% (173). A similar phase transition was previously reported for alpha-lactalbumin (174). Evidence from other studies (175) indicates that the structure associated with these loop regions may be $3_{10}$ helix. Disordered, unfolded structures of proteins have also been studied and characterized by ROA. It is found that the ROA spectra simplify and broaden. For an essentially open-chain structure, there is mainly a broad couplet near 1300 cm$^{-1}$ for lyozyme (176). The further simplification of the ROA upon increasing the temperature has been taken as evidence for water promoting the rapid inter-conversion of backbone structure on the picosecond time scale (177). Most re-cently, the loop structure was studied in human serum albumin and found to decrease in magnitude with decreasing pH as the protein underwent a transition from the normal N state of the native conformation to the F state of the molten globule form (178). Although the band characteristic of the $\alpha$ helix decreased only marginally, the loop band at 1340 cm$^{-1}$ decreased by approximately 40%. Finally, we note that ROA spectra of glycoproteins have been observed, which bear ROA features of both proteins and saccharides (179).

### 5.3.2  Polysaccharides

The ROA of two polysaccharides has been investigated, laminarin and pullulan (180). Here features apparent for smaller saccharide molecules are present, in some cases with enhanced intensity. In a similar vein, the ROA spectra of cyclo-dextrins were reported some time ago, and again enhanced intensities relative to monosaccharides were observed (181).

### 5.3.3  Nucleic Acids

The ROA of nucleic acids in different forms has been reported in recent years. The first of these was a study of different polyribonucleic acids where the solution conformational structures of these polymers were characterized in $H_2O$ and $D_2O$ (182). Included in this work were ROA and Raman spectra for poly(rA), poly(rU), poly(rC), poly(rA)-poly(rU), and poly(rG)-poly(rC). In all of these spectra, the region from ~950 to 1150 cm$^{-1}$ was associated with structural fea-tures in the sugar-phosphate backbone, the region from ~1200 to 1550 cm$^{-1}$ with the base-sugar structure, and the region from ~1550 to 1750 cm$^{-1}$ with base stacking. In a follow-up study (183), evidence for global premelting in poly(rA)-poly(rU) using ROA was found where the magnitude of the ROA bands decreased uniformly across the spectrum with increasing temperature. This pattern is in opposition to the Raman hypochromic effect, where the ordinary Raman intensi-ties increase with increasing temperature. The reason cited for this "antihypo-chromic" effect is the larger number of conformations sampled by the protein with increasing temperature and the fact that the ROA associated with these vari-ous conformations has a tendency to reduce the observed ROA by cancellation. The ROA spectra obtained in these studies of polyribonucleic acids form the empirical basis for the recent report of ROA for calf thymus DNA and phenylala-nine transfer ROA, with and without $Mg^{2+}$ binding (184). These first reports of heterogeneous polynucleic acids are encouraging from the standpoint of the po-tential of ROA to study the structure and conformational dynamics of these mole-cules at a higher level of stereochemical sensitivity than is currently afforded by ordinary Raman scattering.

## 5.4  Theoretical Applications

In this final section on ROA applications, we consider several publications aimed at extending the range of ROA studies to include the resonance Raman effect.

### 5.4.1  Preresonance Raman Optical Activity

Ordinary Raman scattering may be considered far from resonance (FFR) if the frequency of the incident and scattered radiation can be treated in an equivalent way theoretically. Correspondingly, the theory of ROA in the FFR approximation, the original theory of ROA (4,5), is remarkably simple relative to the general theory (160,185). In the FFR limit, ICP and SCP forms of ROA are identical and $DCP_{II}ROA$ vanishes. Also in the FFR limit, backscattering $DCP_I$ and unpolarized ICP, the two most efficient forms of ROA collection, are equal to one another, as has been demonstrated experimentally (20). In order to probe, in a very sensitive way, the onset of preresonance Raman scattering and hence the breakdown of the FFR approximation, study was undertaken of four molecules, *trans*-pinane, α-pinene, verbenone, and quinidine, which have increasing aromatic character and increasingly approach electronic resonance with the argon ion laser line at 514 nm. While *trans*-pinane has no aromatic character and equivalent $ICP_u$ and $DCP_I$ROA spectra, the corresponding spectra for the subsequent three molecules in this series showed increasing sizes of difference. In the case of quinidine, large difference bands were recorded that in fact correspond to the first isolation of $DCP_{II}ROA$ spectra.

### 5.4.2  Resonance Raman Optical Activity

If the incident laser frequency becomes strongly resonant with a single excited state, which is the case if the laser line falls within the electronic absorption band envelope of the sample molecule, the theory of Raman scattering and with it the theory of ROA undergo a remarkable simplification. Recently, the formal theory of ROA in resonance with a single electronic state, the so-called SES limit, was published (186). A striking result of this work is the prediction that the ROA in this case derives its intensity uniformly, for all ROA bands in the spectrum, from the electronic circular dichroism of the resonant electronic state. Thus, all the bands in the resonance ROA (RROA) spectrum have the same sign and the same ratio of ROA to Raman intensity across the spectrum. This ratio is equal to and opposite in sign to the anisotropy ratio of the CD of the resonant electronic state, that is, the ratio of its CD to the intensity of its electronic absorption strength. Subsequent to this theoretical prediction, this effect was observed and its details confirmed for the molecule naproxen in chloroform solution, as shown in Fig. 17 (22). Naproxen sodium salt (high-pH form) has the opposite sign of CD compared to naproxen and its methoxy derivative, and hence the sign reversal seen in the RROA. The resonance interaction in naproxen is supported by the naphthyl

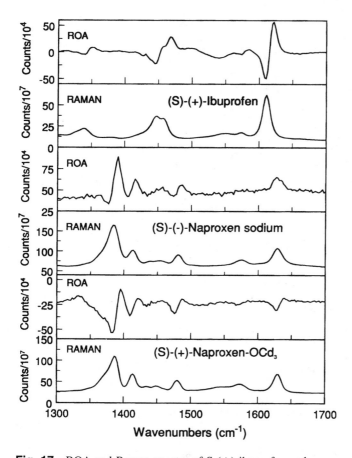

**Fig. 17** ROA and Raman spectra of S-(+)-ibuprofen and resonance Raman and ROA of (S)-(−)-naproxen sodium salt and (S)-(+)-naproxen-OCd₃. All spectra were measured in solution with 514-nm laser-excitation.

group, and if this resonance chromophore is effectively replaced with a phenyl group, as is the case for the similar molecule ibuprofen, a normal nonresonant, or possibly preresonant, ROA spectrum, having both positive and negative bands, is observed.

A final note of theoretical importance relative to resonance ROA is the development of a second approach to the calculation of ROA by ab initio methods (187). In this formalism, the necessary perturbations of the Raman optical activity tensors are generated by a sum-over-states formalism similar to the exciton scheme developed for VCD. An interesting feature of this method is the explicit inclusion of the incident laser frequency, allowing for a theoretical description of an approach to resonance. The starting equations apply more to the FFR approximation but could easily be adapted to a form consistent with the description of strong resonance ROA.

## 6  CONCLUSIONS

We have seen in this review that both VCD and ROA are powerful new spectroscopic tools for the study of various aspects of the stereochemistry of molecules of biological and pharmaceutical interest. VCD and ROA are highly complementary. Each has advantages relative to the other, and there is much to learn about a molecular sample from both techniques. Although VCD has an advantage over ROA in terms of the commercial availability of instrumentation and software for ab initio calculations, ROA will no doubt join VCD in this respect in the not-too-distant future.

## REFERENCES

1.  LA Nafie. Appl Spectrosc 50(5):14A–26A, 1996.
2.  LA Nafie. Ann Rev Phys Chem 48:357–386, 1997.
3.  LA Nafie, TB Freedman, Enantiomer 3:283–297, 1998.
4.  LD Barron, L Hecht, AF Bell. In: GD Fasman, ed. Circular Dichroism and the Conformation of Biomolecules. New York: Plenum Press, 1996, pp. 653–695.
5.  LD Barron, L Hecht, AF Bell, G Wilson. Appl Spectrosc 50:619–629, 1996.
6.  TA Keiderling. In: GD Fasman, ed. Circular Dichroism and the Conformational Analysis of Biomolecules. New York: Plenum Press, 1996, pp. 555–598.
7.  PJ Stephens, FJ Devlin, CS Ashvar, CF Chabalowski, MJ Frisch. Faraday Discuss: 103–119, 1994.
8.  G Holzwarth, EC Hsu, HS Mosher, TR Faulkner, A Moscowitz. J Am Chem Soc 96:251–252, 1974.
9.  LA Nafie, JC Cheng, PJ Stephens. J Am Chem Soc 97:3842, 1975.
10. LD Barron, MP Bogaard, AD Buckingham. J Am Chem Soc 95:603–605, 1973.
11. W Hug, S Kint, GF Bailey, JR Scherer. J Am Chem Soc 97:5589–5590, 1975.
12. LA Nafie, M Diem, DW Vidrine. J Am Chem Soc 101:496–498, 1979.
13. ED Lipp, CG Zimba, LA Nafie. Chem Phys Lett 90:1–5, 1982.
14. LA Nafie, DW Vidrine. In: JR Ferraro, LJ Basile, eds. Fourier Transform Infrared Spectroscopy. Vol. 3. New York: Academic Press, 1982, pp. 83–123.
15. PJ Stephens. J Phys Chem 89:748–752, 1985.
16. PJ Stephens, MA Lowe. Ann Rev Phys Chem 36:213–241, 1985.
17. LA Nafie, F Long, TB Freedman, H Buijs, A Rilling, J-R Roy, RK Dukor. In: J DeHaseth, ed. Fourier Transform Spectroscopy: 11th International Conference. Vol. 430. Woodbury, NY: Amer. Inst. of Phys., 1997, pp. 432–434.
18. MJ Frisch, GW Trucks, HB Schlegel, GE Scuseria, MA Robb, JR Cheeseman, VG Zakrzewski, JA Montgomery, Jr., RE Stratmann, JC Burant, S Dapprich, JM Mil-

lam, AD Daniels, KN Kudin, MC Strain, O Farkas, J Tomasi, V Barone, M Cossi, R Cammi, B Mennucci, C Pomelli, C Adamo, S Clifford, J Ochterski, GA Petersson, PY Ayala, Q Cui, K Morokuma, DK Malick, AD Rabuck, K Raghavachari, JB Foresman, J Cioslowski, JV Ortiz, BB Stefanov, G Liu, A Liashenko, P Piskorz, I Komaromi, R Gomperts, RL Martin, DJ Fox, T Keith, MA Al-Laham, CY Peng, A Nanayakkara, C Gonzalez, M Challacombe, PMW Gill, B Johnson, W Chen, MW Wong, JL Andres, C Gonzalez, M Head-Gordon, ES Replogle, JA Pople. Gaussian 98. A.5 ed. Pittsburgh, PA: Gaussian, 1998.

19. LA Nafie, TB Freedman. Chem Phys Lett 154:260–266, 1989.
20. D Che, L Hecht, LA Nafie. Chem Phys Lett 180:182–190, 1991.
21. M Vargek, TB Freedman, LA Nafie. J Raman Spectrosc 28:627–633, 1997.
22. M Vargek, TB Freedman, E Lee, LA Nafie. Chem Phys Lett 287:359–364, 1998.
23. PK Bose, LD Barron, PL Polavarapu. Chem Phys Lett 155(4,5):423–429, 1989.
24. PL Polavarapu. J Phys Chem 94:8106–8112, 1990.
25. W Hug, H Surbeck. Chem Phys Lett 60:186–192, 1979.
26. LD Barron, L Hecht, W Hug, MJ MacIntosh. J Am Chem Soc 111:8731–8732, 1989.
27. L Hecht, D Che, LA Nafie. Appl Spectrosc 45:18–25, 1991.
28. F Long, TB Freedman, R Hapanowicz, LA Nafie. Appl Spectrosc 51:504–508, 1997.
29. L Hecht, LD Barron. J Mol Struct 347:449–458, 1995.
30. W Hug. In: J Lindon, G Tranter, J Holmes, eds. Encyclcopedia of Spectroscopy and Spectrometry. London: Academic Press, London, 1999, pp. 1966–1976.
31. FJ Devlin, PJ Stephens. J Phys Chem A 101:9912–9924, 1997.
32. LA Nafie, TB Freedman. J Chem Phys 78:7108–7116, 1983.
33. R Dutler, A Rauk. J Am Chem Soc 111:6957–6966, 1989.
34. JL McCann, A Rauk, H Wieser. Can J Chem 76:274–283, 1998.
35. TB Freedman, LA Nafie, D Yang. Chem Phys Lett 227:419–428, 1994.
36. TB Freedman, F Long, M Citra, LA Nafie. Enantiomer 4:103–119, 1999.
37. PL Polavarapu, PK Bose. J Chem Phys 93:7524, 1990.
38. LA Nafie, TH Walnut. Chem Phys Lett 49:441–446, 1977.
39. M Gulotta, DJ Goss, M Diem. Biopolymers 28:2047–2058, 1989.
40. L Hecht, AL Phillips, LD Barron. J Raman Spectrosc 26:727–732, 1995.
41. KM Spencer, RB Edmonds, RD Rauh. Appl Spectrosc 50:681–685, 1996.
42. FJ Devlin, PJ Stephens, JR Cheeseman, MJ Frisch. J Am Chem Soc 118:6327–6328, 1996.
43. LA Nafie. In: MW Mackenzie, ed. Advances in Applied FTIR Spectroscopy. New York: Wiley, 1988, pp. 67–104.
44. J Costante, L Hecht, PL Polavarapu, A Collet, LD Barron. Angew Chem Int Ed 36:885–887, 1997.
45. G-S Yu, TB Freedman, LA Nafie, Z Deng, PL Polavarapu. J Phys Chem 99:835–843, 1995.
46. Z Deng, PL Polavarapu, SJ Ford, L Hecht, LD Barron, CS Ewig, K Jalkanen. J Phys Chem 100:2025–2034, 1996.
47. LA Nafie, ED Lipp, A Chernovitz, G Paterlini. In: H Ishida, ed. FT-IR Characterization of Polymers. New York: Plenum Press, 1987, pp. 81–95.

48. PL Polavarapu. In: HD Bist, JR Durig, JF Sullivan, eds. Vibrational Spectra and Structure. Vol. 17B. Amsterdam: Elsevier, 1989, pp. 319–342.
49. TB Freedman, LA Nafie, TA Keiderling. Biopolymers 37:265–279, 1995.
50. TB Freedman, LA Nafie. In: EL Eliel, SH Wilen, eds. Topics in Sterochemistry. Vol. 17. New York: Wiley, 1987, pp. 113–206.
51. TB Freedman, DM Gigante, MJ Citra, MG Paterlini. In: LA Nafie, HH Mantsch, eds. Biomolecular Spectroscopy III. Vol. 1890. Bellingham, Washington, SPIE, 1993, pp. 40–46.
52. TA Keiderling, SC Yasui, U Narayanan, A Annamalai, P Malon, R Kobrinskaya, L Yang. In: ES Schmid, FW Schneider, F Siebert, eds. Spectroscopy of Biological Molecules: New Advances. Freiburg, Germany, 1988, pp. 73–76.
53. TA Keiderling, P Pancoska, SC Yasui, M Urbanova, RK Dukor. In: V Renugopala-krishnan, PR Carey, ICP Smith, SG Huang, AC Storer, eds. Proteins: Structure, Dynamics and Design. Leiden, Germany: ESCOMP, 1991, pp. 165–170.
54. TA Keiderling, P Pancoska. In: RE Hester, RJH Clark, eds. Biomolecular Spectroscopy, Part B. Vol. 21. Chichester, England: Wiley, 1993, pp. 267–315.
55. TA Keiderling. In: K Nakanishi, ND Berova, RW Woody, eds. Circular Dichroism: Interpretation and Applications. New York: VCH, 1994, pp. 497–521.
56. M Diem, PJ Gotkin, JM Kupfer, AG Tindall, LA Nafie. J Am Chem Soc 99:8103–8104, 1977.
57. M Diem, E Photos, H Khouri, LA Nafie. J Am Chem Soc 101:6829–6837, 1979.
58. M Diem, PL Polavarapu, M Oboodi, LA Nafie. J Am Chem Soc 104:3329–3336, 1982.
59. BB Lal, M Diem, PL Polavarapu, M Oboodi, TB Freedman, LA Nafie. J Am Chem Soc 104, 1982.
60. TB Freedman, M Diem, PL Polavarapu, LA Nafie. J Am Chem Soc 104:3343, 1982.
61. LA Nafie, MR Oboodi, TB Freedman. J Am Chem Soc 105:7449–7450, 1983.
62. WM Zuk, TB Freedman, LA Nafie. J Phys Chem 93:1771–1779, 1989.
63. MR Oboodi, BB Lal, DA Young, TB Freedman, LA Nafie. J Am Chem Soc 107: 1547–1556, 1985.
64. DA Young, ED Lipp, LA Nafie. J Am Chem Soc 107:6205–6213, 1985.
65. TB Freedman, DA Young, MR Oboodi, LA Nafie. J Am Chem Soc 109:1551–1559, 1987.
66. M Diem, PJ Gotkin, JM Kupfer, LA Nafie. J Am Chem Soc 100:5644–5650, 1978.
67. WM Zuk, TB Freedman, LA Nafie. Biopolymers 28:2025–2044, 1989.
68. AC Chernovitz, TB Freedman, LA Nafie. Biopolymers 26:1879–1900, 1987.
69. TB Freedman, AC Chernovitz, WM Zuk, MG Paterlini, LA Nafie. J Am Chem Soc 110:6970–6974, 1988.
70. GM Roberts, O Lee, J Calienni, M Diem. J Amer Chem Soc 110:1749–1752, 1988.
71. M Diem, O Lee, GM Roberts. J Phys Chem 96:548–554, 1992.
72. SS Birke, C Farrell, O Lee, I Agbaje, G Roberts, M Diem. In: RE Hester, RB Girling, eds. Spectroscopy of Biological Molecules. Cambridge, UK: Royal Society of Chemistry, 1991, pp. 131.
73. AC Chernovitz, TB Freedman, LA Nafie. In: J Grasselli et al., eds. 1985 Conference on Fourier and Computerized Infrared Spectroscopy. Proc. SPIE, 1985, pp. 222–223.

74. M Miyazawa, Y Kyogoku, H Sugeta. Spectrochim Acta A 50:1505–1511, 1994.
75. E Tajkhorshid, KJ Jalkanen, S Suhai. J Phys Chem B 102:5899–5913, 1998.
76. WG Han, KJ Jalkanen, M Elstner, S Suhai. J Phys Chem B 102:2587–2602, 1998.
77. C Marcott, HA Havel, J Overend, A Moscowitz. J Am Chem Soc 100:7088–7089, 1978.
78. MG Paterlini, TB Freedman, LA Nafie. J Am Chem Soc 108:1389–1397, 1986.
79. DM Back, PL Polavarapu. Carbohydr Res 133:163–167, 1984.
80. T Chandramouly, DM Back, PL Polavarapu. J Chem Soc, Trans Farad Soc 84(8): 2585–2594, 1988.
81. LA Nafie, TA Keiderling, PJ Stephens. J Am Chem Soc 98:2715–2723, 1976.
82. PL Polavarapu, M Diem, LA Nafie. J Am Chem Soc 102:5449–5453, 1980.
83. RD Singh, TA Keiderling. J Am Chem Soc 103:2387–2394, 1981.
84. L Laux, V Pultz, S Abbate, HA Havel, J Overend, A Moscowitz. J Am Chem Soc 104:4276–4278, 1982.
85. X Qu, E Lee, G-S Yu, TB Freedman, LA Nafie. Appl Spectrosc 50:649–657, 1996.
86. FJ Devlin, PJ Stephens. J Am Chem Soc 116:5003–5004, 1994.
87. D Tsankov, V Dimitrov, H Wieser. Mikrochim Acta: 535–537, 1997.
88. P Bour, J McCann, H Wieser. J Chem Phys 108:8782–8789, 1998.
89. P Bour, J McCann, H Wieser. J Phys Chem A 102:102–110, 1998.
90. DMP Gigante, F Long, L Bodack, JM Evans, J Kallmerten, LA Nafie, TB Freedman. J Phys Chem A 103:1523–1537, 1999.
91. H Wyssbrod, M Diem. Biopolymers 32:1237–1242, 1992.
92. P Xie, QW Zhou, M Diem. J Am Chem Soc 117:9502–9508, 1995.
93. P Xie, QW Zhou, M Diem. Faraday Discuss: 233–243, 1994.
94. P Xie, M Diem. J Amer Chem Soc 117:429–437, 1995.
95. TB Freedman, N Ragunathan, S Alexander. Faraday Discuss 99:131–150, 1994.
96. TB Freedman, S Liu, E Lee, LA Nafie. Biophys J 70:TU435, 1996.
97. LA Bodack, B Chowdry, TB Freedman, LA Nafie. (Unpublished results).
98. PL Polavarapu, AL Cholli, G Vernice. J Am Chem Soc 114(27):10953–10955, 1992.
99. PL Polavarapu, AL Cholli, G Vernice. J Pharm Sci 82:791–793, 1993; ibid. 786: 267, 1997.
100. NP Franks, WR Lieb. Science 254:427–430, 1991.
101. CS Ashvar, PJ Stephens, T Eggimann, H Wieser. Tetrahedron: Asymm 9:1107–1110, 1998.
102. RD Singh, TA Keiderling. Biopolymers 20:237–240, 1981.
103. BB Lal, LA Nafie. Biopolymers 21:2161–2183, 1982.
104. AC Sen, TA Keiderling. Biopolymers 23:1519–1532, 1984.
105. AC Sen, TA Keiderling. Biopolymers 23:1533–1546, 1984.
106. SC Yasui, TA Keiderling. J Am Chem Soc 108:5576–5581, 1986.
107. MG Paterlini, TB Freedman, LA Nafie. Biopolymers 25:1751–1765, 1986.
108. SC Yasui, TA Keiderling, GM Bonora, C Toniolo. Biopolymers 25:79–89, 1986.
109. SC Yasui, TA Keiderling, F Formaggio, GM Bonora, C Toniolo. J Am Chem Soc 108:4988–4993, 1986.
110. P Bour, TA Keiderling. J Am Chem Soc 115:9602–9607, 1993.
111. SS Birke, I Agbaje, M Diem. Biochemistry 31(2):450–455, 1992.

112. VP Gupta, TA Keiderling. Biopolymers 32:239–248, 1992.
113. M Miyazawa, K Inouye, T Hayakawa, Y Kyogoku, H Sugeta. Appl Spectrosc 50: 664–648, 1996.
114. W Mastle, RK Dukor, G Yoder, TA Keiderling. Biopolymers 36:623–631, 1995.
115. G Yoder, TA Keiderling, F Formaggio, M Crisma, C Toniolo. Biopolymers 35: 103–111, 1995.
116. G Yoder, TA Keiderling, F Formaggio, M Crisma, C Toniolo, J Kamphuis. Tetrahedron: Asymm 6:687–690, 1995.
117. V Baumruk, DF Huo, RK Dukor, TA Keiderling, D Lelievre, A Brack. Biopolymers 34:1115–1121, 1994.
118. G Yoder, A Polese, R Silva, F Formaggio, M Crisma, QB Broxterman, J Kamphuis, C Toniolo, TA Keiderling. J Am Chem Soc 119:10278–10285, 1997.
119. G Yoder, P Pancoska, TA Keiderling. Biochemistry USA 36:15123–15133, 1997.
120. R Dukor, TA Keiderling. Biospectroscopy 2:83–100, 1996.
121. MG Paterlini, TB Freedman, LA Nafie, Y Tor, A Shanzer. Biopolymers 32:765–782, 1992.
122. V Baumruk, P Pancoska, TA Keiderling. J Mol Biol 259:774–791, 1996.
123. TA Keiderling. In: I Bainau, H Pessen, TF Kumosinski, eds. New Techniques and Applications of Physical Chemistry to Food Systems (Physical Chemistry of Food Processes. Vol. II). New York: Van Nostrand Reinhold, 1993, pp. 307–337.
124. TA Keiderling, B Wang, M Urbanova, P Pancoska, RK Dukor. Faraday Discuss 99:263–285, 1994.
125. P Pancoska, E Bitto, V Janota, TA Keiderling. Faraday Discuss:287–310, 1994.
126. V Baumruk, TA Keiderling. J Am Chem Soc 115:6939, 1993.
127. P Pancoska, V Janota, TA Keiderling. Appl Spectrosc 50:658–668, 1996.
128. TA Keiderling, P Pancoska. Biophys J 70:MP165, 1996.
129. P Pancoska, H Fabian, G Yoder, V Baumruk, TA Keiderling. Biochemistry USA 35:13094–13106, 1996.
130. P Pancoska, E Bitto, V Janota, M Urbanova, VP Gupta, TA Keiderling. Protein Sci 4:1384–1401, 1995.
131. P Pancoska, E Bitto, V Janota, TA Keiderling. Abstr Pap Amer Chem Soc 210: 151–COMP, 1995.
132. M Urbanova, TA Keiderling, P Pancoska. Bioelectrochem Bioenerg 41:77–80, 1996.
133. BI Baello, P Pancoska, TA Keiderling. Anal Biochem 250:212–221, 1997.
134. P Pancoska, V Janota, J Kubelka, TA Keiderling. Biophys J 74:A65, 1998.
135. C Marcott, HA Havel, B Hedlund, K Overend, A Moscowitz. In: SF Mason, ed. Optical Activity and Chiral Discrimination. New York: Reidel, 1979, p. 289.
136. J Teraoka, K Nakamura, Y Nakahara, Y Kyogoku, H Sugeta. J Am Chem Soc 114: 9211–9213, 1992.
137. RW Bormett, SA Asher, PJ Larkin, WG Gustafson, N Ragunathan, TB Freedman, LA Nafie, S Balasubramanian, SG Boxer, N-T Yu, K Gersonde, RW Noble, BA Springer, SG Sligar. J Am Chem Soc 114:6864, 1992.
138. RW Bormett, GD Smith, SA Asher, D Barrick, DM Kurtz. Faraday Discuss:327–339, 1994.
139. PJ Stephens. Faraday Discuss 99:383, 1994.

140. J Teraoka, N Yamamoto, Y Matsumoto, Y Kyogoku, H Sugeta. J Am Chem Soc 118:8875–8878, 1996.
141. A Annamalai, TA Keiderling. J Am Chem Soc 109:3125–3132, 1987.
142. WX Zhong, M Gulotta, DJ Goss, M Diem. Biochemistry 29:7485, 1990.
143. W-X Zhong, M Gulotta, DJ Goss, M Diem. Biochemistry 29:7485, 1990.
144. L Wang, L Yang, TA Keiderling. In: RE Hester, RB Girling, eds. in Spectroscopy of Biological Molecules. Cambridge, England: Royal Society of Chemistry, 1991, pp. 137–138.
145. T Xiang, DJ Goss, M Diem. Biophys J 65:1255–1261, 1993.
146. L Wang, TA Keiderling. Biochemistry 31(42):10265–10271, 1992.
147. L Wang, L Yang, TA Keiderling. Biophys J 67:2460–2467, 1994.
148. L Wang, TA Keiderling. Nucl Acid Res 21:4127–4132, 1993.
149. L Yang, TA Keiderling. Biopolymers 33:315–327, 1993.
150. L Wang, P Pancoska, TA Keiderling. Biochemistry USA 33:8428–8435, 1994.
151. SS Birke, M Moses, B Kagalovsky, D Jano, M Gulotta, M Diem. Biophys J 65:1262–1271, 1993.
152. BD Self, DS Moore. Biophys J 73:339–347, 1997.
153. BD Self, DS Moore. Biophys J 74:2249–2258, 1998.
154. V Maharaj, D Tsankov, HJ Vandesande, H Wieser. J Mol Struct 349:25–28, 1995.
155. V Maharaj, H Vandesande, D Tsankov, A Rauk, H Wieser. Mikrochim Acta: 529–530, 1997.
156. V Maharaj, A Rauk, JH Vandesande, H Wieser. J Mol Struct 408:315–318, 1997.
157. LD Barron, L Hecht. In: K Nakanishi, ND Berova, RW Woody, eds. Circular Dichroism: Principles and Applications. New York: VCH, 1994, pp. 179–215.
158. LD Barron, SJ Ford, AF Bell, G Wilson, L Hecht, A Cooper. Faraday Discuss 99:217–232, 1994.
159. LD Barron, L Hecht, SJ Ford, AF Bell, G Wilson. J Mol Struct 349:397–400, 1995.
160. LA Nafie, D Che. In: M Evans, S Kielich, eds. Modern Nonlinear Optics, Part 3. Vol. 85. New York: Wiley, 1994, pp. 105–149.
161. LA Nafie, GS Yu, XH Qu, TB Freedman. Faraday Discuss:13–34, 1994.
162. CN Tam, P Bour, TA Keiderling. J Am Chem Soc 119:7061–7064, 1997.
163. CN Tam, P Bour, TA Keiderling. J Am Chem Soc 118:10285–10293, 1996.
164. G-S Yu, D Che, TB Freedman, LA Nafie. Biospectroscopy 1:113–123, 1995.
165. GS Yu, TB Freedman, LA Nafie. J Raman Spectrosc 26:733–743, 1995.
166. G-S Yu, D Che, TB Freedman, LA Nafie. Tetrahedron Asymm 4:511–516, 1993.
167. AF Bell, LD Barron, L Hecht. Carbohydrate Research 257:11–24, 1994.
168. AF Bell, L Hecht, LD Barron. J Mol Struct 349:401–404, 1995.
169. AF Bell, L Hecht, LD Barron. Spectrochim Acta Pt a Mol Bio 51:1367–1378, 1995.
170. AF Bell, L Hecht, LD Barron. J Chem Soc Faraday Trans 93:553–562, 1997.
171. SJ Ford, A Cooper, L Hecht, G Wilson, LD Barron. J Chem Soc Faraday Trans 91:2087–2093, 1995.
172. G Wilson, SJ Ford, A Cooper, L Hecht, ZQ Wen, LD Barron. J Mol Biol 254:747–760, 1995.
173. G Wilson, L Hecht, LD Barron. J Phys Chem B 101:694–698, 1997.
174. G Wilson, L Hecht, LD Barron. J Mol Biol 261:341–347, 1996.

175. G Wilson, L Hecht, LD Barron. J Chem Soc Faraday Trans 92:1503–1509, 1996.

176. G Wilson, L Hecht, LD Barron. Biochemistry 35:12518–12525, 1996.

177. LD Barron, L Hecht, G Wilson. Biochemistry 36:13143–13147, 1997.

178. J Teraoka, AF Bell, L Hecht, LD Barron. J Raman Spectrosc 29: 67–71, 1998.

179. AF Bell, SJ Ford, L Hecht, G Wilson, LD Barron. Int J Biol Molecules 16:277–278, 1994.

180. AF Bell, L Hecht, LD Barron. J Raman Spectrosc 26:1071–1074, 1995.

181. LD Barron, AR Gargaro, ZQ Wen, DD MacNicol, C Butters. Tetrahedron Asymm 1:513–516, 1990.

182. AF Bell, L Hecht, LD Barron. J Am Chem Soc 119:6006–6013, 1997.

183. AF Bell, L Hecht, LD Barron. Biospectroscopy 4:107–111, 1998.

184. AF Bell, L Hecht, LD Barron. J Am Chem Soc 120:5820–5821, 1998.

185. L Hecht, LA Nafie. Mol Phys 72:441–469, 1991.

186. LA Nafie. Chem Phys 205:309–322, 1996.

187. P Bour. Chem Phys Lett 288:363–370, 1998.

# 3

# Vibrational Circular Dichroism of Peptides and Proteins
## Survey of Techniques, Qualitative and Quantitative Analyses, and Applications

**Timothy A. Keiderling**
*University of Illinois at Chicago, Chicago, Illinois*

## 1 INTRODUCTION

Determination of protein secondary structure has long been a major application of optical spectroscopic studies of biopolymers (1–3). In most cases these efforts have been aimed at evaluating the average fractional amount of helix and sheet contributions to the overall secondary structure in a protein or peptide. In some cases further interpretations in terms of turns and specific helix and sheet segment types have developed. This focus on average secondary structure is a consequence of the interactions of primary importance to these techniques and their relatively low resolution, which ordinarily does not provide site-specific information without technologically challenging selective isotopic substitution. Such a limit is in contrast to x-ray crystallography and nuclear magnetic resonance (NMR) spectroscopy, which naturally yield site-specific structural information, due to their very high resolution (at least for NMR), requiring at most uniform labeling. However, these invaluable structural biological techniques are very slow in terms of both data acquisition and completion of the complex interpretive process. Furthermore, the intrinsic time scales of the measurements are slow, thereby not permitting reliable analysis of dynamic structures or of conformations undergoing fast-changing events. Only more limited applications of optical spectra to determination of the tertiary structure or fold of the secondary structural elements have appeared, and these have typically used fluorescence or near-UV electronic circular dichroism (ECD) of aromatic residues to sense a change in the fold rather than to determine its nature (4,5).

For biomolecular structural studies, ECD of transitions in the ultraviolet has been one of the dominant applications of that technique due to its sensitivity to molecular conformation, which is often manifested in complete sign reversals for selected structural changes. Most spectroscopic secondary structure studies of proteins and peptides have used far-UV ECD for the $n-\pi^*$ and $\pi-\pi^*$ transitions of the amide linkage (6–12). Coupling of the involved electronic transition dipoles leads to the extended chirality that is characteristic of the chain conformation and results in the observed CD spectrum, which has a sign/frequency profile that varies for different secondary structures. Since these transitions are broad and fully overlapped, their individual frequencies do not have much impact on the correlation of CD spectral properties with structural features. On the contrary, the usual interpretations employ bandshape-based schemes that are either qualitative in nature or dependent on a statistical fit to a set of (typically protein) spectra that provide the structural reference set (8,10,12). While the earliest methods (13) utilized data from polypeptides for qualitative and quantitative analyses, more reliable results were obtained using a basis set of proteins whose crystal structures were known.

Infrared and Raman analyses of secondary structure historically took a different approach (14,15), due to the natural resolution of the vibrational region of the spectrum into contributions from modes characteristic of different bond types in the molecule. Initial focus was on assigning component frequencies to various secondary structural component types, as has been discussed in several reviews (3,16–20). Most effort focused on the mid-IR amide I (C=O stretch) band with additional use of the amide II (N—H deformation plus C—N stretch) in IR spectra and amide III (oppositely phased N—H deformation plus C—N stretch) with Raman spectral methods. In nonaqueous media, the near-IR amide A (N—H stretch) can also be useful. Since both IR and Raman techniques give rise to single-signed spectral bandshapes that are effectively just the dispersed sum of the contributions from all the component transitions, and those components differ by only relatively small amount in frequency, the bandshapes for different proteins are very similar. However, due to the high signal-to-noise ratio (S/N) of Fourier transform IR (FTIR), this approach can be pushed further with resolution enhancement using second-derivative or Fourier self-deconvolution (FSD) techniques (21–23). These latter methods partially compensate for the relatively low level of bandshape variation available in the IR and Raman spectra, but are subject to abuse by the unwary user and submit the analysis to several assumptions that may well be unwarranted (16,24). Most notably, the frequencies assigned to specific secondary structural types are assumed to be unique, whereas any given type will shift significantly under the influence of different solvents and residues (25). Additionally, nonuniformity and end effects for these segments of a given structural type will have significant impact on the frequencies, resulting in some dispersion of the contribution of a given secondary structure segment over the spectrum (26–28). Finally, most methods assume that the dipole strengths (extinction coefficients) of all the residues, regardless of their conformation, are the same, while they in fact vary (16,29,30). Nonetheless such FTIR methods have proven to be very useful for sorting out differences between proteins with similar IR bandshapes. Bandshape-based analyses similar to those used for ECD have also been applied to FTIR and Raman spectra, with reasonable success, as seen with ECD (26,31–35).

This contrast of FTIR and ECD sensitivities has led to the development of vibrational (or infrared) CD (VCD, or $\Delta A = A_L - A_R$, the differential absorbance for left and right circularly polarized light by vibrational transitions in IR) and its counterpart, Raman optical activity (ROA) in several laboratories. Only the former measurement will be addressed here; but reviews of ROA abound (36–41). The key impetus for moving to the vibrational region of the spectrum is that it is rich with resolved transitions, which are characteristic of localized parts of the molecule, and that by making a VCD measurement one can impart distinct stereochemical sensitivity to each of them (37,39,41–48). The chromophores needed in the molecule for VCD measurement are simply the bonds themselves as sampled by their stretching and bond deformation excitations. Chiral interaction of these bonds (necessary to establish optical activity) will then be manifest in the spectral bandshape that results from a VCD measurement. Furthermore these vibrational excitations are part of the ground state of the molecule (where all thermal processes occur, such as conformational variations), for which one normally wishes to gain structural insight. Though VCD has the three-dimensional structural sensitivity natural to a chiroptical technique, this is manifested in a response distributed over a large number of localized probes of the structure (43,49–53). In other words, VCD is to IR what ECD is to UV absorption spectra; nothing is lost but much is gained.

Of course this benefit comes at a cost, which arises from significantly reduced S/N and some theoretical interpretive difficulty as compared to IR. Developments on the latter front are fast bringing the theoretical capability for prediction of VCD spectra for small molecules to a level that is demonstrably superior to that for ECD spectra (53,54). However, these ab initio quantum mechanical methods are severely limited if one were to apply them to large molecules, such as proteins. Thus most biomolecular applications of VCD use empirically based analyses (45,48,55). Experimentally, instrumentation has reached a stage where VCD spectra for most molecular systems of interest can be measured under at least some sampling conditions (48,51,52,56–59). It is true that most VCD studies of biomolecules in aqueous solution, naturally the condition of prime interest, are restricted to relatively high-concentration samples, much as is characteristic of NMR. (By comparison, ROA measurements demand even higher concentrations (40).)

VCD has the bandshape variability of CD coupled with the frequency resolution of IR, which leads to a demonstrably enhanced sensitivity to secondary structure in proteins (60). Sampling conditions for obtaining experimental VCD spectra of protein and peptide samples are similar to those used in FTIR studies, with the important exception that the data are differential spectra of much smaller amplitude ($\Delta A$ is of the order of $10^{-4}$–$10^{-5}$ of the sample absorbance, $A$) and thus lower S/N. Consequently, to obtain quality VCD spectra, much longer data collection times are required than for FTIR or ECD, and specially designed instruments are used. Experimental methods of biomolecular VCD are summarized in the next section and are fully discussed in separate reviews (41,44,48,51,57,58,61). Theoretical techniques for simulation of small-molecule VCD are also the focus of several previous reviews (37,41,42,52–54,62) and will be only briefly surveyed in a section following that. To date, theory has played a minor role in the interpretation of VCD spectra of most biomolecules due to their large size. However, recent advances in computational techniques and hardware have made larger calculations on sizable (at least, realistic) peptides, ever more feasible.

On the other hand, qualitative analyses can be done with facility utilizing the VCD bandshape and its frequency position to predict the dominant secondary structural type in a peptide or protein. This has become the standard approach for most peptide studies (59). In such smaller, fully solvated, biopolymers, the frequency shifts due to solvent effects and the inhomogeneity of the peptide secondary structure (as well as end fraying) are severe problems for frequency-based analyses. In globular proteins, qualitative estimations of structure remain of interest for determining the dominant fold type, e.g., highly helical, highly sheet, or mixed helix and sheet types, but quantitative estimations of secondary structure content based on empirical spectral analyses are usually of more interest. Quantitative methods for analysis of protein VCD spectra in terms of structure follow the lead of most ECD analyses (8,10) and employ bandshape techniques referenced to a training set of protein spectra (35,63–72). In this respect, globular proteins in aqueous solution are assumed, on average, to have their peptide segments in similar environments and of similar lengths, so the solvent and length effects on the peptide modes will be relatively consistent for the training set and any unknowns studied.

In summary, unlike ECD, VCD can be used to correlate data for several different spectrally resolved features; and, unlike IR and Raman spectroscopies, each of these features will have a physical dependence on stereochemistry. But from another point of view, the combination of these techniques can compensate for each other, providing a balance between accuracy and reliability. The prime questions remaining in the VCD field now relate to application and interpretation of the method. It is clear that, despite claims of fundamental advantages of any one technique, progress in understanding of biomolecular structures will come from synthesizing all the data gathered from various techniques. In our biomolecular work, different types of spectral data are used to place bounds on the reliability of structural inferences that might be drawn from any one technique. Furthermore, such insight into average secondary structure may provide constraints on site-specific structure analyses using NMR and computational methods.

## 2 EXPERIMENTAL TECHNIQUES

This new dimension in optical activity comes at some cost, in that the rotational strengths of vibrational transitions as detected in VCD are much weaker than are those of electronic transitions detected in ECD. Similarly since VCD is a differential IR technique, its S/N can never approach that of FTIR, which represents a summed response. Several research groups have developed instrumentation that makes the measurement of VCD reasonably routine over much of the IR region (51,57,58,73–79). Commercial FTIR vendors are now providing VCD accessories or, in one instance (Bomem-Biotools), a stand-alone VCD instrument that has now been shown by its users to have exceptional S/N and baseline characteristics (80,81). In this section, instrument designs are briefly summarized and compared.

### 2.1 Fourier Transform–VCD versus Dispersive VCD

Available instrumentation makes routine measurement of VCD possible over much of the IR region down to ~700 cm$^{-1}$ on many samples. Development of a VCD instrument is normally accomplished by extending a dispersive IR or an FTIR spectrometer to accommodate, in terms of optics, time-varying modulation of the polarization state of the light and, in terms of electronics, detection of the modulated intensity that results from a sample with nonzero VCD. Our instruments and those of others are described in the literature in detail as referenced in recent reviews (41,49,51,52,57,58). A detailed review contrasting these designs and detailing components needed to construct either type of instrument has been published by this author (51). Here only a brief survey of the important components is given.

VCD instruments share several generic elements with ''normal'' CD instruments, as schematically outlined in Fig. 1. All current instruments use a broadband source of light, typically utilizing black-body radiation from something like a ceramic or graphite-based glower (or tungsten in the near-IR), to allow sampling of a spectrum over the IR region. The method chosen for encoding the optical frequencies divides VCD instruments into two styles. *Dispersive VCD* instruments use a monochromator, based on grating technology, which scans through only the wavelength spectrum of interest, recording the response sequentially. Such an instrument must be optimized for efficient light collection. On the other

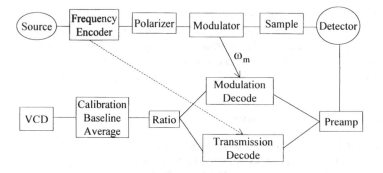

**Fig. 1**  Schematic diagram of a generic circular dichroism spectrometer. The frequency (or wavelength) encoder can be a monochromator or interferometer, whose characteristics (chopping or interferogram, respectively, amplitude modulation) will underlie the detection scheme for transmission. The polarizer, polarization modulator, and detector are optimized for transmission or efficiency in the spectral region of interest; components for VCD in the near- and mid-IR are enumerated in the text. The modulation signal is phase-sensitive detected with a lock-in amplifier referenced to the photoelastic modulator. The decoded intensities are analog ratioed or divided in the computer, which is also used to manipulate and correct the data to yield the final VCD spectrum.

hand, *Fourier transform (FT) VCD* instruments use a Michelson interferometer that encodes the optical frequencies as an interferogram and gains efficiency through the multiplex advantage. With FT-VCD, spectral responses at all wavenumbers of interest are obtained simultaneously, with longer scans serving to improve the resolution of spectral features. Both styles of instrument then are modified to provide for linear polarization of the light beam, normally with a wire grid polarizer, and modulation of it, with a photoelastic modulator (PEM), between (elliptically) right- and left-hand polarization states. The beam then passes through the sample and onto the detector, typically a liquid $N_2$ cooled $Hg_{1-x}(Cd)_xTe$ (MCT) photoconducting diode. After preamplification, the electrical signal developed in the MCT is separated to measure the overall transmission spectrum ($I_{trans}$) of the instrument and sample via one channel and the polarization modulation intensity ($I_{mod}$), which is related to the VCD intensity, via the other.

   These signals are ratioed to yield the raw VCD signal either before or after *A*-to-*D* conversion, depending on the instrumental design. Since VCD is a differential absorbance measurement, $\Delta A = A_L - A_R$ for left/right (*L/R*) circularly polarized light, it is necessary to ratio these two intensities to normalize out any dependence on the source intensity and instrument transmission characteristics. In the limit of small $\Delta A$ values,

$$\frac{I_{mod}}{I_{trans}} = (1.15\ \Delta A)J_1(\alpha_0)g_I \tag{1}$$

where $J_1(\alpha_0)$ is the first-order Bessel function at the maximum retardation of the modulator, $\alpha_0$, and $g_I$ is an instrument gain factor. Evaluation of this term and elimination of the gain factor are obtained by calibration of the VCD using a pseudosample composed of a birefringent plate and a polarizer pair or by measuring the VCD of a known sample (51,56,57,76,82,83). Further processing of the computer-stored VCD spectrum involving calibration, baseline correction and spectral averaging or smoothing, as desired, and conversion to molar quantities, e.g., $\Delta\epsilon = \Delta A/bc$, where $b$ is the path length in centimeters and $c$ is the concentration in moles/liter, completes the process. To give these concepts more substance, some details of the UIC-dispersive and FTIR-based instruments are given next.

Our original dispersive instrument is configured around a 1.0-m focal length, $\sim f/7$ monochromator (Jobin-Yvon, ISA) that is illuminated with a home-built carbon rod source (49,51,58,78). A mechanical chopper provides the modulation necessary for detecting the instrument transmission with an MCT detector. The monochromator output is filtered with a long-wavepass interference filter (OCLI) to eliminate light due to higher-order diffraction from the grating and uses mirrors (achromatic) to focus the beam on the sample. A more recent, compact design has been shown to have advantages in terms of S/N and baseline stability (73,79).

The light is linearly polarized by means of a wire grid polarizer on a $BaF_2$ substrate (Cambridge Physical Sciences) and modulated between left and right circularly polarized states with an Ar-coated ZnSe PEM (Hinds Instruments). Following the sample, a ZnSe lens focuses the light onto an MCT detector chosen in terms of size and shape to match the slit image for optimal operation down to $\sim800$ cm$^{-1}$. Alternatively, very high sensitivity in the near-IR ($\sim5000-1900$ cm$^{-1}$) is possible with an InSb photovoltaic detector, a $CaF_2$ modulator and lens, and a grating optimized for that region. Lower frequencies can be accessed with different detectors but with loss of S/N (51,84).

To process the signal, a lock-in amplifier is used to detect the transmission intensity of the instrument and sample, as evidenced by the signal developed in phase with the chopping frequency. The polarization modulation intensity is measured with a separate lock-in amplifier as that component of the detector signal that is in phase with the PEM frequency. Since the VCD is also modulated by the chopper, the signal can be demodulated again by using another lock-in referenced to the chopper, but now measuring the output of the one referenced to the PEM. Dynamic normalization varies the amplification gain such that the transmission signal is constant. Applying the same gain to the polarization-modulated signal ensures a normalization, much as is accomplished in ''normal'' CD instruments (which vary the high voltage applied to the photomultiplier to vary its gain and yield a constant average, or dc, current). In an alternate design, both signals can be $A$-to-$D$ converted and normalization effected by digital division in the data computer (58,73,79).

Our FTIR-VCD spectrometer uses a Digilab (BIORAD) FTS-60A FTIR as its core (51,61,76,77,82), but the choice of FTIR is fully open to the user, since these optics seem to impart little limitation for practical VCD operation, and very successful instruments in other laboratories have been configured around a number of FTIRs (75,85–90). In a separate compartment, an external beam goes through the same type of linear polarizer, stress optic modulator, lens, and relatively large area MCT detector as described earlier, with only weak focusing at the sample, which leads to flatter baselines. Such MCT detectors can easily saturate to a nonlinear response due to the high light levels in an FTIR. This can be controlled by using optical filters (e.g., 1900 cm$^{-1}$ cutoff low-pass) to isolate the spectral region of interest and by controlling the preamp gain. For aqueous, biological samples, the spectral bandpass is limited by the solvent, so high light level is not a major problem.

The raw VCD is obtained by ratioing the spectrum of the polarization-modulated signal with the normally developed transmission single beam spectrum using the FTIR computer processing software. In a rapid-scan instrument, the detector signal is processed by a lock-in amplifier referenced to the modulator to form an ordinary interferogram of the modulated signal, which the FTIR electronics can process. Most rapid-scan FTIR VCD spectra have concentrated on mid-IR bands, since the near-IR corresponds to higher-frequency sidebands, which the lock-in attenuates. Slow- or step-scan operation yields better response for higher frequency, near-IR, components of the spectrum (76,88–91). The optical frequencies are also encoded through correlation to the mirror position, as measured by use of laser fringe counting, but the time element is removed.

While the instrument throughput (equivalent to $I_{trans}$) is an ordinary IR intensity measurement, for which all instruments are adequately programmed to process, the polarization modulation signal ($I_{mod}$) is not so simply processed. Often the integral of the modulated spectrum is very small, having a rough balance between positive and negative VCD bands. This results in there being only a very weak center burst in the interferogram. For purposes of interferometer alignment and phase correction, this can pose difficulties (51,76,77,82). Normally, specialized software is required to overcome these limits, provide for simultaneous or sequential measurement from two independent detector inputs ($I_{trans}$ and $I_{mod}$), and permit a variety of arithmetic manipulations of the $I_{trans}$ and $I_{mod}$ spectra. Unlike the case for FTIR hardware, software is a central consideration in choosing an instrument for VCD use.

While FT-VCD has many advantages, the restriction to measurement only in the spectral windows of water and the relatively broad bands seen in biopolymer IR spectra can nullify the multiplex and throughput advantages of FTIR and, all other things being equal, favor the use of dispersive VCD. Until now, the expected FTIR advantages have *not* been experimentally realized in terms of the S/N for low-resolution biomolecular (aqueous) FTIR-VCD spectra as compared to what can be measured with the dispersive instrument over a similar time span (51,60,76). Thus, FT-VCD measurement, which always encompasses the full spectrum, can be quite inefficient as compared to concentrating one's effort and maximizing S/N in just one or two spectral bands, such as are accessible for a protein or peptide in $D_2O$.

To get adequate signal-to-noise ratio and determine scan-to-scan reproducibility, the dispersive spectra are averaged for several scans, often using time constants of the order of 10 sec and resolutions of $\sim 10$ cm$^{-1}$. This means a typical IR band can take about a half hour for a single scan. But FTIR-measured VCD spectra can sample a much wider spectral region and take about half an hour to collect an adequate number of scans for detecting the features of interest at higher resolution for a rigid, chiral organic molecule, but would require extensive averaging over much longer times to match the S/N available using the dispersive instrument for single bands in aqueous-phase biopolymers. If, in the end, only one or two adjacent bands are needed for the analysis, much time can be lost with the FTIR-based technique; but if multiple bands are to be studied, FTIR-VCD retains its advantage, even for biological samples (76,92,93). In both cases, these inherently single-beam, though corrected for light-beam intensity by the $I_{mod}/I_{trans}$ ratioing step, VCD scans must be coupled with equally long collections of baseline spectra to correct for instrument- and sample-induced spectral response. Finally in this comparison of FTIR and dispersive-based VCD instruments, it might be noted that the more direct dispersive measurement is intuitively easier to interpret in case something goes wrong, such as noise-on-baseline artifact.

Due to its weak signal size, VCD is subject to artifacts that must be corrected by careful baseline subtraction. The best baseline is determined using racemic material, which is impractical for most biological materials. However, satisfactory baselines for spectral corrections can often be acquired with carefully aligned instruments by measuring VCD spectra of the same sample cell filled with just solvent. There exists no satisfactory theory of these artifacts that can be used to control baselines. Rather there is an empirical body of evidence showing that parallel or slowly converging beams, few reflections, and uniform detector surfaces give the best results. Finally, most artifacts can be minimized by careful optical adjustments, which, at least in our instruments, are very stable, not requiring corrections for months.

## 2.2 Sampling Techniques

Most biomolecular systems are best studied in an aqueous environment. This poses difficulties for IR techniques due to water's being a very strong absorber whose fundamental transitions strongly overlap regions of interest in biomolecules such as the N—H and C=O stretches. Consequently, the peptide amide I′

(primes for N-deuterated amides) band at $\sim 1650$ cm$^{-1}$, which is dominated by the C=O stretch, has normally been measured in D$_2$O-based solution. On the other hand, amide II at $\sim 1550$ cm$^{-1}$ and amide III at $\sim 1300$ cm$^{-1}$ are best studied in H$_2$O.

Protein samples in D$_2$O can be prepared at concentrations in the range of 20–50 mg/ml for VCD. An aliquot of the solution (typically 20–30 µl) is placed in a standard demountable cell consisting of two BaF$_2$ windows separated by a 25–50-µm Teflon spacer. For studies in H$_2$O, concentrations of >100 mg/ml (but <20 µl in volume) and pathlengths of 6 µm are most useful to allow detection of amide I, where the water alone gives absorbances of $\sim 0.9$ at 1650 cm$^{-1}$. This interference causes loss in S/N, but no major artifacts are found in our instruments when properly aligned. For these experiments we find it most useful to use refillable cells (Specac) and to run the H$_2$O baseline first, then replace the solvent with sample without demounting (94).

Final VCD curves are obtained by subtraction of a baseline VCD scan from the sample spectrum and by calibration as noted earlier. Typically, after obtaining baseline and sample VCD scans, single-beam IR transmission spectra of the sample and of the solvent are recorded in the same cell to obtain an absorbance spectrum characteristic of the same instrument and under identical conditions as the VCD spectra. It is often useful to obtain FTIR spectra at higher resolution and optimal S/N on the same samples for purposes of comparison and for resolution enhancement of the absorption spectrum using Fourier self-deconvolution (21). The FTIR spectra can also be used to frequency correct the dispersive VCD spectra. Ideally, VCD should be plotted in molar units such as $\epsilon$ and $\Delta\epsilon$, as is done commonly with ECD measurements. However, since concentration and pathlengths are rarely known to sufficient accuracy, VCD spectra of biomolecules are often normalized to the absorbance, which in this case should be measured on the VCD instrument for consistency. Because the absorbance coefficients for different molecules studied will vary, this is only a first-order correction for concentration.

In our laboratory, ECD spectra are additionally measured for the samples studied using a commercial instrument (Jasco J-600). These spectra are usually obtained under more dilute conditions using strain-free quartz cells (NSG Precision Cells) obtained with various sample pathlengths from 0.1 to 10 mm, the shorter-pathlength cells being somewhat difficult to clean. Since relatively small amounts of biopolymer can give rise to significant ECD signals, it is very important to thoroughly clean sample cells between uses. Concentrations used in our laboratory for ECD are often of the order of magnitude of 0.1–1 mg/ml. For comparison of data obtained under comparable conditions, it is possible to measure ECD on the same samples used for VCD (>10 mg/ml) by employing 6–15-µm path cells constructed with quartz windows and a Teflon spacer (79,95,96).

## 3  THEORETICAL BASIS FOR VIBRATIONAL CIRCULAR DICHROISM

The theory of VCD has been a challenge since before the first VCD was measured for any real samples. This side of VCD research has continued to develop and has yielded valuable tools for the study of small molecules, in particular (53,54,97). Since most such theoretical models do not apply easily to large biomolecules, they will only be touched on here as background for the examples given regarding peptide VCD computations (98–101).

Biopolymer-oriented calculations of VCD were initially based on exciton coupling concepts, whereby local vibrations are coupled via transition dipoles yielding a pattern of oppositely signed VCD for the weakly split coupled modes, whose frequency dispersion results from electric transition dipolar coupling (102). Deutsche and Moscowitz initially simulated polymer VCD (103,104), then Schellman and coworkers (105) modified the exciton method to simulate VCD for polypeptides in $\alpha$ helices and $\beta$ sheets. Holzwarth and Chabay (106) put forth a dipole-based exciton model for the VCD of dimers, much as has been used successfully in ECD studies for biopolymers. This model was revived by Diem and coworkers (107,108), who have termed their result the extended coupled oscillator (ECO) model. By use of comparison to more exact theoretical methods (109), coupled oscillator approaches have been shown to be valid only for weakly interacting (nonbonded) dipolar vibrations. Thus, though often invoked for interpretation of small-molecule VCD, the exciton approach works best for characteristic local modes in DNA (107,108,110,111) and in fact is less useful for peptides (100).

More accurate means of computing VCD spectra have been developed in the last decade. These involve the use of quantum mechanical force fields and ab initio calculation of the magnetic and electric transition dipole moments, usually in the form of parameters termed the atomic polar tensor and atomic axial tensor (APT and AAT, respectively). These computations normally involve the use of relatively large basis sets and some approximation to represent the magnetic dipole term. In effect, the magnetic field perturbation (MFP) method of Stephens and coworkers (42,53,112,113) evaluates transition matrix elements due to the perturbation of the ground-state wave function by a magnetic operator. While successful for a number of small molecules, the MFP model is difficult to scale up to large biomolecules. This is particularly true if one carries out the MFP calculations at higher levels of ab initio theory to incorporate correlation effects (53,114–116) such as MP2 or uses density functional theory (DFT) methods (the latter being much more efficient). These barriers are, however, dropping fast as more computational power becomes available in widely accessible systems (workstations) and as new algorithms lower the intermediate data storage and time requirements of these computations (80,117–120). Nonetheless, even before this, some model calculations for dipeptides had been reported (98), as well as others for tri- and longer peptides that transfer ab initio force field and APT and AAT tensor values from smaller to larger molecules (99,100). Of these advances, the use of density functional theory (93,121,122) for force field development could provide a means of applying ab initio methods to larger molecules and of getting far more accurate results so that the overlapped bands in polymeric molecules can be sorted out.

At this juncture, theory for small-molecule VCD is routinely carried out at a higher level of precision than is even possible for ECD. Thus the younger VCD field can be viewed as progressing very well indeed in terms of both experiment and theory in its efforts to match the status of the older, established spectroscopic tools, such as ECD and FTIR.

## 4  PEPTIDE VCD STUDIES

### 4.1  Empirical VCD of Peptides

In order to build a database for qualitative interpretation of VCD data for peptides and proteins, it is necessary to have spectra of known structures of a single conformational type. The most straightforward method of doing this is to use polypeptides for which structural parameters have been established over several decades of work. Some $\alpha$-helical polypeptides are soluble in nonaqueous environments, which permitted early measurement of VCD for several amide transitions (123–125). For $\beta$-sheets, extensive solution studies are generally not possible due to solubility problems (126–128). On the other hand, there are polypeptides (mostly with charged side chains) that are soluble in aqueous solution that undergo transitions from coil forms to helical or sheetlike forms under perturbations of pH or salt effects (95,129,130).

Aqueous solutions, which are most appropriate for modeling biochemical problems, restrict VCD studies for different bands due to strong solvent absorption interference. $H_2O$ has a major absorption band at $1650 \text{ cm}^{-1}$, directly overlapping the amide I region (primarily C=O stretch), which is of prime importance for peptide and protein studies, and it completely wipes out consideration of the region above $3000 \text{ cm}^{-1}$, including the amide A band (N–H stretch). As noted, amide I VCD can be measured in $H_2O$ using very short-path length cells and relatively high concentrations(94). Unfortunately, these conditions are not compatible with many peptides and some protein systems of interest (if one wants to avoid aggregation complications). By contrast, the amide II and III bands (combinations of N–H deformation and C–N stretch) are more easily measured in $H_2O$-based solutions (69,131). To avoid the interference, many VCD studies have been carried out in $D_2O$, resulting in the exchange of most of the amide protons. Consequently, those studies focus on the amide I (there denoted amide I'), which is the most accessible mode and is only minimally shifted in frequency from its $H_2O$ value. However, in $D_2O$ the amide II′ is strongly shifted and altered in character, and the amide A′ (N–D stretch) and III′ are not detectable with VCD due to the solvent interference.

Strong VCD features are found for right-handed α-helices that consist of a negative couplet in the amide A ($\sim$3300 cm$^{-1}$), positive couplet (+ then −, with increasing frequency) in the amide I ($\sim$1655 cm$^{-1}$), a negative band lower in frequency than the absorption maximum in the amide II ($\sim$1550 cm$^{-1}$), and net positive in the general amide III regions (1350–1250 cm$^{-1}$). The prime example of this behavior is found for the VCD of L (and D, opposite signs) poly-γ-benzyl-glutamates (92,123), which, due to its very long persistence length, has some of the highest-intensity and narrowest-bandwidth α-helical VCD measured. These bandshape patterns are generally characteristic of right-handed α-helices, having been found for numerous systems of varying lengths (92,95,123–125,132–134). Deuteration of the amide N–H (normally for $D_2O$ studies), as illustrated in Fig. 2 for highly α-helical protein albumin, changes the shape of

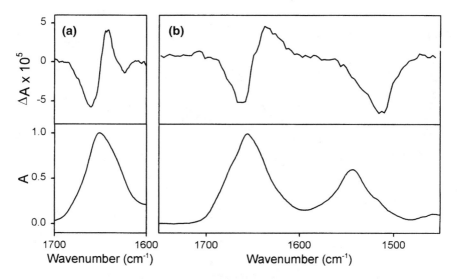

**Fig. 2** VCD (top) and IR absorption (bottom) spectra of highly helical albumin for (a) the amide I′ (N-deuterated) band in $D_2O$ (left) and (b) the amide I and II bands in $H_2O$ (right) showing the highly characteristic positive couplet pattern for the amide I (−,+,− for I′) and the negative amide II VCD shifted down from the absorbance peak. In a soluble polypeptide of uniform α-helical structure in nonaqueous solution, the amide I band would be sharper and more intense (by a factor of 2) than the amide II (see Fig. 4e), but the same shapes would be preserved.

the right-handed α-helical amide I VCD to a three-peaked $(-,+,-)$ pattern (amide I′) (92,94,95,123,124,132,135) and shifts the amide II VCD from its negative maximum at $\sim$1520 cm$^{-1}$ to below 1450 cm$^{-1}$ (amide II′) with retention of its negative sign but with a significant loss of intensity (124).

The β-sheet and coil forms have been shown to have distinctly different amide I′ VCD spectra from that of the α-helix (see Fig. 3) and are also consistent for a variety of polypeptides (95,129,130,132). The β-sheet spectra with very weak negative amide I′ VCD corresponding to the two widely split absorbance features at $\sim$1615 cm$^{-1}$ and 1690 cm$^{-1}$ are characteristic of antiparallel β-sheets with (most probably) appreciable degrees of aggregation (24,136,137). The solu-

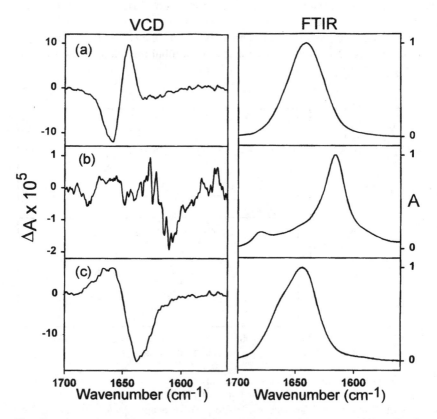

**Fig. 3** Comparison of VCD spectra (left) and IR absorption spectra (right) of polypeptides in α-helical (a, top), β-sheet (b, middle), and random coil (c, bottom) conformations. All spectra were obtained in D$_2$O with (a) poly-(LKKL), (b) poly-(LK) and (c) poly-K, where L is L-Leu and K is L-lys.

bility of β-sheet peptides at IR-compatible concentrations is marginal, which additionally mitigates against obtaining data for them in $H_2O$, where even higher concentrations are required. Hence these split-band, negative VCD patterns relate more to denatured aggregated proteins than to the short sheet segments seen in globular proteins. Studies of oligopeptides that adapt an apparently β-structure in nonaqueous solution evidenced a higher-frequency ($\sim$1635 cm$^{-1}$) couplet VCD whose detailed shape was very influenced by aggregation and solvent (128,138). On the other hand, both polypeptide and protein β-sheet structures give rise to medium-intensity negative couplet amide II VCD and to negative amide III VCD (69,94,131).

The coil form of polypeptides turns out to have a surprisingly intense negative couplet amide I VCD (intense negative followed by a broad weak positive-to-increasing frequency, Fig. 3c). This is often at a somewhat lower frequency and is always opposite in sign pattern to that of the α-helix. Furthermore, the bandshape is insensitive to deuteration. This pattern is exactly the same shape as but smaller in amplitude than the pattern characteristic of poly-L-proline II (PLP II) (139–142), a left-handed 3$_1$ helix of trans peptides. This coincidence of bandshape coupled with extensive oligomer studies and with previous ECD studies (143) has indicated that this "coil" VCD is actually characteristic of a local left-handed twist, which may arise from an extended helix form or from a high propensity for local turn conformations in the coil molecule (141,144). Such a pattern implies that substantial local structure exists in these "random coils," which do lack long-range order. Tests of this property have shown that it is possible to disorder such "coil" molecules by heating or the addition of various salts (144,145). This is consistent with much older proposals by Tiffany and Krimm (143) (and suggested by Paterlini et al. (129)) that these structures can best be viewed as an "extended-helix" conformation that, at least locally, has similarities to that of left-hand helical PLP II. And VCD has the advantage of being sensitive to a much shorter-range interactions, which can highlight such local structures. All of these observations are consistent with the "random coil" having locally ordered regions of a left-handed helical twist sense. It should be realized that, in this situation, VCD data cannot distinguish between the degree of twist in such proposed left-handed turns due to the nonuniformity of any long-range structure in such"coils."

Similarly, other secondary structures could be characterized, especially if magnitude as well as bandshape were considered and if additional bands were analyzed. A number of oligopeptides have been studied (96,146,147) establishing that the 3$_{10}$ helix (Fig. 4e) has VCD of the same sign pattern as the α-helix (Fig. 4a) but gains distinguishability since its amide I VCD has a conservative couplet shape that is much weaker than that of the amide II, while the opposite is true for the α-helix. This provides a means of differentiating these two related helical structures and identifying mixed structures such as occur in longer Aib-Ala oligo-

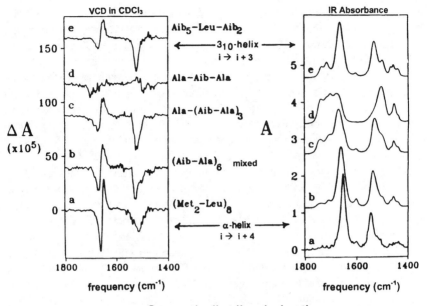

**Fig. 4** Comparison of VCD spectra (left) and IR absorption spectra (right) of α-helical [L-Met$_2$-L-Leu]$_6$ (a, bottom) and 3$_{10}$-helical Aib$_2$-L-Leu-Aib$_5$ (e, top) oligopeptides. Aib (or (αMe)Ala) stabilizes helical structures and favors 3$_{10}$ helix formation in short peptides. The tripeptide, Ala-Aib-Ala (d) lacks any stable secondary structure and gives little VCD, the heptapeptide Ala-(Aib-Ala)$_3$ (c) is 3$_{10}$ helical, and the dodecapeptide (Aib-Ala)$_6$ (b) is a mixed structure, which is also true in its crystal form.

mers shown in Fig. 4. The tripeptide of this series (Fig. 4d) evidences little if any secondary structure, characterized by a VCD barely above the baseline, while the hexapeptide (Fig. 4c) is dominantly 3$_{10}$ helical and the dodecapeptide (Fig. 4b) is probably a mixed structure. Characterizations of the VCD for other minor secondary structure variants—such as β-bend ribbons, alternate cis-trans DL proline oligomers, parallel versus antiparallel strands, and various turns—have also been attempted but are less well established (128,148–151).

Perhaps the most important property of peptide VCD, which was established via oligomer studies, is its general length dependence. Intense VCD signals are found for short oligomers such as 3$_{10}$ helical (Aib)$_2$(L-Leu)(Aib)$_{n-3}$, $n = 3$–8 (Aib = amino-isobutyric acid), (α-Me-L-Val)$_n$, $n = 2$–8, distorted 3$_{10}$ (β-bend ribbon) (Aib-L-Pro)$_n$, $n = 1$–5, and 3$_1$ helical (L-Pro)$_n$, $n = 3$–12 (96,133,140,141,146,148). At short lengths the VCD patterns found for these oligomers closely parallel those of longer chains and thus provide empirical evidence that VCD arises from relatively short-range interactions. For example, for a series of $3_{10}$-helical oligomers, it was found that the VCD bandshape for all three amide bands studied, I, II, and A, was established at least by the $r = 3,4$ length (146). Furthermore, the magnitude of the VCD reached a nearly constant value per subunit value by $n = 5$. This corresponds to only two turns of the $3_{10}$ helix or, alternatively, the coupling of two type III β-turns. Similarly, a comparison of the ECD and VCD of $Pro_n$ oligomers showed that ECD ($\Delta\epsilon$) increased as $1/n$, while the VCD ($\Delta A/A$) had a $1/n^3$ dependence (45,141), indicating that a shorter-range interaction is dominant in VCD.

A comparison of amide I and II VCD strengths vs. chain length for several oligopeptides is shown in Fig. 5. The oligopeptide ECD primarily arises from through-space, electric dipole coupling; but the VCD has a through-bond mechanical coupling, in addition to the through-space dipolar coupling. The added vibrational interaction would normally be represented by force-field mixing for vibrational modes on adjacent subunits (as well as the requirement on all vibrations to conserve the center of mass). Additionally, vibrations naturally interact over only short ranges, so they normally can be characterized by at most near-neighbor residue interactions (aside from interresidue H-bonds). This behavior is supported by our theoretical studies of small oligomers (see later) (98–100). Further proof

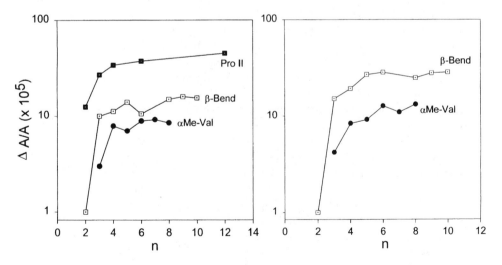

**Fig. 5** Comparison of VCD $\Delta A/A$ values as a function of oligopeptide lengths for amide I (left) and amide II (right) bands in $(Pro)_n$ (left-hand $3_1$ helix), $(Aib-Pro)_{n/2}$ (β-bend ribbon structure), and $(\alpha Me)Val_n$ ($3_{10}$ helix) structures showing the fast rise to a near-constant value for the VCD, which indicates a dominance of the VCD by short-range interactions.

of this concept of short-range dependence is provided by very recent (unpublished) VCD studies of helical peptides containing two adjacent isotopic labels ($^{13}$C on the amide C=O), which show a $^{13}$C-shifted band ($\sim$1600 cm$^{-1}$) with a similar $\alpha$-helical VCD band pattern as seen (much more intensely) for the bulk of the oligopeptide ($^{12}$C at $\sim$1640 cm$^{-1}$) (152).

Because of this length dependence, short peptide fragments that have a stable conformation can give rise to a substantial contribution to the observed VCD in a mixed structure, so for example, the VCD of $\beta$-turns may provide a means of distinguishing the various types or of identifying their presence in a peptide structure. The short Aib peptide results provide examples of type III $\beta$-turn VCD (146). Alternatively, cyclic peptides have been used to study type I and II turns; and those turns, from these results of Diem and coworkers (150,151), are implied to have unique VCD bandshapes.

## 4.2   Theoretical Simulation of Peptide VCD

Before turning to further empirical studies dealing with the more complex conformational problems that are the topic of protein VCD, it is useful to present some computational results that represent the state of the art in terms of theoretical analyses for the VCD of idealized peptides. As has been well established, the magnetic field perturbation (MFP) method of calculating VCD with ab initio wave functions as implemented by Stephens (42,80,112–116,121,153–156) is a fairly reliable theoretical method for simulation of the VCD of small molecules. We first used this calculational approach at the SCF level (4-31G) for a model system (a pseudo di-glycine) containing two peptide bonds whose relative $\phi,\psi$ torsional angles were varied to replicate those in the $\alpha$-helix, $\beta$-sheet, $3_{10}$ helix, and Pro II (left-handed $3_1$) helix polypeptide conformations (98). The computed results for the amide I and II bands of the so-constrained dipeptides qualitatively reflect the experimental H$_2$O VCD patterns found for proteins or peptides in those dominant forms. More recently we have been able to repeat those calculations at the DFT (B3LYP-6-31G**) level for pseudo tri-alanines containing three peptide bonds linked by chiral (Me-substituted) C$_\alpha$ centers (157), as shown in Fig. 6. The computed $\alpha$-helix VCD exhibits a positive couplet in the amide I and two negative bands in the amide II that are lower in energy than the absorption maximum. The Pro II ($3_1$) helix is predicted with the opposite-sign amide I (though positively biased) and a weak amide II VCD. (The sharper intense structure is from overlapped CH$_3$ modes.) The $\beta$-sheet is computed to have a weak, primarily negative amide I shifted down in frequency compared to the others with a relatively strong negative couplet amide II VCD. Due to interference from side-chain modes that is accentuated by errors in the DFT frequencies for the amide modes, the amide II and amide III transitions have more interference as compared to the idealized dipeptide calculations previously published (98). Modified calculations

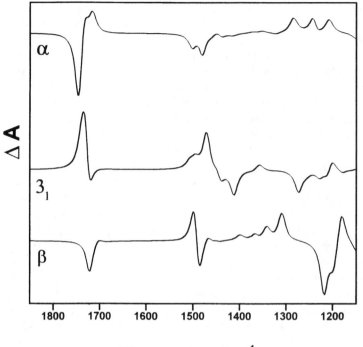

**Fig. 6** Comparison of computed VCD for a pseudo tri-alanine constrained to $\alpha$-helical (top), PLP II–like ($3_1$ helical, middle), and $\beta$-sheet (bottom) conformations over the amide I, II, and III regions. The simulations also include all other vibrations that overlap these regions, primarily from the —$CH_3$ side-chain vibrations. Force fields and APTs were calculated at the DFT level and AAT at the SCF level.

using a variety of isotopic substitutions have verified the source of these spectral distortions as —$CH_3$ modes. These qualitative features are all seen in the experimental data. That this method is successful based originally only on dipeptide calculations (and later for tripeptide) supports the conclusions drawn earlier, based only on empirical observations, that short-range effects strongly influence VCD.

On the other hand, coupled oscillator calculations, which utilize only dipolar coupling (DC) to simulate VCD, yield much poorer representations of the qualitative features seen experimentally (98). MFP-computed VCD represents a wider range of interactions encompassing both DC and other contributions from highly localized interactions. Since the ab initio MFP computations of the VCD for this constrained dipeptide system replicate so many of the observed VCD features, it is clear that these short-range contributions are very important to understanding the overall VCD spectrum. On the other hand, if the DC-based computations were to be used for the long-range interactions, where they are a legitimate representation of the effect (109), and the MFP for the short-range interactions, a sensible theoretical analysis of peptide and protein VCD might result. Recent results by Bour and coworkers (99,157) provide a method of extrapolating the force field and atomic polar and axial tensors obtained via ab initio MFP calculations for a small molecule, such as the dipeptide, onto a larger oligopeptide. This approach permits some reasonable estimate of the larger molecule's spectral properties and can be extended by explicit addition of dipole interactions (J. Kubelka, unpublished). In the Bour et al. (99) published study, tripeptide VCD was also simulated by use of a pseudo-triglycine peptide. We have now also been able to use the results for the tri-alanine model peptides constrained to specific secondary structures (Fig. 6) and have transferred those tensors onto an octa-alanine peptide model for VCD simulation (157). These results also mimic the experiment well. Furthermore, in such calculations it is easy to substitute various positions with $^{13}C$ for comparison to our newest isotopically labeled results (152). Those calculations indicate that with just two isotopically labeled substituents, a VCD pattern emerges resembling what we see experimentally.

## 4.3 Peptide VCD Applications

Some applications of VCD for peptide conformational studies have been alluded to already in the course of describing the qualitative bandshape patterns that can be related to secondary structure. An important aspect of VCD is its resolution of contributions from the amide group (the repeating aspect) from those originating in the side chains or other parts of the molecule. This was vital in determining the secondary structure type and handedness of several aromatic-containing peptides (132,158). A practical issue arose in this regard with respect to a series of β-substituted aryl-alanines in variable alternate sequences with blocked Lys used as model for electron transfer. Understanding the secondary structure was central to determination of the separation of donor–acceptor residues. With VCD it was easy to establish, in those cases, an α-helical conformation (159), which in turn defined the donor–acceptor distance by use of standard molecular bonds and angles.

The sensitivity to handedness and to $3_{10}$-helix formation has been used in a series of studies on peptides containing α-Me substituted residues. The clear preference of L-amino acids for right-handed helices can be distorted for residues with α-substitution. Blocked (α-Me)Phe tetra- and pentapeptides were shown to form left-handed $3_{10}$-like helices in $CDCl_3$ solution (160), in contrast to the right-handed forms seen for ($\alpha$-Me)Val, Aib (i.e., ($\alpha$-Me)Ala), and Iva (isovaline, which has both $\alpha$-Me and $\alpha$-Et substitution) (96,147).

This distinguishability of $3_{10}$- and $\alpha$-helices was utilized to solve a lingering problem in the literature based in part on overreliance on FTIR frequencies for conformational interpretation. Alanine-rich peptides such as Ac-(AAKAA)$_n$-GY-NH$_2$ have a high propensity for helix formation at low temperatures, even in H$_2$O for $n = 3$ or 4. Their amide I′ frequencies are at ~1637 cm$^{-1}$ in D$_2$O, which was used to support EPR-based analyses that a $3_{10}$-helical structure was formed or had a significant contribution to the conformation (161–163), although interpretation of the structural impact of this $3_{10}$-component has varied as more data developed (164). The VCD spectra clearly demonstrate that at high concentration in H$_2$O both the $n = 3, 4$ peptides are dominantly $\alpha$-helical and at lower concentration in D$_2$O the $n = 4$ is largely $\alpha$-helical, while $n = 3$ is a mixed structure (Fig. 7) (134). Temperature variation of these samples and factor analysis (165) of the resulting bandshapes demonstrated that the VCD, due to its short-range length dependence, was able to detect an intermediate structure that could be attributed to growth and decay of the junctions between the central helical portion

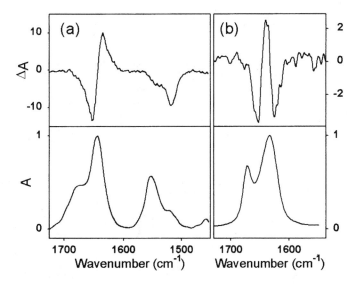

**Fig. 7** VCD spectra (top) and IR absorption spectra (bottom) of highly helical alanine-rich peptides for the amide I and II bands in H$_2$O (a, left) and amide I′ in D$_2$O (b, right). The peptide on the left (a) is Ac-(AAKAA)$_4$ -GY-NH$_2$ and that on the right (b) is Ac-(AAAAK)$_3$ AAAA-Y-NH$_2$. The shoulder (a) and peak (b) at 1672 cm$^{-1}$ are due to a TFA impurity from the solid-phase synthesis. While a common problem in interpreting FTIR spectra of synthetic peptides, it has little effect upon the VCD.

of the molecule and the steadily fraying ends, that were coil-like in VCD (Fig. 8). Recent VCD studies of isotopically labeled molecules of a similar sequence have confirmed our analysis, where the frequency-shifted $^{13}C$-labeled residues have a helical VCD that collapses at a different rate with respect to the bulk of the peptide, depending on where the substitution is placed (152).

Another case of identifying an intermediate formation with VCD comes from a study of the mutarotation of poly-L-proline from the PLP I to PLP II form (142). In this case an intermediate structure developed during the very slow transition and was followed kinetically until it decayed. This intermediate formation could not be detected with ECD, whose analysis had erroneously implied a two-state transition. The key was again the frequency resolution of VCD coupled with its short-range sensitivity. While the PLP I and PLP II VCD bandshapes are only marginally distinct, they undergo a significant frequency shift (which is quite solvent dependent) that allows detection of the intermediate. In this case the magnitude of the signal corresponding to the intermediate (again a junction between the two forms) was quite significant and rose and fell during the transition, which would be consistent with development of several junctions in the

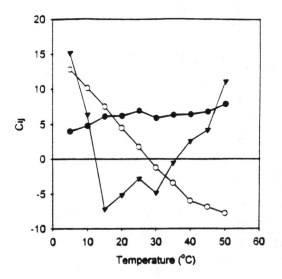

**Fig. 8** Variation of spectral component (factor) loadings with increase in temperature for Ac-$(AAKAA)_4$ -GY-$NH_2$ in $D_2O$. Closed circles represent the loadings for the first factor (nearly constant contribution of the average spectral contribution), open circles are the loadings for the second factor (tracing out the loss of helical contribution), and inverted triangles are the loadings of the third factor (accounting for the rise and fall of the spectral contribution from the helix–coil junctions).

polymer chain, but in contrast to, and more complex than, the previous model of a zipperlike mechanism. Detection of this junction between left- and right-handed helical forms was presaged by a study of DL alternate proline oligomers whose helical preference in terms of handedness was determined by VCD to depend on the chirality of the C-terminal residue (149). While very long DL-Pro polymers have equal populations of optical enantiomers, short oligomers appear to prefer a right-handed helical form for L-residues at the C-terminus (unblocked).

## 5  PROTEIN VCD STUDIES

The foregoing peptide results established general patterns that are indeed apparent in protein spectra. However, most proteins differ from small peptides in terms of the degree of solvation and the uniformity of secondary structure segments. While a helix in a peptide may terminate in a large number of conformations representing the "fraying" of that segment, in a protein this termination is likely to be a relatively well-defined or, at least, conformationally constrained turn or loop sequence. Furthermore, the segment lengths in a protein are determined by the fold rather than the thermodynamics of a structure's stability. Thus interpretation of VCD for proteins has largely been dependent on data for proteins of known structure, with a qualitative dependence on a background of peptide data described earlier. In our laboratory, these interpretations are also constrained to be consistent with ECD and FTIR data of the same system, as is appropriate.

### 5.1  Qualitative Spectral Interpretations

In the amide I' region, if the proteins are first deuterium exchanged before being measured in $D_2O$ or deuterated buffer solution, VCD spectra of proteins are straightforwardly measurable with pathlengths of 25–50 μm and concentrations of 2–5 mg/100 μl (60). Actual amounts of sample needed are lower, since typically only 20–30 μl are needed to fill the cell. Spectra obtained on proteins in $D_2O$ without previous exchange are similar in shape but of lower quality due to interference from absorption of the residual HOD in solution and to the unknown partial degree of exchange. Amide II and III spectra can be obtained for proteins in $H_2O$ if they are prepared at relatively high concentrations (69,131). The VCD of both the amide I and II bands for a single protein sample in $H_2O$ solution can be measured by using very short pathlengths (6 μm), high concentrations (1 mg/10 μl), and extensive signal averaging (>20 hr) (94). While FTIR measurements can be made with less sample and on lower concentrations in $H_2O$, VCD demands an absorbance of the order of 0.1 or greater to get adequate S/N for proteins.

Comparison of the spectra for selected proteins shows that their amide I' VCD spectra are indeed very different from each other. Complete sign pattern inversions and peak frequency variations as large as their bandwidths are found in surveying a set of globular proteins (35,60,64,65,68). By comparison, the IR absorption maximum of these same transitions shifts only a little ($\sim 20$ cm$^{-1}$) within this set of proteins. The degree of variation seen in VCD is also not seen with ECD (6,7,9,12,65,66,94,166). The high variability in the VCD bandshape of proteins arises from the fact that all types of secondary structure give rise to VCD signals of roughly the same intensity. On the other hand, ECD, particularly in the lower-energy UV region, $\lambda > 200$ nm, is dominated by the large contribution from the $\alpha$-helical components.

Figure 9 represents a selected comparison of amide I' VCD, FTIR, and amide ECD for three proteins in D$_2$O solution. Of these, myoglobin (MYO) is in the class of proteins whose secondary structure has a very high fraction of $\alpha$-

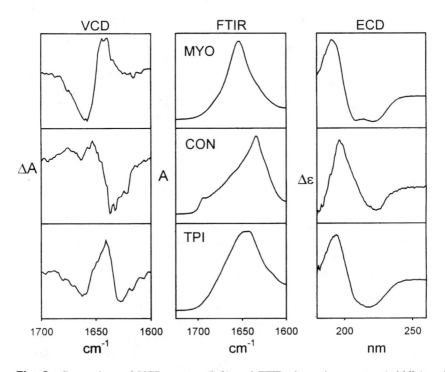

**Fig. 9** Comparison of VCD spectra (left) and FTIR absorption spectra (middle) and ECD spectra (right) of three proteins with dominant contributions from $\alpha$-helical (myoglobin, top), $\beta$-sheet (concanavalin A, middle) secondary structure segments, and from both helix and sheet (triose phosphate isomerase, bottom) conformations. The comparisons emphasize the relative sensitivity to the differences in secondary structure with the three techniques and the distinct bandshapes developed in VCD for each structural type.

helix, while concanavalin A (CON) has almost no helix but does have substantial β-sheet component, and triosphosphate isomerase (TPI) is an example of a protein that has substantial contributions of each structural type. By comparison, all of their ECD spectra are grossly similar in shape, the main differences being in intensity and zero-crossing wavelength, which can in turn be correlated with their helical contents. CON has a big shift of the positive far-UV band, and TPI has a broadened negative band, as compared to MYO. This is the source of the long-recognized success of various simple analyses of ECD data in terms of α-helical fractions (6,7,9,12,65,166). The FTIR spectra of these proteins also change little, with the primary difference being a frequency shift of the peak absorbance that roughly correlates with the amount of β-sheet in the protein. Thus CON has a distorted shape peak at $\sim 1630$ cm$^{-1}$, and TPI is broader and shifted a bit down in frequency from MYO, which is at $\sim 1650$ cm$^{-1}$.

In the VCD spectra, dramatic changes are seen in the bandshapes and significant shifts arise in the frequencies that together cause the spectra to exhibit much more sensitivity to the structural variation in these proteins than do either the ECD or FTIR results (60,66,68,167). The highly helical MYO has an amide I' VCD dominated by a positive couplet with a weak negative feature to low energy, much as seen for model α-helical polypeptides when N-deuterated (Figs. 2, 3). By contrast, the CON amide I' VCD is predominantly negative, with the main feature falling between 1630 and 1640 cm$^{-1}$, which bears only a little similarity to the polylysine antiparallel β-sheet VCD (Fig. 3). Furthermore, globular proteins with a mix of α and β components, such as TPI (Fig. 9), have amide I' VCD spectra resembling a linear combination of these two more limiting types. At least on this qualitative level, the relationship between spectra and structure is clearer in just the amide I' VCD spectra than in the amide I' FTIR alone or even in the ECD measured over the range of 260–180 nm.

Amide II and III VCD for MYO and CON in H$_2$O are also compared in Fig. 10. Highly α-helical MYO has an intense negative amide II VCD lower in energy than the absorbance maximum, and highly β-sheet-containing CON has a weaker negative couplet amide II VCD (131). Proteins with mixed α–β conformations exhibit VCD that is still weaker and representative of a linear combination of these two limiting types. Representative amide III VCD for these same proteins show an overall opposite sign pattern when averaged over the whole 1400–1100-cm$^{-1}$ region, with helices being net positive and sheets being net negative (69). Amide III VCD is very weak, in part reflecting the very low dipole intensity associated with the amide III mode in the IR. Its use for structural analysis is limited at best (131,168,169). The variation in VCD bandshape pattern with change in secondary structure in both the amide II and III regions is less dramatic and less selective than was found for the amide I' spectra. The amide II and III bands are broader than the amide I or I', and all contributions overlap. This is in contrast to the amide I, where frequency as well as bandshape contribute to

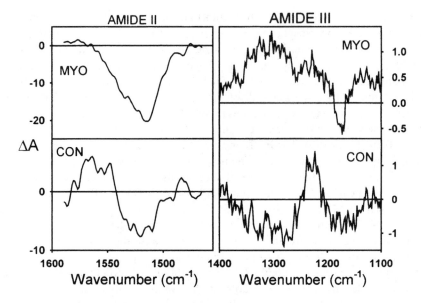

**Fig. 10** Comparison of VCD spectra for the amide II (left) and amide III (right) band regions of two proteins with dominant contributions from α-helical (myoglobin, top) and β-sheet (concanavalin A, middle) secondary structure segments. Comparison to Fig. 9 shows the broader overlapping transitions in these regions in comparison to the more resolved amide I′.

the analysis. It is even in contrast to Raman spectra of the amide III, where each conformer has a characteristic frequency (17). Finally, in summary, it is clear that VCD bandshapes, due to their signed nature, have much more distinctive character and variation than do the respective FTIR absorbance bands for each of the accessible protein amide modes, I, II, and III.

For those proteins for which we have been able to measure amide I VCD in $H_2O$ solution, the differences from the $D_2O$ results are small, aside from the amide I couplet to amide I′ $(-,+,-)$ bandshape change, which parallels that found in the peptide results (Fig. 2). A pattern of identification of the gross fold type emerges. And VCD of proteins with a high helical content usually has an amide I or I′ at ~1650 cm$^{-1}$ dominated by a positive couplet (with a weak low-energy negative component in $D_2O$), an intense negative amide II at ~1530 cm$^{-1}$, and a net positive amide III at ~1300 cm$^{-1}$ (the last being the weakest determinator due to interference from other modes). And VCD of high-sheet-content proteins has a negative amide I at ~1630 cm$^{-1}$, with various weak positive features to higher frequency, a negative couplet amide II, and a net negative amide III.

Mixed helix-sheet proteins are more easily identified in $D_2O$, where they give rise to a distinctive, but relatively weak, W-shaped $(-,+,-)$ amide I' pattern (this is susceptible to modification in the amide I, $H_2O$ case, Fig. 11) (55), accompanied by weak amide II with a typically negative bias and indeterminant amide III spectra. Qualitative appraisal of the VCD cannot yield much beyond establishing the dominant type of native secondary structure in the fold. However, if one studies a single protein system under the influence of some perturbation, such as pH, salt, solvent, or temperature change, much smaller structural variations are detectable using difference spectra techniques (55,170,171) to identify the nature of the change, all other things being constant. To automate such analyses one can do factor analysis of the spectra obtained for varying degrees of perturbation and expect that the second component will represent the major change and its loading will represent its equilibrium shift (134,172). It is this sensitivity to relative change that has long been the forte of optical spectral analyses.

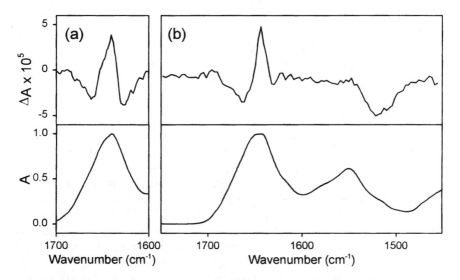

**Fig. 11** VCD spectra (top) and IR absorption spectra (bottom) of the helix- and sheet-containing triose phosphate isomerase for the amide I' (N-deuterated) band in $D_2O$ (a, left) and the amide I and II bands in $H_2O$ (b, right). The spectra of the mixed helix-sheet structure shows the characteristic W-shaped $(-,+,-)$ pattern for the $D_2O$-based amide I' (a), which is less conspicuous for the $H_2O$-based amide I (b). It also evidences (b) the mostly negative amide II VCD (with a small positive lobe to higher frequency), reflecting the reduced contribution of helix to the secondary structure in contrast to the highly helical albumin spectra (Fig. 2).

## 5.2   Quantitative Protein VCD Secondary Structure Studies

Analysis of these protein VCD spectra in terms of the fractional components
(FC) of their secondary structure has centered on the use of the principal component method of factor analysis (165,173,174) to characterize the protein spectra
in terms of a weighted combination of a relatively small number of coefficients
(35,64,65,68). Such an approach has much in common with the methods used
for determining protein secondary structure from ECD data (6–10,12,63,66,
166,175) and methods sometimes used for FTIR spectral interpretation (26,31–
33,35).

In our principal component analysis, the first spectral component (or factor)
represents the most common elements of all the experimental spectra in the training set. The second component represents the major deviations in the set from
the average. Each successive component then becomes less significant, eventually
representing just the noise contributions. By construction, in a function space
spanned by the original set of spectra, these components are orthogonal to each
other, much as would be found in an eigenvector problem. It should be clear that
the orientation of these components in the space (equivalent to the composition
of the eigenvectors) is arbitrary and, due to use of the average spectrum as the
first component, is dependent on the composition of the training set of spectra
used. When complete, the original spectra, $\theta_i$, can be represented as a linear
combination of factors, $\Phi_j$, and loadings $\alpha_{ij}$ as

$$\theta_i = \sum_j \alpha_{ij}\Phi_j \qquad (2)$$

The sum over $j$ can be truncated to sum over just the information-containing
elements by eliminating those $\Phi_j$ corresponding to noise.

For example, the first six subspectra when multiplied by the appropriate
linear coefficients (loadings) determined in the factor analysis (FA) procedure
can be used to reconstruct the experimental amide I' VCD ($D_2O$) spectra (68).
This results in a representation that contains all the spectral information but that
has effectively smoothed out much of the noise contributions. For the amide I
+ II VCD ($H_2O$), more components are needed (35). Furthermore, for the high
S/N ratio FTIR data, even more nonnegligible components can be defined
(31,33,35,70). The loadings of the spectral components making up the experimental spectrum for each protein can be used to create a compact vector descriptor
of each protein's VCD spectral bandshape. This, in turn, can be used to directly
test various regression relationships between the loadings and structural parameters of interest. The measured spectra contain redundant information expressed
as a large number of individual data points; but after factor analysis, the loadings
provide an expression of the maximum derivable information, at least in terms
of independent factors, where each protein is now represented as a handful of

coefficients or as a simple vector. However, these factors may not have a simple relationship to structure, since they are the result of an arbitrary projection out of the training set (experimental spectra). In particular, since their orientation depends on the original input set of protein spectra, if it is biased in some way to favor a specific type of structure (or equivalently, spectral type), the distribution of bandshapes between the component spectra will reflect that bias, and consequently the loadings will have a dependency modified by that bias (8,10,35,65,175,176). Thus attempts to use either a targeted or a broadly based training set of spectra for these analyses will yield different results in detail. To the extent that the training set still reflects the unknowns of interest, the method should develop useful structural insight. Each approach to training set definition has a utility, but a different one, especially in terms of reliability and applicability. Seeking an optimal method of deriving the spectra–structure relationship for protein VCD (as well as ECD, FTIR, and Raman spectra) has been a major research effort in our lab for the past few years (35,64–72,177).

To process all of our spectral data in a parallel manner, we first subject whichever kind of spectra that is being analyzed, i.e., VCD, ECD, FTIR, Raman, etc., for a set of proteins to factor analysis (FA). This step must include the training set of known structure proteins against which one plans to develop a spectra–structure relationship as well as the spectra for the unknown structure one wishes to predict. Treating all the spectral methods with a uniform reduction and correlation method has the advantage of giving us the benefit of experience in terms of the sensitivity of the result to variations in the mathematical methods used.

Regression analysis using the factor analysis loadings ($\alpha_{ij}$) of the spectra and the FC$_\varsigma^i$ values for the $i$ proteins, $\varsigma$ structural types, and $j$ subspectral coefficients can be carried out to establish a first-level quantitative basis for the spectral interpretation. By doing a complete search over for all $\varsigma$ and $j$ factor loadings between the $\alpha_{ij}$ and FC$_\varsigma^i$ sets, we are able to determine which coefficients have a statistically significant dependence on structure and which do not. Thus we investigate the ability of the loadings to fit the FC$_\varsigma$ values in the form:

$$FC_\varsigma^i = \alpha_{ij} C_j C_j^0 \tag{3}$$

by evaluating the goodness of fit between (FC$_\varsigma^i$)$_s$, as determined from spectra and the known value, (FC$_\varsigma^i$)$_x$, determined from the x-ray crystal structure, where $C_j$ and $C_j^0$ are fit coefficients for the loading of the $j$th factor. This can be expanded to a multiple linear regression relationship for more than one loading by considering all pairs, e.g., $\alpha_{ij}$ and $\alpha_{ik}$, triples, etc. There is no a priori reason to treat them in order, as has been done by some (33,178). While the first factor is the most significant component of the spectrum of all proteins in the training set, its loading may or may not have a significant dependence on any particular FC$_\varsigma$ value. Multiple regressions can be evaluated by testing for the statistical significance of adding each new set of coefficients ($\alpha_{ij}$) to the fit for a given set of structural parameters ($FC_\varsigma^i$) (66,68). However, in practice we have now taken a more pragmatic step. Since the critical issue for any analytical scheme is its ability to predict unknown structures, we test all possible regression relations by systematically taking one protein out from the set and redoing the regression (35,64,65). The error of predicting the left-out protein, evaluated as a standard deviation and predicted and known values of $FC_\varsigma$ for the entire set, is used to select the form (number and type of loadings) to be included in the final predictive regression algorithm. We have termed this method to be a selective, or restricted, multiple regression (RMR) approach (35,64,65).

This approach is designed to be conservative, in that it tends to reject contributions from those coefficients that are not significantly correlated to secondary structure. After all, the spectrum observed has contributions from many facets of the protein, its structure, and its environment. There is no reason to feel that each component of the spectrum will relate only to the secondary structure. However, in a multivariable regression approach, leaving in loadings for nondiscriminatory factors will give the illusion of an improved fit with no improvement in information content. The expected problem that will arise is a loss in predictive accuracy for proteins of unknown structure. It is hoped that the subsequent loss in precision in our RMR methods is more than compensated by a gain in predictive accuracy (35,65).

To determine reliable relationships of the spectral data to structure, a comparison of all possible linear regression analyses was undertaken using the FA/RMR approach for the amide I' and II VCD spectra independently (64,65), the amide I + II ($H_2O$) VCD and FTIR (35), as well as those of the ECD (65,66) to find correlations to the $FC_\varsigma^i$ values. All the possible combinations of spectral loadings from these FA sets were tested for which gave the best regression coefficients. At this point known spectra are fit to known structures and nothing new is learned. To apply the method to unknowns, we must evaluate predictive accuracy for which we have used the one-left-out method of testing. The predictions are evaluated for the entire training set and the best predicting regressions are thereby determined. These prediction comparisons show that the most reliable results are obtained with relatively few, selected spectral loadings, no matter which technique is used (35,65).

Our original study (68) focused on just the amide I' and found that the VCD loadings can be correlated at a statistically significant level with the $\alpha$-helical, $\beta$-sheet, and marginally well for the bend and "other" contributions (179) to the secondary structure. An exactly parallel analysis carried out on the ECD data for these same proteins showed that the information content of both techniques is similar, with the VCD analysis being better for $\beta$-sheet, but ECD being much better for $\alpha$-helix (65,66). This probably arises from the fact that the different conformations all contribute to the VCD spectrum at approximately the same level of intensity (short-range dominance), while the α-helix contribution dominates the ECD spectrum (particularly in the near-UV).

Expanding the VCD data set to include amide II data and encompass more proteins led to similar conclusions but has also allowed us to test the model more thoroughly (64). In particular, we focused on the quality of prediction in this extended study. The most far-reaching observation made was that the best secondary structure predictions are found to be correlated with only a few coefficients (35,64,65). Use of all the coefficients worsened the quality of the prediction. In the sense of fitting spectra to structure, the more spectral coefficients used, the better the fit, but this has no real importance. By contrast, use of prediction as a criterion does have real meaning in terms of the ultimate utility for unknowns. Since a number of analyses already in the literature take such an unrestricted approach to spectral coefficients (6,31,63), it is important to realize that this is a fundamental limitation.

Recent ECD-based analyses have taken a more restrictive approach, often altering the training set or optimizing the method for consistent prediction behavior (175,176). It might be seen that those data set reconstructions are complementary to our RMR or other truncated regression (33,178). For ECD this dependence on a few coefficients was dramatic. With our training set, all the secondary structure types were best predicted with just one coefficient (except for ''other,'' sometimes termed ''random coil''). Furthermore, helix and sheet were best predicted with the same coefficient (35,64,65). This results from the interdependence of helix and sheet content that had been previously identified using a neural network study of crystal structure data from the protein data bank (PDB) (48,67). While this might seem to be just a problem of choosing the best training set, we have shown that for 192 nonhomologous structures in the PDB, such a correlation (though nonlinear) still exists. For the VCD of just the amide I′, often two coefficients gave best predictions, but still the second coefficient was most important for both helix and sheet. Changing the training set and modifying the spectral region can modify the dependence on different loadings (as would a rotation of the vector orientation) but in all cases, still only just a few loadings are needed for the most reliable predictions of protein secondary structure. This is borne out by the independent observations of Van Stokkum et al. (33,178) (who truncate the set of loadings) that all the data are not needed for optimal prediction of structure with ECD and FTIR data. On the other hand, we have found that for resolution-enhanced Fourier self-deconvolved (FSD) FTIR data, more coefficients can be useful in predicting minor components, such as turns, due to their unique frequency contributions (70).

Combining the amide I′ and II VCD somewhat improves the predictive ability of the VCD data, making the sheet predictions clearly better than those for the ECD data (65). Use of the $H_2O$ data, amide I + II, gave about the same or a bit worse results, but decoupling the amide I and II made the two analyses equivalent (35). This added flexibility decouples the overall intensity aspect of the amide I and II, thus probably letting the vectors rotate to a more favorable predicting orientation. And FTIR data for the amide I + II was worse than VCD for helix determination, unless the spectra were pretreated by subtracting the average FTIR spectrum from each spectrum in the data set to form what we have termed the difference FTIR (DF) data set (35). Use of difference spectra effectively eliminates the first factor from the analysis and possibly reorients the remainder. It certainly aids the numerical stability of the computations. Analyses of these spectra in terms of structure were comparable to those of the amide I + II VCD. However, in all cases, ECD remained a significantly better source of helix predictions. The best amide I′ + II VCD based predictions were obtained using at least one coefficient from each spectral region, but combining the VCD and the ECD data into a single analysis gave even better predictions. The combination of amide I′ and amide II VCD with ECD gives numerically the smallest error in prediction (65) and the most consistency among the proteins in the data set, but combining the $H_2O$-based VCD or DF data with ECD does about as well (35). We wish to emphasize that it is this combining of the long-range dependence of ECD with the short-range sensitivity of vibrational spectra (FTIR or VCD) in our FA-RMR method that leads to dramatic improvements in prediction error. More importantly, the combined analysis gains stability and has a significant reduction in the errors associated with outliers; in other words it eliminates very poor prediction for specific unknowns. Other researchers, using independent methods, have found similar improvements for combining ECD and FTIR data, thereby also sensing long- and short-range interactions (32,33).

The final predictive capability of this combination of techniques, VCD + ECD, is quite good. Most fractional components are predicted to within 10–15% of their total dynamic range in the training set, such that combining VCD and ECD data sets led to an improvement of the order of 30% over separate predictions. Combining ECD and VCD does give the best of both methods; one gets the superior α-helix prediction of ECD with the special β-sheet sensitivity of VCD, and, in the end, each compensates for the other (35,65).

The standard deviations for the predictions of the fraction of helix or sheet based on the VCD, ECD, or the combined data sets if plotted as a function of the number of spectral coefficients used for the prediction in each case goes through a minimum corresponding to a relatively low number of coefficients (35,64,65). The major prediction capability depends on one or two subspectral coefficients. While adding a few more has little effect, adding many generally makes the predictions worse. Thus prediction is not enhanced by adding more real variation in the optical spectra. This level of error, ~10% of the dynamic range, may be a fundamental limit in the achievable accuracy for spectral predictions of average fractional secondary structure. RMR calculations based on the Levitt and Greer (180) algorithm for secondary structure determination gives similar, but slightly worse, predictive errors. The same is true of more modern algorithms (192). Thus the limits we find cannot be attributed to choosing the wrong algorithm for secondary structure interpretation. This is consistent with these two algorithms' being linearly interdependent (67). It should be clear that the distortions of helices and sheets in a globular protein mean that there are many residues that cannot be uniquely ascribed to helix, sheet, turn, etc. We feel that it is this ambiguity and nonideality of secondary structure segments that imposes a fundamental limitation on the accuracy of determination of the average secondary structure content by optical spectra.

Any training set, such as the ones we used, from the PDB, based on x-ray crystal structures will have strong correlations between the helix and sheet contents. This correlation is general and is independent of the algorithm used to determine the structural components (48,67). The spectrally based ECD determination of β-sheet can be shown to be heavily dependent on the correlation of helix and sheet, whereas the VCD one is more independent (65). In other words, the ECD structural correlations are effectively insensitive to the spectral manifestation of the sheet, while the VCD structural correlations have a direct spectral consequence of sheet content. This explains the apparent ability of ECD to ''predict'' sheet content while being qualitatively insensitive to it. When a protein is normal, i.e., like the others in the set, the ECD does well by utilizing the standard helix–sheet interrelationship from the training set, but the prediction is based on the excellent sensitivity of ECD to helix content only. When a protein structure deviates from the norm, erratic sheet predictions can result. In fact, ECD predicts sheet better based on the helix–sheet correlation than by spectral bandshape alone (65). The field is rife with examples of strange predictions of secondary structure from ECD alone. Certainly others would evolve from FTIR and VCD if also used alone. It is the combination of ECD and VCD that provides some protection in this regard.

## 5.3 Segment Analysis and Higher-Order Structure

Since the RMR for best prediction of average secondary structure necessarily neglects some components of real spectral data, there must be more information content potentially available in the optical spectra, especially ECD and VCD (35,64,65). It is natural to assume that the short-range-dependent techniques, such as VCD, will sense the distortions characteristic of the conformation of residues at the ends of uniform segments of secondary structure in a different manner than does ECD with its longer-range sensitivity. Our first attempt to take advantage of this disparity and to use more of the spectral information was to define a new descriptor of protein secondary structure, one that accounts for the number and connectivity of segments of each type of uniform structure. Our approach to this descriptor is represented by a matrix (64,72,177). This effectively encodes some

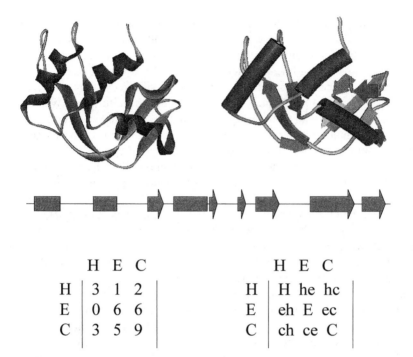

**Fig. 12** Example showing encoding of the segment distribution of secondary structure for ribonuclease A into a matrix form. At this level the number of helices (H), sheet segments (E), and coil segments (C, everything not classified as H or E) are placed on the diagonal and the numbers of interconnections between them as off-diagonal elements. Other types of segments, such as turns or $3_{10}$-helices, can be introduced using the same methodology. The central segment map emphasizes that this descriptor only counts segments and does not encompass tertiary structure or the interactions between the segments in the folded protein indicated on top.

aspects of the length and distribution of such segments, as graphically described in Fig. 12 for ribonuclease A, which has three helical segments and six sheet segments. Other workers, using ECD data, have similarly determined segment length and numbers (181) with a simpler method that disregards the interconnectivity aspects contained in our matrix representation. The rationale for this descriptor is that the helices, for example, are not uniform and deviate from ideal $\phi,\psi$ angles, particularly at the ends. And VCD (particularly due to its short-range length dependence), but other spectra as well, should sense these variations in the conformations of residues making up each type of secondary structure. The contribution of end residues is independent of the total fractional contribution of any structural type; but the higher the number of segments, the higher the number of ends.

Determination of the FC values using the FA/RMR methods described earlier is thus now complemented by determination of a matrix descriptor that is dependent on the number of segments and therefore contains new structural information independent of the average secondary structure. The FA/RMR techniques have proven to be insufficient to accurately determine this matrix descriptor for the training set of proteins (64,72,177). Thus we have developed a novel neural network analysis method utilizing two branches of hidden layers with different optimization methods to devise the values of the matrix descriptor from spectra (72,177). Extensive tests on amide I' VCD, amide I + II difference FTIR (DF) spectra, and ECD have shown that such correlations of matrix descriptor elements and spectra are possible but that, in fact, the lowest errors (again using a one-left-out approach) were obtained with ECD data. This matrix form has redundant terms in that there are sum rules based on the element definitions that restrict the values in any given row (or column). By leaving these in our predictions we can use the sum rules as an additional reliability test. Also, by selectively eliminating some terms, we can improve the overall predictability and recover the missing elements via the sum rule (72). While the relative average errors determined are still of the order seen for the FC values, in terms of the range of matrix values in the training set, such a descriptor moves the analyses conceptually beyond the conventional approaches. Even with such errors, much more is being learned from the spectra about the protein structure. This new insight into secondary structure distribution provided by the matrix descriptor can eventually be coupled to sequence-based structure prediction algorithms, thereby providing an experimental constraint to these highly variable methods. Coupling such methods to spectrally determined structural data may provide a reliable method of surpassing the current accuracy limitations of such primary sequence-based algorithms.

Two applications of the method have appeared to date. In the first, the thermal unfolding of ribonuclease $T_1$ was shown to have an initial loss of the single helical segment (and a fraying of the seven sheet segments) but with some retention of the sheet structure in a hydrophobic core (177). This thermal unfolding is summarized in Fig. 13 for native- and denatured-state spectra, with their analyses of fractional secondary structure for VCD and FTIR and the corresponding segment matrices (as derived from FTIR data). A second analysis used primarily FTIR data of aquaporin and by analogy to bacteriorhodopsin showed it to have six helical segments (182), which was later confirmed by electron diffraction (183).

In summary, the key to the utilization of VCD or any spectroscopic technique is to establish "reliability." We have sought to test thoroughly several different algorithms for the interpretation of data to find their most useful realm

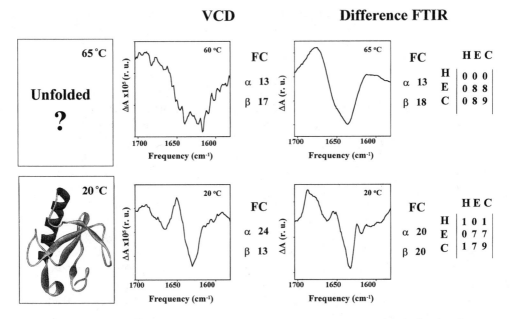

**Fig. 13** Thermal denaturation for ribonuclease T$_1$ as followed by VCD, indicating the VCD determined fractional contributions (FC) to the secondary structure, and by difference FTIR, with its FC values and matrix descriptors. Data and results for the native state (in good agreement with known structures) and the thermally denatured state at 60–65°C are shown, indicating a loss of helix but maintenance of much of the sheet contribution and most of the sheet segments.

of application. Several other groups have taken the same sort of challenge (8,10,12,33,184). These tests demand use of consistent data, access to several spectroscopic probes, and application of a systematic analysis. In one test we have used a neural network to correlate ECD and VCD spectra for the protein training set discussed earlier (167). The ECD could be predicted with reasonable accuracy from the amide I' VCD, but the reverse was not true. Thus, despite its added noise, the VCD was shown to have a higher information content, which undoubtedly arises from its higher resolution.

In an alternate approach to the use of multiple spectroscopic techniques for secondary structure analysis, we have used 2D correlation methods to show which spectral regions are dependent on a given secondary structure type by treating the α-helix or β-sheet content of proteins in a training set as the perturbation variable (71). Heterocorrelation between different types of spectra (ECD with IR and Raman, IR with Raman, VCD with IR and Raman) can be used to identify unique bands with their secondary structure source by using the more easily assigned method, e.g., ECD, to assign the more difficult one (185,186). Such an approach can also be used to identify the spectral regions of highest sensitivity for later analyses using the methods described earlier. Furthermore, the development of reliable secondary structure prediction algorithms also requires secondary testing against perturbed samples to determine the sensitivity of various methods to small structural changes. Such applications (47,55,170,177) continue in our laboratory using factor analysis and 2D correlation methods to identify independent components and trace out mechanistic paths.

## 5.4 Protein VCD Applications

A number of applications have been reported that apply the foregoing qualitative and quantitative methods of VCD analysis to the determination of conformational aspects of specific proteins. An early study of phosvitin (187) at various pH values was used to correct a misinterpreted analysis of ATR-FTIR data and show that this highly glycophosphorylated protein went from a state of low structure (coil-like) to one of antiparallel β-sheet aggregate at pH 1.6. Glycosylation provides problems for ECD analysis but, due to the resolution of vibrational spectra, is generally not a problem for VCD (188). Binding of proteins to ATR plates can shift frequencies and lead to misinterpretation of structures based on FTIR alone (189).

Analyses of unfolding were described earlier for ribonuclease $T_1$. Further studies on ribonuclease A (172) have shown an identifiable intermediate in the thermal unfolding pathway that is highlighted by factor analysis of the entire spectrum and even better by analysis of correlated spectral regions determined with 2D correlation analysis. The initial unfolding step appears to be a loss of secondary structure in the N-terminal helix. This is supported by preliminary spectral studies of ribonuclease S unfolding, which do not show the same pattern due to the break in the sequence at residue 20, the end of this helical segment.

Bovine α-lactalbumin (BLA) is a protein whose structure appears to be unusually malleable and as such has been the focus of many studies of what is termed the "molten globule transition." At low pH, BLA expands and is said to lose tertiary structure. Our VCD analysis showed that it actually gained secondary structure, in particular, α-helix content (55). Furthermore, although its crystal structure is nearly identical to that of hen egg white lysozyme (HEWL), its spectra (ECD, FTIR, and VCD) are noticeably different (170). In fact, only when at a lower pH or in the presence of helix-stabilizing solvent such as TFE or propanol do the spectra of these proteins begin to resemble each other in detail (Fig. 14). It appears that BLA is a dynamically fluctuating structure whose conformation samples many local minima and only under the influence of some perturbation does it find a structural minimum much like that of HEWL. This flexibility of

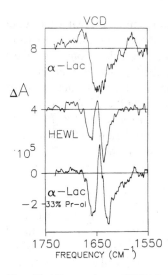

**Fig. 14** Comparison of VCD in the amide I' band for native bovine α-lactalbumin (BLA, top), hen egg white lysozyme (HEWL, middle), and partially alcohol-denatured BLA (bottom). The VCD of the partially denatured BLA protein more resembles that of HEWL than does the native BLA. Both have similar crystal structure, with a domain of helix and a domain of sheet.

structure is not evident in the crystal structure (though it is hinted at), due to the stabilizing forces provided by the packing interactions, and is less evident in NMR, due to the dynamic aspects of many segments. In some cases the helices in BLA may be more $3_{10}$-like and thus yield more detectable differences in VCD than seen in other techniques. This flexibility of BLA serves a function, for it must bind to another protein to become active.

Human chorionic gonadotropin protein consists of two subunits. The β-subunit has three hairpin segments that were thought to fold in an initial step (framework model) and then collapse to form the active state (190). Misfolds are thought to correlate with tumor function. By studying each of the hairpins individually with VCD, we have been able to show that the stability of the β-form is not a property of the sequence alone but is dependent on high concentration or the presence of a micellar environment in sodium dodecyl sulfate (SDS) (138). And VCD was used to distinguish the extended (PLP II–like) structure of the sequence corresponding to hairpin H2-β as opposed to the antiparallel β-form possible with hairpin H1-β and H3-β, the latter being fully dependent on inter-peptide interactions for forming secondary structure. In all of these studies, the correlation of VCD, FTIR, and ECD data over a wide concentration range was essential to deriving a reliable basis for interpretation of folding.

## 6  COMPARISON WITH OTHER TECHNIQUES

It must be stated for the sake of comparison that ECD and FTIR data have S/N advantages over VCD that may have an impact on the relative errors developed in using them for quantitative structural analyses. In particular, deconvolution or derivative techniques (22–24) can enhance the variation seen in the very high-S/N FTIR spectra obtainable for these proteins. The end interpretation of even such high-quality FTIR data still remains dependent on a frequency correlation of band features with secondary structure types. The frequencies of deconvolved FTIR features and those of VCD, after band-fitting deconvolution, are well corre-lated, as one would expect for phenomena arising from the same molecular transi-tions (25). However, it was also shown that for certain-frequency FTIR bands there were sign inversions for the related VCD features over the set of proteins studied. Such sign inversions cannot occur if the features come from the same structural features. The presence of sign pattern changes within a spectral band pass coupled with the high correlation of FTIR and VCD frequency patterns evidences an ambiguity in the assignment of FTIR bands and consequently an ambiguity in the reliability of structures derived by dependence on such assign-ments. An example of this phenomenon was found for two growth hormones where FTIR analysis led to false conclusions regarding the fraction of $\alpha$-helix. The VCD and ECD data in these cases led to consistent interpretations and gave a correct analysis (191).

The sign aspect of optical activity data with its direct dependence on struc-ture gives CD-based measurements another dimension beyond conventional spec-troscopic frequency assignment. Of course, in ECD the resolution of different contributing components is poor, so the overall sign pattern and intensity are the primarily useful properties. The dependence of the traditional far-UV ECD bands on the $\alpha$-helix contribution has been discussed earlier at length. However, there are two other problems with ECD. In the far-UV, contributions from sugar-based transitions (in glycosylated proteins) can distort the spectra. An example of this was noted with glucoamylase, where the IR and VCD gave much better analyses of the thermal denaturation than did the ECD (188). Another conflict arises in the near-UV from contributions of the aromatic groups, which, for example, cause some problems in interpretation of the ribonuclease $T_1$ ECD (177). Both aspects of confusion come from the intrinsically low resolution of ECD. Thus use of VCD and ECD together helps get the benefits of both despite the limitations of both.

In the end, a protein structure is too important and too complex to be left to analysis by a single technique of known (and sometimes hidden) limitations.

Above all, we wish to emphasize the need to take data from multiple spectroscopic techniques and find the model that can satisfactorily encompass all the data. For some questions VCD may give the most important insight, but for most questions it can offer critical data to add to the overall picture of protein conformation and folding.

## ACKNOWLEDGMENTS

This work was supported primarily by a previous grant from the National Institutes of Health (GM 30147), for which we are most grateful. The development of the statistical algorithms for spectral analysis was aided by an international cooperation grant jointly held with Peter Pancoska of Charles University in Prague provided by the National Science Foundation. Instrumentation and theoretical development have been supported in the past by the National Science Foundation. The work described here is the result of the dedicated hard work of a number of talented postdoctoral and graduate student coworkers, whose names are noted in the cited references. Without their efforts VCD would still be addressing small-molecule problems and searching for a theory rather than providing answers. Many of the samples for special applications came to us through generous gifts and/or collaborations with researchers from around the world, to whom we are most grateful and most of whom are identifiable as coauthors in the publications.

## REFERENCES

1. GD Fasman. Circular Dichroism and the Conformational Analysis of Biomolecules. New York: Plenum Press, 1996.
2. HH Havel. Spectroscopic Methods for Determining Protein Structure in Solution. New York: VCH, 1996.
3. HH Mantsch, D Chapman. Infrared Spectroscopy of Biomolecules. Chichester, UK: Wiley-Liss, 1996.
4. RW Woody, AK Dunker. In: GD Fasman, ed. Circular Dichroism and the Conformational Analysis of Biomolecules. New York: Plenum Press, 1996, pp 109–157.
5. E Haas. In: ed. Havel, HA Spectroscopic Methods for Determining Protein Structure in Solution. New York: VCH, 1995, pp 28–61.
6. WC Johnson Jr. Methods Biochem Anal 31:61–163, 1985.
7. JT Yang, CSC Wu, HM Martinez. Methods Enzymol 130:208–269, 1986.
8. WC Johnson Jr. Annu Rev Biophys Biophys Chem 17:145–166, 1988.
9. M Manning. J Pharmaceut Biomed Anal 7:1103–1119, 1989.
10. N Sreerama, RW Woody. J Mol Biol 242:497–507, 1994.
11. RW Woody. In: GD Fasman, ed. Circular Dichroism and the Conformational Analysis of Biomolecules. New York: Plenum Press, 1996, pp 25–67.

12. SY Venyaminov, JT Yang. In: GD Fasman, ed. Circular Dichroism and the Conformational Analysis of Biomolecules. New York: Plenum Press, 1996, pp 69–107.
13. PY Chou, GD Fasman. Methods Enzymol 47:45–148, 1978.
14. HH Mantsch, HL Casal, RN Jones. In: RJH Clark, RE Hester, eds. Spectroscopy. London: Wiley, 1986, pp 1–46.
15. FS Parker. Applications of Infrared, Raman and Resonance Raman Spectroscopy. 1983 ed. New York: Plenum Press, 1983.
16. M Jackson, HH Mantsch. Crit Rev Biochem Mol Biol 30:95–120, 1995.
17. AT Tu. Raman Spectroscopy in Biology. New York: Wiley, 1982.
18. S Krimm, J Bandekar. Advances in Protein Chemistry 38:181–364, 1986.
19. S Krimm. In: TG Spiro, ed. Biological Applications of Raman Spectroscopy. Vol. 1: Raman Spectra and the Conformations of Biological Macromolecules. New York: Wiley, 1987.
20. LG Tensmeyer, EW Keuffman II. In: HA Havel, ed. Spectroscopic Methods for Determining Protein Structure in Solution. New York: VCH, 1996.
21. JK Kauppinen, DJ Moffatt, HH Mantsch, DG Cameron. Appl Spectrosc 35:271–276, 1981.
22. DM Byler, H Susi. Biopolymers 25:469–487, 1986.
23. WK Surewicz, HH Mantsch. Biochem Biophys Acta 952:115–130, 1988.
24. W Surewicz, HH Mantsch, D Chapman. Biochemistry 32:389–394, 1993.
25. P Pancoska, L Wang, TA Keiderling. Prot Sci 2:411–419, 1993.
26. F Dousseau, M Pezolet. Biochemistry 29:8771–8779, 1990.
27. S Krimm, WC Reisdorf Jr. Farad Disc 99:181–194, 1994.
28. NN Kalnin, IA Baikalov, SY Venyaminov. Biopolymers 30:1273–1280, 1990.
29. SY Venyaminov, NN Kalnin. Biopolymers 30:1259–1271, 1990.
30. HHJ De Jongh, E Goormaghtigh, JM Ruysschaert. Anal Biochem 242:95–103, 1996.
31. DC Lee, PI Haris, D Chapman, RC Mitchell. Biochemistry 29:9185–9193, 1990.
32. RW Sarver, WC Kruger. Anal Biochem 199:61–67, 1991.
33. R Pribic, IHM Van Stokkum, D Chapman, PT Haris, M Bloemendal. Anal Biochem 214:366–378, 1993.
34. RW Williams, AK Dunker. J Mol Biol 152:783–813, 1981.
35. V Baumruk, P Pancoska, TA Keiderling. J Mol Biol 259:774–791, 1996.
36. LD Barron. Vibrational Spectra and Structure 17B:343–368, 1989.
37. PL Polavarapu. Vibrational Spectra and Structure 17B:319–342, 1989.
38. LD Barron, L Hecht. In: RJH Clark, RE Hester, eds. Biomolecular Spectroscopy, Part B. London: Wiley, 1993, pp 235–266.
39. LA Nafie. In: M Evans, S Kielich, eds. Modern Nonlinear Optics, Part 3. New York: Wiley, 1994, pp 105–206.
40. LD Barron, L Hecht, AD Bell. In: GD Fasman, ed. Circular Dichroism and the Conformational Analysis of Biomolecules. New York: Plenum Press, 1996, pp 653–695.
41. LA Nafie. Annu Rev Phys Chem 48:357–386, 1997.
42. PJ Stephens, MA Lowe. Ann Rev Phys Chem 36:213–241, 1985.
43. TB Freedman, LA Nafie. Top Stereochem 17:113–206, 1987.
44. M Diem. Vibrational Spectra Structure 19:1–54, 1991.

45. TA Keiderling, P Pancoska. In: RJH Clarke, RE Hester, eds. Biomolecular Spectroscopy, Part B. London: Wiley, Sons, 1993, pp 267–315.
46. TA Keiderling. In: K Nakanishi, N Berova, R Woody, eds. Circular Dichroism Interpretations and Applications. New York: VCH, 1994, pp 497–521.
47. TA Keiderling. In: H Havel, ed. Determination of Protein Structure in Solution by Spectroscopic Methods. New York: VCH, 1995.
48. TA Keiderling. GD In: Fasman, ed. Circular Dichroism and the Conformational Analysis of Biomolecules. New York: Plenum Press, 1996, pp 555–598.
49. TA Keiderling. Appl Spectrosc Rev 17:189–226, 1981.
50. LA Nafie. Adv Infrared Raman Spectr 11:49–93, 1984.
51. TA Keiderling. In: K Krishnan, JR Ferraro, eds. Practical Fourier Transform Infrared Spectroscopy. San Diego, CA: Academic Press, 1990, pp 203–284.
52. LA Nafie, GS Yu, X Qu, TB Freedman. Faraday Discuss 99:13–34, 1994.
53. PJ Stephens, FJ Devlin, CS Ashvar, CF Chabalowski, MJ Frisch. Faraday Disc 99: 103–119, 1994.
54. TB Freedman, LA Nafie. In: M Evans, S Kielich, eds. Modern Nonlinear Optics. New York: Wiley, 1994, pp 207–263.
55. TA Keiderling, B Wang, M Urbanova, P Pancoska, RK Dukor. Faraday Discuss 99:263–286, 1994.
56. LA Nafie, TA Keiderling, PJ Stephens. J Am Chem Soc 98:2715–2723, 1976.
57. PL Polavarapu. In: JR Ferraro, L Basile, eds. Fourier Transform Infrared Spectroscopy. Vol. 4. New York: Acadamic Press, 1985.
58. M Diem In: N Purdie, HG Brittain, eds. Techniques and Instrumentation in Analytical Chemistry. Amsterdam: Elsevier, 1994, pp 91–130.
59. TB Freedman, LA Nafie, TA Keiderling. Biopolymers (Pept Sci) 37:265–279, 1995.
60. P Pancoska, SC Yasui, TA Keiderling. Biochemistry 28:5917–5923, 1989.
61. TA Keiderling, SC Yasui, P Malon, P Pancoska, RK Dukor, RK Croatto, L Yang. 7th International Conference on Fourier Transform Spectroscopy, Proc., SPIE 1989; pp 57–63.
62. PL Polavarapu. Vibrational Spectra: Principles and Applications with Emphasis on Optical Activity. Amsterdam: Elsevier, 1998.
63. JP Hennessey, WC Johnson, JR. Biochemistry 20:1085–1094, 1981.
64. P Pancoska, E Bitto, V Janota, TA Keiderling. Faraday Discuss 99:287–310, 1994.
65. P Pancoska, E Bitto, V Janota, M Urbanova, VP Gupta, TA Keiderling. Protein Sci 4:1384–1401, 1995.
66. P Pancoska, TA Keiderling. Biochemistry 30:6885–6895, 1991.
67. P Pancoska, M Blazek, TA Keiderling. Biochemistry 31:10250–10257, 1992.
68. P Pancoska, SC Yasui, TA Keiderling. Biochemistry 30:5089–5103, 1991.
69. BI Baello, P Pancoska, TA Keiderling. Anal Biochem 250:212–221, 1997.
70. S Wi, P Pancoska, TA Keiderling. Biospectroscopy 4:93–106, 1998.
71. P Pancoska, J Kubelka, T Keiderling. Appl Spectrosc (in press), 1998.
72. P Pancoska, V Janota, TA Keiderling. Anal Biochem 267:72–83, 1999.
73. M Diem, GM Roberts, O Lee, O Barlow. Appl Spectrosc 42:20–27, 1988.
74. P Xie, M Diem. Appl Spectrosc 50:675–680, 1996.
75. CC Chen, PL Polavarapu, S Weibel. Appl Spectrosc 48:1218–1223, 1994.

76. B Wang, TA Keiderling. Appl Spectrosc 49:1347–1355, 1995.
77. P Malon, TA Keiderling. Appl Spectrosc 42:32–38, 1988.
78. CN Su, V Heintz, TA Keiderling. Chem Phys Lett 73:157–159, 1981.
79. G Yoder. Vibrational circular dichroism studies of some helical oligo peptides. PhD dissertation, University of Illinois at Chicago, Chicago, 1996.
80. CS Ashvar, FJ Devlin, PJ Stephens. J Am Chem Soc 121:2836–2849, 1999.
81. PK Bose, PL Polavarapu. Carbohyd Res 323:63–72, 2000.
82. RK Yoo, B Wang, PV Croatto, TA Keiderling. Appl Spectrosc 45:231–236, 1991.
83. P Malon, TA Keiderling. Appl Spectrosc 50:669–674, 1996.
84. R Devlin, PJ Stephens. Appl Spectrosc 41:1142–1144, 1987.
85. CA McCoy, JA de Haseth. Appl Spectrosc 42:336–341, 1988.
86. D Tsankov, T Eggimann, H Weiser. Appl Spectrosc 49:132–138, 1995.
87. RW Bormett, GD Smith, SA Asher, D Barrick, DM Kurz. Faraday Discuss 99: 327–339, 1994.
88. C Marcott, AE Dowrey, I Noda. Appl Spectrosc 47:1324–1328, 1993.
89. F Long, TB Freedman, R Hapanowicz, LA Nafie. Appl Spectrosc 51:504–507, 1997.
90. F Long, TB Freedman, TJ Tague, LA Nafie. Appl Spectrosc 51:508–511, 1997.
91. M Niemeyer, GG Hoffmann, B Schrader. J Mol Struct 349:451–454, 1995.
92. P Malon, R Kobrinskaya, TA Keiderling. Biopolymers 27:733–746, 1988.
93. CN Tam, P Bour, TA Keiderling. J Am Chem Soc 118:10285–10293, 1996.
94. V Baumruk, TA Keiderling. J Am Chem Soc 115:6939–6942, 1993.
95. V Baumruk, DF Huo, RK Dukor, TA Keiderling, D Lelievre, A Brack. Biopolymers 34:1115–1121, 1994.
96. G Yoder, A Polese, RAGD Silva, F Formaggio, M Crisma, QB Broxterman, J Kamphuis, C Toniolo, TA Keiderling. J Am Chem Soc 119:10278–10285, 1997.
97. A Rauk. In: PB Mezey, ed. New Developments in Molecular Chirality. Dordrecht, The Netherlands, Kluwer Academic, 1991, pp 57–92.
98. P Bour, TA Keiderling. J Am Chem Soc 115:9602–9607, 1993.
99. P Bour, J Sopkova, L Bednarova, P Malon, TA Keiderling. J Comput Chem 18: 646–659, 1997.
100. P Bour. Computational study of the vibrational optical activity of amides and peptides. PhD dissertation, Academy of Science, Prague, Czech Republic, 1993.
101. P Bour, TA Keiderling, P Malon. In: HLL Maia, ed. Peptides 1994 (Proceedings of the 23rd European Peptide Symposium). Leiden, Germany Escom, 1995, pp 517–518.
102. I Tinoco. Radiation Res 20:133, 1963.
103. CW Deutsche, A Moscowitz. J Chem Phys 49:3257, 1968.
104. CW Deutsche, A Moscowitz. J Chem Phys 53:2630, 1970.
105. J Snir, RA Frankel, JA Schellman. Biopolymers 14:173, 1974.
106. G Holzwarth, I Chabay. J Chem Phys 57:1632, 1972.
107. M Gulotta, DJ Goss, M Diem. Biopolymers 28:2047–2058, 1989.
108. W Zhong, M Gulotta, DJ Goss, M Diem. Biochemistry 29:7485–7491, 1990.
109. P Bour, TA Keiderling. J Am Chem Soc 114:9100–9105, 1992.
110. LJ Wang, P Pancoska, TA Keiderling. Biochemistry 33:8428–8435, 1994.
111. LJ Wang, LG Yang, TA Keiderling. Biophys J 67:2460–2467, 1994.

112. PJ Stephens. J Phys Chem 89:748, 1985.
113. PJ Stephens. J Phys Chem 91:1712, 1987.
114. FJ Devlin, PJ Stephens. J Am Chem Soc 116:5003, 1994.
115. PJ Stephens, FJ Devlin, KJ Jalkanen. Chem Phys Lett 225:247, 1994.
116. PJ Stephens, KJ Jalkanen, FJ Devlin, CF Chabalowski. J Phys Chem 97:6107, 1993.
117. FJ Devlin, PJ Stephens, JR Cheeseman, MJ Frisch. J Phys Chem 101:6322–6333, 1997.
118. FJ Devlin, PJ Stephens, JR Cheeseman, MJ Frisch. J Phys Chem 101:9912–9924, 1997.
119. CS Ashvar, PJ Stephens, T Eggimann, H Wieser. Tetrahedron Asymmetry 9:1107–1110, 1998.
120. CS Ashvar, FJ Devlin, PJ Stephens, KL Bak, T Eggimann, H Wieser. J Phys Chem 102:6842–6857, 1998.
121. PJ Stephens, FJ Devlin, CF Chabalowski, MJ Frisch. J Phys Chem 98:11623–11627, 1994.
122. P Bour, CN Tam, TA Keiderling. J Phys Chem 100:2062, 1996.
123. RD Singh, TA Keiderling. Biopolymers 20:237–240, 1981.
124. AC Sen, TA Keiderling. Biopolymers 23:1519–1532, 1984.
125. BB Lal, LA Nafie. Biopolymers 21:2161–2183, 1982.
126. AC Sen, TA Keiderling. Biopolymers 23:1533–1545, 1984.
127. U Narayanan, TA Keiderling, GM Bonora, C Toniolo. Biopolymers 24:1257–1263, 1985.
128. U Narayanan, TA Keiderling, GM Bonora, C Toniolo. J Am Chem Soc 108:2431–2437, 1986.
129. MG Paterlini, TB Freedman, LA Nafie. Biopolymers 25:1751–1765, 1986.
130. SC Yasui, TA Keiderling. J Am Chem Soc 108:5576–5581, 1986.
131. VP Gupta, TA Keiderling. Biopolymers 32:239–248, 1992.
132. SC Yasui, TA Keiderling. Biopolymers 25:5–15, 1986.
133. SC Yasui, TA Keiderling, R Katachai. Biopolymers 26:1407–1412, 1987.
134. G Yoder, P Pancoska, TA Keiderling. Biochemistry 36:15123–15133, 1997.
135. RK Dukor, TA Keiderling. In: E Bayer, G Jung, eds. Proceedings of the 20th European Peptide Symposium. Berlin: DeGruyter, 1989, pp 519–521.
136. DM Byler, JM Purcell. SPIE Fourier Transform Spectrosc 1145:539–544, 1989.
137. AH Clark, DHP Saunderson, A Sugget. Int J Peptide Res 17:353–364, 1981.
138. RAGD Silva, SA Sherman, TA Keiderling. Biopolymers 50:413–423, 1999.
139. R Kobrinskaya, SC Yasui, TA Keiderling. In: GR Marshall, ed. Peptides, Chemistry and Biology. Proceedings of the 10th American Peptide Symposium. Leiden, Germany, ESCOM, 1988, pp 65–66.
140. RK Dukor, TA Keiderling, V Gut. Int J Peptide Protein Res 38:198–203, 1991.
141. RK Dukor, TA Keiderling. Biopolymers 31:1747–1761, 1991.
142. RK Dukor, TA Keiderling. Biospectroscopy 2:83–100, 1996.
143. ML Tiffany, S Krimm. Biopolymers 11:2309–2316, 1972.
144. TA Keiderling, RAGD Silva, G Yoder, R Dukor, K. Bioorg Med Chem 6:1–9, 1998.
145. RK Dukor. Vibrational circular dichroism of selected peptides, polypeptides and proteins. PhD dissertation, University of Illinois at Chicago, Chicago, 1991.

146. SC Yasui, TA Keiderling, F Formaggio, GM Bonora, C Toniolo. J Am Chem Soc 108:4988–4993, 1986.
147. SC Yasui, TA Keiderling, GM Bonora, C Toniolo. Biopolymers 25:79–89, 1986.
148. G Yoder, TA Keiderling, F Formaggio, M Crisma, C Toniolo. Biopolymers 35: 103–111, 1995.
149. W Mastle, RK Dukor, G Yoder, TA Keiderling. Biopolymers 36:623–631, 1995.
150. HR Wyssbrod, M Diem. Biopolymers 31:1237, 1992.
151. P Xie, Q Zhou, M Diem. Faraday Discuss 99:233–244, 1995.
152. TA Keiderling, RAGD Silva, SM Decatur, P Bour. In: J Greve, GJ Puppels, C Otto, eds. Spectroscopy of Biological Molecules: New Directions, Dordrecht: Kluwer AP, 1999, pp 63–64.
153. KJ Jalkanen, PJ Stephens, RD Amos, NC Handy. J Am Chem Soc 109:7193, 1987.
154. KJ Jalkanen, PJ Stephens, RD Amos, NC Handy. J Phys Chem 92:1781, 1988.
155. RD Amos, KJ Jalkanen, PJ Stephens. J Phys Chem 92:5571, 1988.
156. RW Kawiecky, F Devlin, PJ Stephens, RD Amos. J Phys Chem 95:9817, 1991.
157. P Bour, J Kubelka, TA Keiderling. Biopolymers 53: 380–395, 2000.
158. SC Yasui, TA Keiderling. In: GR Marshall, ed. Peptides: Chemistry and Biology. Proceedings of the 10th American Peptide Symposium,. Leiden, Germany, ESCOM, 1988, pp 90–92.
159. SC Yasui, TA Keiderling, M Sisido. Macromolecules 20:403–2406, 1987.
160. G Yoder, TA Keiderling, F Formaggio, M Crisma. Tetrahedron Asymmetry 6:687–690, 1995.
161. SM Miick, G Martinez, WR Fiori, AP Todd, GL Millhauser. Nature 359:653–655, 1992.
162. G Martinez, G Millhauser, J Struct Biol 114:23–27, 1995.
163. GL Millhauser. Biochemistry 34:3873–3877, 1995.
164. GL Millhauser, S CJ., P Hanson, KA Bolin, FJM Vandeven. J Mol Biol 267:963–974, 1997.
165. ER Malinowski. Factor Analysis in Chemistry. 2nd ed. New York: Wiley, 1991.
166. CT Chang, CSC Wu, JT Yang. Anal Biochem 91:13–31, 1978.
167. P Pancoska, V Janota, TA Keiderling. Appl Spectrosc 50:658–668, 1996.
168. K Kaiden, T Matsui, S Tanaka. Appl Spectrosc 41:180–184, 1987.
169. F Fu, DB DeOliveira, WR Trumble, HK Sarkar, BR Singh. Appl Spectrosc 48: 1432–1440, 1994.
170. M Urbanova, RK Dukor, P Pancoska, VP Gupta, TA Keiderling. Biochemistry 30: 10479–10485, 1991.
171. M Urbanova, TA Keiderling, P Pancoska. Bioelectrochem Bioenerg 41:77, 1996.
172. S Stelea, P Pancoska, TA Keiderling. In: J Greve, GJ Puppels, C Otto, eds. Spectroscopy of Biological Molecules: New Directions, Dordrecht: Kluwer AP, 1999, p 65–66.
173. RJ Rummel. Applied Factor Analysis. Evanston, IL: Northwestern University Press, 1970.
174. P Pancoska, I Fric, K Blaha. Collect Czech Chem Commun 44:1296–1312, 1979.
175. N Sreerama, RW Woody. Anal Biochem 209:32–44, 1993.
176. P Manavalan, WC Johnson Jr. Anal Biochem 167:76–85, 1987.

177.  P Pancoska, H Fabian, G Yoder, V Baumruk, TA Keiderling. Biochemistry 35: 13094–13106, 1996.
178.  IHM Van Stokkum, HJW Spoelder, M Bloemendal, R Van Grondelle, FCA Groen. Anal Biochem 191:110–118, 1990.
179.  W Kabsch, C Sander. Biopolymers 22:2577–2637, 1983.
180.  M Levitt, J Greer. J Molec Biol 114:181–293, 1977.
181.  N Sreerama, SY Vanyaminov, RW Woody. Prot Sci 8:370–380, 1999.
182.  V Cabiaux, K Oberg, P Pancoska, T Walz, P Agre, A Engle. Biophys J 73:406–417, 1997.
183.  T Walz, T Hirai, K Murata, JB Heymann, K Mitsuoka, Y Fujiyoshi, BL Smith, P Agre, A Engel. Nature 387:624–627, 1997.
184.  AJP Alix. In: G Vergoten, T Theophanides, eds. Biomolecular Structure and Dynamics. Amsterdam: Kluwer AP, 1997, pp 121–150.
185.  J Kubelka, P Pancoska, TA Keiderling. Appl Spectrosc 53:666–671, 1999.
186.  J Kubelka, P Pancoska, TA Keiderling. In: J Greve, GJ Puppels, C Otto, eds., Spectroscopy of Biological Molecules: New Directions. Dordrecht: Kluwer AP, 1999, pp 67–68.
187.  SC Yasui, P Pancoska, RK Dukor, TA Keiderling, V Renugopalakrishnan, MJ Glimcher, RC Clark. J Biol Chem 265:3780–3783 3788, 1990.
188.  M Urbanova, P Pancoska, TA Keiderling. Biochim Biophys Acta 1203:290–294, 1993.
189.  KA Oberg, AL Fink. Anal Biochem 256:92–106, 1998.
190.  RW Ruddon, SA Sherman, E Bedows. Prot Sci 5:1443–1452, 1996.
191.  RK Dukor, P Pancoska, TA Keiderling, SJ Prestrelski, T Arakawa. Arch Biochem Biophys 298:678–681, 1992.

# 4

# Biological Applications of Infrared Microspectroscopy

**David L. Wetzel**
*Kansas State University, Manhattan, Kansas*

**Steven M. LeVine**
*University of Kansas Medical Center, Kansas City, Kansas*

## 1 INTRODUCTION

Spatial resolution, within a microscopic field of view, is the primary achievement of the optically efficient, modern infrared microspectrometer. The chemistry of single cells in biological specimens can be studied *in situ* using an infrared spectroscopic probe of cellular dimensions. Infrared microspectroscopy combines the fields of infrared spectroscopy, microscopy, and computer science. The result enables a comparison of microspectroscopic chemical information to histological structures. In light microscopy, image contrast is produced by the application of stains or fluorescent materials. Electronic microspectroscopy produces spectra of individual pixels or select wavelength images. In vibrational microspectroscopy (infrared or Raman), the use of chemical reagents or stains is not necessary. In mapping procedures with microbeam molecular spectroscopy, the contrast in images produced is from intrinsic infrared absorption bands. From multiple probing, functional-group maps can be established from baseline-corrected absorbance values (peak height or area). In this way, the spectral and chemical integrities of the tissue being studied are not compromised by the elimination of homogenization and chemical modifications via staining. Fourier transform infrared (FT-IR) microspectroscopy does more than perform microanalysis on small samples; it allows spatially resolved localized chemical analysis *in situ* from small portions of the microscopic field, thereby relating localized chemical analysis to the morphology (histology) of the specimen. The presence of certain organic groups is established or excluded by looking for their particular intrinsic absorption bands. Displaying the spectrum in absorbance allows the magnitude of the absorbance reading to be related to the relative concentration of those particular organic groups.

Instrumentation capable of providing excellent spatial resolution was developed primarily for the fields of material and forensic sciences. FT-IR microspectroscopy has been used extensively to characterize defects in silicon wafers (used in the electronics field) or failed hard-drive surfaces, and fibers and various other trace evidences in forensics. The use of FT-IR microspectroscopy for biological applications has been somewhat slower to develop. Our first use of the technique, in 1987, involved comparison of the spectra of different botanical parts of wheat kernels obtained *in situ* with those obtained in physically separated portions of the wheat kernel (1). Subsequently, other plant materials were probed (2). The spectrum of a single red blood cell with and without carbon monoxide was reported by Dong, Messerschmidt and Reffner as early as 1988 (3). This early work was done with an accessory infrared microscope linked to a conventional FT-IR spectrometer, with an interface containing numerous mirrors. Similarly, interfaced equipment was used for the early-cited microspectroscopy in our 1993 review (4), when biological applications were sparse. Activity had increased by 1996 when the review by Victor Kalasinski appeared (5). This chapter is focused (with the exception of select topics) on recent activity, much of which culminated in publications of 1998, including those appearing in our thematic 272-page journal issue (6) and de Haseth's bound volume (7).

Achieving greater spatial resolution and still being able to collect a scan with 2–4-cm$^{-1}$ resolution in 2 minutes or less has been the goal, by design, of modern FT-IR microspectrometers. In order to analyze a small bit of tissue surrounded by neighboring tissue, a high signal-to-noise ratio (SNR) was needed, together with a highly sensitive detector. Most infrared spectrometers are equipped with a dedicated, small-cross-section, liquid-nitrogen-cooled, mercury cadmium telluride (MCT) detector that matches the small-cross-section infrared beam. Optical efficiency is required to maintain the energy of the beam as it goes from the source through the interferometer and the microscope and back to the detector. Such FT-IR instruments, unlike dispersive grating monochromator spectrometers, do not require slits (Jacquinot advantage), and they allow many frequencies to pass through the sample simultaneously (Fellgett or multiplex advantage). The optical efficiency was improved with an integrated microspectrometer design with matching optics that eliminated several mirror bounces to conserve signal. In slightly more than a decade, low-throughput grating-monochromator infrared instruments have given way to high-optical-throughput interferometer optics, with fast Fourier transformation (FFT) used to process data. Low-cost, fast computers for performing the FT, controlling the interferometer and microscope, and handling (receiving, storing, displaying, manipulating, and outputting) data have enhanced FT-IR spectrometry. In the integrated instrument, an infrared microscope and infrared spectrometer were designed as a unit. This contributed to the optical efficiency and favorable signal-to-noise operation. The first sophisticated infrared microscope that had dual remote projected apertures was patented in 1989 by Messerschmidt and Sting (8). Stringent design considerations were required to produce an instrument with a combination of a high signal-to-noise ratio and high spatial resolution in the microscopic field. These considerations have been discussed by Reffner (9), Messerschmidt (10,11), and Wetzel (12). An excellent review was published in 1998 by Reffner (13) that discussed the microscope's optical details and included infinity-corrected as well as focusing reflecting optics.

## 2 INFRARED MICROSPECTROMETERS AND THEIR OPERATION

### 2.1 Operating the Microscope

For transmission work, 6–8-μm-thick frozen sections are prepared. These are mounted onto a barium fluoride disk 2 mm thick and 13 mm in diameter. In some cases, the specimen is sandwiched between two such barium fluoride disks. The infrared microscope is equipped with exclusively front-surface (Schwarzschild) optics for both the objective and condenser. Through these reflective optics, the microscopic field of the specimen is viewed and the portion of the specimen of interest is centered. Typically, the optical path and the infrared path will be exactly coincident. The specimen is viewed via visible radiation passing through the bottom condenser, through the specimen, up into the objective, and finally to the eyepiece or the video camera. The infrared radiation, on the other hand, enters from above, through the objective, before hitting the specimen and subsequently being collected with the condenser that is located below the stage. Once the spot to be analyzed is centered, an image-plane mask is projected onto the field. The mask is adjusted to limit the spot size (aperture) through which infrared radiation will pass. Instruments equipped with image-plane masks both before and after the specimen avoid accidental interrogation of neighboring tissue by way of diffraction. This approach assists in maintaining the desired spatial resolution. During operation, the single-beam infrared spectrum is obtained after coadding a reasonable number of scans. Usually the spectrum is produced in less than 2 minutes, and the resulting single-beam data are ratioed at each wavelength to a background spectrum obtained under the same conditions using the same image-plane masks. The final spectrum is displayed in absorbance or percent transmittance as a result of the ratioing. This instrumentation design has functioned well for many applications used in the decade of the 1990s.

## 2.2  Instrumentation for Achieving Spatial Resolution and Functional-Group Mapping

Distinguishing FT-IR microspectroscopy from infrared microanalysis is important. The infrared microscope is more than a beam condenser for focusing the infrared radiation onto a small specimen, with an eyepiece for incidental viewing. Instead, light microscopy is performed with a research-grade microscope to reveal the fine details of tissue morphology prior to obtaining spectra to explore the complex chemistry of the specimen from well-defined small areas in the field of view. With the patenting by Messerschmidt and Sting in 1989 of a high-performance microscope with all-reflecting optics and dual-remote image-plane masks, the era of modern FT-IR microspectroscopy began. In the words of Reffner (13), "the goal of infrared microspectroscopy is to produce the highest signal-to-noise ratio (SNR) spectrum with the highest spectral resolution from the smallest sample area." The critical research-grade light microscope with substituted front-surface (Cassegrainian–Schwarzschild) mirror lens objective and condenser allowed infrared analysis of localized areas of the field selected by light microscopy. By 1992, most commercial FT-IR instruments could be outfitted with an accessory microscope. Using a dedicated small-area detector and careful alignment of the microscope and spectrometer optics allowed spatial resolution to be achieved. The instrument's performance (13) is limited by several factors, including the detector response, the spectrometer efficiency, source brightness, wavelength range, microscope optics, and sample definition. Three spectral measuring modes include transmission (which requires reflective optics on both sides of the specimen), external reflection (where the same optical device is used both to project the beam onto the specimen and to collect the reflected radiation), and internal reflection with the use of a specialized attenuated total reflectance (ATR) objective. Figure 1 shows the path of radiation using front-surface optics. The two-mirror objective lenses have no spectral absorption, no chromatic aberration, and a high numerical aperture (NA). However, the central obstruction of the small mirror reduces the throughput efficiency. Obviously, the obscurations of both the objective and the condenser must be perfectly aligned, as well as the image-plane masks before the objective and after the condenser. Somewhat less obvious, yet nevertheless important, is the optical matching of the reflecting objective and condenser lenses with the free aperture of the interferometer bench of the FT-IR instrument. Within the free aperture of the interferometer, which is usually considerably larger than the free aperture of the microscope, a detector for the laser creates an obstruction. Unless the interferometer's obstruction is matched optically to the obstructions found in the microscope, a loss of radiation will result. In addition to the classical advantages of FT-IR spectrometers previously mentioned, the circular beam of the interferometer conforms to the apertures of the microscope lenses. This is an important factor.

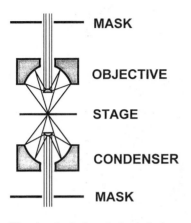

MASK

OBJECTIVE

STAGE

CONDENSER

MASK

**Fig. 1** Optical path through the top remote image-plane mask and reflective objective mirror lens to the microscope stage, and from the stage to the mirror lens condenser through a second remote image-plane mask.

Because the mirror lenses are nonabsorbing, the wavelength range is determined not by the infrared microscope but rather by the source emissivity and detector response and transmission characteristics of the beamsplitter. The mid-infrared spectral range of 2.5–25 μm (4,000–400 cm$^{-1}$) is available on most interferometers. What is generally referred to as a KBr beamsplitter in fact has a thin film of germanium on the surface. The maximum efficiency of all double-pass beamsplitters is 25%. The useful spectral range of the KBr beamsplitter is 1.5–25 μm.

Mercury-Cadmium-Telluride detectors are used because of their high sensitivity, low noise, and wide dynamic operating range of 1.5–25 μm. In general, MCT detectors with a spectral range of 2–16 μm are used, although special detectors may be specified to increase performance for a specific situation. The spatial resolution of an infrared microscope spectrometer system is limited, because the infrared radiation has a long wavelength, and as the aperture size approaches the wavelength, diffraction becomes a serious limitation. The diffraction-limited spatial resolution ($d$) for a microscope with a limiting NA is $d = 0.62/$NA, which is approximately 10 μm in this spectral range (13).

A primary factor in achieving spatial resolution involves careful defining of a sample area. Not only does diffraction limit the wavelength by seriously cutting down the transmission at the longer wavelengths corresponding to the aperture dimensions, but by diffraction for the same image-plane mask, a much broader spread exists for infrared radiation than for visible light. This blurring of the infrared image extends the size of the sample beyond the opening viewed with white light. As a result, the size of the sample increases, and the possibility

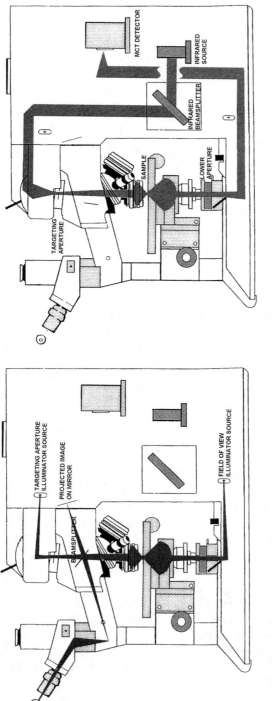

**Fig. 2** Optical diagram of IRµs® integrated FT-IR microspectrometer (Spectra-Tech, Inc., Shelton, CT) showing the path of microscope stage illumination for viewing or video image (left) and the path of infrared radiation (right) from source to beamsplitter to microscope optics, including Schwarzschild objective and condenser mirror lenses, returning to beamsplitter and detector of the spectrometer.

exists of spectral contamination by material outside the sample area defined by the masks. Theoretical studies have described these diffraction effects on sample definition (14). Coates et al. (15) positioned image-plane masks between the illuminated sample and the detector. This design limits only the area of the sample image that is passed onto the detector. Unfortunately, illuminated adjacent sample areas contribute to the radiation coming through the postspecimen mask. Placing an image-plane mask between the source and the specimen limits the illuminated area. In reflection, a single image-plane mask produces a confocal effect by defining both the area of the sample illuminated from the source and the area of the reflected sample directed to the detector. Confocal dual-remote image-plane masks are used in transmission to fix the sample area (Fig. 1).

Spatial resolution was enhanced further with an optically and physically integrated FT-IR microspectrometer. This type of instrument was the workhorse used in nearly all the research reported by the authors of this review. Figure 2 shows the path of the infrared radiation from the beamsplitter to the objective, to the specimen, to the condenser, and returning to the beamsplitter and detector. An optical arrangement allowed viewing the projected mask image onto the full visible field when the bottom image-plane mask was removed. The integrated system is designed as a spectrometer/microscope unit. It has been described in several oral presentations as a microscope containing an embedded interferometer. The implication is that not only is it physically and mechanically secured, but the NA is maintained throughout each optical component of the entire system and not subject to a limiting weak link in the chain. The free apertures of both the spectrometer and microscope are compatible, and the necessary obstructions are matched optically to maximize the microscope's performance. Elimination of as many as eight mirror bounces (by eliminating the usual optical interface) further maintains the signal level, thereby increasing the SNR as compared to an accessory microscope.

## 2.3 Continued Instrumental Developments

### 2.3.1 Further Advances in Instrumentation

In the last half of the 1990s, three major instrumental breakthroughs occurred. On September 12, 1993, at Upton, New York, Reffner, Williams and Carr achieved the first FT-IR microspectrometer data using synchrotron radiation from the vacuum ultraviolet (VUV) storage ring of the National Synchrotron Light Source (NSLS) at Brookhaven National Laboratory (BNL), Upton, NY. A few months later, a workshop on this subject was arranged at BNL, and the authors of this chapter commenced experimentation with the optical efficiency and brightness of this experimental setup applied to biological specimens that required the best possible spatial resolution and efficient detailed mapping capability. In Cincinnati, Ohio, on June 20, 1994, Marcott and Lewis achieved infrared imaging capability enhancement by coupling an indium antimonide (InSb) focal-plane array (camera) detector to a step-scan FT-IR spectrometer equipped with an infrared microscope. For the indium antimonide focal-plane array, the upper wavelength range was limited to the NH, OH, and CH stretching vibration region. Availability of MCT arrays allowed the fingerprint region to be accessed as early as November 1996. More recently, in 1998, an infinity-corrected-optical-design reflection research-grade microscope was introduced. The infinity-corrected design has a nearly columnated beam, which provides two advantages. The placing of infrared beamsplitter, filters, and polarizers does not degrade the image quality of the infrared that contributes to SNR of the spectroscopic function, and visible optics such as differential interference contrast (DIC), filters, and beamsplitters provide enhanced optical viewing convenience.

### 2.3.2  Synchrotron Infrared Microspectroscopy

Prior to 1993, the performance of FT-IR microspectrometers was achieved by conserving intensity as the radiation traversed both the microscope and the spectrometer and utilizing an appropriate, sensitive small-area detector of 0.25 mm. An enhanced source had not been attempted. *Performance* refers to spatial and spectral resolution. When synchrotron radiation was substituted for a conventional thermal (globar) source, the performance was enhanced greatly. Enhanced performance is expected for three excellent reasons with the synchrotron: it has a much greater brightness (approximately 1000-fold), it is free of thermal noise, and the beam is relatively nondivergent. The first two features advantageously affect the SNR of the system. A high SNR allows the production of excellent spectra without excessive coaddition of scans. The spatial resolution is enhanced greatly by the nondivergent characteristic of the synchrotron beam.

In the first day of operation, Reffner and coworkers (16–19) found that they could get most of the flux from the NSLS synchrotron through a 10-μm aperture. In this case, a 1-mm beam size was demagnified by a factor of 100, going from f/100 optics to f/1 optics. The result was significantly better than what was obtained with a conventional thermal source for an aperture of 100μm. In the first test of performance of the synchrotron-illuminated FT-IR microspectrometer, the beam intensity profile in the sample plane established that the infrared beam was focused to a diffraction-limited spot size of less than 12.5 μm. When the beam was centered on the 12.5-μm aperture, the intensity of the beam was more than 60% that of the incident radiation of the same unapertured beam.

In the synchrotron of the NSLS at BNL (Fig. 3), electrons from an electron source (A) are accelerated with a linear accelerator (B) to energy of approximately 75 MeV. They subsequently enter into a booster ring (C), where they are accelerated to a higher energy state before being injected into either an x-ray storage ring or, in this situation, a vacuum ultraviolet (VUV) storage ring. Electrons entering the VUV storage ring have energy of approximately 750 MeV.

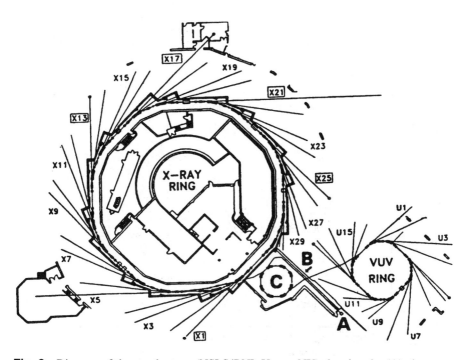

**Fig. 3** Diagram of the synchrotron (NSLS/BNL Upton, NY) showing the (A) electron source, (B) linear accelerator, and (C) booster ring, as well as the VUV storage ring and beamlines.

Electrons that have been accelerated to a velocity near the speed of light give off radiation in a relativistic manner. This radiation is directional and thus concentrated. Synchrotron radiation, besides possessing brightness, also has no thermal noise. Bunches of electrons orbiting in the VUV storage ring emit radiation that is directed from the ring port into the beamlines in proximity to various bending magnets. At beamline U2B, a special interface of the ring and an evacuated tube allowed introduction of the synchrotron beam into the IRμs® FT-IR microspectrometer (Spectra-Tech, Inc., Shelton, CT) several meters downline. The interface at beamline ring U2B was designed by Carr and Williams for other infrared experiments prior to establishing the facility for a synchrotron IR-illuminated microspectrometer (20,21). The microspectrometer was modified only slightly by removing the source mirror.

In the 4000–600-cm$^{-1}$ range, the signal (in watts into a 2-cm$^{-1}$ bandwidth) through a 10-μm pinhole at f/1 has a maximum of slightly more than 10$^{-6}$ for the synchrotron compared to 10$^{-9}$ for a comparable black-body source. The calcu-

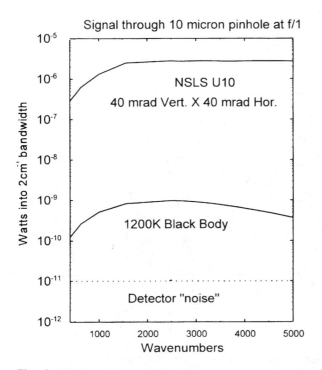

**Fig. 4** Radiant energy profile of synchrotron compared to thermal (globar) source at 1200 K and the usual noise level of a liquid nitrogen–cooled MCT detector showing the SNR advantage of the synchrotron source. (Reproduced from Ref. 22 by permission of *Microchimica Acta*.)

lated profiles of signal vs. wavelength shown in Fig. 4 were based on a 40-milliradian vertical by 40-milliradian horizontal NSLS U10 beamline source and a 1200 K black body. The noise of a typical liquid nitrogen–cooled MCT detector in the same spectral range is estimated at $10^{-11}$. Even without achieving the potential theoretical 1000-fold signal enhancement, the SNR advantage of the synchrotron is obvious (22).

The high SNR of the synchrotron radiation has a great advantage. The synchrotron radiation is not only directional, but also concentrated within a narrow angle into a beam of high flux, so losses of intensity are minimal from the interferometer and infrared microscope optics. Because a large percentage of the beam is concentrated into a small cross-sectional area, aperturing of a field with a microscope does not discard a large percentage of the beam entering the microscope, as it would for a more divergent conventional thermal source. As pointed out by Williams (21) in the case of the synchrotron, the radiation is emitted into an angle of approximately 10 milliradians by 10 milliradians. The emittance is approximately $10^{-4}$ mm$^2$ steradians. The emittance characteristic of a beam closely matches the throughput of the microscope for a properly illuminated 10-μm × 10-μm sample. This is a substantial improvement over the thermal source, which has an emittance of more than $10^{-1}$ mm$^2$ steradians that provides only 0.001% of light from the thermal source for illumination of a 10-μm-size spot on the sample. This is the reason that a conventional source-equipped microspectrometer is acceptable and works well with 24–36-μm spots without the coaddition of a large number of scans. A commercial Spectra-Tech IRμs® instrument identical to the one used by the authors of this review at the Microbeam Molecular Spectroscopy Laboratory at Kansas State University was installed at a specially constructed beamline at the NSLS at BNL (19).

Radiation extracted from the VUV ring at the beamline contains soft x-rays and vacuum ultraviolet radiation in addition to that in the infrared region. The system used between the synchrotron beam port and the IR microspectrometer includes a first mirror that is a standard plane copper laser mirror. This mirror is water-cooled, absorbs x-rays and the VUV flux, and reflects the infrared beam at right angles from the incident radiation to provide Bremsstrahlung shielding. Other mirrors throughout the scheme direct radiation to the infrared microspectrometer. The storage ring operates at a high vacuum, which is maintained through the optical arrangement by an intricate system of gate valves and window valves to a KBr window. The mirror chamber downbeam from the KBr windows is purged continually with nitrogen. An evacuated tube with KBr windows at each end is used to house the beam between the source interface and the microspectrometer. Spatial resolution is important not just for microanalysis but for the isolation of subsamples within the microscopic field to avoid accidental spectral contamination by the neighboring tissue. The synchrotron-illuminated IRμs® microspectrometer (Spectra-Tech, Inc., Shelton, CT) operates near the diffraction limit. Thus, spectral contribution from neighboring tissue is reduced by the use of focal-plane masks after the condenser as well as before the objective to minimize the effect of diffraction on a sample area.

Performance was tested by attempting to obtain the spectrum from each of a series of five laminate layers that were 6, 4, 4, 8, and 32 μm thick. In this test, reported by Reffner and coworkers (16,18), a series of IR absorption spectra was recorded across the layers of the laminated film at 1-μm intervals using a 6-μm × 6-μm dual-confocal-apertured sample area. The variation of the chemical composition across the layers was detected readily. The synchrotron radiation achieved a twofold improvement in spatial resolution over that obtained with a normal thermal emission source. This experiment showed that traversing a sharp boundary between the two phases in adjacent layers produced a chemically specific resolution at or below the optical system's diffraction limit, and the data indicated separation at the 6-μm level. Thus, it was clear that diffraction made a minimum contribution in this experiment. With the synchrotron infrared microspectroscopy, the enhanced spatial resolution enabled routine use of either $6 \times 6$-μm or $12 \times 12$-μm dual image-plane masks. Thus, as will be shown later, single cells or subcellular spots could easily be probed without the necessity of extensive coaddition of scans. Besides probing individual cells or mapping single cells, it was also possible to produce high-density maps to show chemical heterogeneity that was not necessarily visible by viewing with white light. Furthermore, relatively large microscopic areas could be mapped with data interpolated from spot sizes of cellular dimensions obtained in a grid pattern. Throughout the applications section of this chapter, the advantages of synchrotron source will be in evidence.

### 2.3.3 Microscopic Infrared Imaging with a Focal-Plane Array Detector

The recent availability of InSb focal-plane array detectors led to the coupling of such a camera to a step-scan FT-IR microspectrometer. This was first achieved on June 20, 1994, in Cincinnati, Ohio, with a camera furnished by the National Institutes of Health (NIH) and a Bio-Rad step-scan instrument by Proctor and Gamble. Control software was written to operate the instrument and acquire data from a large number of pixels at one time. This has been reported by these and other authors in the following references (23–26). More recently, MCT arrays that are sensitive also in the fingerprint region of the infrared, coupled to the step-scan instrument, provided means of simultaneously obtaining spectra at 4,096 pixels (27). Examples of more recent imaging by this approach follows in the application section. Unlike the raster-scanning procedure, this process takes place in a few minutes when 16-cm$^{-1}$-resolution data are acceptable. With data from these many pixels, the necessity for speed requires that the spectral resolution be limited in comparison to other microspectroscopic procedures. Thus, in our opinion, a trade-off occurs between quantity of simultaneous pixels in a short period of time and the spectral resolution that would require a greater accumulation period. One advantage is that the amount of data from 4,096 pixels simultaneously is sufficient for statistical application of image interpretation. Principal component data treatment and subsequent discriminant analysis allows the categorization of any given pixel in the array. This is accomplished by comparing it with standards obtained by taking spectra at pixels known to be pure with respect to a particular material. From the pattern recognition based on several ''pure'' spectra chosen by coincidental high scores for principal component pairs, each pixel on the image can be classified. Pixels assigned to a particular group are coded in the image by gray scale or color. Objective classification is used to define chemically different species within the field producing shapes and boundaries that then can be compared to electronic photomicrographs to look for coincidence. Some of the applications that follow show the use of focal-plane arrays. When the production of images is the main purpose, rather than higher-resolution spectra, this expedient approach will be used to advantage in the near future. Focal-plane array step-scan FT-IR microspectroscopy has been commercialized by Bio-Rad, Cambridge, MA, and Spectral Dimensions Inc., Olney, MD. In the former case, the standard Bio-Rad microspectrometer model no. FTS-60A/UMA300A was outfitted with a liquid nitrogen–cooled 64 × 64-pixel MCT camera and software. In the latter more recent case, a Nexus (Nicolet Instruments, Madison, WI) step-scan interferometer bench, is coupled to an original equipment manufacturer (OEM) version of the Continuμm® microscope (Spectra-Tech, Inc., Shelton, CT) equipped with infinity-corrected all-reflecting microscope optics but minus the image-plane mask. This instrument with the use of dichroics allows bright-field video observation of the target while providing the display of an infrared image directly from the focal plane on a monitor in real time. The imaging module provided by Spectra-Tech includes optics to direct the microscope-transmitted beam onto the focal-plane array camera (detector) and for 1:1 macro transmission work at a special sampling port. Images from the Spectral Dimensions instrument (Insight IR) were presented at the 1999 Pittsburgh Conference (28). In the laboratory of Chemical Physics at the NIH National Institute of Diabetes and Digestive and Kidney Diseases in Bethesda, MD, the use of focal-plane array cameras for vibrational spectroscopic imaging through a microscope via tunable filters previously had been achieved by Lewis, Tredo, and Levin (29–31). Accumulating interferograms for each pixel via a step-scan FT-IR microspectrometer was a natural progression (24,32).

### 2.3.4  Infinity-Corrected Reflective Optics Infrared Microscope

An infinity-corrected, all-reflecting microscope for infrared microspectrometry has been introduced recently (33). Unlike focusing optics, infinity-corrected optics do not have a fixed tube length, and the rays are parallel. This permits interposing various optical accessories in the beam, including Wolleston prisms for differential interference contrast (DIC) visible light viewing. A better image resulting from the infinity-corrected optics in the visible implies also a better-focused object in the infrared. The result of the improved sharp infrared focus provides a better SNR for the spectra obtained (34). In FT-IR microspectroscopy, the optical necessity to avoid the use of stained sections has made some of the microspectroscopic work inconvenient. Quite often, use of adjacent or neighboring sections is necessary to allow verification of the morphology or histology by staining one section while using the other section for infrared probing. Furthermore, in unstained sections, identifying regions of interest with light microscopy is sometimes difficult. The use of DIC makes that task much easier. With infinity-corrected optics, an infrared polarizer can be placed in close proximity to the actual objective that reduces the risk of depolarization prior to the specimen. In addition, by using beamsplitters and other optical arrangements, viewing can take place while spectroscopic scans are being collected. Thus, scouting a sample both visually and spectroscopically is convenient. With this feature, regions of interest can be identified readily based on their preliminary spectroscopic response that have appropriate specimen thickness. This latest instrumental improvement was introduced very recently, so its use has been limited; however, it is anticipated to increase in the near future. Actually, a routine-use infrared microscope, the InspectIR® (Spectra-Tech, Inc., Shelton, CT) designed for reflection only also employed infinity-corrected optics (35). In this instrument, the spot size is determined by the dimension of the detector. Because infinity-corrected optics have replaced focusing optics for many modern transmission light microscopes, extending this approach to reflecting objectives and condensers was a natural progression. Thus, the advantages of infinity-corrected optics such as DIC have found their way into infrared microspectroscopy, just as they have in visible-light microscopy.

## 3 APPLICATIONS

### 3.1 Nonbiological Applications of FT-IR Microspectroscopy

In the areas of art authentication and art restoration, materials (such as sealer, paint, sizing, glue, varnish, and wax) can be analyzed individually either as layers in a laminate cross section or in successive exposures of layers. By 1993, a collection of spectra of a wide range of these materials had been established at the Getty Conservation Institute by Derrick (36). Authentication of art objects by FT-IR microspectroscopy depends on knowledge of the chemistry of the materials used during the time at which the art object was produced; e.g., acrylic lacquers were not available during early eras. Identities of glues from animal hide, dairy products such as casein and egg (which would probably contain some cholesterol) can be distinguished. Heeren et al. (37) recently reported FT-IR imaging of paint cross sections from 17th–19th century Dutch paintings. Agencies such as the Getty Conservation Institute are very careful when restoring valuable objects. The FT-IR technique was involved in the identification of each layer of the desk of Wilhelm Roentgen. This enabled the same type of material to be used for each layer to produce an authentically valid restoration. Derrick and coworkers also used FT-IR to examine documents such as the Dead Sea scrolls (38).

FT-IR microspectroscopy has been used in archeology for identifying dyes found in fabrics recovered from digs. Work of Martoglio et al. (39,40), for example, established a link, based on dyes found in the wrappings and clothing from digs, between Paracas Necropolis (400 BC to 400 AD) in Peru and the Etowah Mound ca. 1200 AD) located in northwestern Georgia.

In materials science, identifying the composition of polymer laminates is of particular interest (41). In addition, photographic film laminates have been analyzed, including those that have extremely thin layers (18). Laminates used in packaging may include an outside and inside surface and barrier layers for air, moisture, and radiation, as well as binder layers between these other layers. Composites used in the construction of many modern devices can be analyzed by microspectroscopy, and the distribution of the materials that make up the composite can be revealed. In the area of polymers, such as those in films or fibers, the ratio of copolymers can be determined. In addition, in materials science, contamination or failure cause can be determined; for example, subsurface defects can be detected in solid-state electronic devices such as silicon computer chips or photosensitive detectors (42). Surface contamination of objects manufactured under clean-room conditions can be identified, e.g., cellulose lint on the surface of a hard disk drive that has failed. Once the contaminant is identified, then a search for its source proceeds, and elimination of that source from the process can be effected.

The weathering effects of sunlight and rain on polymer coatings have been studied with FT-IR microspectroscopy (43). Layered acrylics applied upon a base coating that has a different composition were analyzed by microtoming a cross section that included all the layers of interest and mapping by FT-IR microspectroscopy with a very narrow aperture starting from the external surface and proceeding to the inner layers of the polymer. By comparing the spectrum of pristine polymer that was beyond the penetration of the suns rays with spectra of each successive layer, moving from the outer surface toward the inside, the depth of penetration of the polymer degrading caused by ultraviolet rays could be determined. Where degradation has occurred, infrared absorption bands of the starting materials disappear and corresponding bands for the degradation products appear. This evidence is found partway into the polymeric film. At the outer edge, the weathering due to rainwater on the degraded surface is also evident by the disappearance of most of the degradation products via solubilization in concert with reduced band intensity from original materials.

Forensic applications of FT-IR microspectroscopy are concerned with "trace evidence." Paint chips from automobiles provide the opportunity, through analysis of the resulting laminate, to reveal not only the composition and origin of the original coating but subsequent aftermarket repainting. The Federal Bureau of Investigation (FBI) has on file paints for virtually all automobiles manufactured in the United States. The chemical composition of a fiber (wool, cotton, cellulose, rayon, acrylic, polyester, or polyamide) is useful for fiber evidence and may be applied to those polymers containing monomers of different types. In addition, the type and ratio of the monomers that make up a copolymer may be of interest (44,45). The finish on the surface of single fibers can be determined by use of an ATR objective with an FT-IR microspectrometer (46–48). ATR is useful in combination with transmission techniques to determine the composition of a single bicomponent fiber. This is done by subtracting the ATR spectrum obtained at the surface from the spectrum obtained by transmission through the fiber. In this way, the identity of the central core can be determined (49).

Single-fiber FT-IR microspectroscopy also has been used with the ubiquitous, generic class, polyethyleneterephthalate (PET) polyester. Nearly all polyester fibers have the same chemical composition. In spite of this, FT-IR microspectroscopy with polarized radiation can be used to determine the processing history of these fibers. Not all manufacturing facilities for PET fibers necessarily use the same processing conditions, and not all polyester fiber products are produced for the same use. A drawing process is applied to heated fibers, elongating them and thereby producing a molecular orientation. Fibers whose macromolecules are oriented exhibit infrared dichroism (50). Polarized infrared spectroscopic data from single fibers have been used to establish 13 subclasses of the formerly single generic class called polyester (51). These subclasses were obtained from a collection of fibers representing polyester manufacturing in North America. Dichroic ratios were obtained for eight readily measurable absorption bands that exhibited dichroic activity for polyester. The dichroic ratio data were used to classify these single fibers based on a discriminate analysis scheme (51).

Geological applications include a recent *Applied Spectroscopy* cover feature article by Guilhaumou, Dumas, Carr and Williams (52). Fluid inclusions produced CH and $CO_2$ bands, which aided in the petrography of diamond-sawed rock samples. Unknown pharmaceuticals or drugs of abuse can be analyzed by FT-IR microspectroscopy (53). Metabolites of heroin or cocaine have been the subjects of analyses of human hair (54). Hair specimens from overdose victims have been used to determine the drug or drugs causing their deaths. Nerve agents have been determined not by microscopy but by FT-IR microsampling of gas chromatograph effluent (55). Microspectroscopy has been used to analyze monosaccharide composition (56).

## 3.2 Biological Applications

### 3.2.1 Neuroscience

The FT-IR microspectroscopic examination of frozen sections of brain tissue from mice was used to explore the potential of this technique for analyzing the *in situ* chemistry of brain specimens. Our initial studies were performed to compare and contrast the chemical differences between white and gray matter. Striking chemical differences were observed between these regions, as described next (57), some of which were reported earlier (58).

White matter was found to have significantly elevated absorbances at 2927 $cm^{-1}$, 1740 $cm^{-1}$, 1469 $cm^{-1}$, 1235 $cm^{-1}$, and 1085 $cm^{-1}$ compared to gray matter (Fig. 5). The $CH_2$ functional group is represented at 2927 $cm^{-1}$ and 1469 $cm^{-1}$. The high concentration of long-chain fatty acids in myelin is responsible for this elevated absorbance. Galactocerebroside and sulfatide represent ~11 and 3%,

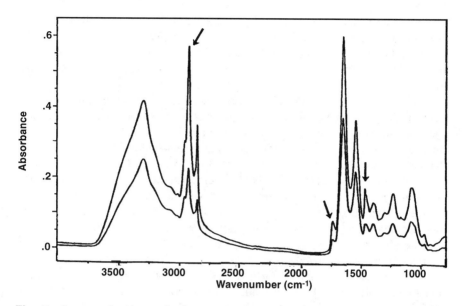

**Fig. 5** Spectra of white matter (top) and grey matter (bottom) obtained by spatially resolved *in situ* probing of a cerebrum section of normal mouse. Infrared bands at 2927 cm$^{-1}$, 1740 cm$^{-1}$, 1469 cm$^{-1}$, 1235 cm$^{-1}$, and 1085 cm$^{-1}$ distinguish normal white matter from grey matter. (Reproduced from Ref. 57 by permission of *Spectroscopy*.)

respectively, of the dry weight of white matter, and the galactose in these molecules is responsible for the greater absorbance at 1085 cm$^{-1}$, which detects HO—C—H. Phospholipids account for 25.2% of the dry weight of white matter, and the P=O in these lipids accounts for the greater absorbance at 1235 cm$^{-1}$. Carbonyl (C=O) at 1740 cm$^{-1}$ also is elevated in white matter compared to grey matter, but the CH$_2$-to-carbonyl ratio is quite high due to the great number of CH$_2$ groups (>35) relative to each carbonyl from the lipids of myelin. The absorbance for amide II at (1550 cm$^{-1}$) was similar between white and adjacent gray matter in the cerebrum.

Psychosine is stored in the brains of twitcher mice, which provide an authentic animal model of Krabbe's disease (globoid cell leukodystrophy). We observed (59) that the peak position of CH$_2$ in psychosine was shifted from its position in white matter (2919 cm$^{-1}$ for psychosine to 2925 cm$^{-1}$ for white matter). A mathematical function was used to determine whether the CH$_2$ peak position in twitcher mice spectra was shifted toward the peak position observed in psychosine. We found that 36% of the spectra collected from brainstem white matter and 19% of the spectra collected from cerebra white matter displayed a shift toward 2919 cm$^{-1}$. The greater percentage of shifted peaks in the brainstem is consistent with our findings of greater pathology in hindbrain compared to forebrain due to the early myelination in the former region and, thus a greater time for toxic accumulation of psychosine to occur (59). Besides advancing the understanding of the pathology in Krabbe's disease, these findings demonstrated that FT-IR microspectroscopy could be used to detect specific compounds that are stored in individual cells.

Chemical changes induced by extravasated blood in cerebral white matter were investigated by FT-IR microspectroscopy (60). In this model, stereotatic injections of blood were given to rats, and then they were sacrificed two or six days later. Analysis of the penumbra region indicated significant chemical changes that caused decreases of H—C=C (3015 cm$^{-1}$), C=O (1740 cm$-1$), P=O (1235 cm$^{-1}$), and HO—C—H (1085 cm$^{-1}$) and increases of N—H stretch (3300 cm$^{-1}$) and amide II (1550 cm$^{-1}$). The chemical changes likely were consequences of digestion by proteases and oxidative tissue damage. The rate of absorbance change as a function of distance from the lesion displayed a trend to be more abrupt at 2 days than at 6 days following lesion initiation, suggesting a spread of chemical changes with time.

Deuterated chemical functional groups can be distinguished from their non-deuterated counterparts by FT-IR microspectroscopy. We showed (61,62) that administration of 30 or 40% D$_2$O to rats for several weeks results in the metabolic incorporation of deuterium into OD, ND, and CD functional groups (Fig. 6). This

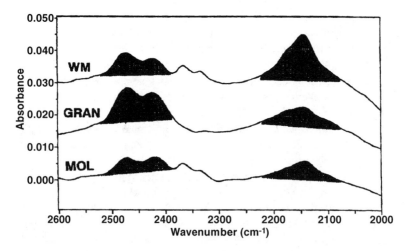

**Fig. 6** Spectra of metabolically incorporated OD, ND (2500–2400 cm$^{-1}$), and CD (2150 cm$^{-1}$) stretching vibrational bands obtained in situ from a 10-μm-thick frozen section of an adult rat cerebellum in the white matter (top), granular cell layer (middle), and molecular cell layer (bottom). (Reproduced from Ref. 62 by permission of *Cellular and Molecular Biology*.)

finding was used to estimate the relative metabolic rates of different layers of the cerebellum. The molecular layer, which contains a high concentration of synaptic connections and a low density of cell bodies, was found to have a high CD:CH ratio compared to that for white matter (Table 1) based on peak areas from 40–80 spectra from each of four animals. This indicates that the relative metabolic rate is greater in the molecular layer than in the white matter, which is supported by active membrane (lipid) synthesis/recycling in presynapses and the slow turn-over of lipids and proteins in myelin reported earlier (63–65). The CD:CH ratio also was greater in the molecular layer compared to the granule cell layer, which has a high nuclear-to-soma ratio. Lewis and coworkers also have studied brain tissue, including cerebellum, using a focal-plane array (FPA) detector (66). Focal-plane array imaging is the subject of an entire chapter (no. 7) authored by Schaeberle et al. in this volume; therefore, no doubt the authors' current and previous work (23–25) are presented there and those results are not discussed further here. In cooperation with Marcott (67), we used the FPA technique on the same actual cerebellum sections that we had previously analyzed by synchrotron and thermal FT-IR microspectroscopy in Refs. 61 and 62. Although the single-pixel SNR was insufficient for analyzing deuterated species, band ratioing and band shifts of other functional groups from 4,096 pixels provided useful images. The term *fast FT-IR imaging* was applied to FPA by Koenig in a recent invited talk (68) at the 1999 Pittsburgh Conference. For those of us who have been imaging all along from spectra obtained by the raster scan mapping procedure, this distinction is appropriate. In this talk, kinetic studies were used as an example of the use of ''fast'' imaging.

In vitro studies by Lamba et al. (69) found that the carbonyl at 1740 cm$^{-1}$ increases when lipids become oxidized, and Signorini et al. (70) reported that the amide I peak at ~1660 cm$^{-1}$ broadens when proteins become oxidized. Analyses of these functional groups were performed on data obtained from cerebral white matter from normal individuals and multiple sclerosis patients (71). The carbonyl:CH$_2$ ratio was elevated in each of the multiple sclerosis patients but not

**Table 1** Deuterated-to-Nondeuterated Ratios Normalized to Those of the Granule Cell Layer for Deuterium Bonded to Carbon as Well as the Combined Nitrogen and Oxygen Bonded Forms

|                    | Bonded to C | Bonded to N or O |
| ------------------ | ----------- | ---------------- |
| Molecular layer    | 1.10        | 1.11             |
| Granule cell layer | 1.00        | 1.00             |
| White matter       | 0.36        | 0.58             |

*Source*: Ref. 62.

in normal individuals. The amide I peak also was broadened in several sites from each of the multiple sclerosis patients but not in normal individuals. These data suggest that oxidation products can be identified in discrete lesion sites of brain tissue.

Synchrotron FT-IR microspectroscopy was used by Choo et al. (72) to characterize the organization of proteins in amyloid plaques. The amyloid plaques from individuals with Alzheimer's disease showed a preponderance of β-sheet structure, indicated by amide I at 1630–1634 cm$^{-1}$. In surrounding tissue, the proteins were mostly in an unordered or α-helical conformation, indicated by the amide I peak frequency at 1652–1654 cm$^{-1}$ (Fig. 7). The β-sheet organization in plaques is in contrast with previous IR studies on synthetic amyloid, which had a peak frequency at 1620–1628 cm$^{-1}$, indicative of aggregated peptides according to Fraser et al. (73–75), Otvos et al. (76), and Fabian et al. (77,78).

### 3.2.2 Plant Material

Although this report is focused on the state-of-the-art FT-IR microspectroscopy at the turn of the century, prefacing the presentation of the present status with the chronology of our plant work is appropriate because it parallels the advances in instrumental capability of the past decade. Our first spectra from wheat sections *in situ* that represented different botanical parts, mentioned briefly in the introduction (1), were accomplished in the laboratory of Messerschmidt at Spectra-Tech

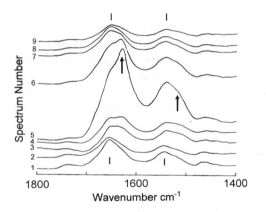

**Fig. 7** Successive spectra obtained (by synchrotron infrared microspectroscopy) at nine equally spaced points along a 108-μm line across a neuritic plaque in the brain section of an Alzheimer patient showing the β-amyloid secondary protein structure (1632 cm$^{-1}$ amide I) in the plaque (spectra 5 and 6) compared to α-helical structure in the adjacent brain tissue (1650–1654 cm$^{-1}$ amide I) before and after the plaque. Note also the shoulder 1519 cm$^{-1}$ amide II band. (Adapted from data presented in Ref. 72.)

in Stamford, CT. A Spectra-Tech model IR-PLAN® infrared microscope acces-sory optically interfaced to a Nicolet FT-IR spectrometer was used to compare parts of wheat kernel sections *in situ* and after physical separation as intermediate fractions of dry flour milling.

Subsequent work with an IR-PLAN® accessory microscope interfaced to a Bowman-Michelson-100 spectrometer was done during a sabbatical of DLW at the Ottawa, Ontario, Agriculture Canada Plant Research Center laboratory of R. G. Fulcher, a botanist, light microscopist, and expert on wheat morphology. Examining the transitions between adjacent botanical parts was the purpose. A motorized stage was not yet available, but for infrared spectra collected from 100-μm spots in a sequence of 50-μm steps, transitions in chemical composition were apparent, to the delight of the investigator, even within two of the botanical parts, the endosperm and the germ. Besides sections of wheat kernels, various ingredients of food were sectioned and their botanical parts analyzed. These in-cluded soybean, barley, vanilla beans, peppercorns, and single starch granules from various sources, and the results, including 65 figures, were reported in a 1990 book chapter (2) resulting from the International Flavor Conference spon-sored by the American Chemical Society Food and Agriculture Division in Crete, Greece, the previous year.

The need for greater spatial resolution and the desire for mapping were met by cooperation with Reffner at Spectra-Tech using the prototype of the first integrated FT-IR microspectrometer, which became the IRμs®. Besides the im-proved spatial resolution by higher SNR from optical efficiency (approximately 8–10 bounces were eliminated) of the integrated instrument, a motorized stage, automated gain, and control software at last made automated mapping (acquiring spectra in a raster scan grid pattern) a reality. To our astonishment, a functional-group map based upon raw absorbance data showed the opposite of what had been expected from the wheat section. The area of endosperm having higher protein had a lower absorbance of amide bands I and II than did the area in the same section where the protein was lower. Clearly, the large index of refraction difference and the resulting scattering of the starchy endosperm overshadowed the absorption of more radiation by the protein present. Software writers came to the rescue with computation of baseline-corrected peak area in time for presen-tation at the International Cereal Chemistry (ICC) Congress in Vienna in 1990. The results of this first wheat-mapping attempt were reported in the ICC proceed-ings (79). Prior to our wheat mapping, Reffner reported ''microspectral'' map-ping (80), and subsequently Kodali et al. (81) used 3-dimensional images from spectral data of arterial tissue.

The Microbeam Molecular Spectroscopy Laboratory at Kansas State Uni-versity was established around an early-model Spectra-Tech IRμs® instrument augmented with high-quality electronic photodocumentation equipment, sample preparation microscopes, and an outboard data processing work station. Extensive

FT-IR microspectroscopy work, including mapping experiments of mostly wheat and corn sections, was reported at various national and international meetings and is detailed, with 110 figures, in a 1993 book chapter (82). Portions of this work have also appeared elsewhere (83), and tracing the migration of water by tempering wheat with $D_2O$ and mapping the OD absorption at 2500 cm$^{-1}$ was accomplished using 20-μm-thick sections to measure this minor band area (84,85). In addition, with polarized infrared, the difference in dichroic ratios in wheat gluten at different positions away from an elongated hole (producing a strain gradient in space) was investigated (86). This work with the spatial resolution capability of the modern FT-IR microspectrometer demonstrated proof of principle. Subsequently, a dedicated instrument for routine optical testing of wheat gluten was produced (87).

Ultraspatial resolution (6 μm × 6 μm) synchrotron infrared microspectroscopic 100% mapping in a small area (100 μm by 100 μm) showed highly localized distribution of carbonyl due to lipid in the central endosperm of corn (88,89). Other probings of cellular and subcellular dimensions with synchrotron radiation were reported on cells and cell walls in a wheat section (90). Mapping whole seeds was performed, in which each data point was taken from a cell-sized aperture with tissue in between, and an image of the entire kernel was produced by interpolation (91).

Other investigations of plant material by FT-IR microspectroscopy include probing of single wheat aleurone cells, wheat primary root cells, barley aleurone cells, corn aleurone cells, and in particular the cell wall portions of oats. Typically in our work of probing a single cell, an actual cell that was positioned in the target area defined by projected apertures was viewed and photographed with visible light and then was probed by microspectroscopy (22). Depending upon the size of the cell, a 6-μm × 6-μm or 12-μm × 12-μm aperture usually was used. In each case, the purpose was to look for compositional similarities or differences between cellular regions in close proximity within the section. In particular, the primary root of hard wheat coleorhiza, the epidermis, cells of the cortex, cells of the central vascular cylinder, and a large cell at the core were probed. Distinct rows of cells occupy the cortex, and a definite line of cellular demarcation occurred between the cortex and the central vascular cylinder, which also had distinct rows of smaller cells. Point-to-point probing allowed determination of when the chemistry of each successive layer of cells had changed. Mapping across the primary root with 6-μm × 6-μm pixels in 6-μm steps produced a series of spectra that showed highly localized chemical detail (89).

Mapping of a single wheat aleurone cell was done as early as 1992 (82,83). The functional-group map of the amide II band showed a pyramid shape, with the higher portion of the pyramid representing the center of the cell. The region of lowest reading of the absorbance for amide II was in the cell walls. The map of a different frequency representing the cellulose of the cell wall produced the opposite kind of image, with a ridge for the cell wall and absence of cellulosic material in the center of the cell. This first work was accomplished by painstakingly collecting many spectra through an aperture of 6 μm × 7 μm using the thermal source integrated IRμs® instrument at the Microbeam Molecular Spectroscopy Laboratory (Kansas State University). Subsequently, a highly detailed mapping experiment with a corn aleurone layer specimen was done using synchrotron radiation. With the synchrotron source, less coaddition of spectra was required, and the spatial resolution was excellent. Therefore, considerably sharper images (Fig. 8) were possible with this mapping procedure (88,89,91). Use of an even smaller pixel size of 3 μm × 3 μm for mapping procedures, including single living cells or diamond-sawed rock sections, with the synchrotron source

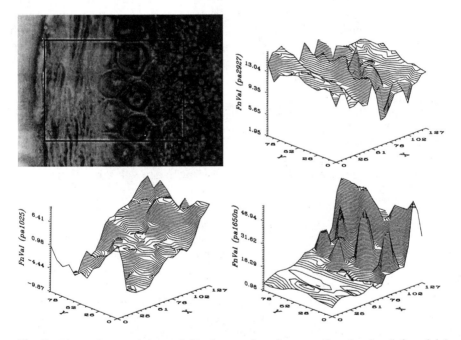

**Fig. 8**  Photomicrograph (upper left) of outer edge of corn section showing (left to right) pericarp, cell walls, cells, cell walls, and subaleurone endosperm. The three-dimensional stacked-contour infrared images resulting from mapping with 6-μm × 6-μm dual-remote image-plane masks and synchrotron infrared microspectroscopy reveal the chemical composition of the photomicrograph feature. Clockwise, the 2927 cm⁻¹ peak area shows the cell wall (based on lipid content), while the 1650 cm⁻¹ peak area shows two intact side-by-side aleurone cells and a broken cell above (based on protein content), and the 1025 cm⁻¹ peak area map reveals the carbohydrate (starch) content of the endosperm. (From Refs. 88 and 89 with permission of *Elsevier Science* and *Cellular and Molecular Biology*.)

is discussed elsewhere in Chapter 7. When the 3-µm aperture was used in the synchrotron work, the cutoff region for usable spectra, based on diffraction, eliminated many of the lower-frequency bands.

FT-IR microspectroscopic mapping of plant material has also been reported with the use of an MCT focal-plane array detector on an instrument illuminated with a conventional globar (thermal) source. Marcott et al. (91) produced functional-group images of several wheat sections with a pixel size of 4.5 µm. Additionally, individual pixel characterization using discriminate analysis and principle component analysis resulted in false color images that contrasted compositional distinctions for comparison with corresponding photomicrographs. Budevska, at a 1999 conference (92), reported FPA images of select botanical parts of rice and corn, and used chemometric statistical tools extensively to maximize the information content from her spectroscopic imaging experiments.

Among grain samples, a particular cultivar of oat known as being high in β-glucan was investigated. β-Glucan was located by a frequency characteristic of cellulosic material and displayed in contour maps, which showed that it tended to be localized in the cell walls from the oat map in the vicinity of a massive cell wall structure (88,89,93).

Grasses used as forage crops in North America have been the subjects of improvement efforts through plant breeding. Digestibility of grasses by cattle is of concern for the efficient use of rangeland. Previous to FT-IR microspectroscopic studies, scanning electron microscopy studies (94) of enzymatically digested grass sections found that the parenchyma bundle sheath that surrounds the vascular bundle is less digestible if it is composed of lignin with a great amount of aromatic character. The frequency at 1508 cm$^{-1}$ is indicative of aromatic character (2,12). Analyzing the parenchyma bundle sheath in the cross section of a leaf of grass represented a challenge to the spatial resolution of the microspectrometer. Early attempts in 1989 with an accessory-type infrared microscope attached to a conventional FT-IR instrument by way of an optical interface were unsatisfying, because the neighboring tissue was sampled accidentally and, thus, the resulting spectra did not accurately represent the parenchyma bundle sheath. When this experiment was repeated with the integrated instrument, where the spectrometer and microscope were built as a unit, the parenchyma bundle sheath alone could be probed. With synchrotron infrared microspectroscopy, the detailed mapping of the vascular bundle and its surrounding tissue was accomplished and the localized aromatic character was revealed (88,89). These results demonstrated that this technique could provide a means for screening various crosses at early stages of a forage-breeding program.

FT-IR microspectroscopic imaging of flax stems was performed by Himmelsbach et al. (95). Waxes, pectin, cellulose, aromatic groups, and acetyl groups were localized in the cuticular and epidermal tissues, fiber bundles, fiber cells, core tissue, and fibers, respectively. In this work, different cultivars of flax were analyzed carefully and evaluated for commercial use. Results of this relatively noninvasive technique were compared with previous studies on flax that used destructive methods.

### 3.2.3 Bone

Two independent research groups have used FT-IR microspectroscopy to examine thin sections of bones using different microspectroscopic instrumentation. Miller and coworkers, at the NSLS, used synchrotron infrared microspectroscopy to examine monkey bone and cartilage (96). Figure 9 shows the spectrum of cartilage versus bone including protein and mineral components, respectively. Particular emphasis was placed on evaluating the ratio of amide I to phosphate, the ratio of amide I to carbonate, and the ratio of carbonate to phosphate. In general, the new bone in close proximity to the osteon had pronounced amide I and amide II bands and relatively little phosphate or carbonate. In the area slightly away from the osteon, the amide I and amide II bands were somewhat diminished, and a prominent phosphate band was exhibited. Still further away, the amide I and II bands diminished more, and the prominent phosphate band gave way to that of the carbonate. New bone contains a high protein/mineral ratio. This decreases as the bone matures and as carbonate ion ($CO_3^{-2}$) is substituted for the phosphate ($PO_4^{-3}$) ion in the hydroxyapatite lattice. An increased carbonate/phosphate ratio also was observed near the marrow space. Related work at the NSLS involving cells, bone, hair, and rock has been reported (97). Recent osteo-

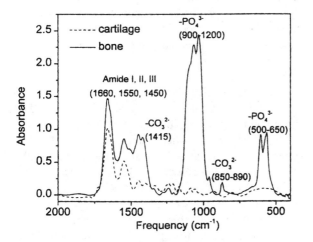

**Fig. 9** Spectra of bone and cartilage showing the amide I and II bands in both, with phosphate and carbonate bands of the bone emphasizing its mineral content. (Reproduced from Ref. 96 by permission of *Cellular and Molecular Biology*.)

porosis studies of subchondral bone from the tibia of ovariectomized monkeys have been done at the newly equipped NSLS beamlines (98).

Marcott et al., working with 5-μm sections of canine alveolar bone tissue, generated spectroscopic images from a locally assembled focal-plane array system using the fingerprint region of the infrared spectrum (99). By ratioing relative intensities of specific bands across the images, they obtained spatial distributions of the mineral-to-matrix ratio and mineral maturity as a function of distance from an osteon. This work was performed independently from the synchrotron infrared microspectroscopy. A conventional thermal source was used, but instead of a single element detector, an MCT focal-plane array detector was used. The arrays were produced from 4096 pixels, and excellent images were obtained. Images from 1740 cm$^{-1}$, 1660 cm$^{-1}$, 1415 cm$^{-1}$, and 1041 cm$^{-1}$ (Fig. 10) clearly defined the different parts of the tissue by their chemical composition. In addition, from mapping, the mineral content in different sections of avelolar bone samples was determined by the peak height of a phosphate band at 1041 cm$^{-1}$. The organic matrix content was determined by the peak height of the amide I band at 1660 cm$^{-1}$. Mineral maturity of the same bone section was determined by dividing the image absorbances at 1030 cm$^{-1}$ by absorbances at 1020 cm$^{-1}$.

An enlarged osteon map showed concentric rings of increasing maturity based on the 1030 cm$^{-1}$/1020 cm$^{-1}$ ratio. Using the focal-plane array detector, with a step-scan FT-IR microspectrometer, enough spectra were obtained to produce images in 5 minutes. Results with the canine specimens using the focal-plane array detector were similar to those produced on monkey specimens with the synchrotron source together with a single-element MCT detector. Prior to the focal-plane work, a tedious process of screening bone specimens from diseased cases had been done with the raster scanning procedure in attempts to understand more about the mechanism of various bone-deficiency diseases (100).

### 3.2.4  Cancer

A number of investigators have used FT-IR microspectroscopic imaging and probing to investigate various cancerous tissues. Lasch and Naumann studied melanoma and colon carcinoma and applied various pattern recognition techniques to the imaging data (101). They used artificial neural networks (ANNs) on a limited number of data channels, which allowed extremely repetitive calculations to take place in a reasonable time. Earlier conventional spectroscopic studies on cancerous tissue by Jackson and Mantsch in 1996 (102), Rigas and Wong in 1992 (103), Wong et al. in 1993 (104), Rigas et al. in 1990 (105), Malins et al. in 1997 (106), and Fabian et al. in 1995 (107) refer to disease-specific spectral signatures in various tissues and suggest that the spectral elements containing relevant information can be extracted. That earlier macrospectroscopic work is not discussed here. More recently, a number of investigators have used FT-IR microspectroscopic imaging and probing to investigate various cancerous tissues.

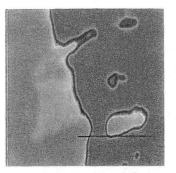
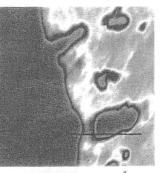
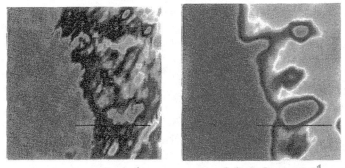

**Fig. 10** Focal-plane array (4096 pixels, MCT detector) maps of bone showing amide distribution (left) in contrast to phosphate (right). (Grey scale images adapted from color originals in Ref. 99 by permission of *Cellular and Molecular Biology.*)

In the laboratory of Naumann, Lasch et al. (108) studied melanoma and colon carcinoma and applied various pattern recognition techniques to image reassembly. They also found that cluster analysis (CA) and principal component analysis (PCA) were useful methods for identifying spectral differences when spectral features of the tissue were unknown. Once these reference spectra were selected and groups were thereby established and defined by the PCA, then reconstitution of the image could be done by assigning each pixel to a particular class or unassigned status. Class or group centers were used to find the origins of the new internal coordinate system. The geometric distances from each point spectrum to all new coordinate origins were calculated. Then all calculated distance values from the origin of the coordinate to any arbitrary point *A* or the distances from the origin of the coordinate to point $A$ were normalized to yield matrices. Normalization was performed by dividing the distance values by the highest distance value of its class. Combining these ratios with the original spatial information resulted in matrices that could be processed further to color or grey scale maps. Each map represented a scaled spatial distribution of one particular spectral pattern. The spatial distribution of the distinct and scaled spectroscopic patterns was found to be in excellent agreement with the pathohistological data. These are illustrated with two-dimensional cases of PCA-based infrared image reassembling. PCA analyses in three or more dimensions could be done as well. These same authors, in addition to the use of PCA and neural networks, have reported more recently the use of hierarchical clustering applied to mapping data from thin sections of colorectal cancer tissues (101).

Other studies on squamous cell carcinoma were done by Schultz and Mantsch (109). These authors produced a keratin map (1305 cm$^{-1}$), a DNA map (968 cm$^{-1}$), and a membrane fluidity map (2852 cm$^{-1}$) of a keratin pearl and surrounding tissue. As expected, keratin was concentrated in the pearl, whereas DNA and membrane fluidity were low in the pearl. Tissue CA and classification was applied to the spectral region from 1350 to 950 cm$^{-1}$ of 315 spectra in a map of a keratin pearl. The result of cluster 1 and cluster 2 maps was that the densities plotted in space were essentially exact opposites, indicating that the two distinct spectral characteristics were mutually exclusive (109). The reassembled maps identified changes in the tissue outside of the pearls, which likely were due to biochemical changes that accompany abnormal growth. This study indicated that FT-IR microspectroscopy together with CA should be able to differentiate between grades of oral carcinomas.

Cancerous and normal biopsied breast tissues were studied by Dukor et al. (110), and cells within the tissues were classified as benign, typical hyperplasia, or malignant. As with other investigators examining suspect tissue specimens, they used multivariate techniques on the spectroscopic data to classify cells. Aside from this, there are two other interesting operational features of Dukor's work. Firstly, unlike most sample treatment for transmission microspectroscopy, histopathological stained specimens mounted on plain glass microscope slides were used. Deparaffinized sections, 5 μm thick, that were stained with hematoxylin and eosin and air-dried were analyzed with an infrared microscope equipped with an ATR objective. This method establishes the possibility of analyzing specimens by infrared microspectroscopy that have been subjected to standard histological methods simply by removing the cover glass and mounting media.

In the laboratory of Diem (111), the dysplastic and neoplastic changes in deparaffinized 6-μm-thick sections of human cervical biopsy tissue with infrared microscopy was studied in collaboration with members of the clinical community. Several chemical changes were detected between normal, benign, dysplasia, and neoplasia, which suggests that FT-IR could be used as a method for screening precancerous tissue in the cervix.

In McNaughton's laboratory, he and coworkers (112) applied artificial neural networks (ANNs) to the diagnosis of cervical cancer from pap smears. Figure 11 shows spectra of benign, dysplastic, and malignant cervical cells. The results from a preliminary investigation indicate that neural networks coupled with infrared spectroscopy may provide an objective and automated screening technique for cervical cancer. This requires developing software for training of ANNs on a large body of data. The goal is to train a network into separations of five individual groups: one normal, one malignant, and three designated as cervical intraepithelial neoplasia (CIN, CIN II, and CIN III).

Lowry (113) has studied exfoliated cervical cells by mapping large areas of 3600- by 4000-μm areas in an overnight automated raster scan procedure. His manipulation of the data resulted in a three-class Mahalonobis discriminate function to produce colored images where each individual pixel has an assignment. From "feature maps" computed from the peak height at the amide II band at 1550 cm$^{-1}$ and from the ratio of the 1028 cm$^{-1}$ band (assigned to glycogen) and the amide II band, a correlation map of a spectral file diagnosed as CIN positive was produced. Ultimately, a classification map was produced using a Mahalonobis distance function.

### 3.2.5 Other Biological Applications

Gallstones with different chemical composition were identified by FT-IR microspectroscopy by Wentrup-Byrne et al. (114,115) and Paluszkiewicz et al. (116). Compounds present included cholesterol and mixtures of cholesterol and bilirubinate salts. By this method, a difference could be demonstrated in the spectra

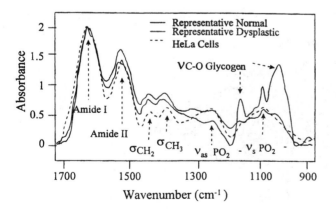

**Fig. 11**   Spectra of benign (top), dyplastic (lower solid line), and malignant (dotted line) cervical cells showing differences in the region of 1244 and 1082 cm$^{-1}$ and of the glycogen at 1047 and 1025 cm$^{-1}$. (Reproduced from Ref. 112 by permission of *Cellular and Molecular Biology*.)

of stones that had different calcium and copper concentration levels. Benedetti et al. (117) studied the spectral changes in megakaryocytic thrombocythemia.

Victor Kalasinsky, who is the author of an excellent review on FT-IR and Raman microspectroscopy of biological materials (5), has done considerable work in the identification of foreign substances found in the human body (118). This research reported on the FT-IR microspectroscopic characterization of breast biopsies from women with silicone breast implants. Polydimethylsiloxane (PDMS), polyesters, and polyurethanes were located and identified in breast biopsies (119). Mapping a large tissue area (e.g., 775 × 725 μm), shown in Fig. 12, compared the distribution of polyester and silicone in the same tissue section, while another pair of maps compared the distribution of polyurethane and silicone. Similar breast biopsies studied in Victor Kalasinsky's laboratory also have been analyzed using the focal-plane array by Lewis et al. of the NIH Laboratory at Bethesda, MD (120,121).

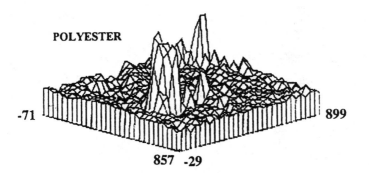

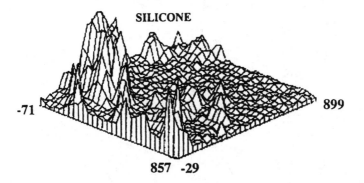

**Fig. 12**  Biopsied breast tissue map showing the location of foreign substances polyester and polydimethylsiloxane. (Reproduced from Ref. 119 by permission of *Cellular and Molecular Biology*.)

Studies by Kathryn Kalasinsky and coworkers (122) examined the distribution of drugs in human hair. Cocaine-doped hair specimens had the highest cocaine carbonyl absorption in the medulla. The corresponding band of the cortex was nearly as low as that of the control medulla. When looking for drugs in hair, it is very important to find the location. Other methods can detect the substances, but confirming their presence inside of the hair and not on the surface is a major feat of investigations of this type at the U.S. Armed Forces Institute of Pathology in Washington, DC. Serious effort has been made to study the process of uptake and deposition of drugs in the hair as well as their distribution. Sample preparation techniques have been developed and other analytical methods have been used in conjunction with the FT-IR microspectroscopy (123).

Detailed mapping of human head hair was accomplished with the use of synchrotron infrared microspectroscopy with a pinhole image-plane mask that produced an aperture of 5 μm (124). Human hair was sectioned longitudinally on a paraffin block, removed from the paraffin section, and then placed between barium fluoride plates in a microcompression cell for mapping. Mapping with a small aperture provided detail of the spatial distribution of chemical composition across the hair and to some degree along the length of the hair. The medulla was the densest region. It contained the most organic matter, as evidenced from the greatest absorption. In addition, the peak areas of 1550 $cm^{-1}$ (amide II) bands showed a higher concentration of protein in the center of the hair, with the maximum being in the region of the medulla. Probing with a 5-μm-diameter aperture at intervals along a 700-μm length of a single control human hair that had been spiked with a heroin metabolite revealed that the metabolite was found at one place along the length. Subsequent mapping experiments across the hair at that point showed an uneven distribution of the metabolite carbonyl band, with a maximum just next to each side of the medulla.

Living cells have been probed by Jamin, Dumas and coworkers with synchrotron infrared microspectroscopy (125) using 3-μm × 3-μm aperturing with an IRμs® (Spectra-Tech, Inc., Shelton, CT) instrument illuminated by synchrotron radiation from the NSLS beamline U2B at the BNL. Lipids and nucleic acids were localized from mapping experiments. Further studies by this same group showed the chemical distribution in a cell undergoing mitosis (126).

Hypercholesterolemic rabbit liver was the subject of the study by Jackson et al. (127). These workers performed functional-group mapping and subsequently treated the data by isodata cluster analysis techniques to produce mid-infrared maps, with each pixel in the 20 × 20 matrix having an assignment denoted by a particular coding or shading. The authors used a common shade for each cluster, and five distinct clusters clearly resulted from five separate groups of spectra. Cluster analysis techniques were used to nonsubjectively identify regions of tissue that have similar spectral properties.

Spectroscopic distinction of bacteria has been studied by macrotechniques with a conventional FT-IR spectrometer equipped with a custom, closed, multiple-sample-handling device. This use of infrared on bacteria has been developed in the laboratory of Naumann (128,129) and promoted effectively and also in the laboratory of Manfait (130). Using an infrared microscope, Lang and Sang (131) developed a simplified way of obtaining the infrared spectra of bacteria colonies directly on agar plates. In this use of an infrared microscope, the aperturing was minimal; nevertheless, a diffuse reflectance approach was taken in order to obtain infrared spectroscopic data. Gram-positive and gram-negative bacteria were differentiated by this process using principal component regression of the data obtained by the direct technique on the agar plates. Subsequent work, using a microscope equipped with an ATR objective, also has been reported by Lang (132) at a recent spectroscopy meeting.

Studies analyzing the secondary structure of proteins by FT-IR microspectroscopy have been reported in 1998. Wentrup-Byrne and coworkers found that insect eggshells had a predominant band in the amide I region at 1634 cm$^{-1}$, which suggested that an antiparallel β-pleated sheet structure was abundant (133). Shoulders at 1662 and 1692 cm$^{-1}$ indicated that β-turn structures also were present. Spectra of chorion from parasitized eggs resembled those from the unparasitized eggs, except for a reduction in the intensities of band absorptions between 2800 cm$^{-1}$ and 3000 cm$^{-1}$ for $CH_2$ and $CH_3$ stretching vibrations.

Single protein crystals have been subjected to infrared crystallographic investigation by Sage and coworkers (134). This exemplary scientific structural work has been made possible by modern FT-IR instrumentation.

## 4  SUMMARY

The field of infrared microspectroscopy is highly technology driven. Advances in instrumentation are directly responsible for infrared microspectroscopic technology and for all applications. Although the concept of infrared microspectroscopy was demonstrated in 1949 by Gore (135), the procedure was not practical until FT-IR spectrometry was developed fully and highly sensitive detectors were available. The decade from 1989 to 1999 brought not only the first high-performance microscope with dual remote image-plane masks and reflecting optics, but also subsequent significant instrumental advances. This progression is discussed in the Foreword, pp. iii–v, of Ref. 6. These instruments have allowed microscopists and other scientists working in the biological sciences to perform *in situ* chemical analysis. This capability provides molecular information that is the basis of the numerous studies and applications that we have cited and of new applications yet to follow.

In a single year alone, more than 100 scientists from more than 65 research institutions worldwide have published articles on infrared microspectroscopy of biological materials. This represents a sharp rise in activity involving this technology in the biological sciences. For the applications cited in this chapter, both integrated microscope/spectrometer and accessory scopes interfaced to FT-IR spectrometers were used. Several of the studies reported used synchrotron infrared microspectroscopy, and others have used either InSb or MCT focal-plane arrays. Most of the work entailed transmission through samples with or without compression cell use. Attenuated total reflectance run on slides prepared for histology may provide useful molecular information on stained sections used for pathological evaluation.

Classical spectroscopic analysis with good spectral resolution is obtained readily while simultaneously achieving localization with excellent spatial resolution. Use of synchrotron radiation produces ultraspatial resolution for situations where this is necessary. Also, infinity-corrected optics enhances the clarity of the infrared image and allows the convenience of DIC viewing microscopy. Infrared microscopic imaging with functional-group maps or spectroscopic pattern-recognition pixel assignment (identification) is made rapid and user friendly with parallel data acquisition at many pixels with a focal-plane array (camera) detector coupled with step-scan FT spectrometer. Such images are produced in minutes when $16$-cm$^{-1}$-resolution spectra are collected. This has been appropriately dubbed *fast FT-IR imaging*; however, the information generated so quickly may require extended time for full interpretation. The multiplicity of pixels allows the use of statistics and image interpretation schemes as objective alternatives to classical data treatment.

Instruments at the present state of the art have opened infrared microspectroscopy to any scientist that could benefit from localized chemical analysis. There is an average of 300 microspectrometers sold each year. The cumulative total for 13 years of manufacturing is at least 3000 systems in use today. It is estimated that less than 10% are devoted to use in the biological sciences. With the recent upsurge in biological applications, we can expect an increase.

**Notes Added in Proof**: In this rapidly emerging technology applied to biological materials it is essential to include late-breaking developments communicated by authors previously referenced. A focal-point article in the A pages of *Applied Spectroscopy* by Max Diem and coworkers showed infrared spectroscopy and histology of human cells and tissue in parallel (136). Spectra were obtained for different parts of a stained cervical biopsy and the contrasts in spectra were discussed. Pathologists on the team thoroughly characterized the histology of the cells and tissues examined to allow correlation of spectra with select parts of the tissue. Spectra from healthy and diseased cells and tissue were compared. Spectra of normal, displastic, and neoplastic cells of the basal layer of cervical epithelium were examined. Infrared spectra of individual myeloid leukemia cells representing six stages of the disease were contrasted. The prospects as well as the pitfalls of pathological diagnosis by way of infrared microspectroscopy are discussed. Sources of contamination of specimens were identified. This same group has published a series of papers (137–140) under the general heading of infrared spectroscopy of human cells and tissues. Paper VI in this series (141) involves a comparative study of histopathology and infrared microspectroscopy of normal, cirrhotic, and cancerous liver tissue.

Microspectroscopy of localized chemistry within bone sections has continued by Miller and coworkers at the NSLS concerning osteoporosis, subchondral osteoarthritis, bone, and the composition of bone protein and minerals (142–144). Fluorescence-assisted infrared microscopy was used. The latest in a series of papers by Mendelsohn and coworkers includes spectroscopy, microscopy, and microscopic imaging of mineralizing tissues from human iliac crest biopsies (145). Combined work of these two research groups at NSLS Beamline U2B dealing with mineral crystal size and phosphate environment in bone has been submitted to *Calcified Tissue International* (146), the journal where Mendelsohn and coworkers previously reported FT-IR microspectroscopic analysis of human untreated osteoporotic bone and normal human cortical and trabecular bone (147,148).

In the area of synchrotron infrared microspectroscopy, Paul Dumas is currently reporting infrared microscopy results form the MIRAGE Beamline developed on an existing synchrotron port (SA5-Super ACO-URE in Orsay, France). A report by Polack et al. (149) describes the optical design and performance of the MIRAGE Beamline, including the IR microscope at that synchrotron location. The group at LURE-CNRS of the Central University of Paris-South in Orsay also experimented with ablation of human tissue with an infrared free-electron laser (FEL) tuned to the infrared frequencies of amide I, amide II, and phosphate vibration bands. Results of the ablation were analyzed by synchrotron infrared microspectroscopy using a 3-$\mu$m $\times$ 3-$\mu$m target size. Results of this localized treatment and analysis showed a frequency dependence (150). At the same facility, a combination of UV and IR work was done. The UV storage ring free-electron laser and the infrared beamline at the same facility were used in the same experiment. The time structure of the synchrotron radiation (SR) in the IR produced short pulses in the subnanosecond range, making it suitable for time-resolved experiments. A storage ring free-electron laser (SR-FEL) at the same storage ring produces intense UV radiation naturally synchronized with the synchrotron radiation that allows interesting two-color pump/probe investigations. Examples are cited for a FEL (UV-pump) + SR (IR-probe) experiments. Alternatively this process may be reversed, where the process may be described as SR (IR-pump) + FEL (UV-probe). With this system, subnanosecond dynamics of intramolecular relaxation processes could be investigated (151).

Infrared beamlines at synchrotrons around the world have expanded from a single one initially to 28. Of these, 15 are equipped with infrared microscopes. A recent paper by Carr et al. (152) is not only an excellent reference on synchrotron infrared hardware, but it also describes six new beamlines at NSLS devoted to infrared. Beamline U12IR is designed for far-infrared, and its pulsed beam allows time-resolved pump-probe experiments. Beamline U4IR is for far- and mid-infrared surface science and spectroscopy. Four other beamlines, U2A, U2B, U10A, and U10B, are set up for mid-infrared spectroscopy. In particular U2B, run by the Albert Einstein College of Medicine, is dedicated to the study of biological systems (153). All of these beamlines, including four equipped with microscopes, are now in operation. Small-pixel-size 100% mapping, as previously reported (52,97,125,126), has continued at NSLS. Chemical imaging of human hair cross sections and studies of organic interaction with hair have recently been reported in experiments using 3-μm × 3-μm apertures (154,155). A summary of applications of infrared microspectroscopy to geology, biology, and cosmetics has also been issued. It deals with fluid inclusions in geological deposits, living cells, hair, bones, and osteoarthritis, complete with color three-dimensional images to illustrate the mapping results as well as the photomicrographs of the original targeted specimens (156).

Recent reports from the Microbeam Molecular Spectroscopy Laboratory at Kansas State University include an invited paper in *Science* (157) and subsequent studies of the chemistry of individual rat retina layers that allowed characterization of each layer from spectra of four pigmented and four albino animals (158). A reduction was observed in the level of unsaturation in the outer segment of the albino rats compared to that of the pigmented rats. Subsequent to that work performed on the Spectra Tech IRμs at Kansas State University, another study was carried out involving chemical changes in the outer segment of the retina caused by oxidative stress. This was performed on the Continuμm™ Spectra Tech instrument equipped with 32 × infinity-corrected mirror lenses. Dramatic changes were in evidence for the oxidative stress, particularly with reference to a reduction of unsaturation and carbonyl groups. An unexpected result was an increase in other functional groups as a result of the trauma associated with injection of either the carrier (saline) or the carrier plus the iron II source of oxidative stress (159).

## REFERENCES

1. DL Wetzel, RG Messerschmidt, RG Fulcher. Chemical mapping of wheat kernels by FT-IR microscopy. Federation of Analytical Chemistry and Spectroscopy Societies, 14th Annual Meeting, Detroit, MI, October 1987, paper no. 151.
2. DL Wetzel, RG Fulcher. In: G. Charalambous Ed. Food Flavors and Off Flavors. Amsterdam: Elsevier, 1990, pp. 485–510.
3. A Dong, RG Messerschmidt, JA Reffner. Biochem Biophys Res Commun 156(2): 752–756, 1988.
4. SM LeVine, DL Wetzel. Appl Spectrosc Rev 28:385–412, 1993.
5. VK Kalasinsky. Appl Spectrosc Rev 31:193–249, 1996.

6. DL Wetzel, SM LeVine. Fourier transform infrared FT-IR microspectroscopy. A new molecular dimension for tissue or cellular imaging and in situ chemical analysis. Cell Mol Biol 44(1), 1998.

7. JA de Haseth. Fourier Transform Spectroscopy: 11th International Conference. Woodbury, NY: American Institute of Physics, 1998.

8. RG Messerschmidt, DW Sting. Microscope having dual remote image masking. U.S. Patent No. 4,877,960, 1989.

9. JA Reffner. Inst Phys Conf Ser No 98 Chapter 13. London: IOP Publ, 1990, pp. 559–569.

10. RG Messerschmidt. In: PB Rousch, ed. The Design, Sample Handling, and Applications of Infrared Microscopes. ASTM STP949. Philadelphia: American Society for Testing Materials, 1987, pp. 12–26.

11. RG Messerschmidt, MA Harthcock, eds. Infrared Microspectroscopy: Theory and Applications. New York: Marcel Dekker, 1988.

12. DL Wetzel. In: HG Charalambous, ed. Food Flavors, Generation, Analysis, Process Influence. Amsterdam: Elsevier Science, 1993, pp. 679–722.

13. JA Reffner. Cell Mol Biol 44(1):1–9, 1998.

14. RG Messerschmidt. In: HJ Huecki, ed. Practical Guide to Infrared Microspectroscopy. New York: Marcel Dekker, 1995, pp. 3–39.

15. VJ Coates, A Offner, EH Siegler Jr. J Opt Soc Am 43:984–991, 1953.

16. JA Reffner, GL Carr, S Sutton, RJ Hemley, GP Williams. Synchrot Radiat News 7(2):30–37, 1994.

17. GP Williams, GL Carr. NSLS Newsletter, November 1993, p. 30.

18. JA Reffner, PA Martoglio, GP Williams. Rev Sci Instr 66(2):1298–1302, 1995.

19. GL Carr, JA Reffner, GP Williams. Rev Sci Intr 66:1490–1492, 1995.

20. GL Carr. In: JA Reffner, GP Williams, eds. Proceedings of the First Workshop on Applications of Synchrotron Radiation to Infrared Microspectroscopy. Upton, NY: National Synchrotron Light Source, 1994, pp. 9–14.

21. GP Williams. Rev Sci Instrum 63:1535, 1992.

22. DL Wetzel, JA Reffner, GP Williams. Mikrochim Acta [Suppl] 14:353–355, 1997.

23. EN Lewis, AM Gorbach, CA Marcott, IW Levin. Appl Spectrosc 50:263–269, 1996.

24. EN Lewis, LH Kidder, IW Levin, VD Kleiman, EJ Heilweil. Optics Lett 22:742–744, 1997.

25. EN Lewis, PJ Treado, RC Reeder, GM Story, AE Dowrey, CA Marcott, IW Levin. Anal Chem 67:3377–3381, 1995.

26. CA Marcott, RC Reeder. In: GW Bailey, JM Corbett, RVW Dimlich, JR Michael, NJ Zaluzec, eds. Microscopy and Microanalysis. San Francisco: San Francisco Press, 1996, pp. 260–261.

27. CA Marcott, GM Story, AE Dowrey, RC Reeder, I Noda. Microchim Acta 14: 157–163, 1996.

28. EN Lewis, LH Kidder, IW Levin, WK Anderson, CJ Manning. New developments in spectroscopic imaging. 50th Annual Pittsburgh Conference on Analytical Chemistry and Applied Spectroscopy, Orlando, FL, 1999, paper no. 897.

29. PJ Tredo, IW Levin, EN Lewis. Appl Spectrosc 46(4):553–559, 1992.

30. PJ Tredo, IW Levin, EN Lewis. Appl Spectrosc 46(8):1211–1216, 1992.

31. EN Lewis, IW Levin. Appl Spectrosc 49(5):672–678, 1995.
32. P Colarusso, LH Kidder, IW Levin, JC Fraser, JF Arens, EN Lewis. Appl Spectrosc 52(3):106A–120A, 1998.
33. JA Reffner, SH Vogel. Confocal Microspectrometry System. U.S. Patent No. 5,864,137, 1999.
34. JA Reffner, SH Vogel. Microscopy Microanal 4 [Suppl 2 proceedings]: 410–411, 1998.
35. JA Reffner, WT Wilborg, Infrared Microspectrometer Accessory. U.S. Patent No. 5,581,085, 1996.
36. MR Derrick. Microbeam Anal 2 [Suppl]: 65–66, 1993.
37. RMA Heeren, J van der Weerd, N Wyplosz. FT-IR Imaging of paint cross-sections from 17th–19th century Dutch paintings. 50th Annual Pittsburgh Conference on Analytical Chemistry and Applied Spectroscopy, Orlando, FL, 1999, paper no. 017.
38. MR Derrick. In: HJ Huecki, ed. Practical Guide to Infrared Microspectroscopy. New York: Marcel Dekker, 1995, pp. 287–322.
39. PA Martoglio. The application of microspectroscopy to dypol fibers and the vibrational assignment of tetronic acid. PhD dissertation, Miami University, Oxford, OH, 1992.
40. PA Martoglio, SP Boufford, AJ Summer. Anal Chem 62:1123A–1128A, 1990.
41. JA Reffner, PA Martoglio, In: HJ Huecki, ed. Practical Guide to Infrared Microspectroscopy. New York: Marcel Dekker, 1995, pp. 41–84.
42. KA Chess. In: HJ Huecki, ed. Practical Guide to Infrared Microspectroscopy. New York: Marcel Dekker, 1995, pp. 111–136.
43. DL Wetzel, RO Carter III. In: JA de Haseth, ed. Fourier transform spectroscopy. 11th International Conference. Woodbury, NY: American Institute of Physics, 1998, pp. 567–570.
44. MW Tungol, EG Bartick, A Montaser. Spectrochim Acta 46B:1535E–1544E, 1991.
45. LL Cho, DL Wetzel. Mikrochim Acta 14:349–351, 1997.
46. PA Martoglio. Microbeam Anal 3 [Suppl]:95–96, 1994.
47. DL Wetzel, JA Reffner, GL Carr, L Cho. In: JA de Haseth, ed. Fourier transform spectroscopy. 11th International Conference. Woodbury, NY: American Institute of Physics, 1998, pp. 657–660.
48. EG Bartick, MW Tungol, JA Reffner. Anal Chim Acta 288:35–42, 1994.
49. LL Cho, Single fiber analysis by FT-IR miscrospectroscopy. PhD dissertation, Kansas State University, Manhattan, KS, 1997.
50. LL Cho, JA Reffner, BM Gatewood, DL Wetzel. J Forensic Sci 44(2):275–282, 1999.
51. LL Cho, JA Reffner, DL Wetzel. J Forensic Sci 44(2):283–291, 1999.
52. N Guilhaumou, P Dumas, GL Carr, GP Williams. Appl Spectrosc 55(8):1029–1034, 1998.
53. JP Kinding, LE Ellis, TW Brueggemeyer, RD Satzger. In: JA de Haseth, ed. Fourier transform spectroscopy. 11th International Conference. Woodbury, NY: American Institute of Physics, 1998, pp. 449–452.
54. KS Kalasinsky, J Magluilo, T Schafer. J Anal Toxicol 18:337–341, 1994.
55. MT Söderström. In: JA de Haseth, ed. Fourier transform spectroscopy. 11th Interna-

tional Conference. Woodbury, NY: American Institute of Physics, 1998, pp. 457–460.

56. RB Melkowits, SP Altomari, SB Levery, JA de Haseth. Monosaccharide composition analysis by infrared microspectroscopy. 50th Annual Pittsburgh Conference on Analytical Chemistry and Applied Spectroscopy, Orlando, FL, 1999, paper no. 793.

57. DL Wetzel, SM LeVine. Spectroscopy, 8(4):40–45, 1993.

58. DL Wetzel, SM LeVine. Proceedings of SPIE—Int Soc Opt Eng 1575:435–436, 1992.

59. SM LeVine, DL Wetzel, AJ Eilert. Inter J Dev Neurosci 12:275–288, 1994.

60. SM LeVine, DL Wetzel. Am J Path 145:1041–1047, 1994.

61. DL Wetzel, SM LeVine, DN Slatkin, MM Nawrocky. In: JA de Haseth, ed. Fourier transform spectroscopy. 11th International Conference. Woodbury, NY:American Institute of Physics, 1998, pp. 294–297.

62. DL Wetzel, DN Slatkin, SM LeVine. Cell Mol Biol 44(1):15–28, 1998.

63. HC Agrawall, K Fujimoto, RM Burton. Biochem J 154:265–269, 1977.

64. CA Fischer, P Morell. Brain Res 74:51–65, 1974.

65. ME Smith. Biochim Biophys Acta 164:285–293, 1968.

66. DS Lester, LM Kidder, IW Levin, EN Lewis. Cell Mol Biol 44(1):29–38, 1998.

67. DL Wetzel, CA Marcott, SM LeVine. Infrared microscopic imaging of brain tissue with a focal-plane array. 50th Annual Pittsburgh Conference on Analytical Chemistry and Applied Spectroscopy, Orlando, FL, 1999, paper no. 791.

68. J Koenig, CR Snivley. Fast FT-IR imaging of multicomponent polymer systems. 50th Annual Pittsburgh Conference on Analytical Chemistry and Applied Spectroscopy, Orlando, FL, 1999, paper no. 016.

69. OP Lamda, D Borchman, WM Garner. Free Rad Biol Med 16:591–601, 1994.

70. C Signorini, M Ferrali, L Ciccoli, L Sugherini, A Magnani, M Comporti. FEBS Lett 362:165–170, 1995.

71. SM LeVine, DL Wetzel. Free Rad Biol Med 25(1):33–41, 1998.

72. L-P Choo, DL Wetzel, WC Halliday, M Jackson, SM LeVine, HH Mantsch. Biophys J 71:1672–1679, 1996.

73. PE Fraser, JT Nguyen, WK Surewicz, DA Kirschner. Biophys J 60:1190–1201, 1991.

74. PE Fraser, JT Nguyen, H Inouye, WK Surewicz, DJ Selkoe, MB Podlisny, DA Kirschner. Biochem 31:10716–10723, 1992.

75. PE Fraser, DR McLachlan, WK Surewicz, CA Mizzen, AD Snow, JT Nguyen, DA Kirschner. J Mol Biol 244:63–74, 1994.

76. L Otvos, GI Szendrei, VMY Lee, HH Mantsch. Eur J Biochem 211:249–257, 1993.

77. H Fabian, LP Choo, GI Szendrei, M Jackson, WC Halliday, L Otvos, HH Mantsch. Appl Spectrosc 47:1513–1518, 1993.

78. H Fabian, GI Szendrei, HH Mantsch, BD Greenberg, LL Otvos. Eur J Biochem 221:959–964, 1994.

79. DL Wetzel, JA Reffner. In: R Lasztity, ed. Proceedings of the International Cereal Chemistry Symposium ICC Congress, Vienna, Austria. Published in Budapest, Hungary: ICC and Technical University of Budapest, 1991, pp. 47–52.

80. JA Reffner. Molecular microspectral mapping with the FT-IR microscope. EMAG-MICRO 89, London, 13–15 Sept. 1989, Inst Phys Conf Ser No 98: Chapter 13(2): 559–569.

81. DR Kodali, DM Small, J Powell, K Krishnan. Appl Spectrosc 45:1310–1317, 1991.

82. DL Wetzel. In: G Charalambous, ed. Food Flavors, Ingredients, and Composition. Amsterdam: Elsevier Science, 1993, pp. 679–728.

83. DL Wetzel, JA Reffner. Cereal Foods World 38:9–20, 1993.

84. DL Wetzel, AJ Eilert. SPIE—Proceedings of Int. Soc. Opt. Eng. 2089:464–465, 1994.

85. JA Sweat. Variations in tempering times for winter wheat by IR microspectroscopic tracking of $D_2O$. MS thesis, Kansas State University, Manhattan, KS, 1996.

86. S Walche. A potential sampling technique of wheat gluten for stress gradient measurement by infrared microspectroscopy. MS thesis, Kansas State University, Manhattan, KS, 1996.

87. DL Wetzel, JA Sweat. Polarized near-IR with an acousto-optic TFS. 50th Annual Pittsburgh Conference on Analytical Chemistry and Applied Spectroscopy, Orlando, FL, 1999, paper no. 991.

88. DL Wetzel. In: HG Charalambous, ed. Food Flavors, Generation, Analysis, Process Influence. Amsterdam: Elsevier Science, 1995, pp. 2039–2108.

89. DL Wetzel, AJ Eilert, LN Pietrzak, SS Miller, JA Sweat. Cell Mol Biol 44(1):145–168, 1998.

90. DL Wetzel. Microscopy and Analysis May: 17–20, 1996.

91. CA Marcott, RC Reeder, JA Sweat, DD Panzer, DL Wetzel. Vibr Spectrosc 19: 123–129, 1998.

92. BO Budevska, Application of infrared microscopy and spectral imaging to biological systems. 50th Annual Pittsburgh Conference on Analytical Chemistry and Applied Spectroscopy, Orlando, FL, 1999, paper no. 018.

93. DL Wetzel, JA Sweat, DD Panzer. In: JA de Haseth, ed. Fourier transform spectroscopy. 11th International Conference. Woodbury, NY: American Institute of Physics, 1998, pp. 354–357.

94. DE Akin. Crop Sci 22:444–446, 1982.

95. DS Himmelsbach, S Khalili, DE Akin. Cell Mol Biol 44(1):99–108, 1998.

96. LM Miller, SC Carlson, GL Carr, MR Chance. Cell Mol Biol 44(1):117–129, 1998.

97. JL Bantignies, GL Carr, P Dumas, LM Miller, GP Williams. Sync Rad News 11(4): 31–36, 1998.

98. LM Miller, R Huang, CS Carlson, MR Chance. Sync Rad News 12(1):21–27, 1999.

99. CA Marcott, RC Reeder, EP Paschalis, DN Tatakis, AL Boskey, R Mendelson. Cell Mol Biol 44(1):109–116, 1998.

100. EP Paschalis, B Jacenko, B Olsen, R Mendelsohn, AL Bosekey. Bone 19:151–156, 1996.

101. P Lasch, D Naumann. Cell Mol Biol 44(1):189–202, 1998.

102. M Jackson, HH Mantsch. In: RJH Clark, RE Mester, eds. Biomedical Applications of Spectroscopy. Chichester, UK: Wiley, 1996, pp. 185–215.

103. B Rigas, PT Wong. Cancer Res 52:84–88, 1992.

104. PT Wong, SM Goldstein, RC Grekin, TA Godwin. Cancer Res 53:762–765, 1993.

105. B Rigas, IS Morgallo, IS Goldman, PT Wong. Proc Natl Acad Sci 87:8140–8144, 1990.

106. DC Malins, NL Polissar, SJ Gunselman. Proc Natl Acad Sci 94:259–264, 1997.

107. H Fabian, M Jackson, L Murphy, PH Watson, I Fichtner, HH Mantsch. Biospectroscopy 1:37–46, 1995.

108. P Lasch, W Wäsche, G Müller, D Naumann. In: JA de Haseth, ed. Fourier transform spectroscopy. 11th International Conference. Woodbury, NY: American Institute of Physics, 1998, pp. 308–311.

109. CP Schultz, HH Mantsch. Cell Mol Biol 44(1):203–210, 1998.

110. RK Dukor, MN Liebman, BL Johnson. Cell Mol Biol 44(1):211–218, 1998.

111. L Chiriboga, P Xie, H Yee, D Zarou, D Zakim, M Diem, Cell Mol Biol 44(1): 219–230, 1998.

112. M Romeo, F Burden, M Quinn, B Wood, D McNaughton. Cell Mol Biol 44(1): 179–188, 1998.

113. SR Lowry. Cell Mol Biol 44(1):169–188, 1998.

114. E Wentrup-Byrne, L Rintoul, JL Smith, PM Frederics. Appl Spectrosc 49(7):1028–1036, 1995.

115. E Wentrup-Byrne, W Chua-Anusorn, TGST Pierre, J Webb, A Ramsay. Biospectroscopy 3:409–416, 1997.

116. C Paluszkiewicz, WM Kwiatek, M Galka, D Sobieraj, E Wentrup-Byrne. Cell Mol Biol 44(1):65–74, 1998.

117. E Benedetti, E Bramanti, F Papineschi, P Vergamini, E Benedetti. Cell Mol Biol 44(1):129–139, 1998.

118. JJ Henton, FB Johnson, RM Przygodzki, VF Kalasinsky, F Al-Dayel, WD Travis. Arch Pathol Lab Med 120(10):967–969, 1996.

119. VF Kalasinsky, FB Johnson, R Ferwerda. Cell Mol Biol 44(1):141–144, 1998.

120. EN Lewis, LH Kidder, IW Levin, VF Kalasinsky, JP Hanig, DS Lester. Ann NY Acad Sci 820:234–246, 1997.

121. LH Kidder, VF Kalasinsky, JL Luke, IW Levin, EN Lewis. Nature Medicine 3(2): 235–237, 1997.

122. KS Kalasinsky. Cell Mol Biol 44(1):81–88, 1998.

123. KS Kalasinsky, J Magluilo, T Schaefer. For Sci Int 63:253–260, 1994.

124. DL Wetzel, GP Williams. In: JA de Haseth, ed. Fourier transform spectroscopy. 11th International Conference. Woodbury, NY: American Institute of Physics, 1998, pp. 302–305.

125. N Jamin, P Dumas, J Moncuit, WH Fridman, JL Teillaud, GL Carr, GP Williams. Cell Mol Biol 44(1):9–13, 1998.

126. N Jamin, P Dumas, J Moncuit, WH Fridman, JL Teillaud, GL Carr, GP Williams. Proc Natl Acad Sci 95:4837–4840, 1998.

127. M Jackson, B Ramjiawan, M Hewko, HH Mantsch. Cell Mol Biol 44(1):89–98, 1998.

128. J Schmidt, T Udelhoven, T Löchte, HC Flemming, D Naumann. In: JA de Haseth, ed. Fourier transform spectroscopy. 11th International Conference. Woodbury, NY: American Institute of Physics, 1998, pp. 260–263.

129. D Helm, D Naumann. FEMS Microbiol Lett 126(1):75–80, 1995.
130. GD Sockalingum, W Boutledja, P Pina, P Allough, C Blouy, M Manfait. Cell Mol Biol 44(1):261–270, 1998.
131. PL Lang, SC Sang. Cell Mol Biol 44(1):231–238, 1998.
132. PL Lang. The attenuated total reflection infrared microspectroscopy of bacterial colonies. Federation of Analytical Chemistry and Spectroscopy Societies, 25th Annual Meeting, Austin, Texas, 1998, p. 122.
133. JM Gentner, E Wentrup-Byrne, PJ Walker, MD Walsh. Cell Mol Biol 44(1):251–259, 1998.
134. H Khachfe, M Mylrajan, T Sage. Cell Mol Biol 44(1):39–52, 1998.
135. RC Gore. Science 110:710–712, 1949.
136. M Diem, S Boydston-White, L Chiriboga. Appl Spectrosc 53(4):148–161A, 1999.
137. L Chiriboga, P Xie, H, Yee, V Vigorita, D Zarrow, D Zakim, M Diem. Biospectroscopy 4:47–53, 1998.
138. L Chiriboga, P Xie, V Vigorita, D Zarow, D Zakim, M Diem. Biospectroscopy 4:55–59, 1998.
139. L Chiriboga, P Xie, W Zang, M Diem. Biospectroscopy 5:253–257, 1999.
140. S Boydston-White, T Gopen, S Houser, J Bargonetti, M Diem. Biospectroscopy 5:219–227, 1999.
141. L Chiriboga, H Yee, M Diem. Appl Spectrosc, 54(1):1–8, 2000.
142. LM Miller, CS Carlson, D Hamerman, MR Chance. Bone 23(5):S458, 1999.
143. LM Miller, J Tibrewala, CS Carlson. Cell Mol Biol, in press.
144. LM Miller, D Hamerman, MR Chance, CS Carlson. SPIE—Int Soc Opt Eng 3775:104–112, 1999.
145. R Mendelsohn, EP Paschalis, AL Boskey. J Biomedic Optics 4:14–21, 1999.
146. LM Miller, V Vairavamurthy, MR Chance, EP Paschalis, F Betts, AL Boskey, R Mendelsohn. Calcified Tissue International, submitted.
147. R Mendelsohn, EP Paschalis, F Betts, E DiCarlo, AL Boskey. Calcified Tissue. Int. 61:480–486, 1997.
148. R Mendelsohn, EP Paschalis, F Betts, E DiCarlo, AL Boskey. Calcified Tissue Int. 61:487–492, 1997.
149. F Polack, R Mercier, L Nahon, C Armellin, JP Marx, M Tanguy, ME Couprie, P Dumas. Optical design and performance of the IR microscope beamline at SUPERACO-FRANCE. SPIE, Eds: P Dumas and GL Carr. 3575:13, 1999.
150. F Glotin, JM Ortega, JL Poncy, F Tourdes, JL Lefaix, P Dumas. Frequency dependence of the ablation of pig tissue with an IR-FEL: A microspectroscopic analysis. SPIE Eds: P Dumas and GL Carr 3575:113, 1999.
151. L Nahon, E Renault, ME Couprie, D Naturelli, D Garzella, M Billandron, GL Carr, GP Williams, P Dumas. Applications of UV-storage ring free electron laser: the case of super-ACO Nuclear Instr Methods A, 429:489–496, 1999.
152. L Carr, P Dumas, CJ Hirschmugl, GP Williams. II Nirovo Cimento 20(4):375–395, 1998.
153. LM Miller, M Sullivan, J Toomey, N Marinkovic, GL Carr, GP Williams, MR Chance. Amer Inst Phys Conf Proc, in press.
154. JL Bantignies, G Fuchs, GL Carr, GP Williams, D Lutz, S Marull. Int J Cosmet Science 20–38, 1998.

155.  JL Bantignies, GL Carr, D Lutz, S Marull, GP Williams, G Fuchs. J Soc Cosmet Chem, submitted.
156.  JL Bantignies, GL Carr, P Dumas, L Miller, GP Williams. Synchrotron Radiation News 11(4):31–36.
157.  DL Wetzel, SM LeVine. Science 285:1224–1225, 1999.
158.  SM LeVine, J Radel, DL Wetzel. Biochemica et Biophysica Acta, 1473:409–417, 1999.
159.  JA Homan, J Radel, DD Wallace, DL Wetzel, SM LeVine. Cell Mol Biol, in press.

## ABBREVIATIONS

| | |
|---|---|
| ANN | artificial neural network |
| ATR | attenuated total reflection |
| BNL | Brookhaven National Laboratory |
| CA | cluster analysis |
| Continuum® | registered trademark of Spectra-Tech, Inc., Shelton, CT |
| DIC | differential interference contrast |
| DSP | digital signal processing |
| FBI | Federal Bureau of Investigation |
| FEL | free electron laser |
| FFT | fast Fourier transform |
| FPA | focal-plane array |
| FT-IR | Fourier transform infrared |
| InspectIR® | registered trademark of Spectra-Tech, Inc., Shelton, CT |
| IRµs® | registered trademark of Spectra-Tech, Inc., Shelton, CT |
| MCT | mercury cadmium telluride |
| NA | numerical aperture |
| NIH | National Institutes of Health |
| NSLS | National Synchrotron Light Source |
| PCA | principal component analysis |
| PET | polyethylene terephthalate |
| SNR | signal-to-noise ratio |
| VUV | vacuum ultraviolet |

# 5

# Biophysical Infrared Modulation Spectroscopy

**Urs P. Fringeli and Dieter Baurecht**
*University of Vienna, Vienna, Austria*

**Hans H. Günthard**
*Swiss Federal Institute of Technology, Zurich, Switzerland*

## 1 INTRODUCTION

Modulation spectroscopy or modulated excitation (ME) spectroscopy can always be applied if a system admits a periodic alteration of its state by the variation of an external parameter, such as temperature ($T$), pressure ($p$), concentration ($c$), electric field ($E$), electric potential ($\Psi$), radiant power ($\Phi$), or mechanical force ($F$). The response of the system to ME will also be periodic, exhibiting the same frequency as the stimulation. In case of a nonlinear system, the response to a sinusoidal stimulation will also contain multiples of the fundamental frequency. After an initial period of stimulation, the system will reach the stationary state, which is characterized by periodic alterations around a constant mean. In case of incomplete reversibility, e.g., the existence of an irreversible exit in the reaction scheme, the signal amplitudes of the initial components and of the intermediate species will decline as the system is approaching its final state.

Phase-sensitive detection (PSD) is used for the evaluation of amplitudes and phase lags of the periodic system response. In a simple view, PSD applied to data from a spectroscopic ME experiment results in a special kind of difference spectra between excited and nonexcited states. Let us consider a system that is stimulated by a sinusoidally oscillating external parameter. During one half-wave, one has excitation followed by relaxation in the other. In the stationary state, this alteration between excitation and relaxation may be repeated as many times as necessary in order to obtain a good signal-to-noise ($S/N$) ratio of the modulation spectra. Moreover, it should be noted that PSD is a narrowband detection; i.e., noise contributes only from a frequency range that is close to the stimulation frequency $\omega$.

Since the periodic system response is evaluated automatically within each period of stimulation, instabilities of the spectrometer, the environment, and the sample are much better compensated than with conventional techniques, where a reference spectrum has to be stored at the beginning of the experiment. As a consequence, the ME technique generally leads to high-quality background compensation with a low noise level, resulting in enhanced sensitivity by at least one order of magnitude.

So far, ME spectroscopy appears as a special type of difference spectroscopy. This is true if the frequency of stimulation is slow compared to the kinetics of the response of the stimulated system. However, if one or more relaxation times of the externally excited process fulfill the condition $0.1 < \omega \cdot \tau_i < 10$, where $\omega$ denotes the angular frequency of stimulation and $\tau_i$ is the $i$th relaxation time of the system, significant phase lags $\varphi_i$ between stimulation and sample responses will occur. As will be derived in Section 6, phase lag and relaxation time are related by $\varphi_i = \arctan(-\omega \cdot \tau_i)$. This phenomenon is paralleled by damping of the response amplitudes $A_i$. Both signify the underlying reaction scheme and the associated rate constants of the stimulated process (1). In this case, selectivity of ME spectroscopy, e.g., with respect to single components in heavily overlapping absorption bands, is significantly higher than that achievable by normal difference spectroscopy. The reason is the typical dependence of phase lags $\varphi_i$ and amplitudes $A_i$ on the modulation frequency $\omega$. If a set of absorption bands of a modulation spectrum exhibits the same phase lag $\varphi_i$, it is considered a correlated population. Such a population consists, e.g., of molecules or parts of them that are involved in the same reaction step. The assignment of a group of absorption bands in a modulation spectrum to a population is considered to be validated if upon changing the stimulation frequency $\omega$ all these bands exhibit further on the same dependence with respect to phase lag $\varphi_i(\omega)$ and amplitude $A_i(\omega)$.

Moreover, the dependence of phase lag and amplitude on $\omega$ may be calculated based on a given reaction scheme. Analytical expressions for a simple reversible reaction $A_1 \rightleftarrows A_2$ are given in Section 6. Obviously, ME spectroscopy enables a very rigorous test of the significance of a reaction scheme, since consistency of experimental data with theory derived from a given reaction scheme must hold over the whole frequency range of stimulation. Theoretically, as will be shown in Sec. 6, modulation spectroscopy and relaxation spectroscopy (2,3) have the same information content regarding kinetic parameters. The access to these parameters, however, is quite different. One relaxation experiment may result in the whole information, because broad band stimulation and detection is applied. As a consequence, a correspondingly high noise level must be expected. On the other hand, ME spectroscopy, as narrow band technique, requires a series of experiments at different excitation frequencies. As a consequence, the signal-to-noise (S/N) ratio is significantly enhanced, and moreover, phase sensitive detection (PSD) of the system response, which is equivalent to Fourier analysis, enables unambiguous experimental detection of non-linearities of the stimulated system. These may result e.g. from 2nd order reactions, phase transitions and/or cooperative phenomena. Therefore, ME spectroscopy enables a more rigorous validation of experimental data, since the system response at any frequency must be consistent with the reaction scheme underlying the kinetic analysis. The price to pay for this advantage, however, are a more complicated theoretical approach for the evaluation of kinetic parameters and more time consuming experiments.

In this chapter we will report on temperature-modulated excitation (T-ME) of poly-L-lysine, on concentration-modulated excitation (c-ME) of acetylcholinesterase (AChE) and arachidic acid (ArAc), on electric field–modulated excitation (E-ME) of poly-L-lysine, and of UV-VIS–modulated excitation ($\Phi$-ME) of pyrocatechol.

Two-dimensional IR spectroscopy (4) as a tool of ME data presentation is discussed in Sec. 4; Sec. 5 deals with signal processing of data obtained by ME. Finally, as an example, the modulation spectroscopic approach to kinetic data of the simple reversible reaction $A_1 \rightleftarrows A_2$ is demonstrated.

## 2 METHODS OF MODULATED EXCITATION

Modulation of external parameters such as temperature ($T$), pressure ($p$), electric potential ($\Psi$), electric field ($E$), concentration ($c$) of a substrate, and excitation of electronic states in molecules by modulated UV-VIS radiation ($\Phi$) may lead to periodic concentration changes of the reactants in chemical, biochemical, or biological systems. Mechanical strain modulation applied to a polymer sample may be used to study strain-induced structural changes in the sample. The mean concentrations corresponding to the stationary state are given by the equilibrium concentrations corresponding to the mean values of the external parameters. Using the general symbol $\eta$ to denote the external parameter used for ME, one obtains in case of a harmonic stimulation

$$\eta(t) = \eta_0 + \Delta\eta_0 + \Delta\eta_1 \cos(\omega t + \theta) \tag{1}$$

$\Delta\eta_0$ denotes the offset of the stationary state with respect to the initial state; i.e., $\eta_0 + \Delta\eta_0$ is the average value of the parameter in the stationary state. $\Delta\eta_1$ is the modulation amplitude of the corresponding parameter. $\theta$ denotes the phase of the stimulation, a parameter that is under experimental control by the operator. The shape for harmonic stimulation with $\theta = \pi$ is shown in Fig. 1. The principle components of an ME spectrometer are depicted in Fig. 2.

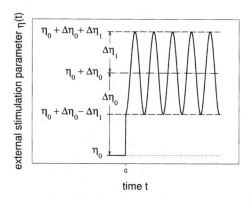

time t

**Fig. 1**   External harmonic modulated excitation according to Eq. (1). $\eta$ denotes any external parameter. $\eta_0$ is the initial value at the beginning of the experiment. $\Delta\eta_0$ is the offset from initial state to stationary state, which corresponds to the equilibrium state at the parameter setting $\eta_0 + \Delta\eta_0$. $\Delta\eta_1$ denotes the modulation amplitude. $\theta$ is the phase angle, determining the onset of the stimulation. With $\theta = \pi$, stimulation starts at the minimum value of the parameter.

## 2.1   Temperature-Modulated Excitation (T-ME)

For an analytical description of the influence of the external parameter $T$ on rate constants, one may use the approximation by Arrhenius, Eq. (2):

$$k = Ae^{-E_a/RT} \tag{2}$$

By introducing Eq. (1) into Eq. (2), one obtains

$$k(t) = Ae^{-E_a/(R(T_0+\Delta T_0+\Delta T_1\cos(\omega t+\theta)))} \tag{3}$$

For small perturbations, i.e., $\Delta T_0, \Delta T_1 \ll T_0$, Eq. (3) may be approximated by the linear part of a Taylor series expansion at $T_0$:

$$k(T(t)) = k(T_0) + k(T_0)\frac{T_A}{T_0^2}(\Delta T_0 + \Delta T_1 \cos(\omega t + \theta)) \tag{4}$$

$T_A = E_a/R$ will be referred to as *Arrhenius temperature*. Under these conditions, Eq. (4) may be used to get the relevant rate equations. An example will be given in Sec. 6.

It should be noted that if the system undergoes a phase transition or is involved in a cooperative process, this approach will still hold. However, more complicated reaction schemes have to be used. Cooperative phenomena have been described, e.g., by Hill (6) and Monod et al. (7).

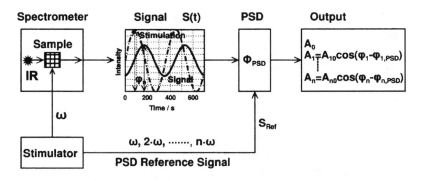

**Fig. 2** Schematic setup for modulated excitation (ME) spectroscopy. A periodic excitation is exerted on the sample with frequency $\omega$. The sample response $S(t)$, as sensed by IR radiation, contains the frequency $\omega$ and higher harmonics at wavelengths that are significant for those parts of the sample that have been affected by the stimulation. Selective detection of the periodic sample responses is performed by phase-sensitive detection (PSD), resulting in the DC output $A_n$ of fundamentals $\omega$ ($n = 1$) and their harmonics $n\omega$ ($n = 2, 3, \ldots$), as well as the phase shifts $\varphi_n$ between the $n$th harmonic and the stimulation. This phase shift is indicative of the kinetics of the stimulated process and of the underlying chemical reaction scheme. Since the PSD output $A_n$ ($n = 1, 2, \ldots, n$; frequency $n \cdot \omega$) is proportional to $\cos(\varphi_n - \phi_{n,PSD})$, absorption bands featuring the same phase shift $\varphi_n$ are considered to be correlated, i.e., to be representative of a population consisting of distinct molecules or molecular parts. $\phi_{n,PSD}$ is the operator-controlled PSD phase setting. Because of the cosine dependence, different populations will have their absorbance maxima at different $\phi_{n,PSD}$ settings, thus enabling selective detection. Moreover, in the case where $0.1 < \omega \cdot \tau_i < 10$ ($\tau_i$ denotes the $i$th relaxation time of the system), $\varphi_n$ becomes $\omega$ dependent, $\varphi_n = \varphi_n(\omega)$. The spectral information can then be spread in the $\phi_{n,PSD}$-$\omega$-plane, resulting in a significant enhancement of resolution with respect to standard difference spectroscopy and conventional time-resolved spectroscopy. (Reproduced from Ref. 5 by permission of Academic Press.)

## 2.2 Pressure-Modulated Excitation (p-ME)

In the Arrhenius model, Eq. (2), the rate constant is considered to be pressure independent. Therefore, a refined calculation of the rate constant is required. Using the activated complex theory (transition-state theory) (8), one obtains for the rate constant $k_+$

$$k_+ = \kappa \frac{kT}{h} K^{\neq} = \kappa \frac{kT}{h} e^{-\Delta G^{\neq}/RT} \tag{5}$$

$\kappa$ is called the *transmission coefficient*. Its value is generally between 0.5 and 1. $k$ and $h$ mean the Boltzmann and Planck constants, respectively. $K^{\neq}$ is the equilibrium constant of the reaction from the initial to the transition state, and

$\Delta G^{\neq}$ denotes the corresponding free enthalpy of reaction (Gibbs energy). The dependence of $k_+$ on pressure $p$ results from the dependence of $\Delta G^{\neq}$ on $p$; i.e., $\partial \Delta G^{\neq}/\partial p = \Delta V^{\neq}$. Since small pressure-induced changes of $\Delta G^{\neq}$ are expected, a linear dependence may be assumed, resulting from Eq. (5):

$$k_+(p) = \kappa \frac{kT}{h} e^{-(\Delta G^{\neq}(p_0) + \Delta V^{\neq}(p(t) - p_0))/RT} \tag{6}$$

$$= k_+(p_0) e^{\Delta V^{\neq}(p(t) - p_0)/RT}$$

Furthermore, since $\Delta V^{\neq}(p(t) - p_0)/RT \ll 1$, a linear influence of pressure on rate constant may be assumed. Therefore, Eq. (6) results after insertion of the oscillating external pressure according to Eq. (1):

$$k_+(\mathrm{p}) = k_+(p_0) \left( 1 + \frac{\Delta V^{\neq}}{RT}(p(t) - p_0) \right) \tag{7}$$

$$= k_+(p_0) + k_+(p_0) \frac{\Delta V^{\neq}}{RT} (\Delta p_0 + \Delta p_1 \cos(\omega t + \theta))$$

## 2.3 Electric Field–Modulated Excitation (E-ME)

Chemical reactions may be influenced by strong electric fields ($E \approx 10^4$ V/m and larger) via dipole-field interaction. If $\vec{M_i}$ denotes the dipole moment per mole of a species $A_i$ that is exposed to the electric field $\vec{E}$, the chemical potential $\mu_i$ will be altered by the energy $-(\vec{M_i} \cdot \vec{E})$ (Stark effect). The corresponding change per extent of reaction is then given by $-(\Delta \vec{M} \cdot \vec{E})$, where $\Delta \vec{M} = \Sigma \nu_i \vec{M_i}$ denotes the reaction dipole moment, which is assumed to be composed of a static part $\Delta \vec{M}_{\mathrm{stat}}$ and an induced part $\Delta \vec{M}_{\mathrm{ind}}$. The latter is given by $\Delta \vec{M}_{\mathrm{ind}} = \Sigma \nu_i \alpha_i \vec{E} = \Delta \alpha \vec{E}$. $\alpha_i$ is the electric polarizability tensor of $A_i$, and $\nu_i$ is the corresponding stoichiometric number, which is defined as a negative or positive number, depending on whether $A_i$ is a reactant or a product, respectively. Thus $\Delta \alpha$ denotes the electric reaction polarizability. Making use of Eq. (5), one obtains for the rate constant

$$k_+(\vec{E}) = \kappa \frac{kT}{h} e^{-(\Delta G^{\neq}(\vec{E}=0) - (\Delta \vec{M}^{\neq}_{\mathrm{stat}} + \Delta \vec{M}^{\neq}_{\mathrm{ind}}) \cdot \vec{E})/RT} \tag{8}$$

$$= k_+(\vec{E} = 0) \cdot e^{\Delta \vec{M}^{\neq}_{\mathrm{stat}} \vec{E}/RT} \cdot e^{\Delta \alpha^{\neq} \vec{E}^2/RT}$$

The two exponentials on the right-hand side of Eq. (8) describe the influence of the electric field on the rate constant via the interaction with static and induced dipole moments of the reactants, respectively. For the sake of simplicity we assume induced dipole moments to have the same direction as the electric field.

In this case, which holds exactly for spherical symmetry, $\alpha_i$ and $\Delta\alpha$ become scalars. The two exponentials may now be expanded into Taylor series until the first electric field–dependent term:

$$e^{\Delta \vec{M}^{\neq}_{\text{stat}} \vec{E}/RT} \approx 1 + \frac{\Delta M^{\neq}_{\text{stat}} \cos(\Delta \vec{M}^{\neq}_{\text{stat}} \vec{E})}{RT} E \quad \text{and} \quad e^{\Delta\alpha^{\neq} \vec{E}2/RT} \approx 1 + \frac{\Delta\alpha^{\neq}}{RT} E^2 \quad (9)$$

Insertion of Eq. (9) into Eq. (8) results in

$$k_+(\vec{E}) \approx k_+(0) \cdot \left( 1 + \frac{\Delta M^{\neq}_{\text{stat}} \cos(\Delta \vec{M}^{\neq}_{\text{stat}} \vec{E})}{RT} E + \frac{\Delta\alpha^{\neq}}{RT} E^2 \right.$$

$$\left. + \frac{\Delta\alpha^{\neq} \Delta M^{\neq}_{\text{stat}} \cos(\Delta \vec{M}^{\neq}_{\text{stat}} \vec{E})}{(RT)^2} E^3 \right) \quad (10)$$

Let's consider now an oscillating electric field according to Eq. (1):

$$E(t) = E_0 + \Delta E_0 + \Delta E_1 \cos(\omega t + \theta) \quad (11)$$

Insertion of Eq. (11) into Eq. (10) results in the time dependence of the alteration of the rate constant due to an oscillating external electric field. It should be noted that according to Eqs. (8) and (9), the interaction of the electric field of frequency $\omega$ with an induced reaction dipole will result in a response of $2\omega$ as the lowest-frequency component. No fundamental frequency $\omega$ should result. However, a response with frequency $\omega$ may be assumed to result from electric field interaction with a static reaction dipole. Since PSD can discriminate between these two signals (see Sec. 5), it offers a powerful tool for reaction mechanistic analysis.

For the sake of completeness it should be mentioned that an exclusive $2\omega$ response to a harmonic stimulation will also be observed when a static molecular dipole reorients synchronously and symmetrically with an electric field oscillating about zero. This is the case because parallel and antiparallel alignment of a molecule results in the same optical absorption.

## 2.4 Concentration-Modulated Excitation (c-ME)

Concentration-modulated excitation (c-ME) is performed by periodic alteration of the concentration of an effector molecule in the spectroscopic flow-through cuvette. In contrast to the three cases discussed earlier, c-ME acts directly on the system by disturbing the chemical equilibrium or stationary state. Again, the time course of the stimulating concentration may be described by Eq. (1):

$$c(t) = c_0 + \Delta c_0 + \Delta c_1 \cos(\omega t + \theta) \quad (12)$$

$c_0 + \Delta c_0$ denotes the stationary concentration, and $\Delta c_1$ is the amplitude of the modulated concentration.

## 2.5 UV-VIS–Modulated Excitation (Φ-ME)

UV-VIS–modulated excitation (Φ-ME) is performed by exposition of the system to a modulated radiant power $\Phi(t)$ [W]. The contribution of Φ-ME to the overall rate of species $A_i$ is then given by

$$\dot{c}_{i,\Phi} = -\Phi(t)w_i c_i \tag{13}$$

$w_i$ denotes the transition probability [$W^{-1}\,s^{-1}$] and $c_i$ is the molarity of $A_i$. According to Eq. (1), the radiant power may be expressed by

$$\Phi(t) = \Phi_0 + \Delta\Phi_0 + \Delta\Phi_1 \cos(\omega t + \theta) \tag{14}$$

For Φ-ME in general, $\Phi_0 = 0$ and $\Delta\Phi_0 = \Delta\Phi_1$, which means that flux modulation occurs around $\Phi_{mean} = \Delta\Phi_0$ between $\Phi_{min} = 0$ and $\Phi_{max} = 2\Delta\Phi_0$. $\Phi(t)w_i$ acts like a rate "constant" of a first-order reaction describing electronic excitation (pumping) of species $A_i$. According to a given reaction scheme, Φ-ME is expressed by the matrix $\mathbf{W}$ of transition probabilities induced by the UV-VIS field. This matrix has to be added to the well-known matrix $\mathbf{K}$ of thermal transition probabilities (matrix of rate constants), as demonstrated by Forster et al. (9) and discussed in Sec. 3.4.

## 2.6 Mechanical Strain–Modulated Excitation (S-ME)

Polymer films change molecular ordering upon stretching, as demonstrated by x-ray diffraction (10). Oscillatory mechanical strain was exerted on thin polymer films in order to detect structural changes in thin polymer films by transmission Fourier transform infrared (FT-IR) modulation spectroscopy (4,11). More recently, it was shown that p-ME of polymer samples on an attenuated total reflectance (ATR) crystal may result in similar structural information as S-ME, however, with more freedom concerning sample size and shape (12).

## 2.7 Polarization Modulation (PM)

Polarization modulation (PM) does not belong to the group of external excitation techniques; however, it also makes use of phase-sensitive detection (PSD) to enhance sensitivity. Phase modulation in combination with infrared reflection absorption spectroscopy (IRRAS) leads to enhanced instrumental stability and background compensation, which is advantageous when working with very thin samples on metal or even at the air–water interface. At about 50 kHz, PM is achieved by means of a photoelastic modulator (PEM). On a metal substrate, the sample will absorb light only in the ∥-polarized half-wave; the ⊥-polarized half-wave of the signal is therefore representative of the background. Subtraction is performed by lock-in technique (PSD) within each PM cycle, i.e., 50,000 times per second. As a consequence, environmental and instrumental contributions are largely compensated (13,14).

## 3 EXAMPLES

### 3.1 Temperature-Modulated Excitation (T-ME) of Poly-L-lysine

Since the pioneering infrared work in the early 1950s (15,16), it has been known that typical amide group vibrations can be used to get information on protein and polypeptide secondary structure. In this context, poly-L-lysine (PLL) played an important role because of its ability to assume helical as well as antiparallel β-pleated sheet conformation, depending on the external conditions. Therefore PLL was used as a reference for protein amide group vibrational studies (17). Although helix and antiparallel β-pleated sheet as boundary states are confirmed by different experimental techniques, little is still known on the nature of the folding process. Application of T-ME has given evidence that intermediate species are accessible by this technique (18).

These experiments were performed with a poly-L-lysine film cast on an ATR plate (internal reflection element, IRE) and hydrated with $D_2O$ (80% rel. humidity, 28°C). The sample was then exposed to a periodic temperature variation of $\pm 2°C$ at a mean temperature of $\overline{T} = 28°C$. According to Eq. (1), these conditions corresponded to $T_0 = 26°C$ and $\Delta T_0 = \Delta T_1 = 2°C$. Slow modulation periods in the range of 2–15 min had to be applied in order to fulfill the optimum condition for modulation spectroscopy, namely, $\omega\tau \approx 1$, where $\omega$ denotes angular frequency and $\tau$ is relaxation time. A schematic drawing of the ATR cuvette used in the experiment is shown in Fig. 3.

The results obtained from T-ME with a 14.7-min period after phase-sensitive detection (PSD) are shown in Fig. 4. Part A shows the stationary spectrum and part B the phase-resolved spectra of the system response at the fundamental frequency $\omega$. The numbers indicated on the spectra denote the phase settings $\phi^{PSD}$ at the phase-sensitive detector, which is a parameter under control of the operator. The ME spectra shown in Fig. 4B may be expressed by Eq. (15), which holds for the fundamental frequency $\omega$. For more detailed information see Sec. 5.

$$
A(\tilde{v}, \phi^{PSD}) = A(\tilde{v}) \cdot \cos(\varphi_{apparent}(\tilde{v}) - \phi^{PSD})
$$

$$
= \sum_{i=1}^{N} A_i(\tilde{v}, \phi^{PSD}) = \sum_{i=1}^{N} A_i(\tilde{v}) \cdot \cos(\varphi_i - \phi^{PSD})
$$

(15)

$A(\tilde{v}, \phi^{PSD})$ is referred to as *modulation spectrum*, which is the superposition of $N$ component spectra $A_i(\tilde{v})$. Each component spectrum $A_i(\tilde{v})$ is characterized by the same phase angle $\varphi_i$ of its absorption bands. Consequently, this set of bands

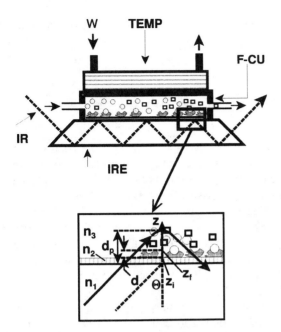

**Fig. 3** IR ATR setup for T-ME experiments under controlled relative humidity. Flow-through cuvette (F-CU) thermostated (TEMP) by water circulation. Alternatively, two thermostats operated at $T_0 = 26°C$, and $T_0 + 2\Delta T_1 = 30°C$, respectively, are connected under computer control to the inlet (W) of the heat exchanger on the cuvette. The modulation period was between 2 to 15 min. The trapezoidal internal reflection element (IRE) was coated by a thin PLL layer spread from a solution in alcohol. $D_2O$ humidified air (80% of r.h.) was circulated through the sealed cuvette.

may be considered to be correlated, i.e., to belong to a population of molecules or functional groups featuring the same kinetic response to the external stimulation. In such a population all absorbance bands exhibit the same dependence on the PSD phase setting $\phi^{PSD}$. The amplitudes become maximum for $(\varphi_i - \phi^{PSD}) = 0°$, minimum (negative) for $(\varphi_i - \phi^{PSD}) = 180°$, and zero for $(\varphi_i - \phi^{PSD}) = 90°$ or $270°$. Obviously, $\phi^{PSD}$ can be used to sense the overall phase angle $\varphi_{apparent}$, which depends on the wavenumber $\tilde{\nu}$. It enables only direct accesss to the phase angle $\varphi_i$ of a population of absorption bands if there is no overlap with bands of a different population. The most accurate way to determine $\varphi_i$ is achieved by line shape analysis of the whole set of phase-resolved spectra shown in Fig. 4B, followed by fitting a phase-resolved set of corresponding bands according to Eq. (15). For an example, the reader is referred to Ref. 18.

   Although the spectral resolution of 4 cm$^{-1}$ was the same for both stationary and modulation spectra, the first impression on comparing Fig. 4A with Fig.

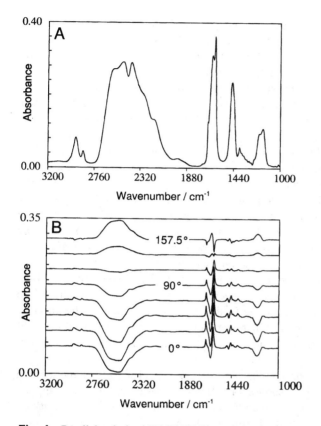

**Fig. 4** Parallel polarized T-ME FT-IR spectra of a poly-L-lysine (PLL) deuterobromide film hydrated with $D_2O$ (80% rel. hum., 28°C). The film was deposited on a CdTe ATR plate. A rectangular temperature stimulation was applied with a period of 14.7 min ($\omega$ = 0.427 min$^{-1}$) at $\overline{T}$ = 28°C ± 2°C. Angle of incidence: $\theta$ = 45°; mean number of internal reflections: $N$ = 9–10. (A) Stationary part of the T-ME-IR spectrum of PLL. (B) Set of phase-resolved T-ME-IR spectra after phase-sensitive detection (PSD) at phase settings $\phi_{PSD}$ = 0°–157.5° (phase resolution 22.5°) with respect to the T-stimulation. $\phi_{PSD}$ = 0° means in phase, with temperature switching from 26°C to 30°C. Heat transfer from the thermostats to the sample resulted in an additional phase lag of $\phi_T$ = 25°. (Reproduced from Ref. 18 by permission of the American Chemical Society.)

4B is that modulation spectra are significantly better resolved. Since modulation spectra reflect only species that are involved in the stimulated process, band overlap may be drastically reduced, as in this case. Furthermore, Fig. 4B shows that not only the intensity but also the shape of phase-resolved spectra is changing with $\phi^{PSD}$-setting. This is an unambiguous indication of the existence of populations of conformational states featuring different phase angles $\varphi_i$, which means having different kinetic responses with respect to the stimulation. Extraction of these populations according to Eq. (15) enables the assignment of intermediate species in the amide I' (1700–1600-cm$^{-1}$) and amide II' (1500–1400-cm$^{-1}$) regions. For more details the reader is referred to Ref. 18.

As an example of a correlated population, the behavior of $CH_2$ stretching of the lysine side chain and the secondary structure of PLL should be mentioned. The weak absorption bands at 2865 cm$^{-1}$ and 2935 cm$^{-1}$ result from symmetric and antisymmetric stretching of these $CH_2$ groups. They are shifted by approximately 3 cm$^{-1}$ toward lower wavenumbers if compared to the corresponding bands in the stationary state (Fig. 4A). This finding is indicative of a conformational change of a hydrocarbon chain from gauche defects into trans conformations (19,20); i.e., the chain is elongated. The phase of the $CH_2$ bands is found to be the same (correlated) as that of the antiparallel β-pleated sheet structure (amide I' bands at 1614 cm$^{-1}$ and 1685 cm$^{-1}$). We conclude, therefore, that the conversion of PLL from helix to β-sheet is paralleled by a conformational change of the lysine side chain from a bent to an extended structure.

The two boundary structures helix and antiparallel β-sheet respond 180° out of phase, which is usual for two species that are reversibly converted into each other. There are, however, three further modulation bands in the amide I' region exhibiting phase angles, which are significantly different from those of the α- and β-structures and which are assigned to intermediate structures of poly-L-lysine (18).

Access to higher harmonics is a further powerful ability of modulation spectroscopy. In general, the response will reflect the same frequency ω as applied with a harmonic stimulation as given by Eq. (1). However, any nonlinearity on the signal path will result in higher harmonics. Excluding instrumental and optical nonlinearities, as admitted in this experiment, we may conclude that the detected 2ω response shown in Fig. 5C must result from a nonlinear chemical reaction scheme, i.e., a reaction scheme containing one or more reaction steps that differ from first order. A further interesting remark should be made: The temperature modulation amplitude $\Delta T_1 = 2°C$, which was applied in this experiment, is so small that only a weak perturbation of the stationary state should be expected. Under such conditions, higher-order kinetics reduce to first-order kinetics (8), producing responses only in the fundamental. The fact that this is not the case in practice, because of prominent 2ω responses, indicates that a magnification by the system must take place, which is typical for phase transitions.

Figure 5 shows the expanded amide I' region of the stationary absorbance spectrum (A) and the parallel polarized modulation spectra in the fundamental frequency (ω) and the first harmonic (2ω). In the stationary spectrum (A), which represents the system at its mean temperature of 28°C, there are three clearly resolved bands, assigned to antiparallel pleated sheet (β-structure) at 1685 and 1614 cm$^{-1}$, respectively. The center band at 1639 cm$^{-1}$ is assigned to the helix. At least four intermediate species may be localized by visual inspection of the

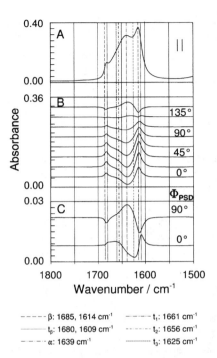

---- β: 1685, 1614 cm⁻¹       ──·──·· t₁: 1661 cm⁻¹
──── tᵦ: 1680, 1609 cm⁻¹      ─ ─ ─ ─ t₂: 1656 cm⁻¹
── ·· ─ α: 1639 cm⁻¹           ──────── t₃: 1625 cm⁻¹

**Fig. 5**  Parallel polarized T-ME FT-IR spectra in the amide I′ region of a poly-L-lysine (PLL) deuterobromide film hydrated with $D_2O$ (80% rel. hum., 28°C). Same experiment as described in the legend to Fig. 4. Expanded amide I′ region with (A) stationary absorbance spectrum, (B) 22.5° phase-resolved modulation spectra in the fundamental ($\omega$), and (C) 90° phase-resolved modulation spectra in the first harmonic ($2\omega$). In the stationary spectrum (A), which represents the system at its mean temperature of 28°C, there are three clearly resolved bands, assigned to antiparallel pleated sheet ($\beta$-structure) at 1685 and 1614 cm⁻¹, respectively, and the center band, which is generally assigned to a helix located at 1639 cm⁻¹. At least four intermediate species may be localized by visual inspection of the modulation spectra: Probably an intermediate antiparallel $\beta$-structure at 1680 and 1609 cm⁻¹, respectively, and three further components at about 1661, 1656, and 1625 cm⁻¹. The latter as well as the antiparallel $\beta$-structure have very prominent bands in the $2\omega$ spectra (C).

modulation spectra: Most probable, an intermediate antiparallel $\beta$-structure at 1680 and 1609 cm⁻¹, respectively, and three further components at about 1661, 1656, and 1625 cm⁻¹. Nearly all modulation bands visible in spectrum B have corresponding bands in spectrum C. This means that a description of the folding process by first-order steps is most probably not correct, provided that the T-stimulation is free of a $2\omega$ component, which was the case in this experiment. As a consequence, we have to assume second-order reactions or more complicated steps due to cooperative phenomena. Second-order steps are surely involved, since PLL is dehydrated upon folding from helix- to β-structure. Finally, it should be noted that the antiparallel β-structure and the component at 1625 cm$^{-1}$ have very prominent components in the 2ω spectra $C$, thus pointing to a distinct nonlinearity.

A detailed analysis based on T-ME experiments performed with different stimulation frequencies will be given elsewhere.

## 3.2 Concentration-Modulated Excitation (c-ME)

### 3.2.1 Experimental Considerations

The performance of a c-ME ATR experiment is based on a very simple concept. Two computer-controlled pumps are connected via a switch to the inlet of the flow-through cell. One pump is connected to a vessel containing solvent (e.g., a buffer solution) with a stimulant of concentration $c_{0S}$, whereas the other feeds solvent with stimulant concentration $c_{0R}$, where $c_{0S} > c_{0R}$ and $c_{0R} \geq 0$. During one half-period, c-ME is active due to pumping a high concentration of reactive compound through the cell, and in the following half-period relaxation occurs by pumping the low stimulant concentration. Under the condition of equal throughput in both channels, the stationary stimulant concentration becomes ($c_{0S}$ + $c_{0R}$)/2. This mean concentration should be adjusted in such a way that the system response is maximum to a change of stimulant concentration (maximum slope). Equation (12) now results in

$$c(t) = c_0 + \Delta c_0 + \Delta c_1 \cos(\omega t + \theta) = \underset{\uparrow}{\frac{c_{0S} + c_{0R}}{2}} + \Delta c_1 \cos(\omega t + \theta)$$

(16)

$$\text{stationary state}$$

While in the steady state, $\Delta c_0$ will be constant at any position in the flow-through cuvette; $\Delta c_1$ depends critically on hydrodynamic and diffusion parameters; and its range is $0 < \Delta c_1 \leq (c_{0S} - c_{0R})/2$. The ratio $2\Delta c_1/(c_{0S} - c_{0R})$ is referred to as *degree of modulation* ρ.

In order to achieve efficient c-ME, it is a prerequisite to take the convection/diffusion problem depicted in Fig. 6 into account. The upper part of Fig. 6 is a schematic vertical sectional view across the SBSR flow-through cell. Single-beam sample reference (SBSR) relates to a commercially available attachment converting a single-beam FT-IR spectrometer into a pseudo-double-beam spectrometer. In the ATR mode, the vertically aligned trapezoidal IRE is subdivided into an upper and lower half-plate, which can be used as reference and sample channel, respectively. Since the SBSR attachment converts the convergent IR beam of the spectrometer into a parallel beam of height $h$, sample ($S$) and refer-

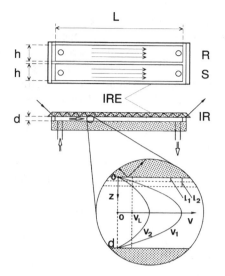

**Fig. 6** Approximate description of the convection/diffusion problem encountered in the ATR flow-through cell. The liquid flow entering the cuvette at the left-hand side is assumed to be laminar in the cell. This results in a parabolic velocity profile $v(z)$, Eq. (17). As a consequence, the flow velocity v is zero at the boundaries $z = 0$ and $z = d$ ($d$: thickness of spacer $S$). Under these conditions, a stimulant molecule can reach the immobilized sample at the surface $z = 0$ of the internal reflection element (IRE) only by diffusion. On the other hand, the diffusion length $l$ depends on the flow velocity v; i.e., the higher the velocity, the smaller the diffusion length $l$. The efficiency of concentration modulation depends critically on the dimensionless quantity $\omega \cdot l^2/D$, where $\omega$ and $D$ denote angular frequency and diffusion coefficient, respectively; see Fig. 7.

ence ($R$) spectra can be scanned alternatively under computer control by moving the whole cell up and down. The main advantage of this technique is enhanced background compensation, which is prominent in the case of long-term measurements. Furthermore, water vapor and carbon dioxide compensation will no longer be a problem. For details the reader is referred to Refs. 5, 21, and 22.

The following assumptions have been made in order to get an approximate solution of the convection/diffusion problem: (a) Laminar flow in the cuvette: As a consequence, the flow rate at the IRE surface ($z = 0$) is zero; i.e., the transportation of the chemical stimulant to a sample immobilized at the IRE surface is not possible by convection but needs an ultimate diffusion step. (b) To make the coupled convection-diffusion problem analytically tractable, we assume a layer $0 \leq z \leq l$ within which diffusion only is considered. For $z > l$, diffusion is neglected; i.e., the stimulant is transported to the boundary $z = l$ by convection only.

It should be noted that unavoidable phase errors in the stimulation process will occur when using a flow-through cell as depicted by Fig. 6, because excitation starts first in the region of the inlet and last at the outlet. The magnitude of this systematic error depends on the time $t$ required to move the wavefront over the distance $L$, namely, $\Delta t(l) = L/v_L$. Where $v_L = v(l)$ denotes the laminar flow velocity at $z = l$. The inset in Fig. 6 illustrates the situation for two flow velocities $v_1 > v_2$. Consider a given velocity $v_L$ at the outer diffusion boundary as marked by a dotted line perpendicular to the v-axis. In this example $v_L$ can be reached by both velocities under consideration. However, one must be aware that the diffusion length becomes longer the slower the flow rate is; i.e., $l_2 > l_1$. Since, in general, diffusion is rate determining, one has to match the dimensions of the flow-through cell and the throughput of solution (i.e., the pumping rate) in order to get a desired response time of the cell and a good efficiency of c-ME.

Let us first consider the convection part. The laminar-flow velocity profile can be described by Eq. (17):

$$v(z) = \frac{6\dot{V}}{hd}\frac{z}{d}\left(1 - \frac{z}{d}\right) \tag{17}$$

$\dot{V}$ denotes the volume throughput (volume per time), and $h$, $d$, and $z$ are height, thickness, and distance from the IRE, respectively, as depicted by Fig. 6.

The maximum velocity reached at $z = d/2$ and the mean flow velocity result, from Eq. (17), as

$$v_{max} = v\left(\frac{d}{2}\right) = \frac{3\dot{V}}{2hd} \quad \text{and} \quad \bar{v} = \frac{\dot{V}}{hd} \tag{18}$$

Now let us consider the diffusion process. The two boundary conditions required to solve the partial differential equation are $(\partial c/\partial z)(z = 0) = 0$ and Eq. (16) at $z = l$ with $\theta = -\pi/2$, which means excitation with a sine function. The initial condition is $c(z, t = 0) = 0$. The general stationary solution for the degree of modulation $\rho(z) = 2\Delta c_1(z)/(c_{0S} - c_{0R})$ is then given after solving the differential equation by

$$\rho(z) = \frac{\cosh\left[\kappa\frac{z}{l}(1 + i)\right]}{\cosh[\kappa(1 + i)]} \tag{19}$$

where

$$\kappa = \sqrt{\frac{\omega l^2}{2D}} = \sqrt{\omega_{red}} \tag{20}$$

$\omega$ is the angular frequency of ME, $D$ is the diffusion coefficient, and $l$ is the thickness of the diffusion layer. We refer to the dimensionless quantity $\omega_{red}$ as the reduced angular frequency, a quantity depending typically on the diffusion parameters. Equation (19) is a complex function, indicating that an additive phase lag results from diffusion besides the amplitude damping. For our purpose the latter is of prime importance. One obtains, from Eq. (19),

$$\left|\rho\left(\frac{z}{l}\right)\right| = \sqrt{\frac{\cosh\left(\kappa\frac{z}{l}\right)^2 \cos\left(\kappa\frac{z}{l}\right)^2 + \sinh\left(\kappa\frac{z}{l}\right)^2 \sin\left(\kappa\frac{z}{l}\right)^2}{\cosh(\kappa)^2 \cos(\kappa)^2 + \sinh(\kappa)^2 \sin(\kappa)^2}} \tag{21}$$

Figure 7 presents the dependence of the degree of modulation $|\rho(z)|$ on the reduced angular frequency $\omega_{red}$ and the relative distance $z/l$ from the IRE surface. The degree of modulation relevant for ATR experiments is obtained by the section parallel to the $\omega_{red}/|\rho(z)|$-plane at $z/l = 0$, i.e., at the surface of the IRE. It follows from Eq. (21) that

$$|\rho(0)| = \sqrt{\frac{1}{\cosh(\kappa)^2 \cos(\kappa)^2 + \sinh(\kappa)^2 \sin(\kappa)^2}} \tag{22}$$

It should be noted that in general the penetration depth $d_p$ of the evanescent field (5,21,23) is small compared to the diffusion length $l$; the assumption $\rho(d_p) \approx \rho(0)$ is therefore a good approximation. The dependence of $\rho(z = 0)$ on the reduced frequency $\omega_{red}$ according to Eq. (22) must now be considered in order to design an efficient ATR flow-through cell for c-ME.

We are aiming to have a cell for c-ME studies with immobilized mono-

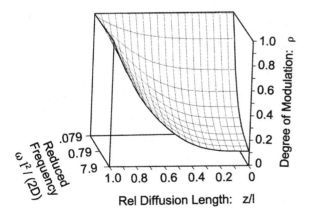

**Fig. 7** Three-dimensional representation of the dependence of the degree of modulation $|\rho|$ on the dimensionless parameters relative diffusion length $z/l$ and reduced frequency $\omega_{red} = \omega l^2/(2D)$ according to Eq. (21).

layers suited for a maximum frequency $f_{max} = 0.1$ Hz, which means that the stimulant is pumped during 5 s through the cell and washed out during the following 5 s, resulting in a stimulation period of $\tau = 10$ s. Typical geometrical parameters of a flow-through cuvette adapted to c-ME are $h = 7.5$ mm, $d = 0.1$ mm, and $L = 40$ mm (22); see Fig. 6. The last is a principal requirement for an adequate number of internal reflections, when working with monolayers and submonolayers on a multiple IRE of thickness 1–2 mm (23).

Intending now to achieve a degree of modulation of $|\rho(0)| = 0.9$ at the maximum frequency $f_{max} = 0.1$ Hz, we have to seek the required volume throughput $\dot{V}$. In order to get this information we first have to determine the diffusion length $l$ required to achieve the conditions stated earlier. This problem can be solved numerically by means of Eq. (22) or graphically by means of Fig. 8.

Aiming for $|\rho(0)| = 0.9$, it follows from Fig. 8 that $\log(\omega_{red}) = -0.23$, resulting in $l = 4.33 \cdot 10^{-2}$ mm for the diffusion length as calculated by means of Eq. (20). The next step in our evaluation is to decide on the maximum systematic phase error $\Delta\theta_{max}$ acceptable in the stimulation process. Setting $\Delta\theta_{max} = 2\pi(\Delta t_{max}/\tau) = \Delta t_{max} \cdot \omega_{max} = 0.314$, which corresponds to 5%, where $\tau$ means the duration of a ME cycle. According to our initial assumptions we have $\omega_{max} = 2\pi f_{max} = 0.628$, resulting in $\Delta t_{max} = 0.5$ s for the maximum time allowed to the wavefront to pass at distance $z = l$ along the length $L$ of the cuvette. The

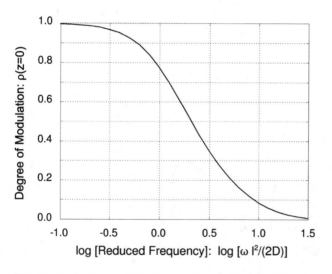

**Fig. 8** Dependence of the degree of modulation $|\rho(0)|$ at the surface of the internal reflection element (IRE) as a function of the reduced frequency $\omega_{red} = \omega l^2/(2D)$ according to Eq. (22).

corresponding minimum velocity is $v_{min}(1) = L/\Delta t_{max} = 8.0$ cm/s. In order to get the corresponding volume throughput, one has to solve Eq. (17) for $\dot{V}$, resulting in

$$\dot{V} = \frac{v_{min}(1)hd^2}{6l\left(1 - \dfrac{1}{d}\right)} \tag{23}$$

Inserting the values from earlier into Eq. (23) one obtains, for the volume throughput, $\dot{V}(d = 100\ \mu m) = 2.44$ cm$^3$/min. For comparison, enhancing the cell thickness to $d = 200\ \mu m$ would require enhancement of the volume throughput to $\dot{V}(d = 200\ \mu m) = 11.3$ cm$^3$/min, which demonstrates that cell thickness is a critical parameter. Although this evaluation is approximate, it has proven to be reasonable in practice.

### 3.2.2 Influence of Ca$^{2+}$ Ions on Structure and Activity of Acetylcholinesterase (AChE)

Little is yet known about structure–activity relationships of proteins and receptors under in situ conditions. ATR c-ME experiments are possible means to get such information on a molecular level. The activity of the enzyme acetylcholinesterase (AChE) is known to depend critically on the concentration of Ca$^{2+}$ (24,25). Thus Ca$^{2+}$-c-ME may be used to get structural details related to the periodic change of AChE activity. The result is shown in Fig. 9.

AChE isolated from *Torpedo marmorata* was immobilized as monolayer at the surface of a germanium (Ge) IRE by adsorption to an aminopropyl triethoxysilane (ATS) layer (26). The activity of the enzyme was determined by means of the Ellman method, the surface concentration of the enzyme was determined spectroscopically (5,21,23). As a consequence, the IR-ATR technique enables in situ measurement of structure and activity. It should be noted at this point, however, that the main problem encountered with experiments of this type is not primarily a technical one, provided high-performance equipment is used, but rather a biochemical one, namely, the preparation of stable immobilized systems. Furthermore, in the context with in situ activity measurements, the adsorption of the enzyme in the tubes and on the rear cover of the flow-through cuvette must also be taken into account.

Figure 9 shows the amide I/II range of the stationary single-channel spectra I and I$_0$, the corresponding transmittance $T = I/I_0$ and the parallel polarized c-ME spectrum in absorbance units. The experiment was performed in H$_2$O buffer, pH 7.0, 0.1 M NaCl. The Ca$^{2+}$ concentrations for stimulation and relaxation were $c_{OS} = 1.25$ mM and $c_{OR} = 0.25$ mM, respectively. Modulation bands with positive sign are typical for high Ca$^{2+}$ concentration; negative bands reflect the AChE response to low Ca$^{2+}$ concentration, i.e., reduced activity. The typical bands for enhanced activity are located at 1652 and 1677 cm$^{-1}$, whereas reduced activity

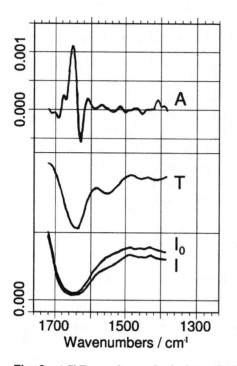

**Fig. 9** AChE monolayer adsorbed to a Ge IRE pretreated by aminopropyl triethoxysilane. Parallel polarized $Ca^{2+}$-c-ME spectrum with $\Delta c_0 = 0.75$ mM and $\Delta c_1 = 0.5$ mM
(see Eq. (16)). 0.1 M NaCl, pH 7. The resulting modulation of the specific activity was
found to be $(650 \pm 150)$ IU/mg. Frequency $f = 0.1$ Hz. Ge IRE, 38 active internal reflections, angle of incidence $\Theta = 45°$. (Reproduced from Ref. 25 by permission of Photo-
Optical Instrumentation Engineers.)

is expressed by three resolved bands at 1630, 1665, and 1690 $cm^{-1}$. Since the
process of c-ME is cyclic, distinct positive and negative bands must correspond
to each other. The most probable combination is 1652 $cm^{-1}$ with the couple at
1630 and 1690 $cm^{-1}$. The former is typical for a helix or random coil, while the
latter reflects an antiparallel pleated sheet structure. The couple 1665/1677 $cm^{-1}$
remains, which probably results from intermediate states with $\beta$-turns involved.

This finding is consistent with pH modulation experiments, which also lead
to a periodic alteration of the enzyme activity (25). Finally it should be noted that
only 0.5–1% of the amino acids of AChE were involved in the dynamic changes
reflected by the modulation spectrum. Therefore one may conclude that activity
control in an enzyme takes place by very small folding/unfolding processes.

### 3.2.3  Influence of the Degree of Protonation on the Structure of an Arachidic Acid (ArAc) Bilayer

The aim of this study was to get information on the influence of electrical surface charges on the molecular structure of an immobilized arachidic acid bilayer, which may be considered as a rough model of the lipid bilayer frame of a bio-membrane. For that purpose, two monolayers have been transferred at a constant pressure of 25 mN/m from the air/water interface of a film balance to the Ge IRE by means of the Langmuir–Blodgett (LB) technique. Since the surface of Ge is hydrophilic, the first layer attached by its polar heads, whereas the second layer attached tail-to-tail. As a consequence, the carboxylic acid groups of the second layer faced the aqueous environment and were therefore accessible for protons or hydroxyl ions. A periodic pH modulation between pH 3 and pH 10 was used to induce a periodic protonation and deprotonation of the carboxylic acid groups. It should be noted that the binding of the ArAc head groups to the Ge surface was so unusual that typically very intense absorption bands of the carboxylic acid group, such as C=O stretching, were completely erased in the spectrum of the inner layer. Therefore, one may assume that the head group signals shown in Fig. 10 result predominately from the second monolayer of ArAc facing the electrolyte.

*3.2.3.1  Assignment of Prominent Bands*

Most assignments given next relate to Ref. 27. The upper traces in Fig. 10A and B show the stationary polarized ATR spectra $A_0$ of the ArAc bilayer in the proton-ated state. They are scaled down for comparison with the modulation spectra. The intense band near 1700 cm$^{-1}$ results from C=O stretching of the COOH group. It consists of at least two components as reflected by the perpendicular polarized spectrum, Fig. 10B. Splitting into two components is also observed with $\nu_{as}$(COO$^-$) near 1570 cm$^{-1}$ and $\nu_s$(COO$^-$) near 1430 cm$^{-1}$. So far it has not been established whether this splitting is typical for the bilayer or whether it is due to local crystallization phenomena (28,29). The narrow peak at 1470 cm$^{-1}$ is assigned to the methylene bending vibration $\delta$(CH$_2$), and the shoulder at 1460 cm$^{-1}$ results most probably from the out-of-phase deformation of the methyl group $\delta_{as}$(CH$_3$). The weak and broad band at 1407 cm$^{-1}$ may be assigned to in-plane COH deformation of the COOH group and the symmetric methyl bending $\delta_s$(CH$_3$). Finally, there is a sequence of nearly equidistant weak bands between 1180 cm$^{-1}$ and 1350 cm$^{-1}$ resulting from the wagging vibrations $\gamma_w$(CH$_2$) of the methylene groups of hydrocarbon chains. This sequence originates from vibrational coupling in a saturated hydrocarbon chain assuming all-trans confor-mation. If $N$ denotes the number of CH$_2$ groups of the chain, the selection rule predicts $N/2$ IR active vibrations for $N$ even (symmetry $C_{2h}$). For $N$ odd, all the wagging vibrations are expected to be IR active, $1/2(N-1)$ are perpendicular

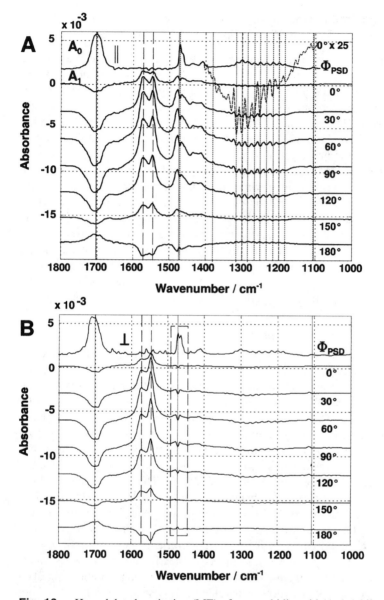

**Fig. 10**  pH-modulated excitation (ME) of an arachidic acid (ArAc) bilayer attached to a germanium internal reflection element (IRE). c-ME was performed by pumping alternatively two buffer solutions (100 mM NaCl, pH 3, and 100 mM NaCl, pH 10) through the ATR cuvette with a modulation period of $\tau = 16$ min, $T = 10°C$. Upper trace $A_0$: Stationary spectrum of a protonated ArAc layer for comparison with modulation spectra. Traces $A_1$: Modulation spectra at PSD phase settings $\phi_{PSD} = 0°, 30°, \ldots, 180°$. The 180° spectrum

polarized (very weak) and $1/2(N + 1)$ are polarized parallel to the symmetry plane (30). A prerequisite for vibrational coupling is all-trans conformation. Therefore, coupling is interrupted by any gauche defect, resulting in shorter all-trans segments with corresponding wagging bands at different wavenumbers. As a consequence, an increasing number of gauche defects at different chain positions will erase the well-resolved wagging sequence due to overlapping and loss of intensity. Finally, the weak band at 1104 cm$^{-1}$ is assigned to C—C stretching $\nu$(C—C).

### 3.2.3.2 Sensitivity of Modulated Excitation Spectroscopy

Taking the pH-modulation spectra shown in Fig. 10 as a typical example, one may check the sensitivity by expanding the lowest intense ME spectrum ($\phi_{PSD} = 0°$) by a factor of 25. The result is shown as the inset in Fig. 10A (top). The wavenumbers of the wagging peaks of the expanded spectrum still coincide with those of the most intense ME spectra ($\phi_{PSD} = 60°$ and $\phi_{PSD} = 90°$; the maximum is expected at $\phi_{PSD} \approx 75°$; so $\phi_{PSD} \approx 165°$ should result in an intensity of zero) and with the stationary spectrum $A_0$. The ordinate scaling factor for the zoomed inset is $4.0 \cdot 10^{-5}$. Comparing it with the other ME spectra (scaling factor $1.0 \cdot 10^{-3}$) one can conclude that bands as weak as $1.0 \cdot 10^{-5}$ absorbance units are still detectable.

Finally, it should be noted that this experiment has been performed in liquid H$_2$O. As a consequence, there was very low spectral energy in the 1640-cm$^{-1}$ region. Nevertheless, a high-quality background compensation is achieved due to the application of modulation technique.

### 3.2.3.3 Selectivity of pH-ME

The highest selectivity of ME spectroscopy is achieved when the stimulation frequency $\omega$ and the kinetics of the stimulated process are matched, i.e., if $0.1 < \omega \cdot \tau_i < 10$, where $\tau_i$ denotes the $i$th relaxation time of the system. In the actual case of pH modulation exerted to the outer monolayer of ArAc, there is no phase resolution achieved, due to the long modulation period of $\tau = 16$ min; therefore no kinetic information is available from these measurements. However, unambiguous discrimination between protonated and deprotonated populations is possible. Only one characteristic example shall be given here. The most promi-

---

corresponds to the 0°-spectrum with opposite sign, because the PSD output is proportional to cos($\varphi - \phi_{PSD}$); see Sec. 5. $\varphi$ denotes the phase difference between the response of a distinct part of the molecule and the stimulation. Due to the long period of $\tau_m = 16$ min, the observed bands in the modulation spectra exhibit only two resolved $\varphi$-values, which are 180° apart, as a consequence of the fact that the chemical relaxation time of protonation/deprotonation of ArAc is much shorter than the stimulation period. In order to demonstrate the excellent S/N ratio, the ordinate of the weakest modulation spectrum has been expanded in the CH$_2$-wagging region by a factor of 25, i.e., the ordinate scaling factor on top of the spectrum results in $4.0 \cdot 10^{-5}$ for the dashed spectrum.

nent band from the protonated state is the C=O stretching vibration $\nu(COOH)$ of the carboxylic acid group near 1700 cm$^{-1}$. Thus all other bands in the ME spectrum that have the same phase belong to the protonated population; they are considered to be correlated. The remaining bands featuring opposite sign are therefore members of the deprotonated population. Consequently, if no phase resolution is achieved, ME spectra reduce to difference spectra, which, however, have a considerably better background and instability compensation than conventional difference spectra, since corresponding sample and reference spectra are measured and evaluated/accumulated within each period $\tau$ of stimulation.

### 3.2.3.4 Spectroscopic Features

Concerning the application of ATR spectroscopy to biomembranes, the reader is referred to Refs. 5, 21, and 23. Let us now consider the wagging region $\gamma_w(CH_2)$ of the spectra shown in Figs. 10A and B. In the stationary absorbance spectrum $A_0$, one can observe nine weak bands resulting from the 18 CH$_2$ groups of the ArAc molecules. We can conclude that the majority of the hydrocarbon chains assume all-trans conformation. Moreover, the wagging sequence in the parallel polarized spectra is significantly more intense than the spectra measured with perpendicular polarized light. This means that hydrocarbon chains are aligned toward the normal of the IRE. This experimental finding is even more prominent in the modulation spectra, which shall be considered now. We look for bands exhibiting the same phase behavior. They represent a population of molecules or parts of them that are correlated, i.e., respond in the same way to the external stimulation. In this experiment the kinetic response to pH change was much faster than the stimulation period of $\tau = 16$ min. Therefore, we have only the protonated and the deprotonated population with phase angles 180° apart. In order to identify a population, it is sufficient to have a reliable assignment of one its bands. Taking, e.g., the modulation spectrum with $\phi_{PSD} = 90°$, we can group the bands into positive and negative ones. Since the band at 1700 cm$^{-1}$ is negative, we know unambiguously that all negative bands of this spectrum belong to the population of protonated ArAc. Consequently, all positive bands at this PSD phase setting belong to the population of the deprotonated ArAc. We can therefore conclude that both $\delta(CH_2)$ and $\gamma_w(CH_2)$ decrease and that $\nu(C—C)$ shifts as ArAc is deprotonated. The population of deprotonated ArAc appears with 180° phase difference. Besides the split $\nu(COO^-)$ bands already mentioned, attention must be drawn to a very strange phenomenon, which probably reflects an electric field effect originating in the negative surface charge density at the planar membrane surface, which is produced by deprotonation. At 1470 cm$^{-1}$ we recognize the sharp vanishing $\delta(CH_2)$ band that is overlapped by an intense and broad positive band. Moreover, this band, which belongs unambiguously to the population of deprotonated ArAc, is completely parallel polarized. The only explanation of this fact is alignment of the transition dipole moment of this band parallel to the normal to the IRE, i.e., parallel to the electric field. We assign this band tentatively to a $\delta(CH_2)$ vibration disturbed by the strong electric field. A more detailed discussion of this phenomenon that also takes into account the behavior of CH-stretching will be given elsewhere (D. Baurecht, G. Reiter, M. Schwarzott, U.P. Fringeli. J Phys Chem, in preparation).

## 3.3 Electric Field–Modulated Excitation (E-ME)

Very strong electric fields are expected in biomembranes, at least in certain regions. The magnitude may be estimated from the knowledge that the transmembrane potential is about 100 mV and the membrane thickness about 100 Å thus resulting in an electric field in the order of magnitude of $10^7$ V/m. Nature uses such high electric field strengths for process control, which means that a coupling to biochemical reactions must occur. Understanding of these reaction mechanisms are therefore of fundamental interest.

So far, considerable theoretical knowledge exists on electric field/biomembrane interaction (31), but only few experimental data are available. The reason is evident, since exposing a biological model system to such extremely high electric field strengths is not an easy operation. However, it is a challenge to try to simulate membrane electrostatic conditions in order to study the response of model compounds on a molecular level.

Indeed, there are several possibilities for exposing model membranes to electric fields of comparable strength. One approach is the direct application of electric potentials across the membrane using a Ge IRE and the rear cover of the cuvette as electrode and counter electrode, respectively. The main problem encountered with this setup is electrochemical degradation of the Ge IRE surface, leading to reduced transmittance and, more critically, to the liberation of protons due to the formation of germanic acid ($H_2GeO_3$) when the Ge IRE is held on a positive potential (32). Experiments of this type may be performed with a hydrated sample sandwiched between two narrowly spaced electrodes. This setup has turned out to require sealed cuvettes and surface-coating-protected Ge IREs.

An indirect method to apply a strong electric field has already been presented in the preceding section—by controlling the surface charge density of a membrane via the pH, whereas a second indirect method makes use of the very strong electric fields occurring in the electric double layer at an electrode surface. The space charge of the double layer can easily and quickly be influenced by very low electrode potentials. However, only very thin immobilized layers can be exposed to these fields, because the penetration depth of the space charge region into the electrolyte is very small. Depending on the ionic strength of the electrolyte, it varies between a few and several hundred angstroms.

It should be noted that the experiments discussed next are preliminary. Nevertheless they deserve to be mentioned, because unambiguous evidence is given that E-ME is feasible and enables a promising approach to get more insight into membrane electrostatics.

### 3.3.1 Direct Application of High Potentials to Hydrated Lipid Multibilayers

In order to study the influence of electric fields to multibilayers of 1,2-dimyristoyl-sn-glycero-3-phosphocholine (DMPC) by IR-ATR spectroscopy, the Ge IRE and the Ge counter electrode were protected by a $SiO_2$ coating. Multibilayers of DMPC were prepared on the IRE simply by evaporation of the organic solvent of a dilute DMPC solution. The IRE and the counterelectrode were separated by a 5-$\mu$m spacer. The lipid was hydrated before sealing of the cell. Potentials up to 200 V could be applied across the lipid phase, resulting in conformational and reorientational effects. The surface coating of the Ge plates was necessary to prevent electrical discharge at elevated electric potentials. Measurements of E-ME using this setup are in progress (personal communication: P. Lasch, D. Naumann, Robert Koch Institute, Berlin).

### 3.3.2 Indirect Application of Modulated High Potentials to an Immobilized Poly-ʟ-lysine Layer

The experimental setup is schematically depicted in Fig. 11. The Ge IRE and the stainless steel rear cover act as electrodes. The latter is grounded, while the former assumes a rectangular potential switched between +50 mV and −1000 mV. Working within these potentials avoids, on the one hand, anodic hydrolysis of Ge (32) and, on the other hand, hydrolysis of water. The modulation period was $\tau = 20$ min. Changing the polarity of the potential at the IRE leads to an

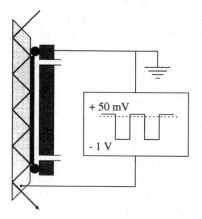

**Fig. 11** Schematic view of a setup for electric field–modulated excitation (E-ME). The sample is immobilized as thin film on the surface of the Ge IRE, which acts as one electrode. The grounded counterelectrode is the rear stainless steel cover of the cuvette, which is filled with electrolyte solution. Stimulation is performed by applying a rectangular potential varying between +50 mV and −1000 mV. These potentials are selected to avoid degradation of Ge at higher positive potentials as well as to avoid electrolytic decomposition of water.

exchange of the ions in the space charge region of the electric double layer at the membrane surface. If the ionic strength is kept low, e.g., about 1 mM as in this case, the Debye length, which determines the thickness of the space charge region, will be in the order of 100 Å. As a consequence, very high electric fields are switched at the surface of the electrode upon changing the polarity (33). Consequently, a thin immobilized layer, such as a model membrane or a protein layer, will be exposed to high electric fields, too. This indirect method may be used to study the influence of an alternating high electric field on a poly-L-lysine (PLL) film immobilized by adsorption to the surface of the Ge IRE (34).

Preliminary results are shown in Figs. 12 and 13. It should be noted that a rectangular ME, as applied in this case, besides the fundamental frequency $\omega$ (1), also results in all odd harmonics, i.e. $3\omega$, $5\omega$, . . . , but with decreasing intensity. However, there is no excitation with even multiples of the fundamental. Therefore, detection of even spectral responses is of special interest, since those result from frequency doubling by the stimulated system, provided the modulated

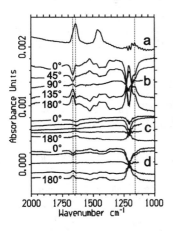

**Fig. 12** E-ME of a poly-L-lysine (PLL) film adsorbed to a Ge IRE in contact with liquid $D_2O$, pD = 10. Exposure of PLL to high electric fields was performed indirectly via the exchange of the space charge in the electric double layer by applying a periodic rectangular potential varying between +50 mV and −1000 mV. Parallel polarized incident light, angle of incidence $\Theta = 45°$, $N = 19$ active internal reflections, $T = 30°C$. (a) Stationary spectrum, (b) phase-resolved modulation spectra obtained by sinusoidal demodulation with the fundamental frequency $\omega$. (c) Phase-resolved modulation spectra obtained by sinusoidal demodulation with the first harmonic frequency $2\omega$. Note that this frequency is not included in a rectangular stimulation with frequency $\omega$. Frequency doubling has occurred by the response of PLL. See also expanded section in Fig. 13. (d) Phase-resolved modulation spectra obtained by sinusoidal demodulation with the second overtone frequency $3\omega$. Note: This response is again initiated by the stimulation, since a symmetric rectangular wave contains all odd multiples of the fundamental. The intensity is damped by the factor $1/(2n + 1)$, with $n = 0, 1, 2, . . . .$ For details see text.

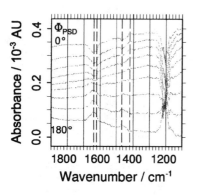

**Fig. 13** Expanded section of the 2ω response modulation spectra shown in Fig. 12(c). Same data but enhanced phase resolution of $\Delta\phi_{PSD} = 22.5°$. Three amide I′ components are resolved, namely, at 1639 cm$^{-1}$, 1620 cm$^{-1}$, and about 1600 cm$^{-1}$. The first two bands overlap; however, the change in shape with varying $\phi_{PSD}$ indicates a slight difference in the response phase angle. Corresponding modulation bands are detected in the amide II′ region, too. Most probably the band at 1470 cm$^{-1}$ correlates with the amide I′ component at 1639 cm$^{-1}$ and may be assigned to helix (18). A further correlation exists between the amide I′ and amide II′ components at 1600 cm$^{-1}$ and 1420 cm$^{-1}$. These two populations are 180° out of phase; i.e. they are periodically converted into each other.

system response is small so that the signal transfer remains linear. In this case, the small signal condition is surely fulfilled, as can be verified from absorbance scaling of Figs. 12 and 13.

Figure 12 presents the downscaled parallel polarized stationary spectrum of the PLL film (a) in liquid D$_2$O environment. Three resolved bands are observed, namely, the amide I′ band near 1639 cm$^{-1}$, the amide II′ band near 1450 cm$^{-1}$, and the band near 1160 cm$^{-1}$, which we assign to the—ND$_3^+$ antisymmetric bending of the lysine side chain. The noise in this region results form low energy due to strong D$_2$O absorption [$\delta$(D$_2$O) at 1200 cm$^{-1}$]. The dotted vertical lines indicate $\delta_{as}$(ND$_3$), helix at 1639 cm$^{-1}$, and a component at about 1665 cm$^{-1}$, which was also observed as intermediate species in T-ME experiments with PLL (see Sec 3.1). No unambiguous assignment of this band is available so far. Traces (b)--(d) of Fig. 12 are modulation spectra with indicated PSD settings (see also Sec. 5). Part (b) represents the PLL response to the excitation with the fundamental ω, while (d) is the response to the 3ω excitation. According to the Fourier series of a symmetric rectangle, the intensity of the 3ω component should be reduced by a factor of 1/3 with respect to the fundamental, which is approximately the case. Spectrum (c) is the 2ω response of PLL to rectangular E-ME. Since no 2ω component exists in the stimulation and all signal amplitudes are very small, which guarantees a linear signal transfer, we can conclude that the weak but significant band at 1639 cm$^{-1}$ has a different origin than the other modulation bands. The signal-to-noise ratio of this measurement is good enough to expand the 1800–1200-cm$^{-1}$ region of Fig. 12(c). This region is shown with enhanced phase resolution of $\Delta\phi_{PSD} = 22.5°$ in Fig. 13. Comparing the shape of the amide I' band along the PSD phase settings, we realize that the shapes are not similar in a mathematical sense; i.e., they cannot be converted into each other by simple multiplication with an adequate factor. Therefore, we have to conclude that at least two components overlap in this band. Visual inspection indicates components at 1639 cm$^{-1}$ and 1620 cm$^{-1}$. Moreover, these two components seem to have a slightly different phase response, which is responsible for the nonsimilarity of the band shapes. At $\phi_{PSD} = 0°$ they are close to their minima (negative bands). A weaker, third band with opposite sign is observed at about 1600 cm$^{-1}$. It vanishes and changes sign at about the same PSD phase setting of $\phi_{PSD} = 90°$; i.e., 1639/1620-cm$^{-1}$ and 1600-cm$^{-1}$ bands have a phase difference of about 180°. A corresponding behavior is also observed in the amide II' region. The bands at about 1470 cm$^{-1}$ and 1639 cm$^{-1}$ are correlated. They have been assigned earlier to the $\alpha$-helical part of PLL (18). On the other hand, the band at about 1420 cm$^{-1}$ correlates with the 1600 cm$^{-1}$ band. For both, there is no reliable interpretation available at present. Additional measurements at elevated excitation frequencies in order to achieve significant time resolution are necessary to get better experimental discrimination.

Concerning the origin of the 2$\omega$ response, a definite answer also cannot be given at present. According to Eqs. 10 and 11, the influence of the electric field on a rate constant via field/induced dipole interaction or spacial reorientation of PLL segments by field/permanent dipole interaction are both possible explanations.

Coming back to the $\omega$ and 3$\omega$ responses shown in Fig. 12, traces (b) and (d). In the amide I' region there is a well-resolved reversible conversion between 1665 cm$^{-1}$ and 1639 cm$^{-1}$. Both bands are known from T-ME experiments (see Sec. 3.1 and Ref. 18). The latter is assigned to the helix; the former probably represents $\beta$-turns. Quite obviously, the rate constant of this conversion seems to be considerably more sensitive to electric fields than rate constants associated with other folding steps occurring in the PLL chain. According to Eq. 10, one may assume a high permanent reaction dipole moment between the initial state and the transition state of this elementary reaction. Therefore, it seems a likely supposition that folding or unfolding of $\alpha$-helical segments is involved in this step.

Let us consider the spectral region around 1200 cm$^{-1}$. The $D_2O$ bending vibration at 1200 cm$^{-1}$ is very well compensated in the stationary spectrum of Fig. 12(a), but it is present in all the modulation spectra of Fig. 12(b)–(d). This means that water molecules are also involved in the cyclic process induced by E-ME. The phase of the $\delta(D_2O)$ modulation band correlates with the 1665-cm$^{-1}$ band. A possible interpretation is a change of hydration of PLL during the 1639/

1665-cm$^{-1}$ transition. However, there might also result significant contributions from cyclic exchange of hydrated anions and cations in the space charge region.

A second band in this region should be mentioned, namely, the antisymmetric bending vibration of $ND_3^+$, $\delta_{as}(ND_3)^+$ at 1160 cm$^{-1}$. It appears only in the $\omega$ and $3\omega$ spectra and correlates with the 1665-cm$^{-1}$ band. Whether the 1639/1665-cm$^{-1}$ reaction step involves a deuteronation of the $\varepsilon$ amino group of the lysine side chain will be the subject of further investigations.

Finally, it should be mentioned that the uneven trace of the $\omega$ and $3\omega$ modulation spectra in the region between 1550 cm$^{-1}$ and 1250 cm$^{-1}$ is not understood up to now. There is no signal-to-noise problem, since the bands exhibit a systematic behavior with respect to the PSD phase setting and, moreover, they correlate in part with the 1665-cm$^{-1}$ and with the 1639-cm$^{-1}$ bands, respectively.

### 3.4   UV-VIS–Modulated Excitation ($\Phi$-ME)

UV-VIS–modulated excitation ($\Phi$-ME) IR spectroscopy could be of significant importance in the investigation of photochemical processes in biology, biochemistry, and chemistry. To our knowledge, no biochemical/biological applications of $\Phi$-ME have been reported so far. Therefore, an example of a sophisticated and pioneering study of the photo-oxidation of pyrocatechol by $\Phi$-ME IR will be presented in short. In this study, $\Phi$-ME IR was supported by complementary $\Phi$-ME ESR experiments, which were very useful for the identification of photochemically produced radicals. For details the reader is referred to Ref. 9.

Pyrocatechol (I, in Fig. 15) was dissolved in carbon tetrachloride with a concentration of $4.0 \cdot 10^{-4}$ mol/L. A transmission flow-through cell of 10-cm path length was used for IR spectroscopy. One side wall of the cell consisted of a quartz window, enabling UV irradiation of the solution. Ultraviolet light from a xenon lamp was interrupted by a mechanical chopper in the frequency range of 2–100 Hz. The IR beam transmitted the cuvette at a right angle to the UV beam. The periodic response of photoexcited species were demodulated by phase-sensitive detection (PSD; see also Sec. 5). The components of the modulation spectra obtained with PSD phase settings 0° and 90° with respect to the excitation were separated in the spectral region of 3200–3700 cm$^{-1}$ by lineshape analysis. This procedure enabled optimum accuracy for the determination of amplitude/frequency and phase/frequency relations. The results of this analysis are presented as Bode diagrams in Fig. 14, which together with a theoretical kinetic analysis formed the basis for the determination of the reaction scheme shown in Fig. 15 as well as of the corresponding rate constants, which were determined by fitting the Bode diagrams. Some interesting key aspects should be mentioned, because they were essential for the postulation of the reaction scheme shown in Fig. 15. From stationary irradiation experiments three end products (III, IV, V) were known. Furthermore, kinetic analysis of the reversible cycle formed by the

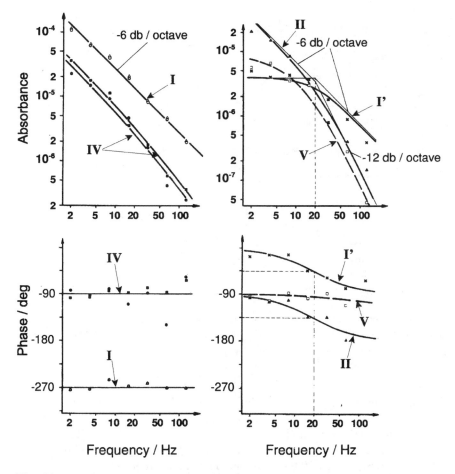

**Fig. 14** Bode diagrams in the frequency range 2–100 Hz from Φ-ME IR absorption bands of pyrocatechol in carbon tetrachloride solution ($4.0 \cdot 10^{-4}$ mol/L, 295 K, $N_2$ purging). The Roman numbers indicate the species and the corresponding frequencies of the modulation bands as indicated in the reaction scheme Fig. 15. (Reproduced from Ref. 9 by permission of North Holland Publishing Company.)

ground state $S_0$ (I), the singlet excited state $S_1$, and the triplet state $T_1$ (see Fig. 15) with slow, irreversible exits resulted in the following finding (9): At low frequencies, ground state I exhibits a phase shift of $-270°$ with respect to the stimulation. Its frequency dependence is a $-6$-decibels (dB) decrease per octave at elevated frequencies. Excited singlet and triplet states are too short-lived to achieve detectable concentrations. Under the condition of a short-lived $S_1$, the

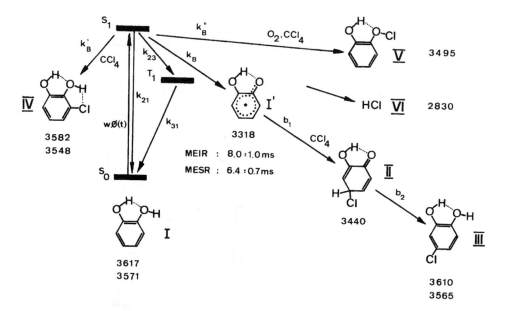

**Fig. 15** Kinetic scheme of the photo-oxidation of pyrocatechol in carbon tetrachloride as determined by Φ-ME IR and Φ-ME ESR spectroscopy. (Reproduced from Ref. 9 by permission of North Holland Publishing Company.)

first particle in an irreversible exit chain of only first-order reactions will exhibit a phase between 0° and −90° as the frequency is increased from zero to infinity. The amplitude decrease at high frequencies is −6 dB/octave. For each succeeding compound in this chain, one has to add −90° to the phase and an additional −6-dB/octave to the slope of the log (amplitude) vs. log (frequency) plot at elevated frequency. However, a second-order reaction step will result in a steeper slope even by itself, e.g., −12 dB/octave as observed with particle V.

In practice, the bode diagrams shown in Fig. 14 have been determined from the experimental results. Three irreversible paths were found: Particles I′ and II behave as first and second compound in a linear chain with first-order kinetics. Particle IV also behaves like the first compound in a chain with first-order reactions. Consequently it must be on a second exit from $S_1$. In addition, it was found independently to be an end product. From the amplitude/frequency behavior, particle V behaves similarly to particle II, which is on the second place in the chain of irreversible first-order reactions. Consequently, particle V must be on a third irreversible exit from $S_1$. Independently it was found that production of particle V depends on the oxygen content of the solution. Therefore it is most probable that the end product V is formed directly from $S_1$ by a second-order reaction. Finally, the OH stretching bands of pyrocatechol (I) could also be assigned due to the phase/frequency behavior, which at low frequency should result in a phase angle of $-270°$.


## 4 TWO-DIMENSIONAL CORRELATION ANALYSIS

Absorption bands of modulation spectra that exhibit equal phase shifts with respect to the external stimulation are considered to be correlated. Two-dimensional correlation analysis is a statistical/graphical means to visualize such a correlation in a 2D plot, which was referred to as a 2D IR spectrum (4). For theoretical background, the reader is referred to Refs. 4 and 18. In general, correlation analysis is made between two modulation spectra with PSD settings 90° degrees apart from each other. However, correlation between any difference of PSD phase settings may be calculated and obviously can lead to a more accurate estimation of the relative phase shift between two bands. Since 2D IR spectra are very susceptible to noise, it was shown that introducing a threshold and plotting only 2D signals above this limit may help disentangling the correlation data (18).

In any case, one may have a different opinion about the scientific value of 2D spectroscopy applied to data obtained via ME. In order to enable the reader to form his own opinion, a synchronous 2D IR spectrum was calculated from the 0° and 90° modulation spectra of Fig. 10A and is presented in Fig. 16. Two interesting features of 2D IR spectroscopy should be mentioned. First, the intensity of cross-peaks is proportional to the product of the intensities of the two bands under consideration. Consequently, a weak band correlating with an intense band can gain considerable intensity in the cross-peak. In our calculation of the 2D FT-IR spectrum shown in Fig. 16, the numerical threshold was set too high for the wagging sequence $\gamma_w(CH_2)$ (see Sec. 3.2.3). Therefore $\gamma_w(CH_2)$ did not result in autopeaks on the diagonal of the 2D FT-IR spectrum. Nevertheless there are two rows of seven cross-peaks correlating with the split $\nu_{as}(COO^-)$ band and the $\gamma_w(CH)_2$ sequence. This demonstrates that 2D correlation analysis is able to detect weak bands, provided they correlate with an intense band. The second example demonstrates a very critical aspect of 2D correlation analysis, namely, the principle impossibility of detecting the correct polarity, i.e., of reporting the correct correlation of a weak band overlapped by a strong band of opposite polarity. This is the case with the narrow $CH_2$ bending band $\delta(CH_2)$ at 1470 cm$^{-1}$ which is assigned to the protonated form of arachidic acid; see Fig. 10A. The broad band with opposite sign around 1470 cm$^{-1}$ determines numerically the sign of $\delta(CH_2)$ and consequently disables the correlation analysis to give the correct answer.

The problem of finding correlated populations is much simpler and considerably more reliably solved by consulting the basic data, namely, the phase-

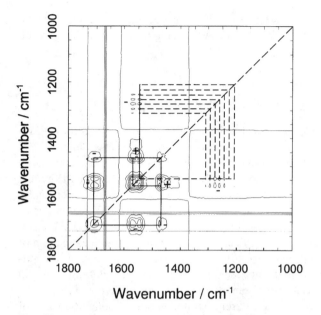

**Fig. 16**  Synchronous 2D FT-IR spectrum of an arachidic acid (ArAc) bilayer attached to a germanium internal reflection element (IRE) influenced by pH-modulated excitation (ME). The 0° and 90° modulation spectra of Fig. 10A were used for calculation of this 2D plot according to Ref. 18. On the dashed diagonal are the positive peaks of the autocorrelation function, which coincide with the wavenumbers of the absorbance spectrum (see Fig. 10). The plus and minus signs in the figure denote the sign of the cross-peak. A positive cross-peak means that the two signals on the diagonal are in phase (positive correlation); a negative cross-peak means out of phase (negative correlation), i.e., exhibiting in this experiment exactly 0° and 180° phase differences (see Sec. 3.2.3). Note the two rows of weak negative cross-peaks, although no autopeaks are visible between 1200 and 1300 cm$^{-1}$. The frequencies coincide with those of the wagging sequence. These cross-peaks gain their intensity from the most intense band in the spectrum, the split antisymmetric carboxylate stretching band $\nu_{as}(COO^-)$. A second phenomenon should be mentioned. Looking at the CH$_2$ bending vibration at 1470 cm$^{-1}$, there is a positive correlation (0°) with $\nu_{as}(COO^-)$ and a negative one (180°) with C=O stretching of the carboxyl group $\nu(COOH)$. This finding is not completely correct, since, as quickly verified from Fig. 10A, there is a weak narrow band at 1470 cm$^{-1}$, which is in phase with $\nu(COOH)$ and out of phase with $\nu_{as}(COO^-)$. This fact cannot be detected by 2D FT-IR plots, since numerically the polarity of the narrow band is falsified due to overcompensation by a more intense band of opposite sign. See also the text.

resolved modulation spectra shown in Fig. 10. In this context, the reader is referred to Ref. 18.

## 5 SPECTROSCOPIC SIGNAL PROCESSING

A modulated perturbation of the sample by an external parameter leads to periodic changes in the concentration $c_i(t)$ of the excited species $A_i$, resulting in modulated alterations of the absorbance of this species. However, changes of the sample thickness $d(t)$, e.g., by stretching or hydration, or changes of the absorption coefficient $\varepsilon_i(\tilde{\nu}, t)$ itself, e.g., due to temperature modulation, can also can also contribute to the time-dependent absorption.

Regarding the common case of a constant absorption coefficient $\varepsilon_i(\tilde{\nu})$ and a constant thickness $d$, Lambert–Beer's law leads to the measured time-dependent absorption of the species $A_i$,

$$A_i(\tilde{\nu}, t) = \varepsilon_i(\tilde{\nu}) \cdot c_i(t) \cdot d \tag{24}$$

Due to Fourier's theorem, the periodic function $c_i(t)$ can always be expressed by a Fourier series in the form

$$c_i(t) = c_{i,0} + \sum_{k=1}^{\infty} c_{i,k} \sin(k\omega t + \varphi_{i,k}) \tag{25}$$

where $k$ denotes the multiple of the fundamental ($k = 1$ fundamental frequency, $k > 1$ higher harmonics) and $c_{i,0}(t)$ is the mean concentration in the stationary state. $c_{i,k}$ and $\varphi_{i,k}$ are the required data for kinetic analysis representing the amplitudes and phase lags of the $k$th harmonic, respectively. To simplify the following algorithm, we use a periodic excitation Eq. (1), with a phase $\theta = -\pi/2$, which leads to the harmonic stimulation

$$\eta(t) = \eta_0 + \Delta\eta_0 + \Delta\eta_1 \cos\left(\omega t - \frac{\pi}{2}\right) = \eta_1 + \Delta\eta_0 + \Delta\eta_1 \sin(\omega t) \tag{26}$$

This stimulation function results in measured phase shifts corresponding directly to the phase lags $\varphi_{i,k}$ introduced by the system itself (Eq. (25)).

Assuming $N$ reacting species, Eq. 24 must be completed to describe the overall absorbance by

$$A(\tilde{\nu}, t) = \sum_{i=1}^{N} A_i(\tilde{\nu}, t) \tag{27}$$

The evaluation of $c_{i,k}$ and $\varphi_{i,k}$ from the measured time-dependent absorption $A(\tilde{\nu}, t)$ is the subject of phase-sensitive detection (PSD).

## 5.1 Principles of Phase-Sensitive Detection

Phase-sensitive detection is used to detect small periodic signals in a noisy background or to separate them from large and noisy background signals, as often encountered in spectroscopy. As stated earlier, the changes in the absorbance due to modulated excitation are periodic, exhibiting the period of the stimulation. Therefore, they can be expressed by a Fourier series:

$$A(\tilde{\nu}, t) = \sum_{i=1}^{N} A_{i,0}(\tilde{\nu}) + \sum_{i=1}^{N} \sum_{k=1}^{\infty} A_{i,k}(\tilde{\nu}) \sin(k\omega t + \varphi_{i,k}) \tag{28}$$

The first part of the right-hand side of Eq. (28) results in the stationary overall absorbance of the system, while the second part contains the modulated response. The amplitude $A_{i,k}(\tilde{\nu})$ is associated with the group of absorption bands exhibiting the same phase angle $\varphi_{i,k}$ at frequency $k\omega$.

Such a group may be assigned to species $A_i$ and is referred to as the $i$th population. In the simplest case, we have to consider at least two populations, one of the reactants, the other of the products.

In order to determine the $k$th amplitude $A_{i,k}(\tilde{\nu})$ and the corresponding phase lag $\varphi_{i,k}$, the periodic function $A(\tilde{\nu}, t)$ is multiplied by a function $\sin(k\omega t + \phi_k^{PSD})$ and integrated over a period of the stimulation $T = 2\pi/\omega$. The factor of $2/T$ is necessary for normalization, i.e., to match the amplitudes before and after PSD. After some straightforward calculation, this leads to the phase-resolved absorbance spectrum $A_k(\tilde{\nu}, \phi_k^{PSD})$ with $\phi_k^{PSD}$ as parameter:

$$A_k(\tilde{\nu}, \phi_k^{PSD}) = \sum_{i=1}^{N} A_{i,k}(\tilde{\nu}, \phi_k^{PSD}) = \frac{2}{T} \int_0^T A(\tilde{\nu}, t) \sin(k\omega t + \phi_k^{PSD})dt$$

$$\tag{29}$$

$$= \sum_{i=1}^{N} A_{i,k}(\tilde{\nu}) \cos(\varphi_{i,k} - \phi_k^{PSD})$$

The parameter $\phi_k^{PSD}$ can be selected by the operator. Obviously the amplitude $A_{i,k}(\tilde{\nu}, \phi_k^{PSD})$ becomes maximum for $\phi_k^{PSD} = \varphi_{i,k}$ resulting in the unknown amplitude $A_{i,k}(\tilde{\nu})$. On the other hand, if $\phi_k^{PSD} = \varphi_{i,k} \pm 90°$, the cosine function and thus the amplitude $A_{i,k}$ become zero. Consequently, maximizing $A_{i,k}(\tilde{\nu}, \phi_k^{PSD})$ by variation of $\phi_k^{PSD}$ results in both the amplitude $A_{i,k}(\tilde{\nu})$ and the phase lag $\varphi_{i,k}$ of the response of the $i$th population to frequency $k\omega$.

## 5.2 Application of Phase-Sensitive Detection in Spectroscopy

As shown by Eq. (29), the time-dependent absorbance has to be multiplied by a trigonometric function and integrated over a period of stimulation to perform the

PSD. It should be noted that the primary signal measured in FT-IR spectroscopy is the interferogram representing the intensity of the IR beam as a function of the interferometer retardation. Therefore, the results of a PSD performed on the IR beam intensity during sampling of the interferogram does not give the same result as a PSD performed in the time-dependent absorbance domain. The reason is the logarithmic relation between intensity and absorbance according to Lambert–Beer's law:

$$I_i(\tilde{v}, t) = I_{0,i}(\tilde{v}, t) \cdot 10^{-A_i(\tilde{v},t)} \tag{30}$$

This nonlinear relation between intensity and absorbance (i.e., concentration) leads to systematic errors as soon as the exponential no longer admits a linear approximation. In this case, a logarithmic amplifier or application of sophisticated mathematical procedures is required. Nevertheless, PSD of many reported FT-IR experiments using modulation spectroscopy were performed in the interferogram domain. Most of these applications used the step-scan mode, with an additional phase modulation by means of the interferometer moving mirror. Consequently, two PSD steps were necessary for complete demodulation (35–38). Up to the time when digital signal processing (DSP) became fast enough, analog lock-in amplifiers were used to perform the PSD. In a first step, lock-in amplifiers were replaced by external digital signal processing (39) or digital lock-in amplifiers. In current developments, the digital signal processor circuits of the spectrometers are used to replace external digital signal processing (12,40).

In our laboratory, we use a digital PSD via software, performed after the sampling of time-resolved absorbance spectra is completed (41). The advantage of this method is the separation of data sampling (including FT-IR phase correction) and the PSD. The data sampling can therefore be done in every time-resolved measurement mode, which is essential for long-period modulation experiments. Application is also possible without step-scan equipment. However, the highest-modulation frequency, which determines time resolution, is then limited by the scan velocity, unless more sophisticated techniques, such as interleaved rapid scan, are applied. Moreover, all of the time-dependent information is still available after the experiment, enabling demodulation at any multiple of the stimulation frequency and at any PSD phase setting. The synchronized sampling and coaddition of time-resolved spectra $A(\tilde{v}, t)$ during the modulation period is a requirement of this method (41,42). Concerning data acquisition, $A(\tilde{v}, t)$ is an array of $n$ absorbance spectra $\{A(\tilde{v}, t_1), A(\tilde{v}, t_2), \ldots, A(\tilde{v}, t_n)\}$ measured with a time interval $\Delta t = T/n$. Accordingly, the basic integral of PSD Eq. (29) becomes discrete:

$$A_k(\tilde{v}, \phi_k^{PSD}) = \frac{1}{C} \sum_{m=0}^{n-1} A(\tilde{v}, t_m) \sin[k\omega(m\Delta t) + \phi_k^{PSD}] \tag{31}$$

where $C$ is a factor for normalization depending on the method of numerical integration. It should be mentioned that $A(\tilde{v}, t_m)$ denotes the $m$th entire spectrum. For that reason, the PSD is performed as a vector PSD and leads immediately to the phase-resolved spectra $A_k(\tilde{v}, \phi_k^{PSD})$.

The modulation amplitude and phase lag at a given wavenumber $\tilde{v}^*$ can now be determined by varying the phase of the PSD until the absorbance at $\tilde{v}^*$ in the phase-resolved spectrum becomes maximum. For non-overlapping bands, the amplitudes and phase lags of all bands can easily be calculated from two orthogonal phase-resolved spectra e.g., $A_k(\tilde{v}, 0°)$ and $A_k(\tilde{v}, 90°)$, which are also called in-phase and out-of-phase spectra, respectively. The modulation amplitude results from

$$A_k(\tilde{v}) = \sqrt{[A_k(\tilde{v}, 0°)]^2 + [A_k(\tilde{v}, 90°)]^2} \qquad (32)$$

To calculate the absolute phase lag in the range of 0°–360°, one has to consider that the result of the arctan function is defined to be only within $-90°$ and 90°, because $\tan \varphi = \tan(\varphi + 180°)$. Table 1 shows how to calculate the absolute phase lag $\varphi_k$ in the range from 0° to 360°.

## 5.3  Phase Resolution of Overlapping Bands

In the case of overlapping bands, modulation spectroscopy can help to discriminate single components if the kinetics and, therefore, also the phase lags of overlapping bands are different. In this case the calculation of a set of phase-resolved spectra with different PSD-phase settings is very helpful (see Figs. 4B and 5B).

Let us consider two absorbing species with different phase lags $\alpha$ and $\beta$. According to Eq. (29), there is always a PSD phase setting resulting in zero absorbance for species $A_i$. For a species with phase lag $\alpha$ this is the case for $\phi_k^{PSD} = \alpha \pm 90°$. If, as assumed, the second species exhibits a different phase

**Table 1**  Calculation of the Phase Lag $\varphi_k(\tilde{v})$ (Absolute Phase Lag) Considering the Non-determined Range of the Arctan Function within 0° and 360°

| $\varphi_{k,\text{absolute}}$ | $A_k(\tilde{v}, 0°)$ | $A_k(\tilde{v}, 90°)$ |
|---|---|---|
| $\varphi_k = \arctan\left(\dfrac{A_k(\tilde{v}, 90°)}{A_k(\tilde{v}, 0°)}\right)$ | $\geq 0$ | $\geq 0$ |
| $\varphi_k = \arctan\left(\dfrac{A_k(\tilde{v}, 90°)}{A_k(\tilde{v}, 0°)}\right) + 360°$ | $\geq 0$ | $< 0$ |
| $\varphi_k = \arctan\left(\dfrac{A_k(\tilde{v}, 90°)}{A_k(\tilde{v}, 0°)}\right) + 180°$ | $< 0$ | $\forall A_k(\tilde{v}, 90°)$ |

lag $\beta$, its modulation spectrum $A_k(\tilde{\nu}, \alpha \pm 90°)$ will not vanish at this phase setting. In this simple case, the remaining absorbance results only from the species with phase lag $\beta$; i.e., separation has been achieved on a completely experimental level. By analogous procedure, the decoupled state of the species with phase lag $\alpha$ is achieved at PSD phase setting $\phi_k^{PSD} = \beta \pm 90°$, resulting in the phase-resolved spectrum $A_k(\tilde{\nu}, \beta \pm 90°)$, which is free of overlapping by absorption bands of the other species.

A typical example should be given: Considering the phase-resolved T-ME spectra shown in Figs. 5B and C, we realize that in the fundamental ($\omega$) spectrum with phase setting $\phi_{PSD} = 135°$, the intense bands of the helix at 1639 cm$^{-1}$ and of the antiparallel $\beta$-sheet at 1614 and 1685 cm$^{-1}$ are virtually absent, indicating that these dominant signals exhibit phase lags of $135° \pm 90°$. However, the remaining modulation spectrum unambiguously shows two weak sigmoidal bands at 1614/1609 cm$^{-1}$ and 1685/1680 cm$^{-1}$, pointing to an intermediate antiparallel $\beta$-structure, possibly a precursor of the final state at 1614 cm$^{-1}$ and 1685 cm$^{-1}$, which is visible in the other modulation spectra and in the stationary spectrum, Fig. 5A. The existence of this intermediate species is confirmed by the first harmonic ($2\omega$) modulation spectrum at the $\phi_{PSD} = 0°$ setting; see Fig. 5C. Moreover, its intense response in the first harmonic, while excitation has been performed in the fundamental frequency $\omega$, gives strong evidence for second-order kinetics with large amplitudes, as expected for intermediates of a phase transition.

## 6 KINETICS OF PERIODICALLY EXCITED SYSTEMS

### 6.1 The Simplest Case of a Reversible First-Order Reaction

As an example of a kinetic analysis using modulation technique we will demonstrate the procedures for a reversible first-order reaction of the type

$$A_1 \underset{k_-}{\overset{k_+}{\rightleftharpoons}} A_2 \tag{33}$$

which may be typical for a simple conformational change, e.g. a cis-trans isomerization.

The rate equations $c_i(t)$ ($i = 1, 2$) are given by

$$\dot{c}_1(t) = -k_+c_1(t) + k_-c_2(t), \qquad \dot{c}_2(t) = k_+c_1(t) - k_-c_2(t) \tag{34}$$

where $c_i(t)$ denotes time-dependent concentrations of the species $A_i$ and $k_+$ and $k_-$ are the rate constants of the forward and the backward reaction, respectively. Introducing the extent of reaction $\xi(t)$ as a relevant parameter that is related to concentrations according to

$$c_i(t) = c_{i,0} + v_i \xi(t) \tag{35}$$

where $c_i$ and $c_{i,0}$ denote the actual and the initial concentration of the $i$th species, respectively, and $v_i$ is the stoichiometric number, which is, according to convention, negative for reactants and positive for products (44). Thus it follows for the concentrations of $A_1$ and $A_2$

$$c_1(t) = c_{1,0} - \xi(t), \qquad c_2(t) = c_{2,0} + \xi(t) \tag{36}$$

The rate equation in terms of the extent of reaction follows from combining Eqs. (34) and (36):

$$\dot{\xi}(t) = -(k_+ + k_-)\xi(t) + k_+ c_{1,0} - k_- c_{2,0} \tag{37}$$

Up to now, no external stimulation was considered. In order to describe a temperature-modulated excitation with small amplitudes, we introduce the temperature-dependent rate constants, as given by Eq. (4). Furthermore, to shorten the notations, abbreviations with distinct physical meaning, summarized by Eq. (38), are introduced:

$$
\begin{aligned}
&\varkappa_1 = k_+ + k_- + \frac{\Delta T_0}{T_0^2}(T_{A_+}k_+ + T_{A_-}k_-) = \overline{k}_+ + \overline{k}_- \\[2mm]
&\varkappa_2 = \frac{\Delta T_1}{T_0^2}(T_{A_+}k_+ + T_{A_-}k_-) = \Delta k_+ + \Delta k_- \\[2mm]
&\varkappa_3 = c_{2,0}k_- - c_{1,0}k_+ + \frac{\Delta T_0}{T_0^2}(T_{A_-}c_{2,0}k_- - T_{A_+}c_{1,0}k_+) = c_{2,0}\overline{k}_- - c_{1,0}\overline{k}_+ \\[2mm]
&\varkappa_4 = \frac{\Delta T_1}{T_0^2}(T_{A_-}c_{2,0}k_- - T_{A_+}c_{1,0}k_+) = c_{2,0}\Delta k_- - c_{1,0}\Delta k_+
\end{aligned}
\tag{38}
$$

$\overline{k}_+ + \overline{k}_-$ denotes the sum of the rate constants at the stationary (mean) temperature $T_0 + \Delta T_0$, while rate constants without overbars relate throughout to the initial temperature $T_0$. $\Delta k_+$ and $\Delta k_-$ denote the change of the corresponding rate constants resulting from a temperature change equal to the modulation amplitude $\Delta T_1$; see also Fig. 1.

The differential equation of the periodically excited system results:

$$\dot{\xi}(t) + [\varkappa_1 + \varkappa_2 \cos(\omega t + \theta)] \cdot \xi(t) + \varkappa_3 + \varkappa_4 \cos(\omega t + \theta) = 0 \tag{39}$$

This equation has no exact analytical solution, because of the time-dependent part $\varkappa_2 \cos(\omega t + \theta)$ in the coefficient of $\xi(t)$. Because we assumed small modulation amplitudes $\Delta T_1$, this part can be neglected, which is confirmed by a numerical solution of Eq. (39). The approximate rate equation thus results in

$$\dot{\xi}(t) + \varkappa_1 \cdot \xi(t) + \varkappa_3 + \varkappa_4 \cos(\omega t + \theta) = 0 \tag{40}$$

The solution of this differential equation is

$$\xi(t) = -\frac{x_3}{x_1} + \left[ \frac{x_3}{x_1} + \frac{x_1 x_4 \cos\theta + x_4 \omega \sin\theta}{x_1^2 + \omega^2} \right] \cdot e^{-x_1 t} \tag{41}$$

$$-\frac{x_4}{x_1^2 + \omega^2} [x_1 \cos(\omega t + \theta) + \omega \sin(\omega t + \theta)]$$

$\xi(t)$ consists of a constant term, an exponentially relaxing term, and a modulated term. The time-dependent trigonometric expression in brackets may be reduced to a cosine function with phase shift, in analogy to Fourier series (43). It follows that

$$x_1 \cos(\omega t + \theta) + \omega \sin(\omega t + \theta) = \sqrt{x_1^2 + \omega^2} \cos(\omega t + \theta + \varphi)$$

where

$$\cos\varphi = \frac{x_1}{\sqrt{x_1^2 + \omega^2}} \quad \text{and} \quad \sin\varphi = -\frac{\omega}{\sqrt{x_1^2 + \omega^2}} \tag{42}$$

$$\text{or} \quad \tan\varphi = -\frac{\omega}{x_1}$$

Denoting the corresponding magnitudes by $\xi_c$, $\xi_r(\omega, \theta)$, and $\xi_m(\omega, \theta)$, respectively, and taking account of Eq. (42), Eq. (41) may be rewritten in the form

$$\xi(t) = \xi_c + \xi_r(\omega, \theta)\, e^{-(\bar{k}_+ + \bar{k}_-)t} + \xi_m(\omega, \theta) \cos(\omega t + \theta + \varphi) \tag{43}$$

$\varphi$ denotes the phase lag introduced by the delayed system response, and $\xi_m(\omega, \theta)$ denotes the amplitude of the modulated part of the extent of reaction. In the notation given by Eq. (43), $\xi_m(\omega, \theta)$ may be positive or negative depending on the sign of $x_4$ (see Eq. (38)). Consequently, the phase angle introduced by the system may assume the value $\varphi$ or $\varphi \pm \pi$. A method to determine the absolute value of $\varphi$ is indicated later.

Direct relations to relevant quantities in kinetics are now obtained by combining Eqs. (38), (41), and (43). In the stationary state of modulated excitation, the time mean of the extent of reaction $\xi(t)$ results in the constant term $\xi_c$:

$$\xi_c = \frac{c_{1,0}\bar{k}_+ - c_{2,0}\bar{k}_-}{\bar{k}_+ + \bar{k}_-} \tag{44}$$

$\xi_c$ describes the extent of reaction between the initial equilibrium state at $T = T_0$ and the stationary state, which corresponds to a new equilibrium state at $T = T_0 + \Delta T_0$. Note that $\xi_c$ becomes zero if the initial concentrations $c_{1,0}$ and $c_{2,0}$ are already equilibrium concentrations at $T = T_0 + \Delta T_0$, as expected and as may be verified by considering the rate Eq. (34) at zero rate, which means equilibrium.

$\xi_r(\omega, \theta)$, from the second term of Eq. (43), denotes the magnitude of the exponentially declining part of the extent of reaction. This part of the general solution is not used in modulation spectroscopy; however, it is of interest for comparison with corresponding results from $T$-jump experiments (3). In a first view, Eq. (43) indicates that the relaxation time results in

$$\tau_m = \frac{1}{\overline{k}_+ + \overline{k}_-} = \frac{1}{(k_+ + \Delta k_+) + (k_- + \Delta k_-)} \tag{45}$$

where $k_+$ and $k_-$ are the rate constants at $T = T_0$ and $\Delta k_+$ and $\Delta k_-$ denote the changes of rate constants induced by the temperature change between initial state $T_0$ and stationary state $T_0 + \Delta T_0$. In relaxation spectroscopy, the relaxation time $\tau_r$ is usually related to the initial temperature $T_0$; i.e., $\tau_r = 1/(k_+ + k_-)$. The difference between $\tau_m$ and $\tau_r$, however, is small and may be calculated via Eq. (38).

Taking the reduction introduced by Eq. (42) into account, one obtains for the magnitude of $\xi_r$

$$\xi_r = \frac{c_{2,0}\overline{k}_- - c_{1,0}\overline{k}_+}{\overline{k}_+ + \overline{k}_-} + \frac{(c_{2,0}\Delta k_- - c_{1,0}\Delta k_+)\cos(\theta + \varphi)}{\sqrt{(\overline{k}_+ + \overline{k}_-)^2 + \omega^2}} \tag{46}$$

The first term on the right-hand side of this equation is a constant that is equal to $-\xi_c$. If we combine it with Eq. (44) we get the portion of $\xi(t)$ corresponding to the $T$-jump relaxation; i.e.,

$$\xi_{jump} = \xi_c[1 - e^{-(\overline{k}_+ + \overline{k}_-)t}] = \xi_c[1 - e^{-t/\tau_m}] \tag{47}$$

This equation demonstrates that relaxation and modulation techniques contain the same kinetic information on relaxation times and amplitudes.

The second term of Eq. (46) depends on modulation amplitude and frequency $\omega$ as well as on the phase angle $\theta$ of the stimulation (see Eq. (1)) and on the phase lag $\varphi$ introduced by the delayed response of the system. This part describes the phase relaxation of the response. Because concentration relaxation ($\xi_{jump}$) vanishes when initial concentrations are equal to stationary concentrations, phase relaxation vanishes when the phase angle $\theta$ of the stimulation differs by $\pm 90°$ from the system phase lag $\varphi$. In practice, however, this condition cannot be fulfilled. Consequently, data acquisition for modulation spectroscopy should be started no earlier than $3\tau_m$ after the beginning of stimulation; otherwise, a systematic error due to a contribution of nonstationary response will occur.

Phase relaxation also vanishes at high modulation frequencies, which means $\omega \gg (\overline{k}_+ + \overline{k}_-)$; however, as will be shown later, the amplitudes of the modulated responses also will vanish under this condition.

Considering now the third term in Eq. (43), we may express the amplitude in the form

$$|\xi_m(\omega)| = \frac{|c_{1,0}\Delta k_+ - c_{2,0}\Delta k_-|}{|\sqrt{(\overline{k}_+ + \overline{k}_-)^2 + \omega^2}|} = \frac{|c_{1,0}\Delta k_+ - c_{2,0}\Delta k_-| \cdot \tau_m}{|\sqrt{1 + (\omega\tau_m)^2}|} \tag{48}$$

As already mentioned (Eq. (45)), $\tau_m = (\overline{k}_+ + \overline{k}_-)^{-1}$ denotes the relaxation time of reaction shown by Eq. (33). Equation (48) results in the absolute value of the modulation amplitude. It should be noted, however, that the amplitude may be positive or negative, depending on the magnitudes of the products between initial concentrations $c_{1,0}$ and $c_{2,0}$ and the change of rate constants under the influence of the modulation amplitude $\Delta T_1$, namely, $\Delta k_+$ and $\Delta k_-$, which depend on the activation energies of forward and backward reactions, respectively. We take the sign of the amplitude via the phase lag $\varphi$ into account, resulting in Eq. (49), which is a modification of the trigonometric functions of the phase lag indicated in Eq. (42):

$$
\begin{aligned}
\cos \varphi &= \frac{\text{sign}(c_{1,0}\Delta k_+ - c_{2,0}\Delta k_-) \cdot (\bar{k}_+ + \bar{k}_-)}{|\sqrt{(\bar{k}_+ + \bar{k}_-)^2 + \omega^2}|} \\
&= \frac{\text{sign}(c_{1,0}\Delta k_+ - c_{2,0}\Delta k_-)}{|\sqrt{1 + (\omega\tau_m)^2}|} \\
\sin \varphi &= -\frac{\text{sign}(c_{1,0}\Delta k_+ - c_{2,0}\Delta k_-) \cdot \omega}{|\sqrt{(\bar{k}_+ + \bar{k}_-)^2 + \omega^2}|} \\
&= -\frac{\text{sign}(c_{1,0}\Delta k_+ - c_{2,0}\Delta k_-) \cdot \omega\tau_m}{|\sqrt{1 + (\omega\tau_m)^2}|} \\
\tan \varphi &= -\frac{\omega}{\bar{k}_+ + \bar{k}_-} = -\omega\tau_m
\end{aligned}
\tag{49}
$$

It should be noted that the right quadrant for the phase angle $\varphi$ can be determined only by considering the signs of two trigonometric functions indicated in Eq. (49), as depicted in Table 2.

Finally, the modulated part of the stationary solution (subscript $m$, stat) of $\xi(t)$ may be written in the form

$$
\xi_{m,\text{stat}}(t) = \frac{|c_{1,0}\Delta k_+ - c_{2,0}\Delta k_-| \cdot \tau_m}{|\sqrt{1 + (\omega\tau_m)^2}|} \cos(\omega t + \theta + \varphi)
\tag{50}
$$

where the phase angle has to be determined from Eq. (49) and Table 2.

As an illustration, the time-dependent concentration $c_2(t)$ of species $A_2$ (see Eq. (36)) was calculated and is shown in Fig. 17A. The frequency dependence

**Table 2** Signs of the Trigonometric Functions

| Quadrant | sin | cos | tan |
|----------|-----|-----|-----|
| I | + | + | + |
| II | + | − | − |
| III | − | − | + |
| IV | − | + | − |

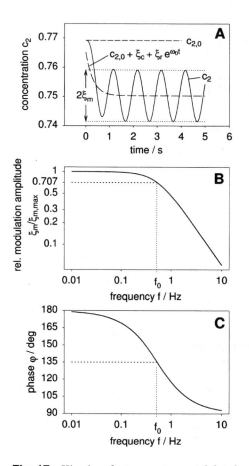

**Fig. 17** Kinetics of a temperature-modulated model system. The concentration $c_2$ of the product in reaction $A_1 \rightleftarrows A_2$ is calculated as a function of time using Eqs. (36) and (50) (A). The mean value of the concentration decreases exponentially, caused by the temperature jump $\Delta T_0$. After a few periods, the stationary state is reached, exhibiting the changes in concentration due to the temperature modulation amplitude $\Delta T_1$. With increasing frequency the modulation amplitude $\xi_m$ decreases, assuming a slope of $-6$ dB per octave at elevated frequency according to Eq. (48) (B), and the phase lag $\varphi$ (C) decreases according to Eq. (49). At the frequency $f_0$, the amplitude $\xi_m$ has decreased by a factor of $1/\sqrt{2}$ ($-3$ dB). The phase lag at the 3-dB point results in $\varphi = 3\pi/4$ (C). The parameters used for calculation were: $E_{a,+} = 35,000$ J/mol, $E_{a,-} = 55,000$ J/mol, $k_+ = 2\ s^{-1}$, $k_- = 0.6\ s^{-1}$, $T_0 = 298$ K, $\Delta T_0 = 5$ K, $\Delta T_1 = 5$ K, $f = 1$ Hz, and $\theta = -\pi$.

of the modulation amplitude $|\xi_m(\omega, \theta)|$ and the phase lag $\varphi(\omega)$ as calculated by Eqs. (48) and (49) are shown in Figs. 17B and C.

## 6.2 Evaluation of Kinetic Parameters from Spectroscopic Modulation Experiments

In Sec. 6.1 the solution in terms of the extent of reaction $\xi$ was derived for a T-ME of the simple reaction scheme depicted by Eq. (33). It is related to the concentration of the reacting species according to Eqs. (35) and (36). Data acquisition, on the other hand, leads to absorbances $A_i(\tilde{\nu}, t)$, which are related to concentrations by Lambert–Beer's law, Eq. (24). Taking into account that data acquisition in modulation spectroscopy is performed in the stationary state of excitation, the relaxation term in Eq. (43) can be omitted, resulting in

$$A_{i,\text{stat}}(\tilde{\nu}, t) = \varepsilon(\tilde{\nu}) \cdot d \cdot c_i(t) = \varepsilon(\tilde{\nu})$$
$$\cdot d \cdot (c_{i,0} + \nu_i \cdot (\xi_c + |\xi_m(\omega, \theta)| \cos(\omega t + \theta + \varphi_i) \tag{51}$$

Note that $\varphi_i$ now denotes the phase of species $A_i$. In this example, $\varphi_i$ and the phase of the extent of reaction $\varphi$ are equal if the corresponding species is the product $A_2$. However, they are out of phase if the $i$th species is a reactant, e.g., $A_1$. Consequently, Eq. (49), which indicates the absolute phase of the extent of reaction, has to be modified as follows in order to get the phase of a distinct species $A_i$:

$$\cos \varphi_i = \frac{\text{sign}(\nu_i) \cdot \text{sign}(c_{1,0}\Delta k_+ - c_{2,0}\Delta k_-)}{|\sqrt{1 + (\omega\tau_m)^2}|}$$

$$\sin \varphi_i = -\frac{\text{sign}(\nu_i) \cdot \text{sign}(c_{1,0}\Delta k_+ - c_{2,0}\Delta k_-) \cdot \omega\tau_m}{|\sqrt{1 + \omega\tau_m)^2}|} \tag{52}$$

$$\tan \varphi_i = -\text{sign}(\nu_i) \cdot \frac{\omega}{\bar{k}_+ + \bar{k}_-} = -\text{sign}(\nu_i) \cdot \omega\tau_m$$

As a consequence of Eq. (52), the absolute value of the stoichiometric number $\nu_i$ has to be used from now on.

After PSD is performed for the fundamental frequency ($k = 1$) on $A_{i,\text{stat}}(\tilde{\nu}, t)$, one obtains, from Eqs. (29) and (51),

$$A_{i,1} = \varepsilon(\tilde{\nu}) \cdot d \cdot |\nu_i| \cdot |\xi_m(\omega, \theta)| \sin(\phi_1^{PSD} - \theta - \varphi_i) \tag{53}$$

Introducing $\theta = -\pi/2$, as in Sec. 5, Eq. (26), which means stimulation with a sine function without phase shift, and solving Eq. (53) for the phase-resolved extent of reaction leads to

$$|\xi_{m,1}(\omega, \theta)| \cos(\varphi_i - \phi_1^{PSD}) = \frac{A_{i,1}(\tilde{\nu}, \phi_1^{PSD})}{|\nu_i| \cdot \varepsilon(\tilde{\nu}) \cdot d} \tag{54}$$

It should be noted that digital PSD results in the phase-resolved modulation spectra $A_{i,1}(\tilde{\nu}, \phi_1^{PSD})$ at any PSD phase setting $\phi_1^{PSD}$. In case of the reversible first-order reaction (Eq. (33)), one can conclude directly from Eq. (52) that the reactant $A_1$ and the product $A_2$ exhibit a relative phase difference of 180°, since $v_1 = -1$, and $v_2 = +1$.

In order to determine the relevant quantities $|\xi_{m,1}(\omega, \theta)|$ and $\varphi_i$, the minimum expenditure is to perform modulation experiments at a frequency in the range $0.1 < \omega \cdot \tau_m < 10$. $|\xi_{m,1}(\omega, \theta)|$ and $\varphi_i$ are then obtained, e.g., by the following procedure:

1. Look for the distinct PSD phase setting $\phi_{1,zero}^{PSD}$, which results in zero amplitude. At this setting it follows that $\varphi_i - \phi_{1,zero}^{PSD} = \pm 90°$, i.e., $\varphi_i = \phi_{1,zero}^{PSD} \pm 90°$. The absolute value of $\varphi_i$ may be determined by means of Table 1 in Sec. 5.
2. Add $\pm 90°$ to $\phi_{1,zero}^{PSD}$. At this setting, $\phi_{1,extreme}^{PSD} = \phi_{1,zero}^{PSD} \pm 90°$, the cosine function on the left-hand side of Eq. (54), assumes $+1$ or $-1$, i.e., an extreme, leading directly to the amplitude $|\xi_{m,1}(\omega, \theta)| \cdot |v_i|$.

It should be mentioned, however, that variation of $\omega$ in the range $0.1 < \omega \cdot \tau < 10$ enables multiple access to $|\xi_{m,1}(\omega, \theta)|$ and $\varphi_i$, which leads to enhanced accuracy and more reliable validation of an assumed reaction scheme. So-called Bode diagrams are presented in Fig. 14 and Figs. 17B and C. The latter depict the amplitude and phase dependence on frequency for a reaction scheme shown by Eq. (33). If the Bode diagrams differ from Figs. 17B and C, as is the case in Fig. 14, it follows that the underlying chemical reaction scheme must be more complicated. For details, the reader is referred to Sec. 3.4 and to Refs. 1 and 9.

In a more complex situation with overlapping modulation bands, it is advisable to proceed with analysis in two steps. First, a lineshape analysis of distinct bands over a complete set of phase-resolved modulation spectra $A_{i,1}(\tilde{v}, \phi_1^{PSD})$ (e.g., at $\phi_1^{PSD} = 0°, 22.5°, 45°, \ldots, 180°$) is performed. In a second step, each distinct modulation band of this phase-resolved set has to be fitted with respect to $|\xi_{m,1}(\omega, \theta)|$ and $\varphi_i$ according to Eq. (54). For a typical example the reader is referred to Ref. 18.

## 6.3 Evaluation of Rate Constants from Phase Lag and Modulation Amplitude

As discussed in Sec. 6.2, ME experiments give access to the relaxation times of a system via phase lag measurements and via the amplitude/frequency dependence. Reaction Eq. (33) exhibits one relaxation time $\tau_m$. The number of relaxation times of a more complicated reaction system equals the number of eigenvalues $\lambda_i$ of the kinetic matrix and is related to them by Eq. (55) (3):

$$\tau_i = -\frac{1}{\lambda_i} \tag{55}$$

$\lambda_i$ is a function of the rate constants and in the case of non-first-order steps also of the initial concentrations of the corresponding reacting species. A number of relaxation times of distinct reactions have been calculated and summarized in Ref. 3 and 45.

In the case of reaction Eq. (33), the relaxation time is given by Eq. (45). In order to evaluate the single constants $\bar{k}_+$ and $\bar{k}_-$, a second, independent relation is necessary, e.g., $\xi_c$, the extent of reaction between the initial equilibrium state at $T = T_0$ and the stationary state at $T = T_0 + \Delta T_0$ as described by Eq. (44). It follows from Eqs. (44) and (45) that

$$\bar{k}_- = \frac{c_{1,0} - \xi_c}{(c_{1,0} + c_{2,0})\tau_m} \quad \text{and} \quad \bar{k}_+ = \frac{1}{\tau_m} - \bar{k}_- \tag{56}$$

Finally, some features concerning the Bode diagrams shown in Figs. 17B and C should be mentioned. Plotting the relative extent of reaction $|\xi_m(\omega)|/|\xi_m(0)|$ as calculated from Eq. (48) versus frequency, it follows that the frequency fulfilling the condition $\omega\tau_m = 1$ has a special meaning, we call it $\omega_0$ or $f_0 = 2\pi\omega_0$. According to Eq. (48), at this frequency the amplitude has decreased by a factor of $1/\sqrt{2}$, which corresponds to $-3$ decibels (dB). At elevated frequency, the amplitude declines with a slope of 6 dB per octave, i.e., by a factor of 2 upon frequency doubling. Moreover, the phase lag of species 2 (Eq. (33)) as calculated by Eq. (52) is shown in Fig. 17C. In order to understand the traces and the scaling of coordinate axes of Fig. 17, we have to look at the input data used for calculation of this example. The initial equilibrium concentrations at $T = T_0 = 298$ K were $c_{1,0} = 0.231$ mol/L and $c_{2,0} = 0.769$ mol/L. Temperature displacement and modulation amplitudes were equal, $\Delta T_0 = \Delta T_1 = 5\ K$, and the phase shift of temperature stimulation according to Eq. (1) was $\theta = -\pi$. The following activation energies have been used for calculation: $E_{a+} = 35$ kJ/mol and $E_{a-} = 55$ kJ/mol. A relevant quantity for amplitude and phase determination is $(c_{1,0}\Delta k_+ - c_{2,0}\Delta k_-) = -0.0624$, i.e. sign $(c_{1,0}\Delta k_+ - c_{2,0}\Delta k_-) = -1$ as calculated by Eq. (38), using the data indicated earlier. Now considering the product $A_2$ of reaction Eq. (33) it follows that sign$(v_2) = +1$. Insertion of these data into the phase equations (52) results in the signs of the trigonometric functions; i.e. sign(cos $\varphi_2$) = $-1$, sign(sin $\varphi_2$) = $+1$, and sign(tan $\varphi_2$) = $-1$. Taking two of these signs and consulting Table 2 results in the second quadrant to be relevant for $\varphi_2$. As follows now from Eq. (52), the phase of species 2 alters from 180° at $\omega = 0$ to 90° at $\omega = \infty$, which is shown by Fig. 17C. The absolute phase lag at the 3-dB point, where $\omega\tau_m = 1$, results in $\varphi_2 = 3/4\pi = 135°$. The phase behavior of the reactant $A_1$ differs from the product $A_2$ only insofar that sign$(v_1) = -1$. Consideration of Eq. (52) and Table 2 assigns the forth quadrant to the phase $\varphi_1$, i.e., from $3/2\pi$ to $2\pi$. It should be noted that multiples of $2\pi$ may be added to $\varphi$ without changing the absolute character.

Introducing $\varphi_1 = 7/4\pi$ and $\varphi_2 = 3/4\pi$ into Eq. (51) and setting $t = 0$

demonstrates that the response of the system is ahead of the stimulation. Although mathematically correct because of the $2\pi$ periodicity of the process, it is usual in practice to subtract $2\pi$ from the phases determined by the procedure just described. Consequently, we indicate the absolute phase lags of the species of reaction Eq. (33) as $\varphi_1 = -\pi/4$ and $\varphi_2 = -5/4\pi$.

## 7  CONCLUSIONS

Modulation spectroscopy, or modulated excitation (ME) spectroscopy, has been presented as an experimental tool for the very sensitive detection of periodic concentration changes in a system that has been stimulated by a periodic modulation of an external parameter. The high signal-to-noise ratio typically achieved with ME techniques is based on the narrowband characteristic of phase-sensitive detection (PSD). If a large and noisy signal contains weak amounts of a periodic component with distinct frequency, e.g., the frequency of excitation, PSD selectively demodulates this modulated part and converts it into a direct current (DC) output. On the other hand, the dominant rest of the signal will be canceled, because PSD converts it to an output with mean value zero.

Since demodulation occurs within each period, there is minimal influence of apparative and environmental instabilities.

There is a close analogy between ME spectroscopy and relaxation spectroscopy. The latter is initiated by a jumplike change of an external parameter. This leads to a relaxation of the equilibrium concentrations before jump to the new equilibrium concentrations after jump. In ME spectroscopy, on the other hand, relaxation from the initial equilibrium state to a stationary state is observed. This state is characterized by a periodic concentration modulation about the stationary mean values. Furthermore, if the modulation frequency is matched to system kinetics, which is the case if the modulation frequency is close to the reciprocal relaxation time, significant phase shifts between stimulation and system response will be observed. This phenomenon is paralleled by increased amplitude damping with increasing frequency. For a given reaction scheme, both phase shift and amplitude damping depend systematically on the relaxation times of the system.

Obviously, both methods, relaxation and modulated excitation, lead to system relaxation times, which can be related to basic kinetic constants. Two significant differences should be noted, however, because we think that they mean an advantage of the ME technique. The first one is frequency doubling by nonlinear systems. If a sinusoidal ME is exerted under adequate conditions to a chemical system, producing a response in the first harmonic, one can unambiguously conclude that non-first-order kinetics are involved in the stimulated process. A nonlinear system response may result from 2nd order kinetics or more complex reactions like cooperative phenomena encountered with phase transitions or protein folding. Of course, a non-linear system response is also expressed by the shape of a relaxation curve. The difference to ME techniques is, however, that PSD enables a selective detection of higher harmonics without application of any fitting model. Validation of a given model is then performed by variation of the stimulation frequency, since both phase shifts and amplitudes of the concentrations of reacting species depend critically on the underlying reaction scheme. In practice, procedures developed for electronic circuit analysis, such as Bode and Nyquist diagrams (46), may be applied.

Finally, it should be noted that the advantage of narrow-band measurements just mentioned must be paid for by extended data acquisition time.

## REFERENCES

1. HsH Günthard. Ber Bunsenges Phys Chem 78:1110–1115, 1974.
2. M Eigen, L De Maeyer. In: SL Friess, ES Lewis, A Weissberger, eds. Techniques of Organic Chemistry. Vol. 8. Part 2. New York: Wiley-Interscience, 1963, pp. 895–1054.
3. H Strehlow, W Knoche. Fundamentals of Chemical Relaxation. Weinheim, Germany: Verlag Chemie, 1977.
4. I Noda. Appl Spectrosc 44:550–561, 1990.
5. UP Fringeli. In: Encyclopedia of Spectroscopy and Spectrometry. New York: Academic Press, 1999, pp. 58–75.
6. TL Hill. Thermodynamics for Chemists and Biologists. Reading, MA: Addison-Wesley, 1968.
7. J Monod, J Wyman, JP Changeux. J Mol Biol 12:88–118, 1965.
8. AA Frost, RG Pearson. Kinetics and Mechanism. 2nd ed. New York: Wiley, 1961, pp. 77–102.
9. M Forster, K Loth, M Andrist, UP Fringeli, HsH Günthard. Chem Phys 17:59–80, 1976.
10. O Kratky. Kolloid Z 64:213–222, 1933.
11. C Marcott, AE Dowrey, I Noda. Anal Chem 66:1065A–1075A, 1994.
12. C Marcott, GM Story, I Noda, A Bibby, CJ Manning. Pressure-modulation dynamic attenuated-total-reflectance (ATR) FT-IR spectroscopy. In: JA deHaseth, ed. Proceedings of the 11th International Conference on Fourier Transform Spectroscopy. Woodbury, NY: AIP Conference Proceedings 430, 1998, pp. 379–380.
13. D Blaudez, JM Turlet, D Dufourcq, D Bard, T Buffeteaux, B Desbat. J Chem Soc Faraday Trans 92:525–530, 1996.
14. D Blaudez, T Buffeteaux, JC Cornut, B Desbat, N Escafre, M Pezolet, JM Turlet. Thin Solid Films 242:146–150, 1994.
15. A Elliot, EJ Ambrose. Nature 165:921–922, 1950.
16. H Lenormant, ER Blout. Nature 172:770–771, 1953.
17. F Dousseau, M Pézolet. Biochemistry 29:8771–8779, 1990.
18. M Müller, R Buchet, UP Fringeli. J Phys Chem 100:10810–10825, 1996.
19. M Maroncelli, SP Qi, HL Strauss, RG Snyder. J Am Chem Soc 104:6237–6247, 1982.
20. H Casal, RN McElhaney. Biochemistry 29:5423–5427, 1990.
21. UP Fringeli, J Goette, G Reiter, M Siam, D Baurecht. Structural investigations of oriented membrane assemblies by FTIR-ATR spectroscopy. In JA deHaseth, ed. Proceedings of the 11th International Conference on Fourier Transform Spectroscopy. Woodbury, NY: AIP Conference Proceedings 430, 1998, pp. 729–747.

22. OPTISPEC, Rigistrasse 5, CH-8173 Neerach, Switzerland.

23. UP Fringeli. In: FM Mirabella Jr, ed. Internal Reflection Spectroscopy, Theory and Application. New York: Marcel Dekker, 1992, pp. 255–324.

24. P Hofer, UP Fringeli, WH Hopff. Biochemistry 23:2730–2734, 1984.

25. UP Fringeli, P Ahlström, C Vincenz, M Fringeli. Structure-activity relationships in enzymes: an application of IR-ATR modulation spectroscopy. Fourier and Computerized Infrared Spectroscopy, Montreal. Bellingham, WA: SPIE 553, 1985, pp. 234–235.

26. P Hofer, UP Fringeli. Biophys Struct Mech 6:67–80, 1979.

27. NB Colthup, LH Daly, SE Wiberley. Introduction to Infrared and Raman Spectroscopy. 3rd ed. New York: Academic Press, 1990.

28. F Kopp, UP Fringeli, K Mühlethaler, HsH Günthard. Z Naturforsch 30c:711–717, 1975.

29. W Münch, UP Fringeli, HsH Günthard. Spectrochim Acta 33A:95–109, 1977.

30. R Zbinden. Infrared Spectroscopy of High Polymers. New York: Academic Press, 1964.

31. G Cevc, D Marsh. Phospholipid Bilayers: Physical Principles and Models. New York: Wiley Interscience, 1987.

32. H Gerischer. In: P Delahay, ed. Advances in Electrochemistry and Electrochemical Engineering. New York: Wiley-Interscience, 1961, pp. 142–232.

33. P Delahay. Double Layer and Electrode Kinetics. New York: Wiley, 1965.

34. M Schwarzott, D Baurecht, UP Fringeli. Structural changes of PLL by electric field stimulation: an FTIR ATR modulation spectroscopic study. Structural Heterogeneity and Dynamics of Biological Macromolecules, Joint Meeting of the Dutch and German Biophysical Societies and the Biochemistry and Molecular Biology Society, May 13–16, 1999, Hünfeld, Germany, 1999, p. 88.

35. I Noda, AE Dowrey, C Marcott. Appl Spectrosc 42:203–216, 1988.

36. RA Crocombe, SV Compton. The Design, Performance and Application of a Dynamically Aligned Step-Scan Interferometer. FTS/IR Notes 82, Bio-Rad, Digilab Divison, Cambridge, MA, 1991.

37. RA Palmer. Spectroscopy 8:26–36, 1993.

38. RA Palmer, JL Chao, RM Dittmar, VG Gregoriou, SE Plunkett. Appl Spectrosc 47: 1297–1310, 1993.

39. CJ Manning, PR Griffiths. Appl Spectrosc 47:1345–1349, 1993.

40. R Curbelo. Digital signal processing (DSP) applications in FT-IR. Implementation examples for rapid- and step-scan systems. In JA deHaseth, ed. Proceedings of the 11th International Conference on Fourier Transform Spectroscopy. Woodbury, NY: AIP Conference Proceedings 430, 1998, pp. 74–83.

41. UP Fringeli. Int Pat Publ, PCT, WO 97/08598, 1997.

42. D Baurecht, W Neuhäuser, UP Fringeli. Modification of time-resolved step-scan and rapid-scan FTIR spectroscopy for modulation-spectroscopy in the frequency range from Hz to kHz. In: JA deHaseth, ed. Proceedings of the 11th International Conference on Fourier Transform Spectroscopy. Woodbury, NY: AIP Conference Proceedings 430, 1998, pp. 367–370.

43. CRC Standard Mathematical Tables. 18th ed. Cleveland, OH: Chemical Rubber Co, 1970.

44. I Mills, T Cvitas, K Homann, N Kallay, K Kuchitsu. Quantities, Units and Symbols in Physical Chemistry. Oxford: Blackwell Scientific, 1988, p. 38.

45. DN Hague. Fast Reactions. New York: Wiley-Intersciences, 1971.

46. HW Bode. Network Analysis. London: Van Nostrand Company, 1955.

# 6

# Time-Resolved FT-IR Difference Spectroscopy: A Tool to Monitor Molecular Reaction Mechanisms of Proteins

**Klaus Gerwert**

*Institute of Biophysics, Ruhr University Bochum, Bochum, Germany*

## 1 INTRODUCTION

A challenge in the life sciences today is understanding processes in living organisms at the molecular level. A deeper insight into these processes enables progress in medicine and biotechnology. The two essential players at the molecular levels are DNA and proteins (1). DNA stores the genetic information. Proteins are the nanomachines that perform the biochemical processes of living organisms. Understanding biological processes at the molecular level therefore means understanding the function of proteins at the atomic level. An essential step in the elucidation of the functioning of a protein is the determination of its three-dimensional structure. Structural models of proteins are provided today almost routinely by x-ray structure analysis and for smaller water-soluble proteins by NMR spectroscopy (2,3). However, both techniques usually deliver structural models of the quiescent protein ground state. Understanding how proteins function needs in addition time-resolving techniques to determine transient occupied states with atomic resolution. Spectroscopic techniques fulfill both conditions (4). Infrared spectroscopy especially is capable of monitoring absorption bands of single molecular groups of a protein and their absorbance changes down to femtosecond time resolution. Time-resolved Fourier transform infrared (FT-IR) difference spectroscopy has proved itself as a powerful method for studies of molecular reaction mechanisms of proteins larger than 100,000 daltons and time resolutions up to nanoseconds (5). Using monochromatic IR spectroscopy, the time resolution is even pushed into the picosecond time range (6). However, dispersive techniques provide absorbance changes only at selected single wavenumbers, but not complete spectra as does the polychromatic FT-IR technique. In order to assign bands, which is important for interpretation, complete spectra are needed. In addition, pulsed IR sources in combination with the FT-IR technique will allow monitoring even in the picosecond time range in the future.

In this chapter various time-resolved FT-IR and FT-Raman techniques are explained, and then the application of these techniques to the light-driven proton pump bacteriorhodopsin, the redox-driven proton pump cytochrome-$c$-oxidase, and the photosynthetic reaction center will be described. Finally, the use of photolabile trigger compounds to initiate GTPase activity of H-$ras$ p21 will be presented.

This chapter does not represent a complete review in which all relevant publications are cited; only instructive examples are referenced. For a more detailed literature overview, the cited reviews and original publications are recommended.

## 2   METHODS

### 2.1   Difference Spectroscopy

The infrared spectrum of a protein is dominated by its peptide backbone amide I ($C=O$) and amide II (mostly NH) vibrations. In addition, water, which is indispensable for protein function, is also a strong absorber in the mid-infrared. In Fig. 1, upper part, a typical spectrum of a hydrated protein is shown. It is taken from the membrane protein bacteriorhodopsin (more details will be given later). The band positions at 1650 cm$^{-1}$ (amide I) and 1550 cm$^{-1}$ (amide II) indicate the predominant $\alpha$-helical structure (7). The water absorption causes the shoulder at the 1650 cm$^{-1}$ band. The protein backbone and water absorption overrule the much smaller absorption bands of single residues. In order to select these small absorption bands of functionally active single residues from the large background absorption of the entire protein, difference spectra are performed between different active protein states. The difference provides absorption bands only of groups undergoing reactions and cancels the quiescent background. The absorbance changes that are to be determined are on the order of 10$^{-3}$ beyond a background absorbance of up to 1 (Fig. 1). To monitor such small changes, highly sensitive instrumentation is required. Fourier transform infrared (FT-IR) spectroscopy is able to reliably detect such small changes due to the multiplex and the Jacquinot advantages (8–10). The multiplex advantage allows shorter measuring times in comparison with dispersive instruments. The Jacquinot advantage reflects the higher light throughput of interferometers as compared to dispersive instruments. This increases the signal-to-noise ratio (SNR) remarkably.

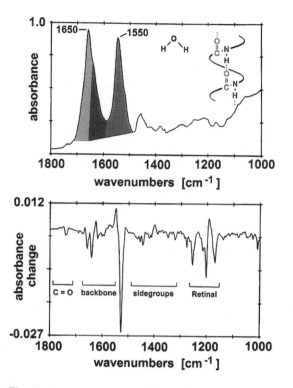

**Fig. 1** Upper part: a typical absorption spectrum of a hydrated protein film (bacteriorhodopsin, BR). Lower part: a typical difference spectrum (BR-L of BR) is shown.

## 2.2 Principles of FT-IR

Here, only a very brief introduction is given. For more details, Refs. 8–10 are recommended.

In general, an FT-IR instrument consists of a Michelson interferometric arrangement (see Fig. 2) with a light source G (globar), a beam splitter BS, a fixed mirror (FM), a movable mirror (MM), and a mercury cadmium telluride detector (MCT). The IR beam is split by the beam splitter and reflected by the fixed and movable mirrors. The beams are recombined and the intensity is measured on the detector. Due to the movement $X$ of the movable mirror, the split beams travel different distances and therefore have, after recombination, different phase delays. This causes an interference pattern called an *interferogram*. This effect is more easy to understand for monochromatic light: if instead of a polychromatic source (as the globar) a monochromatic source (such as a HeNe laser) is used, the movement $X$ produces a cosinewave interferogram. The Fourier transformation of the cosinewave provides a single peak at the frequency of the HeNe

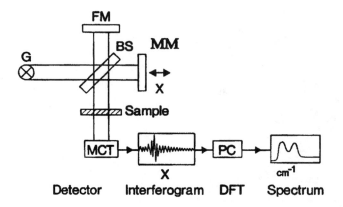

**Fig. 2** Schematic diagram of an FT-IR instrument.

laser in the spectrum. In case of a polychromatic source, all single cosine contributions overlap in an interferogram. Nowadays, the Fourier transformation is performed on a conventional PC in less than one second to yield the absorption spectrum.

In contrast to dispersive instruments, the FT-IR technique drastically reduces the measuring time due to the multiplex advantage, since all spectral elements are measured parallel instead of consecutive. The advantage increases with the width of the measured spectrum.

The second advantage of FT-IR instruments is the Jacquinot advantage, which increases the light throughput as compared to dispersive instruments, because no entrance and exit slits need be used. This significantly improves the performance, because a linear increase in the light intensity increases as the square root of the SNR.

The length $X$ of the interferogram determines the spectral resolution $\Delta v$:

$$\Delta v \propto \frac{1}{X}$$

The sampling points $x_i$ at which the interferogram intensity is measured are determined by the zero crossings of the cosine caused by a monochromatic HeNe laser beam, which is transferred colinearly with the IR beam. The resulting high wavenumber accuracy constitutes the third advantage of FT-IR.

Artifacts in the FT-IR spectrum can arise primarily because digital computers perform a discrete rather than a continuous Fourier transformation of the interferogram $I(x)$, an approximation that requires care to avoid errors. For a detailed description, Refs. 8–10 are recommended. Here, only a brief overview is given.

The discrete interferogram is Fourier transformed to yield the spectrum $S$ with a discrete algorithm (DFT). As a result of DFT, the continuous variables, i.e., the scan length $X$ and the frequency $v$, become the discrete variables $n \cdot \Delta x$ and $k \cdot \Delta v$:

$$S(k \cdot \Delta v) = \sum_{n=0}^{N-1} I(n \cdot \Delta x) e^{i2\pi n \cdot k/N}$$

In practice, one uses a less redundant fast Fourier transform algorithm, e.g., the Cooley–Tukey algorithm, rather than the expression shown. This discrete instead of continuous Fourier transformation over a limited instead of an infinite pathlength requires care to avoid errors such as the picket fence effect, aliasing, or leakage: the truncation of the interferogram at maximum optical pathlength results in the convolution of the true interferogram with a boxcar function, which causes "leakage" of the band intensities into side lobes. Leakage can be avoided by employing the appropriate apodization function instead of the boxcar function. This seems to be a disadvantage; but with dispersive instruments, one has to deal with the instrumental line shape, while FT instruments offer a free choice of the apodization function and thus allow optimizing the apodization for a specific application. For our purposes, the Blackman–Harris apodization function appeared to be most useful.

In contrast to the theoretical expectation, the measured interferogram $I(x)$ is typically not symmetric about the center burst ($x_0 = 0$), and the maximum peak at $I(x_0)$ is not at $x = 0$. This is a consequence of experimental errors, e.g., frequency-dependent optical and electronic phase delays. In order to compensate for these phase errors, a short part with length PIP (phase interferogram points) of the interferogram is measured double-sided (PIP/2 points on either side of the center burst). Since the phase is a weak function of the wavenumber, one can easily interpolate the low-resolution phase function and use the result later for phase correction. If there is considerable background absorption, phase errors may falsify the intensities of bands in the difference spectra. To avoid such phase errors for difference spectroscopy, the background maximum absorbance should therefore be less than 1.

## 2.3  Sample Preparation

### 2.3.1  Transmission Cells

Due to the considerable absorption of water in the mid-infrared spectral region, meaningful spectra of hydrated proteins are obtained by transmission measurements only through very thin (2–10 μm) films. This involves placing a drop of a protein suspension or solution on an IR window and then carefully concentrating it under a nitrogen stream or under vacuum. Alternatively, a protein suspen-

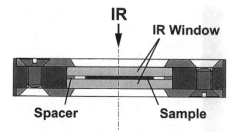

**Fig. 3** Diagram of the sample chamber for transmission measurements.

sion of a membrane-bound protein is centrifuged and the pellet squeezed between two IR windows. A typical measurement requires about 100–150 μg of protein. The initial concentration is about 10 mg/ml. The final concentration is in the millimolar range. The sample chamber is closed by a second IR window, which is separated from the basic IR window by a thin spacer of several micrometers. Figure 3 shows an outline of the IR sample chamber. In the case of bacteriorhodopsin, it was explicitly shown that such carefully hydrated thin protein films afford the same rate constants as the suspension (11). Sufficient hydration is crucial for biological samples.

### 2.3.2 Attenuated Total Reflectance Technique

The attenuated total reflectance (ATR) technique can be used to measure solutions. This ATR spectroscopy utilizes the phenomenon of total internal reflection (12). A beam of radiation entering an ATR crystal will undergo total internal reflection when the angle of incidence at the interface between the sample and the crystal surface is greater than the critical angle. However, the beam penetrates a fraction of a wavelength beyond the reflecting surface. The resultant attenuated radiation reflects the absorption characteristics of the fluid sample on the ATR crystal. In principle, this technique allows solution measurements, because the beam enters the solution on the ATR crystal only few micrometers.

On the other hand, if transmission cells can be used they are clearly superior to ATR cells, because the signal-to-noise ratio is much better.

## 2.4 Time-Resolved FT-IR Techniques

This section describes the state of the art of fast time-resolved FT-IR techniques, together with the advantages and limitations of each method.

### 2.4.1 Trigger Techniques

The difference technique requires a clear-cut initiation of the protein reaction. Furthermore, the sample cannot be moved out of the apparatus, because such a change induces larger absorbance changes than those induced by the protein activity.

### 2.4.1.1 Photobiological Systems

Ideally suitable are light-induced reactions in photobiological systems like bacteriorhodopsin and the photosynthetic reaction centers that carry intrinsic chromophores. In these systems, the chromophore can be directly activated by light, which induces isomerization or redox reactions of the prosthetic groups.

### 2.4.1.2 Caged Compounds

A much broader applicability can be achieved by the use of caged compounds. In this case, biologically active molecules are released from inactive photolabile precursors. They allow the initiation of a protein reaction with a nanosecond UV laser flash. Caged phosphate, caged GTP, caged ATP, and caged calcium especially have been established as suitable trigger compounds (13).

2.4.1.2.1. CAGED PHOSPHATE. This is given as an example for the caged compounds. The 1-(2-nitrophenyl)ethyl moiety is used to protect phosphate, nucleotides, and nucleotide analogs (see Fig. 4). The application of UV flashes leads to the release of the desired phosphate compound. The mechanism of photolysis of compounds containing the 2-nitrobenzyl group was the topic of several investigations (14).

The scheme in Fig. 4 shows the generally accepted reaction pathway (15). After photolysis, the caged compound decays to orthophosphate and the by-product 2-nitrosoacetophenone. Typical FT-IR difference spectra of this photolysis reaction are shown in Fig. 5 (16). At first, a spectrum is measured as reference before the photolysis. After photolysis of the caged phosphate, further spectra are taken and the difference spectra are calculated.

Only those vibrational modes that undergo reaction-induced absorbance changes cause bands in the difference spectra. Negative bands in the difference spectrum are due to the caged phosphate, whereas positive bands are due to the photolysis products, orthophosphate and 2-nitrosoacetophenone (Fig. 4). Characteristic bands of 1-(2-nitrophenyl)ethyl derivatives are the disappearing antisymmetric and the symmetric $NO_2$ stretching vibrations [$\nu_{as}(NO_2)$ = 1525 cm$^{-1}$;

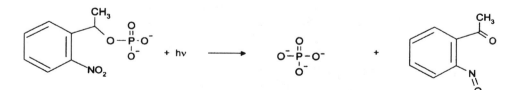

**Fig. 4** Reaction scheme for the photolysis of caged phosphate.

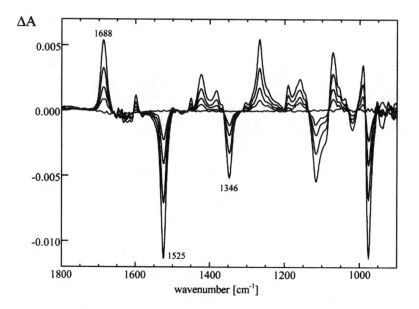

**Fig. 5**  Photolysis difference spectra of caged phosphate: 50 mM caged phosphate, 0.1 M Mops, pH 7.0. The baseline shows as a control the difference between two spectra measured before the flash is applied. Then difference spectra taken after the first, second, fourth, and ninth flash was applied are shown.

$v_{sy}(NO_2) = 1346$ cm$^{-1}$]; and the positive band is due to the carbonyl group of 2-nitrosoacetophenone 5 [$v(C=O) = 1688$ cm$^{-1}$].

Important compounds for the investigation of biological systems are caged nucleotides such as caged GTP and ATP. The photolysis reaction of caged GTP has been investigated in detail by FT-IR spectroscopy. Bands were assigned by the use of $^{18}O$ phosphate labeling by Cepus et al. (16). In addition, in Ref. 16 FT-Raman spectra and photolysis spectra of caged Ca$^{2+}$ are presented.

### 2.4.2  Static FT-IR

In the static FT-IR mode, which is the simplest technique to realize, data acquisition takes several seconds. At such low time resolution, it is necessary to stabilize the transient intermediates for few minutes, for instance, by cooling, by pH variation, or by photostationary accumulation. As an example, the so-called BR-L difference spectrum of bacteriorhodopsin is shown in the lower part of Fig. 1. The L intermediate of bacteriorhodopsin decays at room temperature in about 50 μsec. However, its decay can be slowed down to hours by cooling the sample to 170 K. In the first step, the sample is cooled down in the dark, and a spectrum of the ground state is taken. Then for several minutes the sample is illuminated within the FT-IR apparatus by a halogen lamp equipped with an interference filter. Finally, the spectrum of the light-activated state is taken.

### 2.4.3 Band Assignments

A huge number of bands appears in the difference spectra in Fig. 1. The frequency range in which a band appears allows a rough tentative assignment of the band. For example, the retinal vibrations are expected in the fingerprint region between 1300 cm$^{-1}$ and 1100 cm$^{-1}$, the carbonyl vibrations of aspartic or glutamic acids between 1700 cm$^{-1}$ and 1770 cm$^{-1}$. In order to draw conclusions from the spectra, the bands have to be assigned to individual groups. A clear-cut band assignment can be performed by using isotopic-labeled proteins or by amino acid exchange via site-directed mutagenesis.

Isotopic labeling shifts the frequency of the labeled group due to the increased reduced mass $\mu$:

$$ \nu = \frac{1}{2\pi} \sqrt{\frac{k}{\mu}} $$

Isotopic labeling can be performed on prosthetic groups such as retinal (17) or nucleotides such as GTP (16,18). These compounds can be chemically synthesized. As an example for the site-specific isotopic-labeled prosthetic groups, caged GTP is presented later.

Isotopic labeling of all amino acids of one kind can be achieved by biosynthetic incorporation of isotopic-labeled amino acids, e.g., aspartic acid (19). This is relatively easy to perform. On the other hand, site-directed exchange of an amino acid by mutagenesis eliminates the absorption band of the exchanged group. This is schematically illustrated in Fig. 6.

As an example of an assignment based on site-directed mutagenesis, the BR-L difference spectra of the wild type and the Asp-96-Asn mutants are shown in Fig. 7 (20). Absorbance changes in the spectral range between 1500 cm$^{-1}$ and 1000 cm$^{-1}$ are highly reproducible. This indicates that this specific mutagenesis is noninvasive and does not disturb the protein structure. Only the carbonyl band shift from 1742 cm$^{-1}$ to 1748 cm$^{-1}$ is absent in the spectrum of the mutant, as shown on an enlarged scale in the inset. Thus, the missing difference band is caused by the 4-carbonyl vibration of the exchanged Asp 96 (20). The carbonyl band appearing around 1700 cm$^{-1}$ is caused by the new Asn 96.

Mutation of an amino acid changes the structure of the protein more or less, but it is easy to achieve by site-directed mutagenesis, a standard molecular biological method (1). On the other hand, isotopic labeling has the advantage of marking the molecular group noninvasively.

However, the most suitable method, site-directed isotopic labeling of an

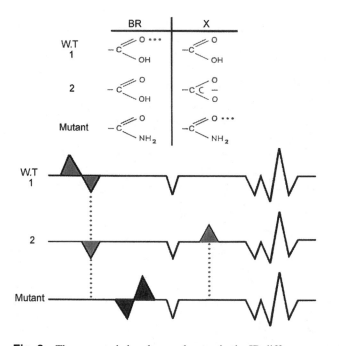

**Fig. 6** The expected absorbance changes in the IR difference spectra are presented schematically. If in the transition of BR to intermediate X (case 1), a H-bond change of a
protonated carboxylic acid in wild type (WT) takes place, a difference band in the difference spectra (lower part 1) is expected due to the frequency upshift of the carbonyl vibration. For a deprotonation (case 2), a negative carbonyl band should disappear and a carboxylate band appear (lower part 2). If the amino acid causing the absorbance changes is
exchanged and the mutant protein is measured, the carbonyl band (in cases 1 and 2) and,
in case 2, the carboxylate band should disappear (marked by dots) as compared to the
WT difference spectrum. In addition, in the mutant a new carbonyl band might appear
for an Asp (Glu) to Asn (Gln) mutation. It is important to notice that all other bands
remain the same (as shown), indicating that most of the structure is not changed and that
the mutation is noninvasive.

amino acid, demands not only the exchange of one amino acid against another
natural amino acid, but also the substitution by an isotopic-labeled amino acid.
This is an expensive molecular biological method and only very rarely is successfully applied (21,22). Today, chemical synthesis of a protein allows site-directed
isotopic labeling and may be the method of the future.

### 2.4.4 Rapid Scan

The principle of the rapid-scan mode is simple: after taking a reference spectrum
of the protein in its ground state, one activates the protein (e.g., by a laser flash)

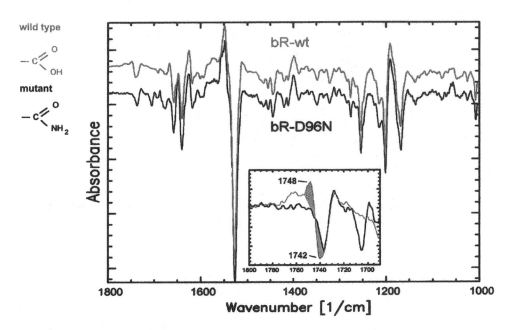

**Fig. 7** BR-L difference spectra of (a) wild type and (b) Asp-96-Asn. Inset: expanded carbonyl region.

and obtains interferograms in much shorter times than the half-lives of the reactions (11,23). The time pattern of such an experiment (Fig. 8) shows in the upper trace the "Take Data" signal. Its "high" state indicates interferogram recording. The lower trace shows schematically a reaction pathway. The first two reference (R) interferograms reflect the ground state (A), and the following three interferograms are taken during the reaction pathway (B, C, and D).

The velocity of the scanner $V_{max}$ and the desired spectral resolution $\Delta v$ determines the scan duration $\Delta t$ and thereby the time resolution:

$$\Delta t = \frac{1}{2V_{max}} \cdot \frac{1}{\Delta v}$$

Today, state-of-the-art spectrometers typically yield a time resolution of 10 msec at 4-cm$^{-1}$ spectral resolution. The Fourier transform technique allows the resolution of processes whose half-lives are on the order of the scan time for first-order reactions or even below (24). If the half-life of the observed process is shorter than the duration of the scan, the intensity of the interferogram is convoluted by the absorption change of the sample. In the case of first-order reactions, the interferogram is convoluted with exponential functions, resulting after Fourier transformation in Lorentzian line shape broadening only.

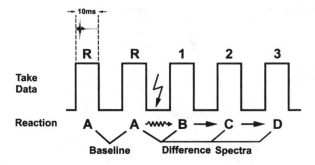

**Fig. 8** Diagram of the rapid-scan method. Two interferograms are taken as reference, the reaction is initiated, and interferograms are recorded during the reaction from A to D. Then the difference spectra are calculated. The first difference spectrum is a control for the baseline.

Significant improvements in state-of-the-art spectrometers yielding increased scanner velocity are unlikely to occur. This has practical reasons, e.g., the extreme acceleration of the scanner at the turning points. Without a radical change of design, the time resolution for the rapid scan is limited to the millisecond range. A typical experimental setup for rapid-scan measurements is described in Fig. 9. The sample is activated by a laser flash. In parallel, the absorbance change in the visible is measured by a conventional photolysis setup.

### 2.4.5  Stroboscope Technique

Merely because of software improvements, the stroboscope method afforded a time resolution of a few microseconds even if the rapid-scan technique is used to collect the interferograms (25). The idea is this: If it takes a time period $T$ to monitor a complete interferogram with the length $X$, it requires only a time period $T/N$ to record a segment $X/N$ of the interferogram. In successive experiments, all $N$ segments of the interferograms are recorded. The software sorts and combines these segments to a new interferogram with an increased time resolution of $T/N$. A time resolution of 20 μsec is realized with the stroboscope technique (25). Typical measuring times are 10–15 hours. Because the Jacquinot advantage is preserved, this instrumentation still has improved SNR as compared to dispersive instruments. Nowadays, the step-scan technique is quite well established, so it seems more useful to employ this technique for microsecond time resolution.

### 2.4.6  Step Scan

In the step-scan mode, the scanner (the movable mirror MM) is kept constant at the interferogram position $x_n$ (Fig. 10), the protein activity is initiated, for exam-

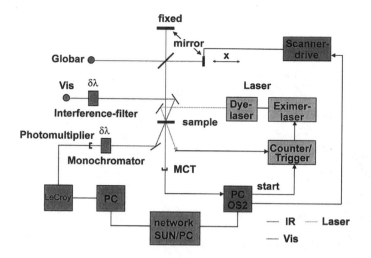

**Fig. 9** Typical experimental setup, consisting of a FT-IR instrument with globar, beam-splitter, mirrors, detector (MCT); a photolysis setup with light source, interference filters, monochromator, photomultiplier, and transient recorder; and an excimer pumped-dye laser system to activate the sample. The Vis and IR data are transferred to a workstation (SUN) network for kinetic analysis.

ple, by a laser flash, and the time dependence of the intensity change at this interferogram position $x_n$ is measured. Then the scanner "steps" to the next interferogram position, $x_{n+1}$, and the reaction is measured again. This is repeated at each sampling position of the interferogram. The position of the scanner must be kept accurately down to about 1–2 nanometers at $x_n$ while the intensity change of the interferogram during the reaction is measured. Therefore, the method is very sensitive to external disturbances (e.g., noise!). The time resolution is usually determined by the response time of the detector, which is about a few nanoseconds. After the measurement the data are rearranged to yield time-dependent interferograms $I(t_i)$. Using pulsed IR sources instead of the globar, the time resolution is determined by the time duration of the probe pulse. This can give in principle femtosecond resolution with broadband femtosecond IR lasers.

### 2.4.6.1 Experimental Setups

For more details on the step-scan technique, see Refs. 26–30. A typical experimental setup is described by Rammelsberg et al. (30) and is shown schematically in Fig. 11. Except for the home-built sample chamber, the FT-IR instrument is evacuated to 3 mbar during the measurement. This increases the stability and reduces the sound sensitivity of the movable mirror. Furthermore, the setup is

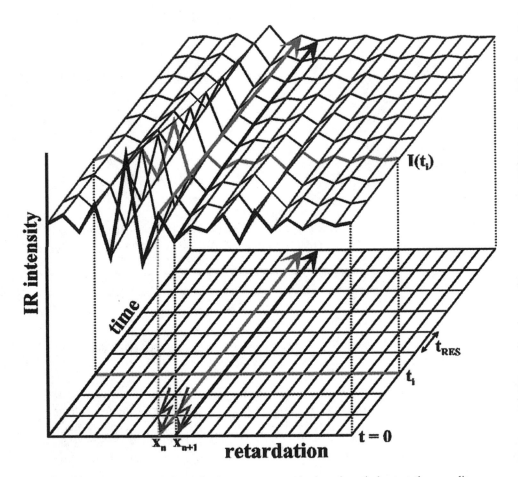

**Fig. 10** Step-scan technique: In the step-scan mode the mirror is kept at the sampling positions $x_n$. Then the reaction is initiated and the time dependence of the IR intensity during protein activity is measured. The time resolution is determined by the detector. After relaxation of the system, the scanner steps to the next position, $x_{n+1}$, and the reaction is again started. After measuring the time dependence of all sampling positions, the computer rearranges the data to yield the time-dependent interferograms $I(t_i)$. Fourier transformation yields the time-dependent spectra.

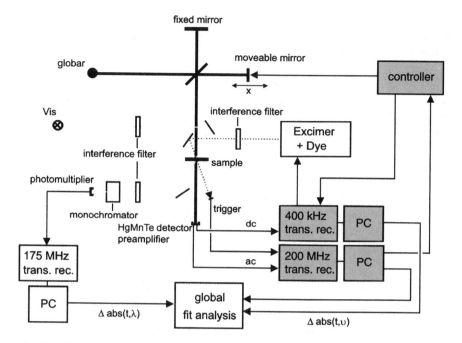

**Fig. 11** Typical experimental setup, consisting of an FT-IR instrument (Bruker IFS66v) with globar, beamsplitter, mirrors, controller, detector, and preamplifier connected to a 400-kHz and a 200-MHz transient recorder; a photolysis setup with light source, interference filters, monochromator, photomultiplier, and transient recorder; and an excimer pumped-dye laser system to activate the sample. The Vis and IR data are transferred to a workstation (SUN)–PC network for kinetic analysis.

positioned on a vibration-isolation table (Newport) in a temperature-controlled laboratory. We have determined the residual spatial fluctuations of the movable mirror to be ±0.5 nm. The sample chamber is purged with dry air (dew point −70°C, Balston 75-62 FT-IR purge gas generator). The IR absorbance changes are detected by a photovoltaic HgMnTe detector (Kollmar Tech., KV100-1-B-7/190, cutoff 850 cm$^{-1}$). The detector's signal is amplified in a home-built two-stage preamplifier with an ac- and a dc-coupled output. The bandwidth of the dc part is limited to 400 kHz, whereas the bandwidth of the ac part is 200 MHz. Controls ensure that the output signal of the preamplifier depends linearly on the IR intensity. The dc-coupled output of the preamplifier is digitized by a 12-bit, 400-kHz transient recorder (Keithley, ADWIN Gold) connected to a PC. The offset of the input signal can be compensated to zero. This allows subsequent amplification of the signal and allows use of the full dynamic range of the transient recorder. In order to prune the huge amount of data from the nanosecond to the millisecond time domain, the 400-kHz acquisition rate is slowed down to 100 kHz at 100 μsec. This allows time averaging. The ac-coupled output of the preamplifier is recorded by a 200-MHz, 8-bit transient recorder (Spectrum, PAD 82).

At every sampling position of the interferogram the correct positioning of the movable mirror is checked before data acquisition starts. A TTL output signal then triggers the excimer laser to initiate the reaction. A fast photodiode (rise time 10 nsec) is activated by a reflex of the dye laser flash and starts data acquisition of the 200-MHz transient recorder.

The spectral range is limited below the Nyquist-wavenumber 1975 cm$^{-1}$ by an interference filter to reduce the number of sampling points of the interferogram. This filter also shields the IR detector from scattered light of the dye laser and the heat emitted by the sample. (The dye laser's pulse causes a small warming of the sample.) The resulting interferogram contains 780 positions. It is multiplied with the Norton–Beer weak apodization function and then zero-filled by a factor of 2. The phase spectrum $\varphi(v)$ is calculated with a spectral resolution of 50 cm$^{-1}$, whereas the difference spectra are taken with a spectral resolution of approximately 3 cm$^{-1}$. Time-resolved measurements proceed as follows (Fig. 12):

1. The scanner moves to the first acquisition point of the interferogram. The dc value of the IR intensity is measured. Afterwards the offset of the dc signal is set to zero.

2. The reaction is started by a laser flash. Both transient recorders simultaneously measure the time-dependent infrared intensity changes at the respective sampling positions. The 200-MHz transient recorder measures the time domain from 30 nsec to 20 μsec, whereas the 400-kHz transient recorder monitors from 2 μsec until the end of the reaction in the millisecond time range. At every sampling position of the interferogram, several photocycles are averaged to improve the signal-to-noise ratio. Typical data are shown later for bacteriorhodopsin in Fig. 18. In this case, 100 photocycles were accumulated at each sampling position. This measurement took about 2 hours. After measuring the time courses of all interferogram sampling positions, the data were rearranged to yield time-dependent difference interferograms $\Delta I(t_i)$. Because these difference interferograms contain positive as well as negative spectral features, the usual Mertz phase correction cannot be directly applied. Therefore the stored phase $\varphi(v)$ from the first measurement in step 1 was used. The phase does not change between both measurements, because the movable mirror stops exactly at the same sampling points and only small absorbance changes take place.

Possible errors due to transient heating of the sample by the actinic laser flash, baseline distortions, and nonlinearity of the IR detectors are discussed in detail by Rammelsberg et al. (30).

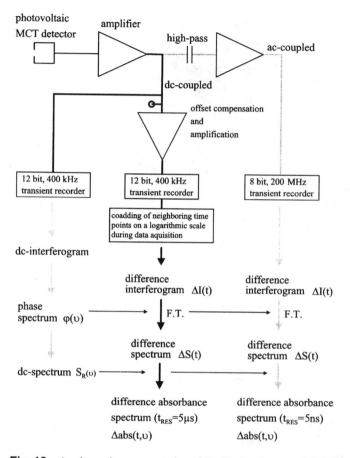

dc-interferogram

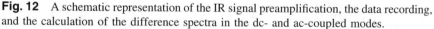

**Fig. 12** A schematic representation of the IR signal preamplification, the data recording, and the calculation of the difference spectra in the dc- and ac-coupled modes.

## 2.4.6.2 Step Scan of Noncyclic Reactions

The step-scan technique cannot be applied directly to noncyclic reactions, because the investigated process has to be repeatedly initiated at about 1000 sampling positions of the interferogram. Consequently, to investigate noncyclic systems, the sample has to be renewed at every sampling position. In the flow cells used so far, the optical pathlength is too large to perform difference spectroscopy of hydrated biological samples. We have to use 4-µm-thin films to depress the water background absorption of biological samples.

In a novel approach, the IR beam and the excitation laser beam are focused to a very small diameter of 200 µm (Fig. 13). Thereby, only a small segment of the sample, which has a diameter of 15 mm in total, is excited and probed. By

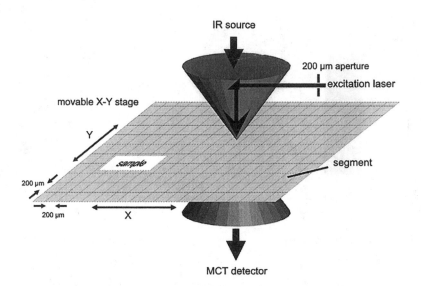

**Fig. 13** The IR beam and the excitation laser beam are focused to a diameter of 200 μm. The sample is mounted on a movable X-Y stage and divided into a few thousand 200-μm × 200-μm segments. The sample is renewed by moving the next segment into the spot of the IR beam and laser beam.

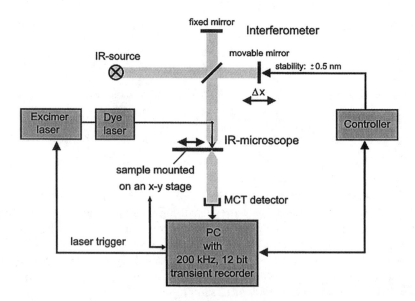

**Fig. 14** Experimental setup with IR source, Michelson interferometer, detector, transient recorder, and controller. The sample is mounted on a movable X-Y stage in the focus of an IR microscope. An excimer pumped-dye laser flash initiates the reaction.

moving the sample, which is mounted on a movable X-Y stage, to different non-excited segments, the reaction can be repeated until a complete interferogram data set has been recorded (Fig. 14).

The technique was successfully applied to a noncyclic reaction, the photolysis of caged ATP (31). The difference spectra are shown in Fig. 15. By this technique the transiently formed aci-nitro anion complex is also measured (compare also to Fig. 5). The successful performance with 20-μsec time resolution

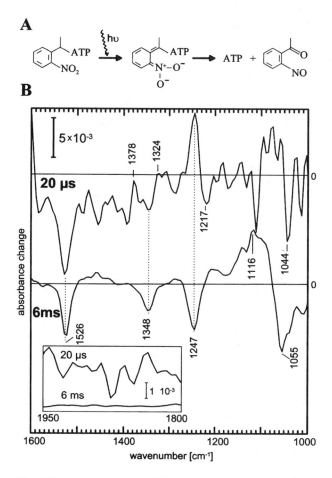

**Fig. 15** Time-resolved difference absorbance spectra of the caged ATP photolysis. The reaction scheme is shown in (A). It is measured with the setup shown in Fig. 14. (B) The 20-μsec spectrum (top) shows the *aci*-nitro intermediate. The 6-msec spectrum (bottom) shows free ATP and by-products. In the inset, the noise of the spectra is compared.

now opens the door to many new applications of step-scan FT-IR measurements to noncyclic reactions.

### 2.4.6.3  Fourier Transform–Raman Spectroscopy

Near-infrared excited FT-Raman spectroscopy has recently begun to show promising results. In contrast to resonance Raman spectroscopy, the fluorescence is drastically reduced (32,33). Furthermore, it has the Jacquinot advantage over resonance Raman spectroscopy, affording a better signal-to-noise ratio. Thus, FT-Raman is an excellent technique to supplement FT-IR difference spectroscopy in investigations of intramolecular protein reactions. The FT-Raman spectra are not affected by water; they may therefore be employed to study protein solutions or suspensions. This method has been used to investigate bacteriorhodopsin as well as bacterial reaction centers (34,35). Once the step-scan technique is established, it can also be used for the FT-Raman technique, because the same Michelson interferometer is used (Fig. 16). The time resolution is in principle limited to the microsecond time range due to the near-IR germanium detector. However, problems arise in triggering the reaction by light because the sample has to be highly concentrated. The actinic light is already absorbed at the surface layers of the sample. In case of bacteriorhodopsin, a time resolution for the step-scan Raman technique of few milliseconds is achieved (36). Fluorescence of the sample inhibited a better time resolution. For solid-state samples, a time resolution of nanoseconds was obtained (37). As an example, in Fig. 17 two typical FT-Raman spectra of bacteriorhodopsin are shown.

### 2.4.7  Kinetic Analysis

For the interpretation of time-resolved data, adequate kinetic analysis is important. A so-called global-fit analysis affords the apparent rate constants of the analyzed processes (38). The global analysis not only includes fitting the absorbance change at a specific wavenumber, but at up to 250 wavenumbers simultaneously. All reactions are assumed to be first order and can therefore be described by a sum of exponentials. The fit procedure minimizes the difference

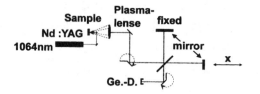

**Fig. 16**  FT-Raman setup. The 1064-nm line of a Nd:YAG laser is scattered on the sample. The Raman line is blocked by the plasma lens. The Raman radiation travels through the interferometer and is detected by a germanium detector.

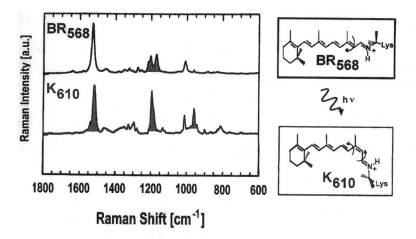

**Fig. 17** FT resonance Raman spectra of the BR ground state and of the K intermediate stabilized at 77 K are shown. The different band pattern in the fingerprint region indicates an all-trans to 13-cis isomerization from BR to K.

between the measured data $\Delta A_{\text{measured}}$ and the theoretical description $\Delta A$, weighted according to the noise $w_{ij}$ at the respective wavenumbers, and summarized not only over time ($t$) but also over the wavenumbers ($i$). In global-fit analysis, the absorbance changes $\Delta A$ in the visible and infrared are analyzed with sums of $n_r$ exponentials with apparent rate constants $k_l$ and amplitudes $a_l$:

$$\Delta A(\nu, t) = \sum_{l=1}^{n_r} a_l(\nu) e^{-k_l t} + a_o(\nu)$$

In this analysis, the weighted sum of squared differences $f$ between the fit with $n_t$ apparent rate constants $k_l$ and data points at $n_w$ measured wavelengths $\nu_i$ and $n_t$ time-points $t_j$ is minimized:

$$f = \sum_{i=1}^{n_w} \sum_{j=1}^{n_t} (w_{ij})^2 \left( \Delta A_{\text{measured}}(\nu_i, t_j) - \sum_{l=1}^{n_r} a_l(\nu_i) e^{-k_l t_j} + a_o(\nu_i) \right)^2$$

For unidirectional forward reactions, the determined apparent rate constants are directly related to the respective intrinsic rate constants describing the respective reaction steps (2,39). However, if in addition significant back-reactions occur, the analysis becomes more complicated. Then the reaction has to be modeled until the guessed intrinsic rate constants fulfill the experimentally observed time course described by the apparent rate constants. Because the number of intrinsic rate constants in the model is larger than the number of experimentally observed apparent rate constants, the problem is experimentally underdetermined and the solution is not unequivocal.

An alternative method is the singular value decomposition or principal component analysis (38). Thereby, the basis difference spectra are calculated from all difference spectra measured. This procedure allows the determination of transient spectra independent of specific kinetic models and independent of the temporal overlap (38).

## 3  APPLICATION TO PROTEINS

### 3.1  Bacteriorhodopsin

Bacteriorhodopsin is a large (27000 daltons) membrane protein, located in the purple membrane of *Halobacterium salinarum* (41). A structural model based on x-ray analysis has been published (42). After light excitation, protons are pumped across the membrane (43). The protein spans the membrane by seven α-helices (see Fig. 18a, b). The chromophore retinal is embedded deeply within the protein, shielded by the helices. Retinal connects to Lys 216 of the protein via a protonated Schiff's base (Fig. 18b).

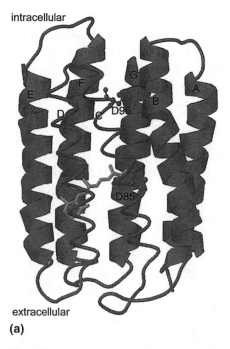

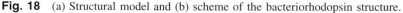

**Fig. 18**   (a) Structural model and (b) scheme of the bacteriorhodopsin structure.

# cytoplasmic

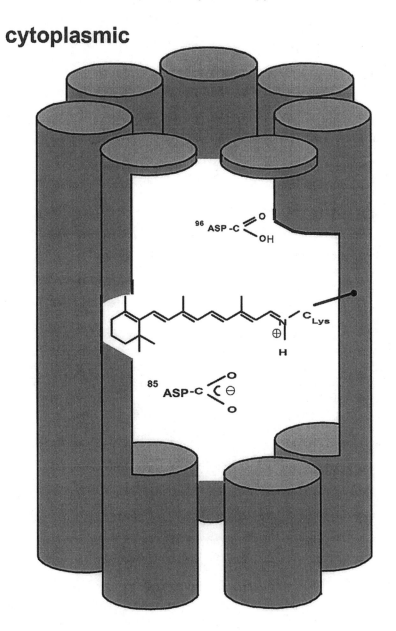

# extracellular

(b)

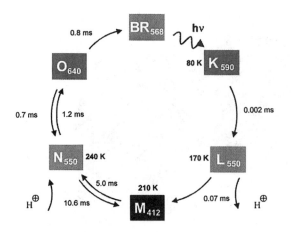

**Fig. 19** Photocycle of bacteriorhodopsin. The different intermediates are characterized by different absorption maxima, as indicated by the subscripts.

Light excitation of ground state $BR_{568}$ (the subscripts refer to the absorption maxima) initiates a photocycle with the intermediates $K_{590}$, $L_{550}$, $M_{412}$, $N_{550}$, and $O_{640}$ (Fig. 19). Upon light excitation, the chromophore retinal undergoes an all-trans to 13-cis isomerization in 450 fsec to J, which decays in 1.5 ps to K (not shown). The Schiff's base is deprotonated during the L-to-M reaction and is reprotonated again in the course of the M-to-N conversion. M is generally believed to be the essential intermediate in the proton pumping mechanism. During the L-to-M reaction, the protein releases a proton to the extracellular side; another proton is taken up from the cytoplasmic side during the M-to-BR reaction (Fig. 19) at pH 7.

The IR absorbance changes during bacteriorhodopsin's photocycle taken in a typical step-scan measurement are shown in Fig. 20 in a three-dimensional graph. The negative disappearing bands represent the BR ground state, and the positive bands correspond to the respective intermediates. In the fingerprint region between 1300 and 1100 $cm^{-1}$, retinal C—C vibrations are observed. They were assigned using isotopic-labeled retinal (44,45). The disappearing bands at 1202 $cm^{-1}$ and 1167 $cm^{-1}$ and the appearance of the C—C stretching vibration band at 1190 $cm^{-1}$ indicates the all-trans to 13-cis isomerization of retinal. The isomerization takes place within 450 fsec and is not time resolved here. The first difference spectrum shows the BR-to-K transition; compare also the FT-Raman spectra of BR and K in Fig. 17. The disappearance of the band at 1190 $cm^{-1}$ at about 200 μsec indicates the deprotonation of the Schiff's base. The loss of charge at the Schiff's base greatly reduces the IR absorbance of the chromophore,

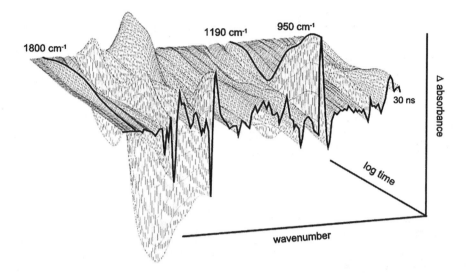

**Fig. 20** A three-dimensional representation of the IR absorbance changes between 1800 cm$^{-1}$ and 1000 cm$^{-1}$ with 30-nsec time resolution and 3-cm$^{-1}$ spectral resolution accompanying bacteriorhodopsin's photocycle as revealed by a global-fit analysis. The time axis has a logarithmic scale in order to show a complete bacteriorhodopsin photocycle in one representation.

leading to the disappearance of this band. (The Schiff's base connects the retinal to Lys 216 of the protein (Figure 18b)).

The detailed kinetic can be followed by the absorption change at 1187 cm$^{-1}$ versus time (Fig. 21). Beside the absorbance change, the fitted curve and the contributions of the individual rate constants as obtained by global fit are shown. In addition to the absorbance change at 1187 cm$^{-1}$, the absorbance change at 1758 cm$^{-1}$ is presented. It reflects the carbonyl vibration of Asp 85. Its appearance indicates the protonation kinetics of this residue. It is nicely seen that Asp 85 is protonated with the same apparent rate constants as the Schiff's base becomes deprotonated. From this result it was concluded that it accepts the Schiff's Base proton in the rate-limiting step (11,20). Asp 85 represents the catalytic proton-binding site on the release pathway (see Fig. 22). The further proton release pathway involves internal water molecules. A proton delocalized in a H-bonded network spanned by internal water molecules is identified as participating in the transport of a proton in a Grothuss-like proton transfer mechanism to the surface of the protein (47). The reappearance of the band at 1187 cm$^{-1}$ in the millisecond time domain in Fig. 21 indicates the reprotonation of the Schiff's base. It is reprotonated by Asp 96 (11,20). The final disappearance at 1187 cm$^{-1}$ shows

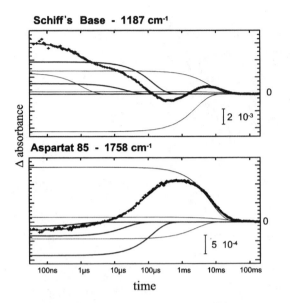

**Fig. 21**  As examples, time courses at specific wavenumbers are shown. The absorbance change at 1187 cm$^{-1}$ represents a C—C stretching vibration of the 13-cis protonated Schiff's base retinal. Its appearance indicates the all-trans to 13-cis isomerization (not time resolved). The disappearance within about 200 μsec indicates the deprotonation of the Schiff's base, its reappearance within a few milliseconds the reprotonation, and the final disappearance the back-isomerization to all-trans retinal. The absorbance change at 1758 cm$^{-1}$ indicates the protonation of Asp85. It is protonated with the same kinetic as the Schiff's base becomes deprotonated.

**Fig. 22**  The current model of the proton pump mechanism of bacteriorhodopsin, to which many groups have contributed (for references, see text). After the light-induced all-trans to 13-cis retinal isomerization in the BR-to-K transition, the Schiff's base proton is transferred to Asp 85 in the L-to-M transition. Simultaneously, an excess proton is released from an icelike H-bonded network of internal water molecules to the extracellular site. This network is controlled by Glu 204, Glu 194, and Arg 82. Asp 85 reprotonates the network in the O-to-BR reaction. The Schiff's base is oriented in the $M_1$-to-$M_2$ transition from the proton release site to the proton uptake site and determines thereby the vectoriality of the proton transfer. A larger backbone movement of the helix F is observed in the M-to-N transition as compared to the $M_1$-to-$M_2$ transition. Asp 96 reprotonates the Schiff's base in the M-to-N transition. Asp 96 itself is reprotonated from the cytoplasmic site in the N-to-O transition.

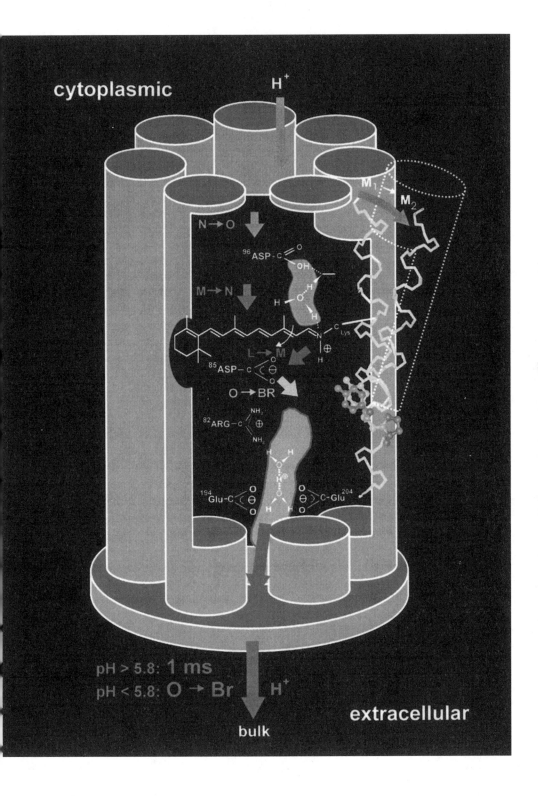

the relaxation of the chromophore from 13-cis to the all-trans BR ground state configuration.

Based on the FT-IR experiments, a detailed model of the light-driven proton pump bacteriorhodopsin is elucidated and presented in Fig. 22.

## 3.2 Bacterial Photosynthetic Reaction Centers

The bacterial photosynthetic reaction centers are the first membrane proteins whose structures were solved at nearly atomic resolution (Fig. 23) (see color plate) (48). In the photosynthetic reaction center (RC) of *Rhodobacter sphaeroides*, light excitation induces a transmembrane charge separation originating at the primary donor P (bacteriochlorophyll a-dimer), proceeding via $B_A$ (bacteriochlorophyll a-monomer), $H_A$ (bacteriopheophytin a), and $Q_A$ (ubiquinone-10 = $UQ_{10}$) and terminating at $Q_B$ ($UQ_{10}$) (49).

A considerable amount of IR spectra from the different charge-separated states was monitored using the static FT-IR method (50–54).

The step-scan method with a 10-μsec time resolution was used to investigate the $PQ_A$-to-$P^+Q_A^-$ and the P- to-$^3$P transitions (55,56). The absorbance changes of this transition are much smaller than for the $P^+Q_A^-$ transition, demanding an improved SNR. The experimental result shows convincingly that even when the relaxation time is extremely long, it is still possible to apply the step-scan technique. However, the sample has to be stable over about 20 hours. The results are shown in Fig. 24.

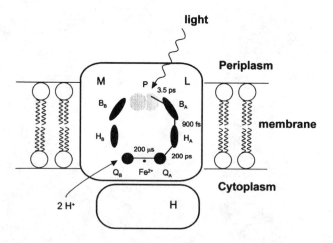

**Fig. 23**  (b) Structural model of the bacterial photosynthetic reaction center of *Rhodobacter sphaeroides* with schematic representation of the electron pathway and the proton uptake.

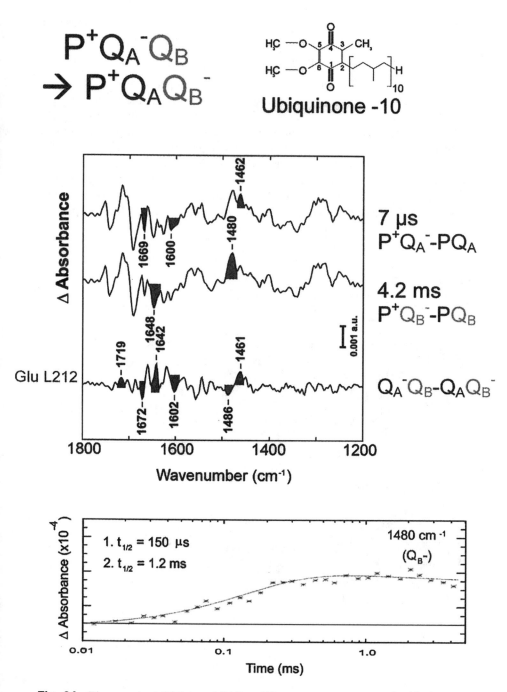

**Fig. 24** Time-resolved $P^+Q_A^-$ and $P^+Q_B^-$ difference spectra measured with the step-scan technique. In addition, the double difference is shown, which contains only absorbance changes due to the $Q_A$–$Q_B$ transition. The partial protonation of Glu 212 is seen for example at 1719 cm$^{-1}$. In the lower part, the rise of the $Q_B^-$ band at 1480 cm$^{-1}$ is shown. It represents the 1- and 4-C=O stretching vibrations of $Q_B^-$.

### 3.3   Cytochrome *c* Oxydase

As an example for a redox-driven proton pump, haem-copper oxidase is presented
(57). Haem-copper oxidases are redox-driven proton pumps in the aerobic respiratory chain of microorganisms and mitochondria (1). The enzymes are integrally
located in energy-transducing membranes. They use either cytochrome *c* or quinols
as donors of electrons transferred to the terminal acceptor oxygen (see Fig. 25).
During the chemical process of oxygen reduction and water formation, protons are
taken up from the cytoplasm (bacteria) or matrix (mitochondria). The free energy
of the chemical reaction is sufficient to drive additional movement of protons across
the membranes. Recently the 3D structure of cytochrome *c* oxidase from the proteo-
bacterium *Paracoccus denitrificans* was solved to a level of 2.8 Å (58).

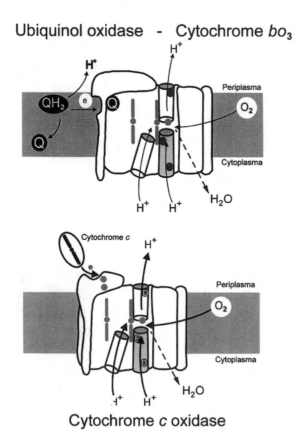

**Fig. 25**   Schematic representations of ubiquinol oxidase and cytochrome *c* oxidase. The
proton transfer pathways are indicated.

In order to initiate the reaction, a ''caged electron'' technique was developed (59,60), which employs flavinemononucleotide (FMN) as an intermediate electron transmitter (Fig. 26). After photochemical activation by ultraviolet laser light, the caged compound FMN injects electrons liberated from the sacrificing donor EDTA into the redox centers of the cytochrome oxidase. This approach has led to the detection of absorbance differences in various haem-copper oxidases (59,60).

In Fig. 27, typical redox difference spectra in the carbonyl region are shown. In the E286Q mutated quinol oxidase, the difference band at 1745/1735 cm$^{-1}$ is no longer observed. This is nicely elaborated in the double differences in the lowest trace. The frequency shift indicates an environmental change of E286 in the wild type. The movement of E286 seems to connect two proton transfer pathways. Therefore, it seems to be crucial for the proton transfer in the proton uptake pathway of the protein (Fig. 28) (59). Alternatively, electrochemical cells can be used to monitor these redox-induced reactions (61).

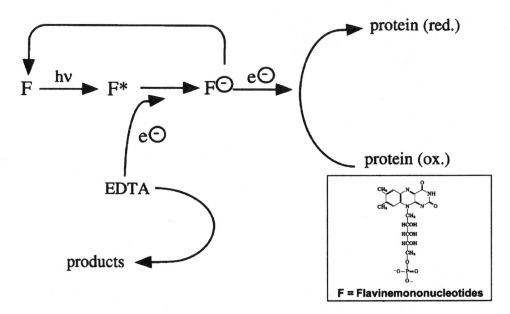

**Fig. 26** Reaction scheme of the photoreduction mechanism with the caged electron compound flavinemononucleotide (F), simplified for the sake of clarity. The asterisk (*) represents the photoactivated state.

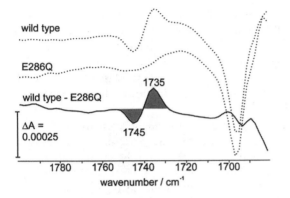

**Fig. 27**  Redox FT-IR spectra of wild-type cytochrome $bo_3$ and mutant E268Q. The difference bands with a minimum at 1745 cm$^{-1}$ and maximum at 1735 cm$^{-1}$ in the double difference can be assigned to Glu 268.

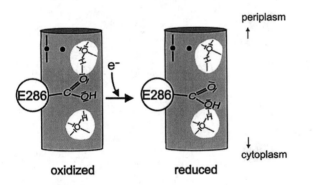

**Fig. 28**  The model shows the redox-induced conformational changes of E286, by which the carboxylic side chain is positioned into environments of distinct hydrophobicity. The side-chain conformational change connects both hydrogen-bonded networks and thereby facilitates proton transfer.

## 3.4   H-*ras* P21

The small GTP-binding protein H-*ras* p21 plays a central role in signal transduction leading to cell growth (62). In Fig. 29 (see color plate), the structural model is shown (63). P21 appears to act as a switch in the signal transduction: in the GTP-bound form a signal is given to an effector molecule that is part of a cascade leading to cell proliferation and differentiation (62). The intrinsic GTPase activity is accelerated by GAP (GTPase-activating protein) or NF1. The transition from

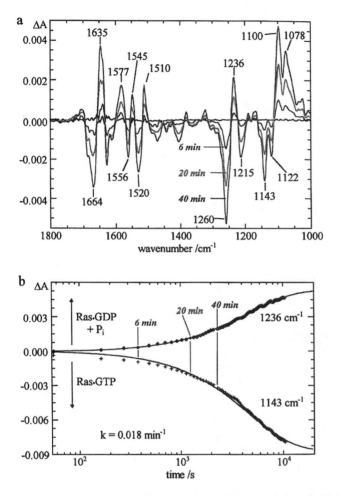

**Fig. 30** Absorbance changes during the GTPase activity of p21. GTP is released from caged GTP using several UV flashes and GTPase activity starts. Difference spectra are taken between the first spectrum measured directly after the GTP is released and spectra measured during the following GTPase reaction (a). In (b), two representative kinetics are shown, the disappearing γ-phosphate band at 1143 cm⁻¹ and an appearing GDP band at 1236 cm⁻¹.

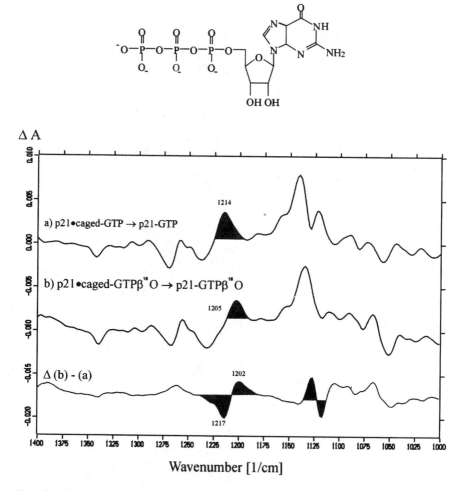

**Fig. 31** Photolysis difference spectra between p21-caged GTP and p21-GTP. The GTP bands are positive. Due to the $^{18}O$-β-GTP labeling, the band at 1214 cm$^{-1}$ is downshifted to 1205 cm$^{-1}$, as also seen in the double difference spectra in the lowest trace. The band at 1214 cm$^{-1}$ is assigned to the β-GTP vibration.

the active GTP-bound form to the inactive GDP-bound form is accompanied by a conformational change.

The reaction can be studied using caged GTP complexed to p21 instead of GTP, to prevent hydrolysis. Using an intense UV flash (308 nm, 100 mJ), the caged group is photolysed, GTP is released, and GTPase activity is precisely started. In Fig. 30 difference spectra between p21-GTP and finally p21-GDP are shown in dependence on time in part (a) (18). The region between 1400 and 1000 cm$^{-1}$ is dominated by phosphate bands. For example, the band at 1143 cm$^{-1}$ is assigned to the γ-phosphate vibration (16,18). Its absorbance change can be used as marker band for the GTPase kinetic. The time course of a disappearing GTP band and an appearing GDP band is shown in part (b).

Using isotopic-labeled caged GTP, the phosphate vibrations were assigned. In Fig. 31, photolysis difference spectra between p21-caged GTP and p21-GTP are shown as examples of the band assignment. In Fig. 31b, the $^{18}$O-β-labeled caged GTP difference spectra are shown. Based on the downshift from 1214 cm$^{-1}$ to 1205 cm$^{-1}$, this band was assigned to the β phosphate vibration (18). This result shows that the β phosphate is unusually frequency shifted when bound to the protein, indicating strong H-bonding of β phosphate to the protein environ-

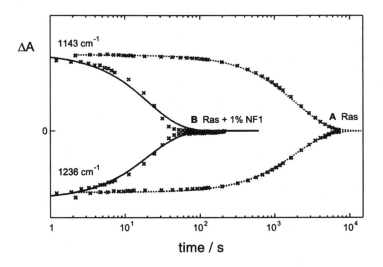

**Fig. 32** Time course of the absorbance changes for the measurement of the GTPase reaction without NF1-333 at 310 K (A) and with 1% NF1-333 at 283 K (B) at 1143 and 1236 cm$^{-1}$, indicating the disappearance of GTP and the appearance of GDP. Global-fit analysis yields $k = 4.7 \times 10^{-4}$ sec$^{-1}$ (A) and $k = 4.6 \times 10^{-2}$ sec$^{-1}$ (B). In contrast to Fig. 30, the last and not the first spectrum is taken as reference. Therefore, the absorbance changes relax to zero.

ment. By the H-bonding, the $\beta\gamma$ connecting oxygen bond is weakened and the splitting of the $\gamma$-phosphate during GTPase activity is preprogramed by the protein environment (18). The results on p21 show that molecular reaction mechanisms of proteins without an intrinsic chromophore but with caged compounds can be nicely investigated with time-resolved FT-IR. Beside the intrinsic mechanism of p21, the GAP-catalyzed GTPase mechanism can also be studied. In Fig. 32, the acceleration of the GTPase via the addition of NF1 in catalytic amounts is shown. This demonstrates that protein–protein interactions can be studied.

## 4 CONCLUSION

The presented results demonstrate that time-resolved FT-IR difference spectroscopy in combination with molecular biological methods allows detailed analysis of the molecular reaction mechanism of proteins from the nanosecond time range up to several minutes. The use of caged compounds provides a powerful tool also for non-photobiological proteins.

## REFERENCES

1. L Stryer. Biochemistry. New York: Freeman, 1995.
2. CR Cantor, PR Schimmel. Biophysical Chemistry. Vols I–III. New York: Freeman, 1980.
3. K Wüthrich. NMR of Proteins and Nucleic Acids. New York: Wiley, 1989.
4. NB Colthup, LH Daly, SE Wiberley. Introduction to Infrared and Raman Spectroscopy. 3rd ed. San Diego, CA: Academic Press, 1990.
5. K Gerwert. Curr Opin Struct Biol 3:769–773, 1993.
6. PO Stoutland, RB Dyer, WH Woodruff. Science 257:1913–1917, 1992.
7. S Krimm, J Bandekar. Adv Prot Chem 38:181–364, 1986.
8. PR Griffiths. Chemical Infrared Fourier Transform Spectroscopy. New York: Wiley, 1975, pp. 21–30.
9. PR Griffiths, JA deHaseth, Fourier Transform Infrared Spectroscopy. New York: Wiley, 1986.
10. K Gerwert. In: B Schrader, ed. Infrared and Raman Spectroscopy. Weinheim: VCH Verlagsgesellschaft, 1995, pp. 617–637.
11. K Gerwert, G Souvignier, B Hess. Proc Natl Acad Sci USA 87:9774–9778, 1990.
12. NJ Harrick. Internal Reflection Spectroscopy. Ossining, NY: Harrick Scientific, 1987.
13. JA McCray, DR Trentham. Ann Rev Biophys Chem 18:239–270, 1989.
14. JET Corrie, DR Trentham. In: H Morrison, ed. Biological Applications of Photochemical Switches. New York: Wiley, 1993, pp. 243–305.
15. JW Walker, GP Reid, JA McCray, DR Trentham. J Am Chem Soc 110:7170–7177, 1988.
16. V Cepus, C Ulbrich, C Allin, A Troullier, K Gerwert. Methods Enzymol 291:223–245, 1998.

17. J Lughtenburg, RA Mathies, RG Griffin, J Herzfeld. Trends Protein Sci 13:388–393, 1988.
18. V Cepus, RS Goody, AJ Scheidig, K Gerwert. Biochemistry 37:10263–10271, 1998.
19. M Engelhard, K Gerwert, B Hess, W Kreutz, F Siebert. Biochemistry 24:400–407, 1985.
20. K Gerwert, B Hess, J Soppa, D Oesterhelt. Proc Natl Acad Sci USA 86:4943–4947, 1989.
21. WB Fischer, S Sonar, T Marti, HG Khorana, KJ Rothschild. J Biol Chem 269:28851–28858, 1994.
22. JA Ellman, BF Volkmann, D Mendel, PG Schulz, DE Wemmer. J Am Chem Soc 114:7959–7961, 1992.
23. K Gerwert. Ber Bunsenges Phys Chem 92:978–982, 1988.
24. K Gerwert, G Souvignier. Proceedings of the 8th International Conference on FTIR Spectroscopy. SPIE-Int Soc Opt Eng 1575:431–432, 1992.
25. G Souvignier, K Gerwert. Biophys J 63:1393–1405, 1992.
26. RA Palmer, CJ Manning, J Chao, I Noda, AE Dowrey, C Marcott. Appl Spectrosc 45:12–17, 1991.
27. RA Palmer, J Chao, RM Dittmar, VG Gregoriou, SE Plunkett. Appl Spectrosc 47:1297–1310, 1993.
28. O Weidlich, F Siebert. Appl Spectrosc 47:1394–1400, 1993.
29. W Uhmann, A Becker, C Taran, F Siebert. Appl Spectrosc 45:390–397, 1991.
30. R Rammelsberg, B Hessling, H Chorongiewski, K Gerwert. Appl Spectrosc 51:558–562, 1997.
31. R Rammelsberg, S Boulas, H Chorongiewski, K Gerwert. Vib Spectrosc 19:143–149, 1999.
32. P Hendra, C Jones, G Warnes. Fourier Transform Raman Spectroscopy. Chichester, UK: Ellis Horwood, 1991.
33. B Schrader. In: JR Ferraro, K Krishnan, eds. Practical Fourier Transform Infrared Spectroscopy. San Diego, CA: Academic Press, 1990, pp. 167–202.
34. J Sawatzki, R Fischer, H Scheer, F Siebert. Proc Natl Acad Sci USA 87:5903–5906, 1990.
35. TA Mattioli, A Hoffmann, B Robert, B Schrader, M Lutz. Biochemistry 30:4548–4554, 1991.
36. M Brandt. Diploma thesis, Ruhr-Universität Bochum, Bochum, Germany, 1997.
37. T Johnson, S Burke. Bruker Optics, TRVS IX, Tucson, AZ, 1999.
38. B Hessling, G Souvignier, K Gerwert. Biophys J 65:1929–1941, 1993.
39. A Fersht. Enzyme Structure and Mechanism. New York: Freeman, 1985.
40. B Hessling, J Herbst, R Rammelsberg, K Gerwert. Biophys J 73:2071–2080, 1997.
41. R Henderson, JM Baldwin, TA Ceska, F Zemlin, E Beckmann, KH Downing. J Mol Biol 213:899–929, 1990.
42. H Lücke, HT Richter, JK Lanyi. Science 280:1934–1937, 1998.
43. D Oesterheld, W Stoeckenius. Methods Enzymol 31:667–678, 1974.
44. MS Braiman, K Rothschild. Ann Rev Biophys Chem 17:541–570, 1988.
45. T Kitagawa, A Meada. Photochem and Photobiol 50:883–894, 1989.
46. J le Coutre, J Tittor, D Oesterhelt, K Gerwert. Proc Natl Acad Sci USA 92:4962–4966, 1995.

47. R Rammelsberg, G Huhn, M Lübben, K Gerwert. Biochemistry 37:5001–5009, 1998.
48. J Deisenhofer, O Epp, K Mikki, R Huber, H Michel. Nature 318:618–624, 1985.
49. J Deisenhofer, H Michel. Science 245:1463–1473, 1989.
50. W Mäntele. In: J Deisenhofer, JR Norris, eds. The Photosynthetic Reaction Center. Vol II. New York: Academic Press, 1993, pp. 239–283.
51. B Robert, E Nabedryk, M Lutz. In: RJH Clark, RE Hester, eds. Time-Resolved Spectroscopy. Chichester, UK: Wiley, 1989, pp. 301–334.
52. S Buchanan, H Michel, K Gerwert. Biochemistry 31:1314–1322, 1992.
53. R Brudler, HJM de Groot, WBS van Liemt, WF Steggerda, R Esmeijer, P Gast, AJ Hoff, J Lugtenburg, K Gerwert. EMBO J 13:5523–5530, 1994.
54. R Brudler, HJM de Groot, WBS van Liemt, P Gast, AJ Hoff, J Lugtenburg, K Gerwert. FEBS Letters 370:88–92, 1995.
55. JR Burie, W Leibl, E Nabedryk, J Breton. Appl Spectrosc 47:1401–1404, 1993.
56. R Brudler, K Gerwert. Photosynth Res 55:261–266, 1998.
57. M Wikström. Curr Opin Struct Biol 8:480–488, 1998.
58. C Ostermeier, A Harrenga, U Ermler, H Michel. Proc Natl Acad Sci USA 94:10547–10553, 1997.
59. M Lübben, K Gerwert. FEBS Lett 397:303–307, 1996.
60. M Lübben, A Prutsch, B Mamat, K Gerwert. Biochemistry 38:2048–2056, 1999.
61. P Hellwig, J Behr, C Ostermeier, OMH Richter, U Pfitzner, A Odenwald, B Ludwig, H Michel, W Mäntele. Biochemistry 73:7390–7399, 1998.
62. F Wittinghofer, L Wiesmüller. Cell Signalling 6:247–267, 1994.
63. A Wittinghofer, EF Pai. Trends Biochem Sci 16:382–387, 1991.

# 7
# Biological Vibrational Spectroscopic Imaging

**Michael D. Schaeberle, Ira W. Levin, and E. Neil Lewis**
*National Institutes of Health, Bethesda, Maryland*

## 1   INTRODUCTION

The utility and increasingly extensive roles of conventional Raman, mid-infrared (IR), and near-infrared (NIR) vibrational spectroscopic methods in the study of biomolecular assemblies are well documented (1–27). With the vast array of sampling procedures currently available for implementation, the capabilities now exist for researchers to readily acquire vibrational spectra of complex, heterogeneous materials. Although often labyrinthine, these spectra reflect both qualitative and quantitative information characteristic of the structural, morphological, and environmental properties of the system under scrutiny. Since the observed vibrational signal is intrinsic to the molecular system, no additional perturbants such as dyes, tissue stains, or excessive labeling methods are required for either signal generation or identification, obvious advantages to be considered carefully when clarifying subtle conformational effects and assessing functional molecular relationships.

Although vibrational spectroscopic imaging methods are relative newcomers to the panoply of vibrational techniques, they inherit the versatility of the traditional single-point infrared and Raman approaches and, in combination with digital imaging technology, allow for new perspectives in the interpretation of the spectra of biomolecules. Imaging advantages are derived, in part, from the ability of an investigator to process and discern effectively a complex two-dimensional spatial representation of a sample in terms of a variety of spectrally related molecular parameters. In particular, a single spectroscopic image, or chemical map, is capable of summarizing and conveying a wealth of spatial, chemical, and molecular data of benefit to both specialists and nonspecialists.

Regardless of the specific method used to collect data during a vibrational spectroscopic imaging experiment (see Sec. 2), the final representation is typically expressed as a three-dimensional image cube, or hypercube, consisting of two spatial dimensions and one spectral dimension. The image cube concept, shown in Fig. 1, is interpreted in one of two ways, either (1) a series of images collected as a function of wavenumber (Fig. 1A) or (2) a vibrational spectrum corresponding to each spatial location, or pixel, within the image plane (Fig. 1B). The ability to merge and recall, within a single analytical technique, information corresponding to the spatial and spectral axes of the image cube provides extraordinary flexibility in probing and extracting structural and compositional information from biological samples. Thus, a data set may be either summarized as a single image or expressed as a series of images derived from one or more spectroscopic parameters, such as spectral peak intensity, band frequency, or linewidth. The chemical and morphological interpretations of the resulting images are then based on the spatial variations of the intrinsic molecular property being highlighted and their relevance to the biological questions being pursued. In contrast to images generated by conventional point mapping, which generally suffer from low spatial resolution due to relatively few sampling points, direct imaging approaches using the advantages inherent in two-dimensional array detectors derive an enormous benefit from the highly parallel nature of the data collection. That is, a single data set, which typically contains many tens of thousands of independent spectral channels, may be analyzed and interpreted using a variety of statistical algorithms, allowing subtle spatial and spectral variations that are often overlooked or misinterpreted in traditional spectroscopic studies to be tested and revealed.

Although vibrational spectroscopic imaging has been widely applied to diverse materials, including polymers, semiconductor, pharmaceuticals, cosmetics, and consumer products, this chapter will focus primarily on a review of selected biological applications. The strengths and adaptability of vibrational spectroscopic imaging rest not only on the ability to determine chemical compositions and component distribution within a sample, but also on being able to extract localized molecular information relevant to the sample architecture. For example, vibrational spectra are sensitive to alterations in protein secondary and tertiary structures. Spectral shifts could, therefore, be used to image specifically the distributions of specific structural moieties within a biological sample in contrast to measuring simply an overall distribution of a general class of molecules. Multivariate approaches may also be used to generate a composite metric indicative of either disease progression or biological function. Although these metrics cannot be readily interpreted in terms of biochemical changes, they may often be statistically correlated with disease. In addition, by simultaneously recording data

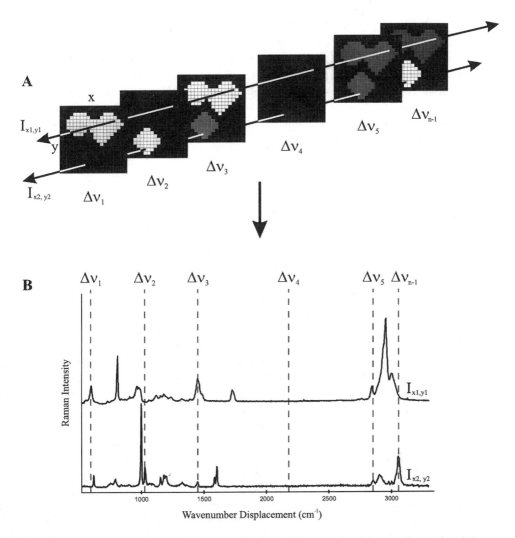

**Fig. 1** Conceptualization of a three-dimensional hyperspectral image data cube. (A) Images as a function of wavelength. (B) Selected spectra associated with pixels $(i, j)$ in the image.

on a two-dimensional focal-plane array detector from a myriad of spatial locations within a biological matrix, pooled spectra may be treated statistically. In this manner, analysis of variance can be employed to more robustly test the significance of observed spectral changes appearing, for example, as a consequence of a perturbed or diseased state. In this chapter, we will present a range of applications, including, for example, the imaging of functioning cells, the detection of tissue inclusions, a mapping of bone maturity, and the identification of diseased tissue. These applications will also emphasize the combining of vibrational spectroscopic mapping and imaging techniques with increasingly complex levels of data treatment. Because various aspects of the vibrational spectroscopic imaging technology are still relatively new and unfamiliar to the general readership, we shall also include a brief discussion of the instrumentation generally employed and their modes of implementation.

## 2 METHODS

### 2.1 Raman Imaging Techniques

Raman spectroscopic imaging methods can be generally divided into three groups: mapping, spatial encoding, and wide-field techniques. Mapping techniques, currently the predominant method of acquiring Raman spectroscopic image data, began with the introduction of point mapping and the Raman microprobe in 1975 (28). Point mapping, shown in Fig. 2A, employs a lens or microscope objective to focus the laser beam to a small point at the surface of the sample. The Raman scattered light from that point is collected and a spectrum is recorded using a standard dispersive spectrograph and detector. In order to generate an image, a motorized stage is employed to raster-scan the sample in both the $X$ and $Y$ directions, while the instrument records a spectrum at sequential points. The complete spectral set can then be used to reconstruct an image corresponding to each spectral frequency (or wavenumber).

One limitation experienced by many optical imaging techniques is a relatively poor axial, or depth, resolution capability. This can be overcome by employing a confocal approach in which the Raman scattering is focused through a pinhole aperture. In this manner, contributions to the overall image from Raman scattering arising from planes above and below the plane of interest are eliminated and radiation from only a specified sample plane is passed to the spectrograph. Although this approach significantly decreases the signal intensity compared to standard mapping techniques, confocal Raman mapping (29,30), as well as direct confocal imaging (31), is often desirable.

Another Raman spectroscopic mapping technique is represented by line-scan imaging, illustrated in Fig. 2B as an extension of point mapping. This approach employs either cylindrical optics or scanning mechanisms for elongating the exciting laser beam in one dimension (32,33). The line, now defining the excitation radiation, is focused through an objective and onto the sample so that the collected, scattered radiation is parallel to the entrance slit of a dispersive spectrograph. Raman spectral signals are collected using a two-dimensional charge-coupled device (CCD) detector, in which one dimension of the CCD rec-

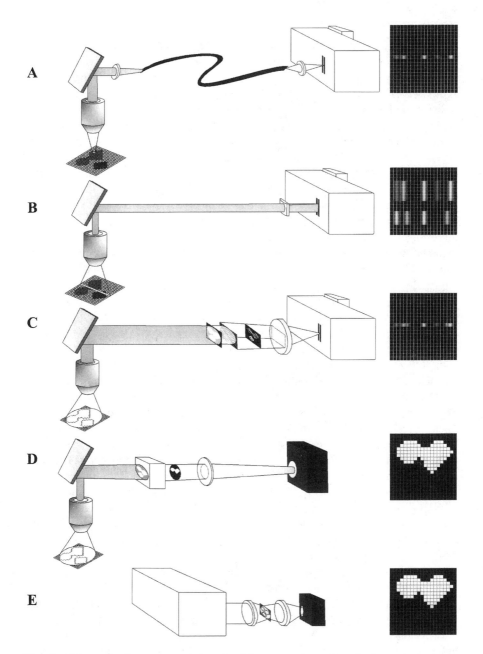

**Fig. 2** Schematic of spectroscopic imaging instrumentation and methods. (A) Point mapping. (B) Line-scan imaging. (C) Hadamard transform imaging. (D) Generic wide-field imaging. (E) Infrared step-scan imaging.

ords spectral information while the other registers a spatial dimension. The second spatial dimension is recorded by either scanning the excitation beam or moving the sample along an axis perpendicular to the line of focused radiation.

Spatial encoding methods, such as Hadamard transform (HT) Raman imaging, which is shown in Fig. 2C, have also been demonstrated for collecting spectroscopic image data. In one arrangement, the entire sampling area is irradiated with wide-field epi-illumination (34–36). A mask with a sequence of apertures is placed in the emission path, allowing the Raman scattered radiation to be encoded prior to being focused onto the slit of a dispersive spectrometer. A two-dimensional CCD preserves one spatial dimension and the one spectral dimension, similar to line scan imaging. A second spatial dimension is recorded by changing the pattern of aperture openings in the mask according to the Hadamard encoding sequence. The prescribed manner to change the mask settings involves using either a stepped linear mask or a rotating disk mask (37,38). Unfortunately, masks involving moving parts suffer from inherent mechanical failures and misalignments that decrease the achievable spatial resolution of the technique. Because of the moving-mask limitations and the additional expertise required for implementation, conventional HT Raman imaging systems are currently employed mainly as academic research tools and are not commercially available.

Attempts have been made to improve the mask technology used in Raman and IR HT spectroscopic imaging instruments by avoiding the traditional mechanical movements (39,40). One of the new technologies, for example, employs a 127-element solid-state liquid crystal spatial light modulator (LC-SLM) (39). The apertures defining the mask are individually addressable and can be opened (clear) or closed (opaque) by varying the voltage applied to the LC elements. Although the LC-SLM performs adequately in an HT spectrometer, the signal-to-noise ratio of the instrument is degraded due to stray light through the opaque elements (transmission $\neq 0$) and less than full transmission through the clear elements (transmission $\neq 1$). The multiplexing of elements, however, minimizes this defect in performance (39). Another recently developed mask technology for HT imaging involves the use of digital micromirror arrays (DMAs) (40). The micromirrors are constructed of highly reflective aluminum and are mounted directly on top of the integrated circuits, which are used to control their positions. Such DMAs exhibit many of the properties required for HT imaging, including fast element-switching speed (on/off), reliable positioning, variable mask dimensions, and high radiation tolerance. This currently represents the greatest potential for a commercialization of HT spectrometers and imaging systems.

Wide-field Raman imaging is generally shown in Fig. 2D. In this approach, either the laser beam is defocused or scanning mechanisms are employed to rapidly scan the beam in both the $X$ and $Y$ directions (41), thus allowing, in principle, all sampling points in the field of view to be simultaneously Raman excited. The

Raman scattering from all points is collected by a microscope objective and is presented to an imaging tunable filter for wavelength selection. Many types of imaging tunable filters, including rotating dielectrics (42), fixed filter/tunable source (43), Fabry–Perot filters (44), acousto-optic tunable filters (AOTFs) (45–51), and liquid crystal tunable filters (LCTFs) (52–56), have been used in Raman spectroscopic imaging systems. The resulting filtered Raman image is collected by a two-dimensional CCD detector, which captures the two spatial dimensions; the spectral dimension is recorded by sequentially tuning the filter and then acquiring images. Although the spectral resolution of imaging tunable filters is generally inferior to that of mapping techniques, the spatial resolution is often much improved, approaching the diffraction limit.

## 2.2 Infrared Imaging Techniques

Many of the methods employed for IR imaging are analogous to those used for collecting Raman images. These include point scanning (57) and many of the wide-field imaging techniques that utilize tunable filters, such as fixed bandpass filters (58), rotating dielectric filters (59–61), and AOTFs (46,62,63). Although they are not widely available, it has also been reported that LCTFs capable of operating in the infrared region can be constructed (P. Miller, personal communication, 1998). A novel imaging method not often employed in Raman imaging approaches involves interferometric techniques, as shown in Fig. 2E. The instrument consists of a commercially available mid-infrared Michelson step-scan interferometer coupled to an IR microscope and a focal-plane array (FPA) detector (64–67). Modifications have been made to the microscope optics and interferometer electronics to provide an efficient integration with the FPA. The stepping of the interferometer is synchronized to FPA acquisition so that, as the interferometer mirror is stepped, a series of image frames is acquired, averaged, and stored. An image data set is assembled by repeating this process at each mirror step.

Thermal imaging also makes use of the infrared region of the electromagnetic spectrum; however, since the images are not generated from intrinsic molecular vibrations within the sample, this technique will not be discussed here. For a discussion of the use of thermal imaging in biological research, the reader is directed to Ref. 68.

## 3 RAMAN AND INFRARED IMAGING APPLICATIONS

### 3.1 Biochemical Composition

One active area of vibrational spectroscopic imaging research applies these techniques to visualizing the spatial distributions of the biochemical components within a given sample. Since the Raman and/or IR spectra act as fingerprints for a particular molecule, spectra from the image cube are used to identify uniquely and unambiguously sample components. Using spectral parameters, one can generate chemically specific images for assessing the distribution of components within the sample matrix and for identifying the morphological properties of the sample. Imaging techniques have been recently applied in a wide variety of biological, medical, and agricultural studies, including visualization of cells (69), drug analysis in human hair (70), detection of lipids, proteins, and nucleic acids (55,71,72), polytene chromosomes in cells and/or tissue (73,74), and the examination of flax (75) and other plant tissues (76,77).

Biochemical vibrational spectroscopic imaging allows the identification and chemical distribution of cellular materials and tissue components, information essential to understanding biological activity. Raman imaging, using an LCTF-based microscope assembly with two-dimensional CCD detection in conjunction with ratiometric and multivariate image processing, has proved useful in visualizing the spatial distributions of protein and lipid components in an intact animal breast tissue model (55). In the extraction of useful information from a complex biological sample, this study highlights the potential of Raman spectroscopic imaging evolving from a basic research tool into a viable diagnostic technique for use in a clinical setting. In addition to tissue applications, confocal Raman microspectroscopy has been utilized to map both the cytoplasm and nuclei of intact cells (73). Raman line imaging data have also been derived from polytene chromosomes, which exhibit of patterns of alternating bands and interbands reflecting different Raman scattering activities (74).

While Raman spectroscopic imaging methods have benefited from technical advances in visible detectors, fiber optics, and tunable filters, their use in routine biomedical diagnostics is hampered by low signal levels and long exposure times. Additionally, the effectiveness of global Raman imaging is complicated by the necessity of the high-powered lasers that are required to maintain a reasonable power density over the entire sampling area. Many of these limitations can be overcome by employing the complementary techniques embodied in IR spectral mapping and imaging methods. The higher signal levels inherent in the IR approaches result in shorter experimental acquisition times, an attractive advantage when high-throughput scenarios are contemplated. Infrared spectral mapping, in the form of synchrotron infrared microscopy, for example, has been used to map the cellular distributions of lipids, proteins, and nucleic acids within living hybridoma B cells (72) as well as cells in the final stages of cell division and necrosis (71). Synchrotron radiation, in place of a standard IR broadband source, allows one to achieve improved spatial resolution and signal-to-noise characteristics, thus enabling the mapping of single cells (72).

The identification of specific cellular constituents is often inhibited by the complications arising from the superposition of their vibrational modes. It is possible, however, to assign major chemical classes based upon vibrational bands associated with their characteristic functional groups. For example, in the mapping of hybridoma B cellular components, the 1240 cm$^{-1}$ band serves as a marker for the phosphodiester (PO$_2^-$) of the nucleic acids, the bands at 1650 and 1545 cm$^{-1}$ reflect the amide I and II modes of the protein, and the band at 2929 cm$^{-1}$ identifies the methylene (CH$_2$) groups of the lipids. The reliable use of the antisymmetric CH$_2$-stretching peak to map the lipid content requires that the protein CH groups do not contribute or interfere significantly with the lipid signal at this frequency (72).

The hybridoma B cell work demonstrates the ability of IR spectroscopic mapping using a synchrotron source to obtain the molecular distributions of individual cells at the micron level (72). To accomplish this mapping, data from a living cell consisted of a 14 × 14 spectral matrix collected by raster scanning a 3 × 3-µm aperture across the sample in 3-µm steps. Maps of the cell were generated using the integrated intensities from the nucleic acid, protein, and lipid marker bands, as shown in Fig. 3A–C, respectively. An optical image is presented in Fig. 3D for comparison. The nucleic acid and protein maps, shown in Fig. 3A and B, respectively, are relatively homogeneous, with the greatest intensity appearing at the center of the cell. The lipid map in Fig. 3C, on the other hand, is more heterogeneous, with an independent sample distribution in comparison to the protein map.

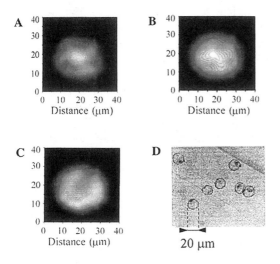

**Fig. 3** Chemical distribution of components within an in vivo mouse UN2 hybridoma B cells. (A) Nucleic acid distribution image based on the PO$_2^-$ antisymmetric stretching band. (B) Protein distribution image based on the amide II band. (C) Lipid distribution image based on the hydrocarbon CH$_2$ antisymmetric stretching band. (D) Optical image showing the cell imaged in (A)–(C). (From Ref. 72. © 1998 *Cellular and Molecular Biology*.)

With the increasing concern regarding corporate cost and effect of illicit drug use in the workplace, screening agencies, including law enforcement, are searching for additional methods for detecting and documenting the presence and use of illegal narcotics. One new avenue that is being explored is the IR spectroscopic imaging of human hair. Hair has become an increasingly amenable sample for drug testing because of the long-term history contained within a fiber and the ease of sample collection. Point-scan IR spectral imaging employing both conventional and synchrotron IR sources has been used to investigate the localization of drug metabolites in human hair. By mapping the distribution of the metabolites, it was shown that drug affinity might be linked to either the hair's medulla or its central cortex (70). Hydrophobic molecules bind to the medulla, while hydrophilic molecules tend to be spread throughout the hair's cortex. Although the data presented in this study were not considered to be quantitative, the relative infrared intensities of the bands were proportional to drug concentrations as determined by independent testing methods (70). This IR mapping study provides a graphical means for unraveling the pharmacokinetics of drug incorporation in hair and may lay the necessary groundwork for validating the use of hair as an efficient drug-testing methodology.

The applications of vibrational spectroscopic mapping and imaging have also been extended to problems involving cellular structure and morphologic changes occurring in a variety of agricultural products (75–77). For example, the renewed desire to use natural fibers for textile and industrial purposes has generated an interest in improving the quality and methods of obtaining fibers from flax plants. To realize this improvement, it is necessary to increase our understanding of the biochemistry and structure of plant cell walls and how these affect the integrity of the plant fibers. One IR spectroscopic mapping study examined the chemical morphology of cross-sectional samples of two different types of flax stems, *Ariane* and *Natasja*. Images of each flax stem were generated from a standard IR mapping system by raster-scanning the sample in 10-µm increments over a $140 \times 140$-µm area of the sample. By mapping baseline-corrected peak heights, integrated intensities, or band ratios for specific vibrational modes, images based on the lipid, pectin, cellulose, aromatic, and acetyl group distributions were generated (75). While the authors state that direct comparisons between the two different varieties of flax are difficult, this work demonstrates that FT-IR spectroscopic mapping is capable of detecting differences in location and concentration of components within a given flax stem sample. The ability to detect spatial variations provides botanical and agricultural researchers with a new tool for the noninvasive in situ assessment of a sample's chemical composition and distribution (75).

In a separate study of the cellular structures of grain seed, differences in spatial resolution determined from either conventional IR globar radiation or synchrotron radiation in FT-IR mapping was investigated (76). Unlike globar emission, synchrotron radiation enables nearly diffraction-limited spatial resolution to be achieved, thus allowing the interrogation of single cells or integral cellular components in situ. Spectra with reasonable signal-to-noise ratios were achieved using an infrared aperture of 5–6 µm and a minimum number of coadded spectra, allowing differences between neighboring tissues in the grain seeds to be distinguished

A greatly improved spatial resolution of 4.5 µm was achieved in the analysis of wheat kernels by using a step-scan FT-IR spectrometer coupled to an IR microscope and a two-dimensional FPA (77). Statistical data-processing methods were applied to the entire data set, which consisted of 4096 simultaneously obtained spectra, one for each detector pixel. High-fidelity IR images, which differentiated individual cell layers in the outer portion of the wheat kernel as well as in the primary root within the germ, were generated (77). Step-scan FT-IR imaging using a two-dimensional FPA could provide genetic engineers with valuable information concerning chemical changes occurring as a result of selective breeding methods.

## 3.2 Foreign Inclusions

An imaging area receiving recent prominence involves the incorporation and migration of foreign inclusions into human tissue. For example, considerable emphasis has been placed on the identification of silicone gel inclusions in breast tissue arising from failed breast implants. Other foreign components have also been incorporated within the area surrounding implants, namely, the biocompatible polyester fixative patches used to anchor the implant to the chest muscle wall. Wide-field Raman spectroscopic imaging, employing an AOTF and two-dimensional CCD detection, has been used to visualize inclusions that were found in sectioned breast tissue from a biopsy sample (Fig. 4) (50). The depiction in Fig. 4B, showing the distribution of the inclusion, is a ratio image generated by dividing an image corresponding to an inclusion peak at $1615$ cm$^{-1}$ by a nonspecific background image at $1670$ cm$^{-1}$. The spectroscopic component associated with the image pixels unambiguously identifies the inclusions as Dacron polyester, likely originating from the fixative patch (50).

Infrared spectroscopic mapping and imaging have also been employed to determine the chemical composition, distribution, and size of inclusions in human breast tissue. An IR point-mapping approach has been used to map the distribution of several implant components, including silicone, polyester, and polyurethane (78). Some of the most striking images were generated with a wide-field imaging approach using a step-scan interferometer and a two-dimensional FPA detector (Fig. 2E). This technique has been employed to visualize silicone gel and Dacron polyester inclusions found in separately sectioned biopsy samples (79). The Dacron detection was analogous to that of the Raman study mentioned previously and shows the complementary nature of the two imaging techniques. Using unique spectroscopic markers for silicone gel ($2963$ cm$^{-1}$), lipids and paraffin ($2927$ cm$^{-1}$), and proteins ($3350$ cm$^{-1}$), it is possible to generate the specific

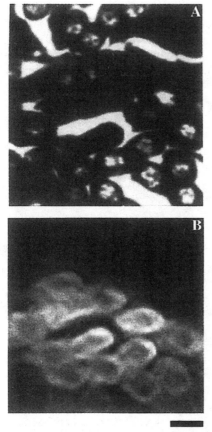

15 μm

**Fig. 4**  High-fidelity Raman image of Dacron polyester inclusions in human breast capsular tissue. (A) Brightfield reflectance image. (B) Inclusions appearing as bright areas in the background-ratioed Raman image (1615 cm$^{-1}$/1670 cm$^{-1}$). (From Ref. 50. © 1996 American Chemical Society.)

images shown in Fig. 5. The distribution of silicone gel, shown in Fig. 5B, reveals the inclusions as several bright domains. The lipid distributions of lipid/paraffin and protein are shown in Fig. 5C and D, respectively, and reflect a moderately homogeneous distribution of each of the components. By using the imaging approach, instead of point spectroscopy, it was also possible to readily identify a small silicone inclusion on the order of ~10 μm (see box in Fig. 5B) that had remained undetected by conventional pathology techniques (79).

These applications accentuate the capabilities of vibrational spectroscopic

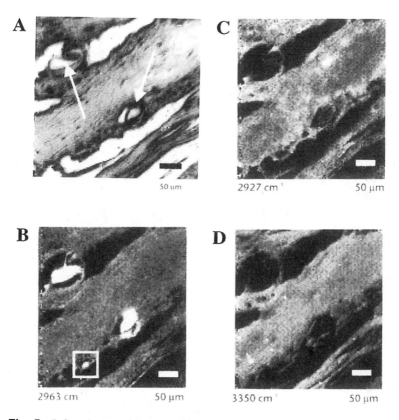

**Fig. 5** Infrared spectral images of human breast tissue. (A) Multiwavelength, infrared brightfield image of a 5-µm-thin section of human breast implant capsular tissue. The bright areas within the image are vacuoles in the tissue sample. (B) Image obtained at 2963 cm$^{-1}$ highlighting silicone gel inclusions in the breast tissue. (C) Image obtained at 2927 cm$^{-1}$ corresponding to the $CH_2$ and $CH_3$ stretching modes of lipid and paraffin. (D) Image obtained at 3350 cm$^{-1}$ showing a moderately homogeneous distribution of protein throughout the tissue sample. (From Ref. 79. © 1997 *Nature Medicine.*)

imaging to complement traditional histopathologic determinations. The techniques aid the researcher by not only characterizing the two-dimensional sample morphology (size, shape, and distribution) of foreign inclusions, but also by providing a means for unambiguous chemical identification.

## 3.3 Mineralized Tissue

Although bone as a composite material is relatively well understood, the basic chemical mechanisms regulating its formation and regeneration are not. Investigators are now beginning to use Raman and IR spectroscopic imaging methodologies to elucidate these complex mechanisms. Raman line-scan imaging has been used to study the chemistry of bone growth and regrowth by monitoring the spatial distribution of phosphate species in mature and newly generated bone samples (80,81). The phosphate ($PO_4^{3-}$) band at 958 cm$^{-1}$ was used as a marker for completely mineralized bone (mature), while the hydroxyphosphate ($HPO_4^{2-}$) band at 1000 cm$^{-1}$ was used to identify incompletely mineralized, newly generated bone. By combining Raman imaging with factor analysis, images representative of the Raman signal from $(PO_4)^{3-}$ and from $(HPO_4)^{2-}$ were generated. Examination of these images revealed that even in mature bone regions, new bone formation and regeneration occurred. A similar analysis was performed on newly generated bone. In these samples it was not possible to separate the two categories by factor analysis, leading the authors to conclude that the $(PO_4)^{3-}$ and $(HPO_4)^{2-}$ were more evenly distributed in the newly generated bone than in mature bone (80).

The chemical composition of bone has also been analyzed using a synchrotron radiation source (82). Two infrared spectroscopic mapping methods were used to obtain high-resolution (5-μm) infrared spectroscopic images. In the osteon method, linear maps are generated from the center of the osteon, which is composed of new bone, to the periphery, consisting of older bone. The transverse method, which was specifically applied to subchondral bone, involved collecting linear maps extending from the older bone at the edge of the particular cartilage to the newer bone at the marrow space. Age-dependent comparisons for both methods were made based on the protein/mineral and phosphate/carbonate integrated area band ratios. It was observed that most of the chemical changes seen in the bone samples occurred in the first 20 μm from the site of new bone growth. The analysis of both methods showed that the protein/mineral ratio, generated by dividing the integrated area of the amide I band (1600–1700 cm$^{-1}$) by the integrated area of either the phosphate (925–1180 cm$^{-1}$) or the carbonate (1395–1425 cm$^{-1}$) bands, was higher in new bone and decreased as the bone aged and mineralized (82). The decrease in the phosphate/carbonate ratio observed as bone ages suggests that during the aging process carbonate is substituted into the phosphate sites of the hydroxyapatite lattice. High-resolution synchrotron infrared spectroscopic mapping, when combined with conventional light microscopy, provides an in vitro method to correlate observed microstructures with their chemical composition (82).

Infrared spectroscopic imaging using a conventional radiation source has been employed to study the mineralized bone tissue (83,84). Using a mercury-cadmium-telluride (MCT) FPA detector coupled to a Fourier transform infrared microscope, it has become possible not only to image the mineral and organic matrix content, but also to image bone maturity (83). As an example, the maturity image of a canine alveolar bone sample is shown in Fig. 6. This image was

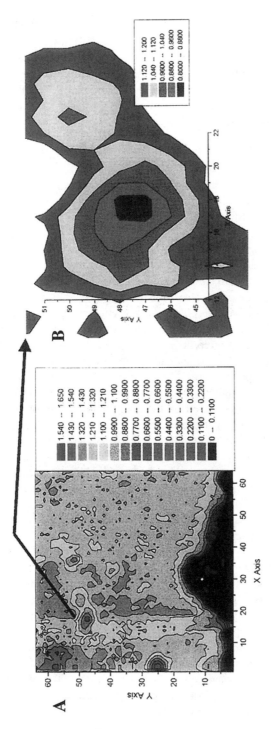

**Fig. 6** Imaging bone mineral maturity. (A) Mineral maturity of a bone section determined by dividing the image absorbance associated with the stoichiometric substituted apatites (1030 cm⁻¹) by the absorbance of the nonstoichiometric ones (1020 cm⁻¹). (B) An expanded view of two osteons from (A) (From Ref. 83. © 1998 *Cellular and Molecular Biology*.)

generated by ratioing the absorbance at a band indicative of stoichiometric substituted apatites ($1030$ cm$^{-1}$) by the absorbance at a band representing poorly crystalline, nonstoichiometric substituted apatites ($1020$ cm$^{-1}$). The ratio of these bands is a measure of the mineral crystallinity/maturity and is consistent with the results of previous studies (85).

## 3.4  Diseased Tissue

Vibrational spectroscopy has been used extensively to examine differences in spectral signatures between diseased and normal tissue. These studies may represent the greatest potential for vibrational spectroscopic imaging to impact medical research in the diagnosis of disease (23,24,86–93). Representative examples of these studies in pathological determinations are discussed in this section.

A study of the effect of a cholesterol-rich diet on the livers of hypercholesterolemic rabbits was undertaken to understand the spectral changes that occur between normal and hypercholesterolemic livers (94). A high-cholesterol diet causes the level of serum cholesterol, detected as cholesterol esters in lipoprotein particles, to increase in the hypercholesterolemic rabbits. Functional-group mapping and cluster analysis were employed to generate chemically significant images that could be compared to the stained tissue. A map of the integrated intensity of the amide I absorption band ($1600–1700$ cm$^{-1}$) exhibited a homogeneous distribution of protein throughout the tissue. This is not the case for the lipid distribution, which is mapped based on the acyl chain absorption band at $2925$ cm$^{-1}$, suggesting that the two categories, lipids and proteins, are differently dispersed and not colocalized. In this study, the researchers demonstrated that functional-group mapping based on IR microspectroscopy can be used to identify specific biological materials (protein, lipids, etc.) within the tissue at a spatial resolution of 40 μm. They have also shown that regions of the tissue with similar spectra may be nonsubjectively classified and mapped through cluster analysis, an analytical technique that creates groups, or clusters, based on comparisons between either the entire spectrum or a particular section of each spectrum and all other spectra (94).

Infrared spectroscopic mapping has also been investigated as an adjunct to the conventional Pap smear test for cervical cancer (95). The current, conventional protocol requires a highly trained technician to make an informed, but subjective decision between normal and abnormal cells based on visual inspection using an optical microscope; it is recognized that erroneous test results often occur. The IR microscopic mapping technique employed in this comparative study recorded spectra at 100-μm intervals across the surface of a 5-mm BaF$_2$ window onto which the cervical-cell films were cast (95). The objective of this work was to relate observed spectral changes to the disease state of the cells and to determine whether these spectral differences were evenly distributed in the sample or whether they arose from specific regions of morphologically different cells. A variety of postprocessing methods was applied to the data in an attempt to establish a method for cancer screening (95). These methods included baseline-corrected peak height, band ratioing to compensate for sample thickness, correlation metrics, similarity matching, and a Mahalanobis distance classification. The results from each processing method were compared with samples that had been diagnosed as either "within normal limits" or being cervical intraepithelial neoplasia positive by a pathologist observing Pap smears taken before the samples were prepared for FT-IR analysis. It was found that an adequate qualitative-feature map could be generated by dividing the glycogen image of the data set by the amide II image to account for differences in sample thickness (95). In order to create a quantitative image, a correlation method was used that compared the spectroscopic image data to reference spectra from pure components. It was observed during this study that spectral features attributed to either normal or abnormal samples were relatively constant across the entire sample. This indicated that spectral changes do not arise from individual abnormal cells, but instead are attributed to chemical differences in a majority of the cells obtained from the Pap smear (95). A larger goal of the study was to determine the feasibility of FT-IR spectroscopic imaging as a biodiagnostic for early cancer screening. It is speculated that the demonstration of FT-IR microscopy as a biodiagnostic tool may eventually lead to the development of a rapid, easy-to-use, automated technique for early cervical cancer detection.

A study by Lasch and Naumann explored FT-IR spectroscopic mapping to distinguish human melanoma and colon carcinoma tissue from normal tissue (96). They applied several postprocessing methods to the data, including functional-group mapping, cluster analysis, principle component analysis, and artificial neural networks. While assessing the applicability of each of these, it was discovered that although the chemical images produced by functional-group mapping showed the spatial distribution of different tissue structures, the technique was unable to establish a reliable reference set for tissue differentiation (96). By applying methods based on pattern recognition, it was possible to improve the discrimination between different tissue types as well as to increase the contrast between normal and cancerous tissue. Unsupervised methods, such as cluster analysis and principal component analysis, were useful in cases where the exact spectra or number of components in the tissue were unknown. Use of artificial neural network processing, which is a supervised classification method, produced erroneous results if not all spectral components were identified in the training set (96). This work demonstrates that high-quality IR images based on pattern recognition methods can be obtained for certain types of diseased tissue and shows that spectral information can be used to differentiate biological tissues.

Mantsch and coworkers have made many contributions to biological vibrational spectroscopy and mapping procedures (8,21,27,89–94,97,98). Two particularly interesting applications employ IR image mapping to determine the distribution of collagen in diseased hearts (8) and to examine the distribution of biochemical components within cancerous tissue (98). The goal in the first study was to evaluate IR spectroscopic mapping as an alternative method to in vivo detection of collagen deposition in heart tissue. To accomplish this, the researchers used sections from cardiomyopathic and normal hamster hearts to compare patterns of cardiac collagen deposition (8). The integrated intensity of the band at 1204 cm$^{-1}$ was used to generate chemical maps of collagen distribution. The ability of the IR maps to visualize collagen apportionment was verified in comparisons with tissue sections stained with trichrom, a dye identifying the interstitial matrix proteins in cardiac myocytes. The images reflected a detection limit within a single volume element in the map (50 × 50 × 10 μm) of 25 ng and showed that the collagen concentration in 200-day cardiomyopathic animals was elevated relative to that of the control samples (8).

In another application, these authors examined cellular variations of proteins, specifically keratin, in squamous carcinomas of the oral/oropharyngeal epithelium by using FT-IR point microspectroscopy to map the distribution of components in the tissue (98). A typical feature found in well-differentiated squamous cell carcinomas (SCCs) is the existence of keratin pearls. An image of an unstained keratin pearl sample is shown in Fig. 7A. Maps showing the distribution of keratin (1305 cm$^{-1}$) and DNA (968 cm$^{-1}$), as well as a map of membrane fluidity based on the lipid band (2852 cm$^{-1}$) are presented in Figs. 7B–D, respectively. The keratin biochemical map, in Fig. 7B, not only shows a high concentration of keratin in the center of the pearl but also reveals additional entities not apparent in the visible image (Fig. 7A), including the extension of the pearl toward the lower right-hand corner of the image and the existence of a substructure within the pearl. The map in Fig. 7C shows a significant decrease of DNA in the pearl and its surrounding area, supporting the theory of keratin pearl development (98). The membrane fluidity map in Fig. 7D shows a significant decrease in the cells in the center of the pearl and serves to add additional information to the cell differentiation model (98). The IR spectroscopic images in Fig. 7 suggest that the keratin pearl can be described as a densely folded and highly compressed superficial layer that extends far beyond that seen in the visible image (98). With the application of additional image analysis modalities based on two-dimensional tissue classification (98), IR spectroscopic mapping provides valuable information for SCC diagnosis and prognosis.

Vibrational spectroscopic imaging is being applied to a wide variety of areas of brain research, including strokes (7), neurotoxicity (99), genetic disease (100,101), Alzheimer's disease (AD) (97), and metabolism (102). It has been suggested that amyloid deposition is important in the pathogenesis of the most common form of dementia, Alzheimer's disease (103,104). Synchrotron FT-IR spectroscopic mapping has been employed to study β-amyloid peptide in AD

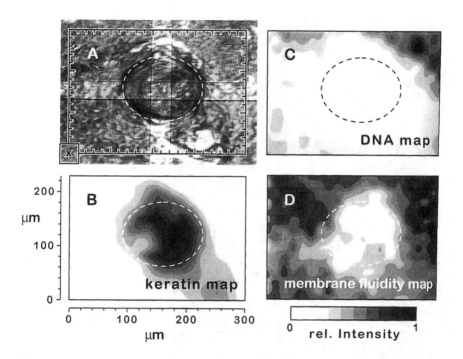

**Fig. 7** Biochemical mapping of a keratin pearl. (A) Unstained image of the keratin pearl. The frame overlaying the image outlines the area where the 315 individual IR spectra used to create (B)–(D) were collected. The square in the lower left corner indicates the image scale and is 30 × 30 μm. The double cross in the middle of the image displays the aperture system and is transparent only to visible light. (B) Gray-scale contour map with relative infrared intensities for protein based on the keratin peak at 1305 cm⁻¹. (C) Gray-scale contour map with relative infrared intensities for DNA using the band at 968 cm⁻¹. (D) Gray-scale contour map with relative membrane fluidity derived from the lipid band at 2852 cm⁻¹. (From Ref. 98. © 1998 *Cellular and Molecular Biology.*)

brain tissue in an attempt to further the understanding of the structure–function relationship of this peptide and its relationship to the pathogenesis of the disease (97). Results of the in situ study on intact slices of AD brain tissue confirmed that conformation of proteins in the gray matter are predominately α-helical and/ or unordered, while the conformation in the amyloid is mostly a β-sheet configuration. By generating images based on the shift in the frequency of the amide I peak, it is possible, as shown in Fig. 8, to monitor the conformational changes across the sample. A three-dimensional surface plot of the integrated intensity of the area under the amide I band is shown in Fig. 8A. It can be seen from the plot that a substantial increase in the integrated intensity occurs in the region of

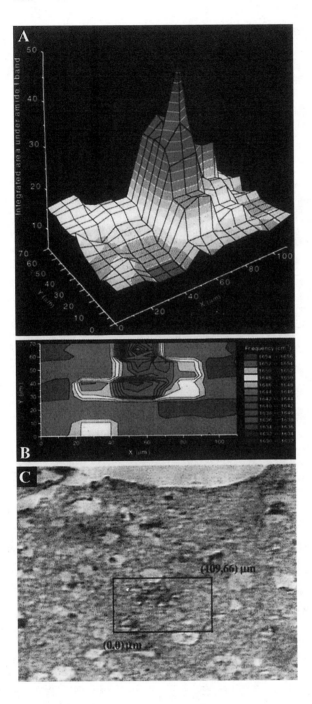

the plaque compared to that of the surrounding tissue. Figure 8B presents a map of the frequency of the amide I band, where a frequency around $1652–1654 \ cm^{-1}$ is expected for proteins in the $\alpha$-helical and/or unordered state and a shift to lower frequency indicates a greater contribution from the $\beta$-sheet conformation. These results indicate that the feature highlighted in Figs. 8A and B is composed of proteins in the $\beta$-sheet conformation, and suggests that it is an amyloid core of a mature neuritic plaque. Figure 8C shows the AD brain slice after it has been stained with Congo red, further corroborating the conclusions derived from the IR maps.

Lewis and coworkers have used wide-field FT-IR spectroscopic imaging with FPA detection in several studies to examine biochemical changes in the brain due to genetic disease and neurotoxicity (25,99–101). One study used a macroscopic imaging approach to examine the biochemical modifications that occur in the cerebellum of a Niemann–Pick type C (NPC) mouse (100). This NPC is representative of a genetic lipid-storage disease exhibiting neuropathologic effects. In this study, 10-micron sections from both diseased and control samples were examined and compared to respective contiguous sections that had undergone standard Nissl staining to reveal the protein and lipid distributions within the sample. Infrared spectroscopic images of the lipid/protein distribution were created by rationing a predominately lipid image collected at $2927 \ cm^{-1}$ to a protein image at $3290 \ cm^{-1}$. The resulting images for both the control and diseased brain slices revealed that the white matter contained the highest relative amount of lipid, followed by the molecular and granular layers. It was also observed that the lipid in the white matter in the NPC sample had decreased relative to the control. Because of the large sample population ($>4000$/sample), the statistical significance of the lipid decrease in the NPC samples compared to the controls could be tested. An analysis of variance showed that the decrease in the lipid content observed predominately in the white matter was statistically significant at the 95% confidence interval. It was also determined that the effect in the molecular and granular layers was not statistically significant at this confidence level. It is easily appreciated that by applying statistical methods to the thousands of spectra that make up the spectroscopic imaging data set, one can augment standard histopathologic methods to simultaneously provide a visualization scheme

---

**Fig. 8** Synchrotron IR mapping of $\beta$-amyloid in Alzheimer's disease tissue. (A) The integrated area under the amide I band is used to generate a three-dimensional surface map of the region marked in (C). (B) Frequency shifts are mapped using two-dimensional contour image of the amide I peak frequency of the region marked in (C). (C) Photomicrograph of the tissue of interest after staining with Congo red for amyloid. A deposit appears as a dark area in the upper center portion of the boxed region. (From Ref. 97. © 1996 Biophysical Society.)

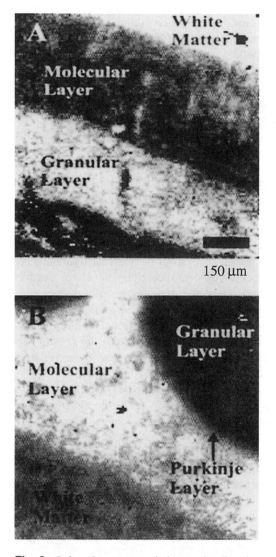

**Fig. 9**  Infrared spectroscopic image showing the spatial distribution of lipid and protein in thin cerebellar sections from a control and a cytarabine-treated animal. Ratio images are generated by dividing the lipid-absorbance image at 2927 cm$^{-1}$ by the protein-absorbance image at 3350 cm$^{-1}$. (A) Ratio image for a control rat treated with saline. (B) Ratio image for a drug-treated rat. (From Ref. 99. © 1998 *Cellular and Molecular Biology*.)

with a quantitative means of describing biochemical heterogeneity within a sample.

Wide-field FT-IR spectroscopic imaging is not limited solely to macroscopic work, and it has been employed to study brain neurotoxicity on the microscopic scale (99,101). Brain tissue sections from rats treated with the antineoplastic drug cytarabine, along with control samples, were imaged in order to correlate lipid and protein distributions within specific cell types in the cerebellum. Images of cerebellum sections showing the lipid protein distribution of the control and cytarabine-treated rats were generated by ratioing the 2927 cm$^{-1}$ lipid band to the protein band at 3380 cm$^{-1}$ and are shown in Fig. 9A and B, respectively. These images emphasize the variation in biochemical composition between tissue cell layers. Changes between the control and cytarabine-treated samples can readily be seen. For example, the boundary region between the Purkinje and molecular cell layers is represented by a minimum in the lipid/protein ratio in both the control and treated samples, although the transition is less noticeable in the control samples. Use of a scatter plot comprised of two spectral bands, such as those for lipids and proteins, provides an unsupervised method of examining the populations present in the image. By selecting specific populations from the scatter plot, spatial location maps can be obtained. An example of this is presented in Fig. 10, in which three distinct populations in a drug-treated sample are ob-

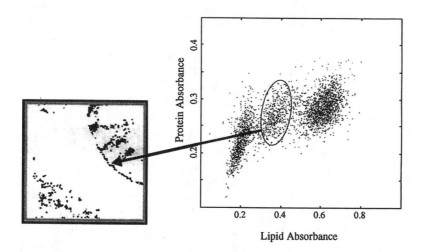

**Fig. 10** Spatial plot of lipid and protein peak absorbance values observed in a cytarabine-treated animal (see Fig. 9B for original lipid/protein image). The population encompassed by the ellipse is mapped back onto the sample image, creating a location map that demonstrates the localization of this population to both the Purkinje and granular layers. (From Ref. 99. © 1998 *Cellular and Molecular Biology*.)

served. The values encompassed by the ellipse are localized within the Purkinje and granular layers.

## 4  CONCLUSIONS

Vibrational spectroscopic imaging techniques based on Raman, mid-infrared (IR), and near-infrared (NIR) have the potential to become indispensable tools for biological and medical research because of their broad applicability and seamless integration with conventional imaging approaches. These techniques provide researchers with the ability to obtain both qualitative and quantitative molecular and structural information for individual sample components. Imaging methods also intrinsically maintain the visualization capabilities necessary for many histological and pathological protocols while at the same time building upon the extensive diagnostic capabilities of vibrational spectroscopy. A large sampling population, such as that obtained with two-dimensional FPA detection, aids in the generation of meaningful two-dimensional maps based on the heterogeneity in a sample's chemistry, morphology, or physical properties. Furthermore, with the immense amount of data implicit in the hyperspectral representation, statistical methods for testing the significance of observed spectral changes significantly enhance the ability to meaningfully interpret the spectral data. This is particularly important in working with complex, heterogeneous biological systems.

To realize its full potential in clinical diagnostics, vibrational spectroscopic imaging must develop interactive systems and methodologies that can be implemented in clinical settings to provide in situ analyses in real time. Although vibrational spectroscopic imaging is ideal for in situ situations due to the noninvasive nature of the technique, several operational points need to be considered. Since most biological samples contain a high percentage of water, which is strongly absorbing in the mid-infrared, imaging techniques employing this spectral range would be expected to have limited functionality in "real-world" clinical settings. In contrast to the mid-infrared, Raman spectroscopy is relatively unaffected by the presence of water in tissue. This suggests that Raman methods may be amenable to real-time clinical studies once the ubiquitous contaminating strong fluorescence signals have been minimized and the inherently low signal levels are amplified to useful levels. A currently underutilized technique, NIR spectroscopic imaging, however, may have more extensive clinical utility (105). Near-infrared spectral signal levels, corresponding to overtone and combination bands, are orders of magnitude stronger than Raman signals, potentially allowing rapid, real-time data acquisition.

From this summary of both IR and Raman vibrational spectroscopic imaging approaches, it can be appreciated that a consideration of the capabilities and limitations of each technique should, in the future, make it possible to design instrumental assemblies optimized for a particular research or clinical application.

Developments in both the hardware and software utilized in vibrational spectroscopic imaging methodologies are rapidly advancing toward achieving the goal of automated, clinical diagnostics for incorporation into standard clinical and pathological protocols.

## REFERENCES

1.  FS Parker. Applications of Infrared, Raman, and Resonance Raman Spectroscopy in Biochemistry. New York: Plenum Press, 1983, pp 421–479.
2.  RNAH Lewis, RN McElhaney. In: HH Mantsch, D Chapman, eds. Infrared Spectroscopy of Biomolecules. New York: Wiley-Liss, 1996, pp 159–202.
3.  JJ Duindam, GFJM Vrensen, C Otto, J Greve. Opthalmic Res 28:86–91, Suppl 1, 1996.
4.  P Lasch, CP Schultz, D Naumann. Biophys J 75:840–852, 1998.
5.  R Mendelsohn, DJ Moore. Chem Phys Lipids 96:141–157, 1998.
6.  FS Parker. Applications of Infrared, Raman, and Resonance Raman Spectroscopy in Biochemistry. New York: Plenum Press, 1983, pp 83–153.
7.  RJ Dempsey, LA Cassis, DG Davis, RA Lodder. Ann NY Acad Sci 820:149–169, 1997.
8.  KZ Liu, IMC Dixon, HH Mantsch. Cardiovasc Pathol 8:41–47, 1999.
9.  PA Terpstra, C Otto, GMJ Segersnolten, JS Kanger, J Greve. Biospectrosc 1:255–263, 1995.
10. C Marcott, RC Reeder, JA Sweat, DD Panzer, DL Wetzel. Vib Spectrosc 19:123–129, 1999.
11. DL Wetzel, JA Reffner. Cereal Food World 38:9–20, 1993.
12. DL Wetzel, RG Fulcher, RG Messerschmidt. Cereal Food World 33:663–663, 1988.
13. D Neumann, CP Schultz, D Helm. In: HH Mantsch, D Chapman, eds. Infrared Spectroscopy of Biomolecules. New York: Wiley-Liss, 1996, pp 279–310.
14. D Helm, D Naumann. FEMS Microbiol Lett 126:75–79, 1995.
15. HC Vandermei, D Naumann, HJ Busscher. Arch Oral Biol 38:1013–1019, 1993.
16. D Naumann, D Helm, H Labischinski. Nature 351:81–81, 1991.
17. D Naumann, V Fijala, H Labischinski, Mikrochim Acta 1:373–377, 1988.
18. D Naumann, V Fijala, H Labischinski, P Giesbrecht. J Mol Struct 174:165–170, 1988.
19. D Naumann, H Labischinski, W Ronspeck, G Barnickel, H Bradaczek. Biopolymers 26:765–817, 1987.
20. FS Parker. Applications of Infrared, Raman, and Resonance Raman Spectroscopy in Biochemistry. New York: Plenum Press, 1983, pp 155–185.
21. M Jackson, HH Mantsch. In: HH Mantsch, D Chapman, eds. Infrared Spectroscopy of Biomolecules. New York: Wiley-Liss, 1996, pp 311–340.
22. L Chiriboga, P Xie, H Yee, V Vigorita, D Zarou, D Zakim, M Diem. Biospectrosc 4:47–53, 1998.
23. L Chiriboga, P Xie, V Vigorita, D Zarou, D Zakim, M Diem. Biospectroscopy 4:55–59, 1998.

24. L Chiriboga, P Xie, W Zhang, M Diem. Biospectroscopy 3:253–257, 1997.
25. EN Lewis, AM Gorbach, C Marcott, IW Levin. Appl Spectrosc 50:263–269, 1996.
26. SM LeVine, DL Wetzel, SW Dickson. Mikrochim Acta 14:463–466, Suppl, 1997.
27. MF Stranc, MG Sowa, B Abdulrauf, HH Mantsch. Brit J Plast Surg 52:210–217, 1998.
28. M Delhaye, PJ Dhamelincourt. J Raman Spectrosc 3:33–43, 1975.
29. CJ de Grauw, C Otto, J Greve. Appl Spectrosc 51:1607–16–12, 1997.
30. CJH Brenan, IW Hunter. J Raman Spectrosc 27:561–570, 1996.
31. NM Sijtsema, SD Wouters, CJ de Grauw, C Otto, J Greve. Appl Spectrosc 52:348–355, 1998.
32. M Bowden, GD Dickson, DJ Gardiner, DJ Wood. Appl Spectrosc 44:1679–1684, 1990.
33. M Bowden, DJ Gardiner, G Rice, DL Gerrand. J Raman Spectrosc 21:37–41, 1990.
34. PJ Treado, MD Morris. Appl Spectrosc 44:1–4, 1990.
35. PJ Treado, MD Morris. Appl Spectrosc 43:190–192, 1989.
36. PJ Treado, MD Morris. Appl Spectrosc 42:897–901, 1988.
37. P Hansen, J Strong. Appl Opt 11:502, 1972.
38. RD Swift, RB Wattson, JA Decker Jr, R Paganetti, M Herwit. Appl Opt 11:1596, 1972.
39. DC Tilotta, RM Hammaker, WG Fateley. Appl Spectrosc 41:727–734, 1987.
40. RA DeVerse, RM Hammaker, WG Fateley. Vib Spectrosc 19:177–186, 1999.
41. SR Goldstein, LH Kidder, TM Herne, IW Levin, EN Lewis. J Microsc Oxford 184:35–45, Part 1, 1996.
42. DN Batchelder, C. Cheng, GD Pitt. Adv Mater 3:566–568, 1991.
43. GJ Puppels, M Grond, J Greve. Appl Spectrosc 47:1256–1267, 1993.
44. KA Christensen, NL Bradley, MD Morris, RV Morrison. Appl Spectrosc 49:1120–1125, 1995.
45. PJ Treado, IW Levin, EN Lewis. Appl Spectrosc 46:1211–1216, 1992.
46. PJ Treado, IW Levin, EN Lewis. Appl Spectrosc 46:553–559, 1992.
47. EN Lewis, PJ Treado, IW Levin. Appl Spectrosc 47:539–543, 1993.
48. MD Schaeberle, JF Turner II, PJ Treado. Proc SPIE—Int Soc Opt Eng 2173:11–20, 1994.
49. MD Schaeberle, CG Karakatsanis, CJ Lau, PJ Treado. Anal Chem 67:4316–4321, 1995.
50. MD Schaeberle, VF Kalasinsky, JL Luke, EN Lewis, IW Levin, PJ Treado. Anal Chem 68:1829–1833, 1996.
51. HT Skinner, TF Cooney, SK Sharma, SM Angel. Appl Spectrosc 50:1007–1014, 1996.
52. HR Morris, CC Hoyt, PJ Treado. Appl Spectrosc 48:857–866, 1994.
53. HR Morris, CC Hoyt, P Miller, PJ Treado. Appl Spectrosc 50:805–811, 1996.
54. HR Morris, B Munroe, RA Ryntz, PJ Treado. Langmuir 14:2426–2434, 1998.
55. NJ Kline, PJ Treado. J Raman Spectrosc 28:119–124, 1997.
56. JR Schoonover, F Weesner, GJ Havrilla, M Sparrow, P Treado. Appl Spectrosc 52:1505–1514, 1998.
57. MA Harthcock, SC Atkin. Appl Spectrosc 42:449–455, 1988.

58. JA Majewski, DH Matthiesen. J Crystal Growth 137:249–254, 1994.
59. D Wienke, K Cammann. Anal Chem 68:3987–3993, 1996.
60. D Wienke, K Cammann. Anal Chem 68:3994–3999, 1996.
61. EN Lewis, IW Levin. Appl Spectrosc 49:672–678, 1995.
62. NJ Chanover, DA Glenar, JJ Hillman. J Geophysical Res Planets 103:31335–31348, 1998.
63. PJ Treado, IW Levin, EN Lewis. Appl Spectrosc 48:607–615, 1994.
64. EN Lewis, PJ Treado, RC Reeder, GM Story, AE Dowrey, C Marcott, IW Levin. Anal Chem 67:3377–3381, 1995.
65. Spectral Dimensions Inc. Website: http://www.spectraldimensions.com.
66. LH Kidder, IW Levin, EN Lewis, VD Kleinman, EJ Heilweil. Opt Lett 22:742–744, 1997.
67. EN Lewis, LH Kidder, JF Arens, MC Peck, IW Levin. Appl Spectrosc 51:563–567, 1997.
68. H Kaplan. Practical Applications of Infrared Thermal Sensing and Imaging Equipment. 2nd ed. Bellingham, WA: SPIE, 1999.
69. LH Kidder, AS Haka, PJ Faustino, DS Lester, IW Levin, EN Lewis. SPIE 3257:178–186, 1998.
70. KS Kalasinsky. Cell Mol Biol 44:81–87, 1998.
71. N Jamin, P Dumas, J Moncuit, WH Fridman, JL Teillaud, GL Carr, GP Williams. Proc Natl Acad Sci USA 95:4837–4880, 1998.
72. N Jamin, P Dumas, J Moncuit, WH Fridman, JL Teillaud, GL Carr, GP Williams. Cell Mol Biol 44:9–13, 1998.
73. GJ Puppels, FFM de Mul, C Otto, J Greve, M Robert-Nicoud, DJ Arndt-Jovin, TM Jovin. Nature 347:301–303, 1990.
74. C Otto, CJ de Grauw, JJ Duindam, NM Sijtsema, J Greve. J Ram Spectrosc 28:143–150, 1997.
75. DS Himmelsbach, S Khalili, DE Akin. Cell Mol Biol 44:99–108, 1998.
76. DL Wetzel, AJ Eilert, LN Pietrzak, SS Miller, JA Sweat. Cell Mol Biol 44:145–167, 1998.
77. C Marcott, RC Reeder, JA Sweat, DD Panzer, DL Wetzel. Vib Spectrosc 19:123–129, 1999.
78. SR Ali, FB Johnson, JL Luke, VF Kalasinsky. Cell Mol Biol 44:75–80, 1998.
79. LH Kidder, VF Kalasinsky, JL Luke, IW Levin, EN Lewis. Nature Medicine 3:235–237, 1997.
80. JA Timlin, A Carden, MD Morris, JF Bonadio, CE Hoffler II, KM Kozloff, SA Goldstein. J Biomed Optics 4:28–34, 1999.
81. JA Pezzuti, MD Morris, JF Bonadio, SA Goldstein. SPIE 3261:270–276, 1998.
82. LM Miller, CS Carlson, GL Carr, MR Chance. Cell Mol Biol 44:117–127, 1998.
83. C Marcott, RC Reeder, EP Paschalis, DN Tatakis, AL Boskey, R Mendelsohn. Cell Mol Biol 44:109–115, 1998.
84. C Marcott, RC Reeder, EP Paschalis, DN Tatakis, AL Boskey, R Mendelsohn. 19th Annual International Conference of the IEEE Engineering in Medicine and Biology Society. Paper #753, Oct. 31, 1997.
85. EP Paschalis, E DiCarlo, F Betts, P Sherman, R Mendelsohn, AL Boskey. Calcif Tissue Int 59:480–87, 1996.

86. TCB Schut, GJ Puppels, YM Kraan, J Greve, LLJ VanderMaas. Int J Cancer 74: 20–25, 1997.
87. SM Levine, DL Wetzel. Free Radical Bio Med 25:33–41, 1998.
88. SM Levine, DL Wetzel. Appl Spectrosc Rev 28:385–412, 1993.
89. M Jackson, JR Mansfield, B Dolenko, RL Somorjai, HH Mantsch, PH Watson. Cancer Detect Prev 23:245–253, 1999.
90. CP Schultz, KZ Liu, PD Kerr, HH Mantsch. Oncol Res 10:277–286, 1998.
91. MG Sowa, JR Mansfield, GB Scarth, HH Mantsch. Appl Spectrosc 51:143–152, 1997.
92. CP Schultz, KZ Liu, JB Johnston, HH Mantsch. Leukemia Res 20:649–655, 1996.
93. H Fabian, M Jackson, L Murphy, PH Watson, I Fichtner, HH Mantsch. Biospectrosc 1:37–45, 1995.
94. M Jackson, B Ramjiawan, M Hewko, HH Mantsch. Cell Mol Biol 44:89–98, 1998.
95. SR Lowry. Cell Mol Biol 44:169–177, 1998.
96. P Lasch, D Naumann. Cell Mol Bio 44:189–202, 1998.
97. L Choo, DL Wetzel, WC Halliday, M Jackson, SM Levine, HH Mantsch. Biophys J 71:1672–1679, 1996.
98. CP Schultz, HH Mantsch. Cell Mol Biol 44:203–210, 1998.
99. DS Lester, LH Kidder, IW Levin, EN Lewis. Cell Mol Biol 44:29–38, 1998.
100. LH Kidder, P Colarusso, SA Stewart, IW Levin, NM Appel, DS Lester, PG Pentchev, EN Lewis. J Biomedical Optics 4:7–13, 1999.
101. EN Lewis, LH Kidder, IW Levin, VF Kalasinsky, JP Hanig, DS Lester. Ann NY Acad Sci 820:234–247, 1997.
102. DL Wetzel, DN Slatkin, SM Levine. Cell Mol Biol 44:15–27, 1988.
103. DJ Selkoe. J Neuropathol Exp. Neurol 55:438–447, 1994.
104. SS Sisodia, DL Price. FASEB J 9:366–370, 1995.
105. RJ Dempsey, DG Davis, RG Buice, RA Lodder. Appl Spectrosc 50:18A–34A, 1996.

# 8

# Clinical Applications of Near- and Mid-Infrared Spectroscopy

**H. Michael Heise**
*Institute for Spectrochemistry and Applied Spectroscopy at Dortmund University, Dortmund, Germany*

## 1  INTRODUCTION

Nowadays, medical diagnostics is efficiently supported by clinical chemistry. Many different human biosamples have been investigated by chemical, biochemical, and physical methods to provide information to the physician on the pathophysical state of the patient. Traditionally, infrared (IR) spectroscopy has been one of the most important physical methods in the chemical laboratory, since it plays an important role in the elucidation of structures and in the identification of all categories of compounds. It has also been accepted for the study of biological molecules (1,2) of different categories, such as proteins and peptides, chromophores, and chromophoric proteins, nucleic acids, carbohydrates, lipids, and biomembranes, as presented within a recent conference (3). Despite this fact, for many years the application of IR spectroscopy to clinical chemistry was only rudimentary. One field of early application worth being mentioned was the analysis of urinary calculi based on the KBr pellet technique (4). A previous review (5) marked the entry of IR spectroscopy for the quantitative analysis of biofluids, of which whole blood and blood fluids are of greatest importance. An obstacle to the analysis of biofluids was the high water content in such samples, so new, emerging techniques, such as attenuated total reflectance (ATR), or the exploitation of near-infrared (NIR) spectroscopy were necessary to access high-quality spectral data from such specimens, leading to a new dimension in clinical chemistry.

Establishing infrared spectroscopy as a diagnostic tool certainly presents significant challenges. It seems to run contrary to the trend of reducing costs for clinical assays. On the other hand, the investment for a small spectrometer, serving for the mid-infrared spectral range with medium spectral resolution, is currently comparable to that for a gas chromatograph, although several approaches are under development to reduce the price and some IR photometers are already on the market (6).

The versatility of infrared spectroscopy is certainly its great advantage, because it is applicable to gaseous, liquid, and solid samples. The combination of IR spectrometry with chromatography for the analysis of drugs, toxic compounds, and their metabolites provides splendid possibilities, as the detection limits continue to be lowered (7,8). An example for precision trace gas analysis by Fourier transform infrared (FT-IR) spectroscopy is given for the determination of the $^{13}C/^{12}C$ isotope ratio of $CO_2$, which can be used for a variety of human breath tests (9). It was recently discovered that the bacterium *Helicobacter pylori*, when colonizing the inner stomach lining, is responsible for most stomach ulcers. The bacterial urease secreted hydrolyses ingested $^{13}C$-labeled urea to $^{13}CO_2$ and $NH_3$, which enter the blood and are exhaled (10). A quite different application is the investigation of drugs localized in the hair of drug-abusing persons by infrared microscopy (11). There are many different measurement techniques, in particular for solid samples, either homogeneous or inhomogeneous, and the latter types can again be studied advantageously by infrared microscopy.

Conventionally, samples are taken by biopsies, punctuation, or using syringes. For diabetic patients, simple puncturing of the skin by using lancets to draw a blood sample from the capillary bed of the dermis is a routine procedure for self-monitoring of blood glucose, although some minimally invasive techniques have been developed for harvesting body fluids, in particular, interstitial fluid from dermis or subcutaneous tissue. One emerging technology uses, for example, reverse iontophoresis to extract biofluids electro-osmotically. Thermal microporation and suction techniques have also been proposed, especially for blood glucose testing; for a recent review see Ref. 12. Sample volume can be reduced to a few microliters, so appropriate measurement techniques have to be developed for IR spectroscopic assays.

Continuous measurements are often necessary in intensive care units, which can be realized with extracorporal sensing devices. Ex vivo sensors using IR spectrometry can be combined with microdialysis probes or systems that rely on open tissue perfusion. Noninvasive technology is based mainly on near-infrared spectroscopy. One established field is the measurement of oxygen saturation in blood using pulse-oximetry, although some deficiencies still have to be resolved (13). Recent applications tackle the imaging of larger heterogeneous areas, e.g., of skin flaps, to study the viability of critical regions (14).

One trend is certainly toward decentralization in the hospital, with the associated demand for small-sized instrumentation close to the patient. Similar requirements are claimed for assays, which can be performed in the physician's laboratory, thus avoiding central laboratories for special analysis parameters. Test strips based on solid-phase chemistry were developed for such purposes. In a similar way, however, less complicated and reagent-free assays are being tested employing IR spectroscopic dry-film measurements giving immediate test results. This move is parallel to the development of noninvasive measurement techniques, particularly aimed at blood glucose self-monitoring in diabetic patients (15).

The quantitative analysis of biofluids in the clinical chemistry laboratory using IR spectroscopic techniques is certainly attractive, due to the ease of sample handling, without the need for expensive reagents, and its potential for the simultaneous assay of several parameters. The number of applications has grown tremendously within the last decade, so these will be presented in the following sections. Although solid-phase measurements are also involved, particularly from dry-film measurements of biofluids, the analysis of solid samples from patients submitted for inspection and analysis to the clinical laboratory will be discussed separately.

## 2  MEASUREMENT TECHNIQUES AND CHEMOMETRICS

Advances in IR spectrometry during the last few years are due to improvements in measuring techniques and to refinements in data evaluation, which provide a significant impact on analytical assays for biomedical applications. In general, the mid-IR spectral range contains more selective information, in particular with respect to the fingerprint region, than does the near-IR, where most intensive vibrations stem from molecular moieties with hydrogen atoms involved, such as O—H, C—H and N—H. The short-wave near-infrared (SW-NIR) carries higher overtone bands with even smaller absorptivities, so sample pathlength for transmission measurements can be increased using cuvettes of up to centimeter pathlength. The visible and the SW-NIR spectral ranges also exhibit significant bands, due to electronic transitions.

The analysis of biofluids using mid-IR spectroscopy was severely hindered in the past, because water-compatible materials were needed, transparent for IR radiation within a broad spectral range. Water is nature's biosolvent, but its bands are intensive and cover most of the interesting mid-infrared regions; see Fig. 1. $CaF_2$ and other expensive materials (16) were needed as window material for transmission cells, which were used for standard measurements. The identification of major compounds in human urine is an example of an early investigation (17). Our first quantitative experiments were carried out with aqueous glucose solutions, measured in $CaF_2$ cells of about 50-$\mu$m optical pathlength, which was still tolerable to obtain good quantitative spectral information within the so-called fingerprint region. On the other hand, we noticed that reaching out for the lowest physiological concentrations, the rather high refractive index of the window material led to interference fringes in the spectrum, complicating the baseline. We tried

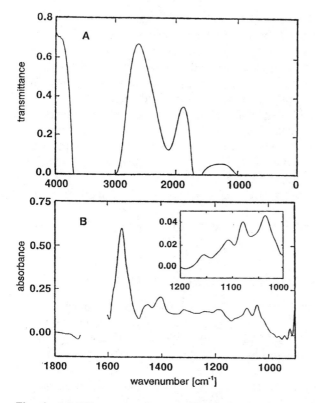

**Fig. 1** Mid-IR spectra of aqueous fluids measured in transmission (CaF$_2$-cell, sample pathlength 50 μm). A: Spectrum of water. B: Absorbance spectra of a reference serum (Seroquant-AU) and of an aqueous glucose solution (concentration 5 g/L; see inset) with water absorbance compensation. (Reproduced from Ref. 34.)

several defringing methods based on software implementations (18); however, essential advances were made by using the attenuated total reflection technique. The Circle cell from Spectra-Tech became commercially available, which is also still widely used in its microcell configuration, often equipped with a ZnSe crystal and advantageously employed as flow-through cell. In general the latter requires sample volumes of about 50 μl and more but still renders appropriate transmission-equivalent optical pathlengths for precise quantitative spectroscopy. A schematic diagram of the accessory, including a "ray tracing" for the attenuated total reflection experiment, is shown in Fig. 2A. A discussion of the measurement parameters important in ATR spectroscopy and optical penetration depth into the sample medium can be found, e.g., in Ref. 19.

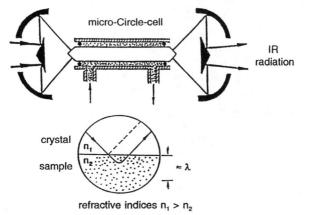

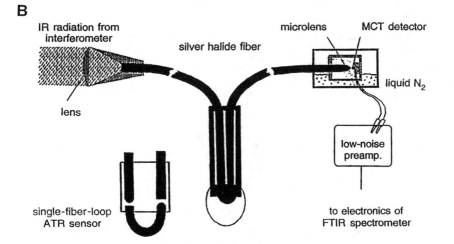

**Fig. 2** Biofluid assays using the ATR technique. A: Micro-Circle cell accessory with schematics of the electromagnetic wave leakage into the sample medium. B: Simple and inexpensive setup for evanescent wave spectroscopy using fiber optics. (Reproduced from Ref. 22.)

Recent achievements can be noticed in the field of mid-IR fiber-optic sensors for chemical analysis. With the availability of crystalline silver halide fibers, which are excellent optical materials due to their inertness and optical losses lower than 0.5 dB-m$^{-1}$ in the wavelength region between 3 and 16 μm (20), simpler, miniaturized, and less expensive accessories can be constructed, allowing even remote-sensing applications; see Fig. 2B and, e.g., Refs. 21–23.

The near-IR spectral range was originally serviced by extending the operating range of ultraviolet/visible (UV/VIS) instrumentation using scanning grating monochromators. Nowadays, many instruments are based on polychromators, with diode arrays to measure the whole spectrum simultaneously. Other measurement technologies are, for example, based on acousto-optical tunable filters or interferometers, which are the heart of Fourier transform spectrometers; for more information see Refs. 15 and 24. Optical materials in the near-IR for routine use can be of glass or quartz. For fiberoptic applications, one affordable material is especially low-OH-grade quartz, with high transmission reaching, in part, into the mid-IR range. Measurement accessories for remote sensing, such as required for catheters, but also for routine transmission or reflection experiments can be constructed from such fibers (15,25).

The diffuse-reflectance measurement technique has been very successful for near-IR spectroscopy. In Fig. 3, an efficient, custom-made accessory is shown that can be used for a variety of different samples, such as scattering fluids (whole blood), solid films, and powders (26). In addition, high-quality skin tissue spectra have been recorded, being essential for noninvasive blood glucose monitoring (27,28).

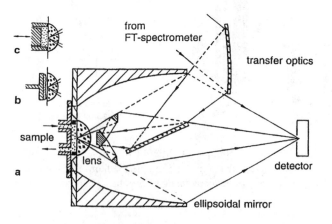

**Fig. 3** Diffuse-reflectance accessory used for near-IR spectrometry of scattering samples: (a) liquid samples, (b) solid films, and (c) powders. (Reproduced from Ref. 26.)

Because water is the main constituent of biofluids, its absorbance spectra in the different spectral ranges measured under appropriate conditions are shown in Fig. 4. The mid-IR spectral features in the fingerprint region, with band maximum at 1640 cm$^{-1}$, arise from the water-bending vibration. Interestingly, the major effects from temperature variations on the ATR absorbance spectrum are due to changes in density and refractive indices, giving rise to a baseline offset over a large interval (29).

For the bands of the fundamental OH-stretching vibrations of water covering broad sections above 3000 cm$^{-1}$ and corresponding combination and overtone bands in the near-IR, the temperature dependencies produce significant changes in the spectrum. Because of this effect, the hydrogen-bonding network of the water molecules in the liquid phase must be discussed, for this leads to a

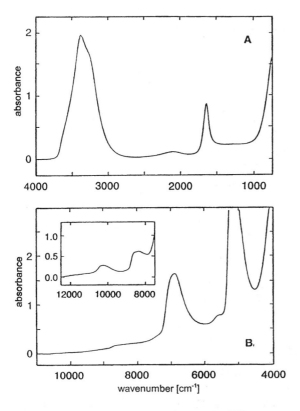

**Fig. 4** Absorbance IR spectra of water. A: ATR spectrum measured using a micro-Circle cell. B: Spectra measured by transmittance (respective cell pathlengths 1 mm and 10 mm) (SW-NIR, see inset). (Reproduced from Ref. 73.)

broadening of the absorption bands due to large variations in the molecular force field and structures resulting from intermolecular interactions. The association equilibrium between the water molecules is sensitive not only to temperature changes (30) but also to changes in concentrations of different electrolytes (31) and to changes in pH, allowing titrations to be followed by near-IR spectroscopy (32); and strong interactions with biopolymers such as proteins or carbohydrates can also be noticed. Recently, the water in skin has been studied with respect to such effects (33). In Fig. 5, the effects on the near-IR water absorbance spectrum are illustrated for changes in sample temperature and concentrations of NaCl and

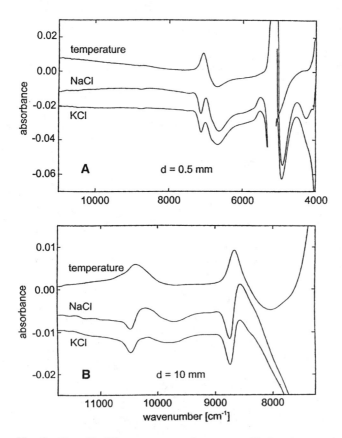

**Fig. 5** Near-IR difference spectra for water at 32.5°C, for an aqueous solution of NaCl and of KCl (5% each) versus a spectrum of water at 30°C. A: Near-IR spectra recorded with a cell of 0.5-mm pathlength. B: SW-NIR spectra recorded at 10-mm pathlength. (Reproduced from Ref. 41.)

KCl, respectively. For such reasons, near-IR spectroscopic in vitro experiments are usually carried out under strict cell thermostating conditions.

A discussion of different measurement techniques for the analysis of biofluids was recently published (34). To avoid the large spectral variations from water in the analysis of aqueous biofluids, dry-film measurements have been proposed after evaporation of the water content. In Fig. 6A, the ATR spectra were

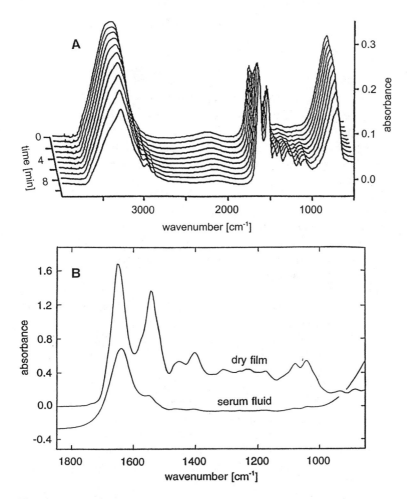

**Fig. 6** A: ATR spectra acquired during the process of drying of film of whole blood with single reflection. B: Absorbance spectra from a serum sample (Seroquant-AU) obtained with a single-loop fiber sensor (spectra of the serum fluid as well as from a dry film obtained by serum dipping and following water evaporation).

acquired during the process of drying a film of whole blood on a circular ZnSe crystal with a flat sample surface and single reflectance. Similar spectra can be recorded by using silver halide fibers dipped into the biofluid. After withdrawal of the fiber from the fluid, the aqueous film dries down within a few minutes. The composition of the liquid serum sample can be advantageously displayed after scaled subtraction of a water-absorbance spectrum, but an enlargement by a factor of 10 is needed to reach similar spectral absorbances as obtained with the dry-film measurement (see Fig. 6B).

Another technique for recording dry-film spectra is transmission, which still needs water-compatible but IR-transparent substrates, or reflection-absorption spectroscopy. The latter technique was used by us in combination with gold-coated, diffusely reflecting substrates, which has the advantage that the IR radiation passes through the sample twice to double the absorbance signals compared to a corresponding transmission experiment (see also Fig. 7). It is necessary to achieve homogeneous layers; otherwise, linearity as described by Lambert–Beer's law cannot be guaranteed (26). Dried biofilm preparations allow the analysis of compounds at trace concentrations due to the lower detection limits principally reached. Another advantage is their inherent potential for long-term archiving, which enables clinical IR fingerprinting of the patient.

A more complex approach has been to use cryoenrichment at temperatures of about $-2°C$ in combination with ATR measurements, where band intensities of biochemical components in an aqueous solution could be enhanced by a factor of 100 compared to measurements at room temperature (35).

Another technique, which is suited to optically ''thick'' solid samples such as found for dry blood films, is photoacoustic spectroscopy (34). Here, the optical sample pathlength depends on several measurement parameters and thermal properties of the sample. One advantage is certainly the depth-profiling capability of photoacoustic spectroscopy, which can be exploited for the mid-IR spectra of different biopsy samples (36). Additionally, skin tissue has also been studied, in particular for blood glucose concentration (37).

Before clinical assays are presented, some fundamental spectroscopic aspects should be discussed. It is the wealth of information within the IR spectra that allows a direct reagent-free multicomponent assay. The selection of appropriate spectral intervals best suited for quantitative analysis is important. Some spectra are presented providing an impression of similarities and striking differences. In Fig. 8, the mid-IR spectra of three different blood proteins are shown, prepared with the KBr pellet technique and measured as aqueous solutions, respectively. There are other factors, such as pH and ionic strength of a solution or the adsorption process from the aqueous phase, that lead to alterations in the structure of proteins and that can be followed by changes in time of the corresponding spectra (38).

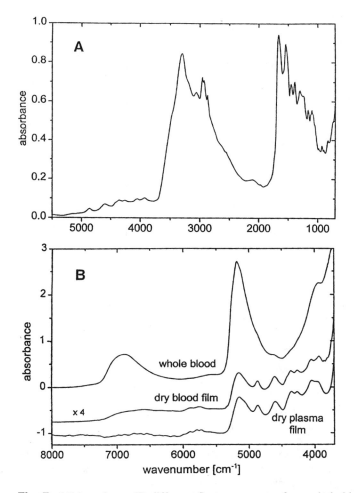

**Fig. 7** Mid- and near-IR diffuse reflectance spectra from whole blood and plasma. A: Diffuse reflectance spectrum of a dry whole-blood film prepared on a gold-coated substrate of sandpaper. B: Absorbance spectrum of a layer of whole blood (200 µm) and of the corresponding dry-film preparation on a gold-coated substrate of sandpaper (for comparison, a spectrum from a dry plasma film is also shown). (Reproduced from Ref. 26.)

Because glucose plays an important role in medical diagnostics, the spectra of this compound, which is an aldohexose, are compared with those of another important sugar species, D-fructose, which is found in fruits. This ketohexose could be used, for example, by diabetic patients as an alternative caloric sweetener. The crystalline compounds of glucose and fructose show many sharp spec-

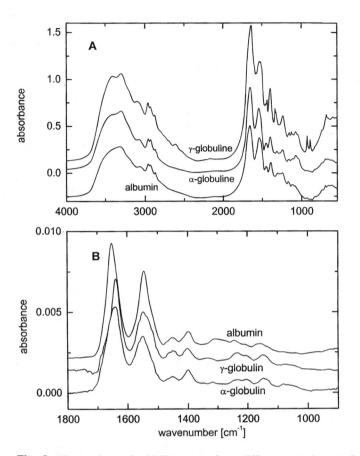

**Fig. 8** Comparison of mid-IR spectra from different proteins. A: Spectra prepared by using the KBr pellet technique. B: Spectra from aqueous solutions with water-absorbance compensation (for the globulin solutions, an isotonic 0.9% NaCl solution was used).

tral features, so discriminating between these components is not a problem; see Fig. 9. In Figs. 10A and B, the fingerprint spectra of the crystalline species and their aqueous solutions are presented; the latter were recorded by using a micro-Circle cell. Differences in both spectral types can be noticed, due to their specific fundamental molecular vibrations of the functional groups. This can be the basis for a quantitative assay, e.g., for a ternary mixture of sucrose, fructose, and glucose in aqueous biological samples (39). In addition, there is enough structural information in the spectra to follow the mutarotation of D-glucose in water (40), by which an equilibrium between two different anomer conformations, the α- and β-forms, is established. With D-fructose in aqueous solution, two stereoiso-

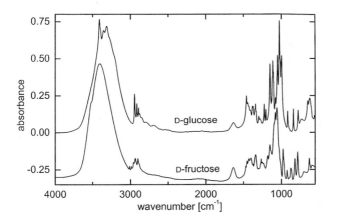

**Fig. 9**  Mid-IR spectra of crystalline D-glucose and D-fructose prepared by using the KBr pellet technique.

meric pairs (two pyranose and two furanose forms) are present in significant concentrations.

The near-IR spectra of some components recorded by using diffuse-reflectance techniques are shown in Fig. 11, displaying tremendous differences in their spectral signatures. The phospholipid lecithin (phosphatidylcholine) was chosen as an example for a special lipid category. For completeness, further lipid classes are fatty acids and their derivatives, triglycerides, and sterols. On the other hand, the diffuse-reflectance spectra of crystalline glucose, fructose, and galactose are also shown, now measured as pure powders without any preparation (see Fig. 12A). For comparison, the spectra of glasslike samples prepared from a syrup and measured in reflection-absorption after drying—the spectra are equivalent to aqueous-solution spectra (41)—are also shown, elucidating that differences can be found only for the combination bands just above 4000 cm$^{-1}$. This underlines the fact that many blood glucose assays favor the use of the narrow transmission window around 4500 cm$^{-1}$, which is accessible with transmission cells up to 1-mm pathlength. Despite this, a simultaneous analysis of an aqueous two-component solute system containing glucose and fructose, using the range between 4660 and 4320 cm$^{-1}$, failed due to their spectral similarities (42).

Traditional quantitative spectroscopy is based on the Lambert–Beer law. Liquid samples have generally been measured in fixed-pathlength transmission cells or ATR cells, as discussed earlier. Other alternatives are the diffuse-reflectance technique for powders, films, or in vivo transcutaneous measurements; for signal linearization the Kubelka–Munk transformation (see, e.g., Ref. 43) or negative log (reflectance) values may be used (44). Linear modeling of the spectra

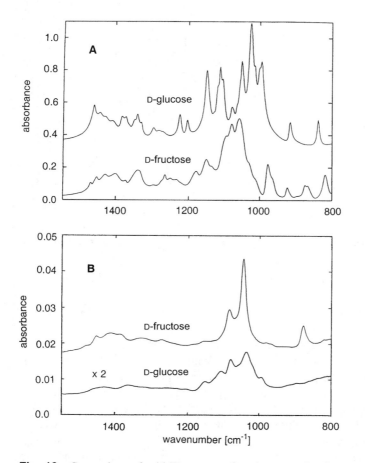

**Fig. 10** Comparison of mid-IR spectra of D-glucose and D-fructose in the fingerprint region. A: Crystalline substances prepared with the KBr pellet technique. B: aqueous solutions of same concentration (250 mg/dl) measured by using the micro-Circle cell with water-absorbance compensation.

is the fundamental basis for their quantitative analysis. A multicomponent evaluation requires at least the same number of wavelengths as components, and the classical multivariate approach for determining the concentrations from several compounds is to model the sample spectrum with all component spectra, which can be pure compound spectra; or, in case of molecular interactions, the component spectra can be derived from mixture spectra of similar composition, such as for the sample to be analyzed (calibration step). The component concentrations are usually estimated by least squares of the linear equation systems, overdeter-

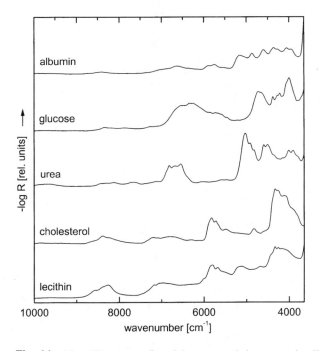

**Fig. 11**  Near-IR spectra of model compounds important in clinical blood analysis. Apart from lecithin (bottom), measured as thin layer in transflectance, the other substances were measured in diffuse reflectance from powders (for clarity, the individual spectra were offset). (Reproduced from Ref. 15.)

mined in wavelengths. An example is the challenging quantitative analysis of plasma proteins in aqueous solution, requiring robust numerical algorithms for matrix inversion due to the similarity of compound spectra (45).

For assays of biosamples, it is nearly impossible to have quantitative information on all components contributing to the spectrum. However, regressing concentrations against spectral data, the so-called soft modeling approach, works well. For statistical calibrations, it is essential that the calibration data span the range of variations, which can influence the spectra of future unknown samples. There are a number of different multivariate calibration algorithms available. Due to the so-called collinearity within the calibration spectra, ill-conditioned linear equation systems are often found, so factor methods such as partial least squares (PLS) are applied, providing slightly biased solutions (46,47). With wavelength selection for achieving robust calibration models, multiple linear regression based on least squares solutions is possible. A different approach, which is able to model nonlinearities explicitly, is the application of neural networks; for example,

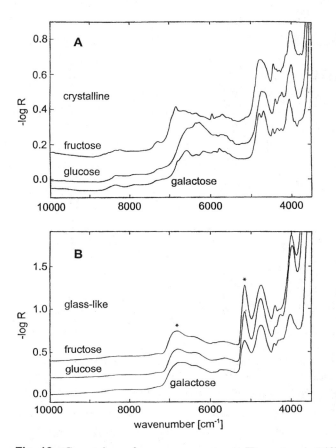

**Fig. 12** Comparison of pure component near-IR spectra. A: Diffuse-reflectance spectra of different crystalline monosaccharides of biochemical importance. B: Spectra measured from glasslike samples obtained from dried syrup preparations and measured in transflectance (residual water bands are marked by asterisks). (Reproduced from Ref. 41.)

see Refs. 48 and 49. As in other multivariate methods, calibration by "learning" is essential. However, such an approach tends to be more susceptible to overfitting and more difficult to handle and to interpret than the linear approaches discussed before.

When statistical multivariate calibration methods are used, spurious correlations in the data set must be avoided. Under certain conditions, high correlations existing between chemical component concentrations or with time can easily be modeled by such algorithms, in particular when many spectral variables are used in the regression. Therefore, a randomization in the data collection is essential, avoiding such time correlations. Effects on the prediction performance of spectroscopic calibration models from such correlations were recently studied by different groups (50–52). This underlines the necessity of a sound calibration design for such applications. For in vitro studies on cell culture media, McShane and Coté (50) used a combination of sample spectra from the biotechnical process and from appropriate artificial aqueous mixtures. Their results for glucose, lactate, and ammonia indicate that a selective calibration model can be set up under such an approach. Similar experiences with the same procedure, called ''matrix-enhanced calibration'' by the authors, were reported by Riley et al., who required only a small number of culture medium samples for improving the calibration transfer from aqueous mixtures to culture medium samples (53).

## 3  BIOFLUID ASSAYS

### 3.1  Fluids Derived from Blood

For clinical diagnostics, blood and urine samples are most important, although a variety of other biofluids, such as cerebrospinal fluid, synovial fluid, interstitial fluid, aqueous humor, and saliva, can be mentioned. Often, the identification of extravasal fluids is necessary or quantification for testing their clinical significance. The serous fluids (i.e., pleural, pericardial, and peritoneal) must be added to this list. Blood can be measured as whole blood with different additives or as plasma after centrifugation and elimination of the cellular components or as serum after cell and fibrin separation. Even whole blood can be capillary, venous, or arterial, which can show different compositions.

Different classes of analytes have to be considered. Substrates, enzymes, electrolytes, hormones, metabolites, and vitamins have to be quantified; in addition, different immunological assays are in use. From the substrate category, glucose and creatinine are most often demanded. Total protein is found at high concentrations (serum reference range 62–82 g/L), whereas other substrates are determined in per-mill concentrations and below. Due to its importance, the most frequently analyzed substrate in the clinical laboratory is D-glucose, owing to the large number of disorders within the carbohydrate metabolism, especially related to diabetes mellitus. Self-monitoring of blood glucose is part of the daily routine for such patients, and research activities concentrate on the development of minimally invasive or noninvasive methodology, possible with skin spectroscopy within the near-IR range, allowing penetration of the radiation to probe blood constituents.

There is a large number of studies on artificial mixtures in aqueous solutions, but the real test of the methodology is to use clinical specimens. For most parameters, the reference methodology, which is needed for the calibration step, requires plasma or serum, because the cellular components usually disturb the assays, and concentration values are reported for such fluids. The concentration values in the plasma phase are most significant for diagnosis, so IR assays are carried out with the use of such samples.

Recently, we studied the effect of inhomogenously distributed analytes, found only in the plasma phase, on the quantitative ATR spectroscopy of whole blood (54). The model compound was hydroxyethylstarch (HES), which is regularly applied as plasma expander for the purpose of hemodilution or as blood substitute after severe blood losses. Whole-blood and plasma samples were spiked to give an HES concentration range between 0.2 and 4.5 g/dl. The hematocrit, the relative blood volume fraction of erythrocytes, influences the spectral absorption features of such inhomogenously distributed analytes significantly (see Fig. 13). The limitations of our spectroscopic HES assay using a horizontal ATR accessory with aqueous and plasma solutions were manifested by relative standard deviations of the concentration estimates of 2% and lower. The precision—for example, obtained for a concentration of 0.2 g/dl—was similar.

The measurement of whole-blood samples is affected by two effects, the sedimentation of red blood cells and the adsorption kinetics of blood proteins onto the ATR crystal, so a strict time protocol is generally needed for achieving a high reproducibility in recording whole-blood spectra; see also Ref. 55. For an isotonic suspension of erythrocytes, 3 minutes after the start of the sedimentation experiment a 4% relative reduction of the HES absorbances was observed by us. Despite a proper timing of the spectral measurements, PLS calibrations on blood samples of different hematocrit levels, spiked with HES, yielded a cross-validated average error (RMSEP) of 6.9% for the HES concentration values, due to the hematocrit-dependent absorbances.

With knowledge of the hematocrit value, it is possible to model the absorption spectra, if the component absorptivities in aqueous phase (e.g., plasma) are given. The model of a step-shaped cell—within the cellular compartment the analyte of interest is absent—was adapted for the case of a matrix absorption-free region. Due to the special disk shape of human erythrocytes (see, for example, Ref. 56), a blood cell shape factor was introduced, because the evanescent field is actually probing a larger plasma volume at the crystal interface than given by the hematocrit value. The quantitative results based on a two-wavelength measurement (absorption band maximum at 1022 cm$^{-1}$ and baseline close by at the lower wavenumber 960 cm$^{-1}$; see also Fig. 13) were improved based on such a model's including hematocrit values obtained from centrifugation (the relative mean prediction error was 5.2%; compare with the earlier PLS-multivariate HES concentration results).

Hematocrit is certainly also an important clinical parameter, because deviations from the normal ranges allow the recognition of fluid shifts between blood and the extravascular fluid space. In particular, anemia is a disease related to the loss or reduced formation of red blood cells. Other pathological conditions affect-

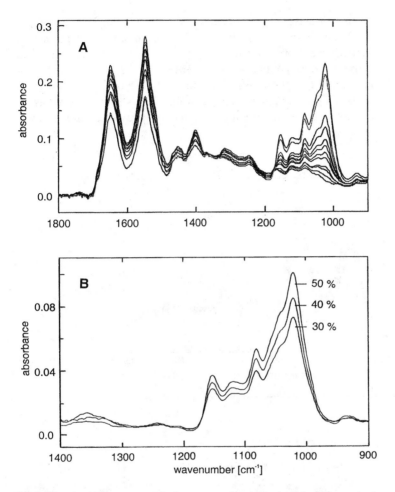

**Fig. 13** A: ATR spectra of plasma samples spiked with different hydroxyethylstarch (HES) amounts versus a background spectrum with a water-covered horizontal ATR crystal. B: Difference spectra of HES in blood at different hematocrit values (in %) versus the corresponding unspiked blood spectrum. (Reproduced from Ref. 54.)

ing the hematocrit also exist. Noninvasive spectroscopic methodology based on near-IR photoplethysmography has been proposed (57), similar to the techniques used in pulse oximetry, but this needs further clinical testing. Recently, we investigated the possibility of hematocrit measurement by analyzing mid-IR spectra of whole blood measured by using the ATR technique, complementing the earlier assay on plasma-phase-residing analytes (55). Blood samples from 15 different volunteers were used and in part diluted to give 109 samples spanning the hematocrit range between 30% and 50%. Reference values were obtained by centrifugation of blood-filled capillaries to give the packed cell volume. Best prediction results for hematocrit were obtained for a spectral interval between 1600 and 1200 cm$^{-1}$ [relative average prediction error (RMSEP) of 2.7% obtained from cross-validation]. Figure 14A shows ATR spectra of whole blood; part B provides a scatter plot of the predicted versus reference values. This spectroscopic assay can principally enable the determination of a variety of parameters directly from

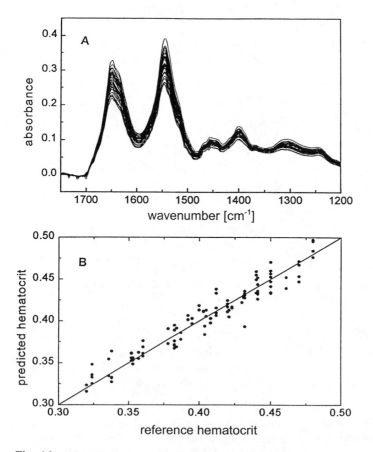

**Fig. 14**   A: ATR spectra of whole-blood samples of different hematocrit from 15 different patients and with additional hemodilution as measured versus a background provided by a water-covered ATR crystal. B: Scatter plot of hematocrit values calculated from a PLS calibration model using IR spectra between 1600 and 1200 cm$^{-1}$ versus reference values obtained by centrifugation. (Reproduced from Ref. 55.)

a whole-blood measurement without preparation of blood plasma, which reduces the amount of sample handling considerably.

Generally, an assay for the determination of the patient blood volume, which is a critical parameter after large blood losses from an accident or surgery, is possible with the HES assay discussed earlier; even simple two- or three-wavelength mid-IR photometers are feasible for such a purpose. A blood sample can be taken from the patient to provide an optional spectral reference background of blood, before giving a bolus of a concentrated HES solution to the patient's vascular system. The dilution factor can be obtained from a concentration reading from a spectrum of a blood sample taken after about 15 min of distribution and mixing within the body blood volume.

For glucose, clinical methods also are available for whole blood, which is hemolyzed during the analysis. Glucose also diffuses into the red cells, but still a slight difference between whole-blood and plasma concentrations exists (the plasma glucose level is about 10% higher) (58). For completeness, it is necessary to state the type of blood fluid, such as for deproteinized samples, for example, the volume-displacement effect from the separated proteins is relevant, increasing analyte concentrations; see, e.g., Ref. 59. For large biomolecules from the substrate class, it is evident that those are distributed in the plasma volume.

Whole blood was investigated by us for glucose with mid-IR spectrometry, whereas blood plasma was used for a list of substrates, employing different spectral ranges and measurement techniques. The first measurements for glucose in the mid-IR using the ATR technique were published in 1989 (60). Parallel studies were carried out on whole protein, glucose, triglycerides, total cholesterol, and urea in blood plasma (58). However, substantial improvements were necessary to reach clinical acceptance limits for precision and accuracy. With improved sample handling, including regular cell cleaning and enabling a better spectral signal-to-noise ratio, compared to the initial studies, the assay performance was good enough for routine application; for the composition of the plasma samples studied, see also Table 1. A variance analysis, including estimates on the vari-

**Table 1**  Concentration Intervals $c_{min}-c_{max}$, Average Values $c_{av}$, and Standard Deviations $\sigma_{pop}$ for Population of 126 Blood Plasma Samples Studied by Different IR Spectroscopic Assays

|  | Glucose | Total protein | Total cholesterol | Triglycerides | Urea | Uric acid |
|---|---|---|---|---|---|---|
| $c_{min}-c_{max}$ | 36–482 | 59–83 | 125–329 | 47–640 | 14–69 | 2.1–9.8 |
| $c_{av}$ | 207.5 | 70.5 | 219.0 | 163.1 | 36.8 | 5.2 |
| $\sigma_{pop}$ | 90.4 | 4.4 | 44.0 | 98.6 | 10.0 | 1.4 |

Total protein in g/l; others in mg/dl.

ances of the reference assays, the spectroscopic random noise, and baseline instability, allowed the actual limitations of the spectroscopic multicomponent assay
to be predicted (61). Results with respect to assay performances are summarized
in Table 2. Further investigations were carried out on quantifying experimental
errors in the analysis of glucose in plasma with respect to spectrometer drift, cell
temperature, sample loading into the micro-Circle cell, and protein adsorption
(29).

We studied several chemometric approaches using different intensity data,
either absorbance spectra or logarithmized single-beam data. The latter possess
the advantage of a lower noise level compared to absorbance data, which require
a two-step measurement, including the background single-beam spectrum. The
background spectra were usually recorded with a water-filled sample cell to use
the full dynamic range of the data-acquisition system, because sample and reference spectra are recorded with a similar radiant energy flux falling onto the detector. In this way, photometric errors from the nonlinear response of the sensitive
mercury cadmium telluride (MCT) detector during interferogram recording could
also be compensated for absorbance spectra. Furthermore, the PLS prediction
performance of parsimonious calibration models with respect to a reduced num-

**Table 2**  Comparison of Standard Errors of Prediction for Different Infrared
Spectrometric Assays Based on Logarithmized Single-Beam Data of the Same
Blood Plasma Sample Population[a]

|  | Root-mean squared prediction errors | | |
| --- | --- | --- | --- |
| Compound | SW-NIR[b] | NIR[c] | MIR[d] |
| Total protein | 1.08 | 1.07 | 0.90 |
| Glucose | 47.3 | 16.2 | 9.8 |
| Total cholesterol | 15.4 | 7.7 | 8.2 |
| Triglycerides | 23.4 | 12.1 | 10.3 |
| Urea |  | 4.7 | 2.6 |
| Uric acid |  |  | 1.0[e] |

*Source*: Ref 71.
[a] See also Table 1; results for protein are given in g/L, for other substrates in mg/dl.
[b] Spectral calibration interval for all substrates 11015–7621 $cm^{-1}$, transmission measurements with cell pathlength of 10 mm.
[c] Spectral range for protein 6001–5508 $cm^{-1}$, for cholesterol and triglycerides 6001–5508 and 4520–4212 $cm^{-1}$, for glucose 6788–5461 and 4736–4212 $cm^{-1}$, cell pathlength of 1 mm.
[d] Spectral range for protein 1700–1350 $cm^{-1}$, for cholesterol 3000–2800, 1800–1700, and 1500–1100 $cm^{-1}$, for triglycerides 1800–1700 $cm^{-1}$ and 1500–1100 $cm^{-1}$, for glucose 1200–950 $cm^{-1}$, for urea 1800–1130 $cm^{-1}$, for uric acid 1800–1700 and 1600–1150 $cm^{-1}$, measured by attenuated total reflection using a micro-Circle cell.
[e] *Source*: Ref. 61.

ber of spectral variables was studied under different wavenumber spacing in the spectra. Multiple models were set up for plasma substrates, with the starting point of the data interval being shifted, so complete spectral data exhaustion was performed stepwise. The individual models gave slightly different predictions due to the effect of mainly random spectral noise in the calibration data, and the results could be used for estimating the uncertainty of the mean-squared prediction errors of the PLS models of different rank (62).

Additionally, blood plasma samples were spiked gravimetrically with glucose, and optimal PLS calibrations produced a root mean-squared prediction error (RMSEP) of 3.7 mg/dl (1.7% compared to the population mean value of 218 mg/dl). Using the reference data from the clinical laboratory, the relative average prediction error increased to 3.7%), illustrating the effect of the raised imprecision existing for the hexokinase/glucose-6-phosphate dehydrogenase method installed on a routine analyzer (29).

The efficiency of the mid-IR spectroscopic assay led us to carry out experiments with samples obtained from microdialysis probes (22,63). Microdialysis simplifies the biotic fluid matrix considerably, because compounds of low molecular weight can be separated from proteins, which otherwise cause problems for continuous monitoring of body fluids due to protein adsorption processes occurring at the ATR crystal surface. This separation, however, is at the cost of a dilution of the analyte under investigation through the continuous elution process within the implanted probe. Such devices have been used for developing a glucose sensor for continuous in vivo monitoring, which can improve the glycemic control in diabetic patients with disorders of their carbohydrate metabolism; for a review, see Ref. 12. Such a device will be particularly useful in an artificial pancreas for such individuals. Because the lifetime of the amperometric enzyme biosensors is limited and frequent recalibration is needed, an alternative IR spectrometry–based glucose sensor is most promising. Prediction errors achieved with PLS calibration models were 6.1 mg/dl (population average concentration 39.9 mg/dl for dialysates obtained from blood plasma of different patients) and 1.3 mg/dl for aqueous glucose solutions, which are equivalent to a constant-biofluid matrix case (population average concentration 39.7 mg/dl).

For the development of multiwavelength IR photometers it is essential to know how many spectral variables are needed for a satisfactory calibration. This is similar to the case of searching for optimal conditions using simple near-IR filter spectrometers, as is routinely used for the analysis of agricultural products, pharmaceuticals, etc. Special selection strategies have been followed in multiple linear regression (MLR) methods, which generally rely on a reduced number of spectral variables. The number of wavelengths optimally required for spectroscopic calibration has been discussed intensively in the past. In fact, once one uses as many variables as there are independent spectral constituents, the addition of further wavelengths should serve to reduce effects from noise. On the other hand, as more wavelengths are used, the probability of encountering additional spectral interferences increases. This progression eventually leads to a situation where the use of more variables starts to degrade the accuracy of the result.

Recently, considerable effort was placed toward evaluating procedures that identify spectral variables carrying useful information for the setup of robust calibration models. A selection of papers dealing with these aspects, considering search strategies such as genetic algorithms, simulated annealing, and artificial neural network approaches are cited in an interesting paper by McShane et al. (64). One of their examples involves the choice of glucose calibration models based on near-IR spectra. We developed a rapid and reliable variable selection algorithm for statistical calibrations based on PLS regression vector choices, which has been tested for various calibration scenarios (44,65,66). Because the optimum regression vector obtained from statistical calibrations contains the weights for the spectral variables needed for concentration prediction, spectral variables were chosen pairwise, providing the minima and maxima of the regression vector, but in a ranking order decided on their coefficient weight size. The analytical performance of the calibration models, based on a consecutively increasing number of spectral variable pairs, is tested by cross-validation. The minimum RMSEP was taken for the selection of the optimum calibration model with a reduced number of variables under this scheme. For mid-infrared studies with clinical calibrations for glucose in blood plasma or whole blood, fewer than 10 variables were needed to reach the equivalent analytical performance obtained with broad-spectrum evaluation (65).

An example is presented in Fig. 15, where the mean prediction errors are presented in part A for 126 blood plasma samples from different patients, dependent on the number of spectral variables selected under the scheme discussed earlier. The spectral data were logarithmized single-beam sample spectra. The minimum prediction error (9.6 mg/dl) is achieved with 18 spectral variables, which are displayed in Fig. 15B. Here, the regression vector weights, each based on the optimum PLS calibration matrix rank, are presented for a PLS model using all 193 spectral variables within the interval 1200–950 cm$^{-1}$ (wavenumber spacing 1.29 cm$^{-1}$) and for another model based on 22 spectral variables equidistantly spaced, giving a minimum prediction error of 10.6 mg/dl, similar to a result from absorbance spectra containing all the spectral variables from the same interval at a spacing of 5.1 cm$^{-1}$.

The validation process for testing the prediction performance of PLS calibration models of different rank (equivalent to the number of orthogonal PLS factors) has often been discussed. For calibration testing, the data are usually divided into two subsets, the first for calibration modeling and the second for prediction validation with independent data. This approach has the advantage that it does not rely on any statistical assumptions derived from solving the calibration equation system; see also Ref. 47. Another strategy is to have a subset of calibra-

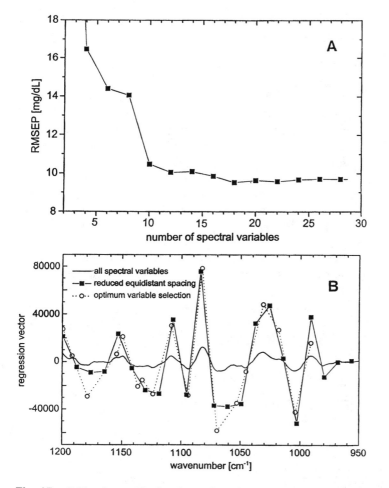

**Fig. 15** Calibration results for glucose in 126 blood plasma samples using logarithmized single-beam ATR spectra between 1200 and 950 cm⁻¹. A: Minimum root-mean squared errors of prediction (RMSEP) of PLS models calculated using an increasing number of relevant spectral variables. B: Optimum PLS regression vectors based on different number of spectral variables (with 18 variables, the global RMSEP minimum was achieved).

tion samples (size $n-m$) and a validation subset of size $m$, but putting the data back after validation for repeat calibrations and selecting another subset of samples previously not implemented in the validation stage, until all samples have been used in the validation process. The root-mean squared error of prediction (RMSEP) is calculated from the prediction residual sum of squares. When the validation subset size is 1, this is called the *leave-one-out* strategy, which is a reasonable approach when only a limited number of samples is available for model training.

In Fig. 16 the RMSEP values for calibration models, which were all based on the same 18 spectral variables optimally selected from negative log (single-beam) data, as discussed earlier, are plotted for various validation subset sizes versus an increasing number of PLS factors (total sample population size of pairs of blood plasma spectra and glucose reference concentration values was 126). Throughout, the optimum rank is 8, and the RMSEP values have a maximum spread of 0.2 mg/dl, with the result from the leave-one-out strategy lying in the middle of all prediction error estimates. From Fig. 16 it is evident that when halving the sample population (equivalent to the leave-64-out case), the least squares estimate obtained by solving the matrix of full rank is giving less precise glucose predictions on average by 0.5 mg/dl compared to cases with a larger number of samples as input for the calibration stage.

The performance of a mid-IR spectroscopic assay of comparable quality is demonstrated in Fig. 17A. The same scatter was obtained for single repeat reference measurements, carried out in the clinical chemistry laboratory at a later stage, to show that storage of the plasma samples in a freezer at −35°C over some months had no effect on glucose concentration (the abscissa concentration values were averages from triplicate glucose determinations using the enzymatic hexokinase method) (67).

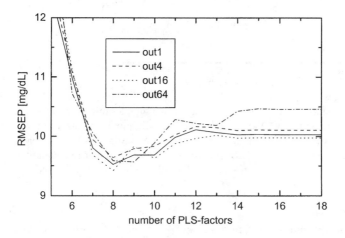

**Fig. 16**  Glucose calibration results, i.e., RMSEP values obtained for models calculated from a different number of PLS factors, based on 18 spectral variables from logarithmized blood plasma ATR spectra between 1200 and 950 cm⁻¹, with a different subset size used for cross-validation.

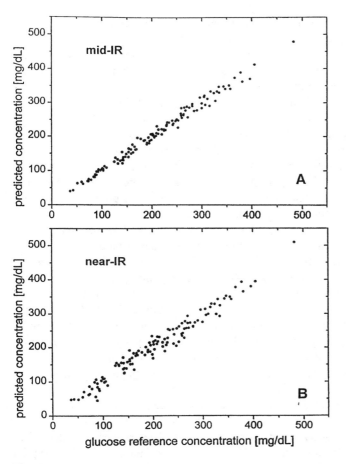

**Fig. 17** A: PLS calibration predictions for glucose concentration in blood plasma based on 10 specially selected variables between 1200 and 950 cm$^{-1}$ using ATR absorbance data. B: Results based on 24 optimally selected variables from logarithmized single-beam spectra within the spectral intervals of 6800–5450 cm$^{-1}$ and 4750–4200 cm$^{-1}$ using a 1-mm quartz window cell. (Reproduced from Ref. 22.)

Activities to replace conventional ATR accessories by mid-IR fiber probes have already been mentioned (21–23). A feasibility study for quantitative analysis of blood serum samples was published by Simhi et al. (68), but significant improvements have to be achieved to reach assay performances, for example, as obtained with the micro-Circle cell. One reason for the deficiencies was the use of a room-temperature DTGS detector (studies from other groups were carried out with more sensitive liquid nitrogen–cooled MCT detectors). To improve the stability of the fibers, avoiding the deterioration in transmission caused by interaction with the biological fluids, a dip-coating procedure was recently presented. Polystyrene and silicone-elastomer films of about 3-μm thickness on the fibers were found to guarantee 70% transmittance at 2500 cm$^{-1}$ after 150 hours of contact time with a saline solution (69).

Contrary to the ATR measurements of blood and plasma performed by us and other groups, transmission measurements with whole-blood samples using a flow-cell (CaF$_2$ windows, 50-μm spacer) were recently carried out by the group around the late Robert Kellner, and mid-IR spectroscopic calibrations, based on spectral intervals between 1500 and 900 cm$^{-1}$, were presented for glucose (70). The lowest RMSEP value reported was 13.1 mg/dl, which is about one-third higher than the results of our studies with blood plasma. Calibration experiments with dry-film measurements will be discussed later.

Similar investigations on the same population of 126 blood plasma samples, which were used for mid-IR multicomponent assay development, were carried out by us using the near-IR (71) and the SW-NIR spectral range (72,73). For the initial study, a quartz window transmission cell of 1-mm pathlength was used; for the later experiments, the pathlength was extended to 10 mm by using a different spacer. A comparison of mean assay prediction errors (RMSEP) for the substrates studied is given in Table 2. For glucose, the optimum PLS model performance is illustrated in Fig. 17B with a scatter plot of predicted versus reference values. A similar variance analysis was performed as described earlier for the mid-IR study using the micro-Circle cell to test the ultimate limitations of the near-IR spectroscopic assay (71). Our selection scheme for most significant spectral variables was also applied to the near-IR spectral data recorded with the 1-mm-pathlength cell; see Ref. 66. In contrast to the mid-infrared studies, based on the same clinical blood plasma population, up to 26 variables were needed to reach the clinical acceptance for glucose. An overview on the various calibration models and their performance with respect to mid-IR and near-IR spectral range exploitation for glucose in blood plasma is given in Ref. 22. For the whole set of substrate parameters obtainable from the FT-IR spectra, which were recorded either with a 10-mm cell (SW-NIR, transmission), with a 1-mm cell (near-IR, transmission), or with a micro-Circle cell (mid-IR, ATR) under optimum spectral interval selection, the RMSEP values are presented in Table 2.

The activities of another group in this clinical area must be mentioned, which culminated in a recent publication (74). Extensive calibrations were performed on about 240 undiluted human serum samples for total protein, serum albumin, globulin protein, triglycerides, cholesterol, urea, glucose, and lactate based on the spectral range between 5000 and 4000 cm$^{-1}$. Raw and digitally filtered data were used as PLS calibration input. For the latter spectra, Gaussian-shaped filter response functions for different digital Fourier filters were calculated to simplify the PLS calibration modeling, since it decreases the number of PLS factors required to achieve the same level of performance as accessible with spectral raw data. Except for lactate, the concentrations of which were below the detection limit, satisfactory near-IR assays could be developed (for glucose: RMSEP = 23.3 mg/dl). Calibration transfer to data recorded on a modified spectrometer after nearly two years resulted in a slight but constant positive prediction bias (after offset correction, the RMSEP value for the glucose concentrations of the 50 new samples was still a factor of 2 larger than previously obtained: 45 mg/dl).

There is a significant number of other clinical studies based on near-IR spectroscopy of fluids derived from blood. Hall and Pollard provided an extensive study of human sera using the NIR approach for the clinical analysis of proteins, triglycerides, and glucose (75). A further study from these authors was devoted to total protein and protein fractions, e.g., albumin and globulin, as well as to urea (76). They recently provided a summary of their previous NIR validation results for human sera; in addition, they presented spectra of whole blood as measured in transmission and reflectance (77). Rapid transmission measurements of whole blood were carried out by Norris and Kuenstner, who evaluated different multivariate calibration models for several blood substrates, such as total protein, albumin, cholesterol, and urea (78). The robustness of multivariate calibration models for six analytes, including glucose, based on absorbance spectra of human sera, was tested by Gatin et al. employing two data sets recorded by two operators at different times (79). Cross-validation produced very similar prediction errors compared to those obtained from calibration models built from one data set, respectively, and applied to the second set for validation. A recent investigation was aimed at the determination of protein and β-lipoprotein in blood serum using transflectance measurements (80). Further near-IR studies on lipoproteins were reported by Demsey et al. (81), who were developing an in vivo assay for apolipoproteins immobilized in the walls of human arteries (carotic plaque). The different particles are usually classified based on density by ultracentrifugation, for example, low-density lipoproteins (LDL) and high-density lipoproteins (HDL). Among the major lipid components, triglycerides and cholesterolesters are important markers for enlightening the metabolism of lipoproteins. For the interested reader, a recent mid-IR study of very low-density lipoproteins (VLDL) and LDL compounds may be mentioned, compared to triolein and cholesteryl oleate spectra (82).

Human albumin and γ-globulin measurements have attracted clinical spectroscopists, because their ratio (A/G ratio) is usually employed as an indicator for various infections or nutritive conditions. These two proteins were measured in phosphate buffer solutions at physiological concentrations over the spectral range of 1300–1850 nm (83), and RMSEP values were 1.5 g/L and 1.2 g/L for albumin and γ-globulin, respectively. A similar study was carried out by the same group, but using a control serum solution as matrix for both proteins with equivalent results (84).

Another clinically interesting parameter is hemoglobin, and visible and near-IR transmittance spectra of unlysed blood samples were recorded in an open cell and a vertical IR radiation path. The hemoglobin content could be determined within a standard error of 0.43 g/dl using second-derivative data at 1740 and 1346 nm (85). A noninvasive measurement of hemoglobin is certainly advantageous. Thus spectral data for different hemoglobin species, including oxy-, deoxy-, carboxy-, and methemoglobin, were presented for the visible and the near-IR regions from 620 to 2500 nm by Kuenstner and Norris (86). An extension of that work was carried out to test different filter calibrations, based on three different wavelengths chosen for a modified pulse oximeter, which could be employed for noninvasive hemoglobinometry (87). An even more extensive study was performed by Yoon et al., who tested spectra of whole-blood samples from 165 patients (88). Calibration models were based on the evaluation of whole spectra (500–900 nm) and special three-wavelength schemes with a similar assay performance (521 nm, 615 nm, and 894 nm; RMSEP = 0.35 g/dl). For evaluating noninvasive methodology, the optical properties of circulating human blood are of great interest; results within the spectral range between 400 and 2500 nm were recently reported (56).

A different aim was to monitor changes in human blood during storage by near-IR spectroscopy through their original plastic bags as used in surgery for blood transfusion. Reflectance spectra were recorded for different blood groups, and global calibrations could be established for predicting storage time using first- and second-derivative data between 1000 and 1600 nm (89).

## 3.2 Dry-Film Measurements

Several measurement techniques for dry biofluid films have already been described in Sec. 2. Some attractive features, in particular for routine measurements, are involved here: There is no need for cuvettes, samples of a few microliters and less can be handled, and the drying step eliminates a significant obstacle to the conventional spectroscopic assay of biofluids, i.e., the absorption of water. Furthermore, the methods are suited for automation, dry-sample handling is straightforward, and the substrate is a fascinating analog to a test strip, but the assay is reagent-free and based on "infrared colors." An early interesting study was carried out using dried human serum samples on fiberglass filters (90). The performance of the urea assay employing diffuse-reflectance techniques was comparable to the analysis of unmodified serum samples. However, improvements in dry-film preparation should lead to even better assay performance than using aqueous biofluids, because the signal-to-noise ratio for the substrates can be increased. Much wider spectral intervals are available due to the elimination of water, as illustrated in Fig. 7, and hence it follows that especially the near-IR spectral variances arising from the sample water need not be modeled.

A first feasibility study on the determination of glucose and cholesterol using mid-IR spectra of dried whole-blood and serum samples on a polyethylene film carrier (PE cards) was presented by Budinová et al. (91). Potassium thiocyanate with an absorption band at 2059 cm$^{-1}$ served as an internal standard for spectrum normalization throughout the experiments, which was necessary because of the film inhomogeneities. The sample application onto the PE cards was more reproducible for blood serum than for whole blood. Average prediction errors were presented for glucose (22 mg/dl) and cholesterol (31 mg/dl), calculated as the mean values obtained for different calibration models. However, these results were not significantly better than our early blood plasma assays (58).

A large population of 300 sera spectra was used as the basis of a recent investigation using dry films, which also contained a constant amount of potassium thiocyanate, added for spectrum normalization (92,93). The films were prepared on BaF$_2$ windows from serum samples of 3.5-μl volume, diluted by the same volume of a KSCN solution. Whereas the calibration models were satisfactory for total protein, albumin, triglycerides, cholesterol, glucose, and urea, the mid-IR assay using the dry-film absorbance spectra proved to be less suited for uric acid and creatinine due to their low concentrations. The methodology proposed is well suited for routine analysis; however, the results must again be compared to conventional techniques using liquid biofluids (see also Table 2). The multicomponent assay based on mid-IR spectra of dry films gives slightly worse results for cholesterol, and, apart from glucose, the other substrates give prediction errors (RMSEP) that are worse by a factor of 2 or more.

In another paper published by the same group, different biofluids (serum and amniotic and synovial fluids) were prepared as dry films on inexpensive glass substrates, which limits the accessible spectral range down to wavenumbers above 2400 cm$^{-1}$. The most prominent absorption bands correspond to the CH and NH groups of the constituents, which are clearly observable due to the drying process and suited for calibration input. The results for triglycerides, using just the spectral range of 2800–3000 cm$^{-1}$ for a PLS calibration of undiluted-serum films without compensation for variability in film thickness, were equal to the previous study, leading to a standard prediction error of 0.22 mmol/L (19 mg/dl) (94). The potential of IR spectroscopy for the analysis of amniotic and synovial fluid will be discussed later.

Approaching the same goal of a reagent-free multicomponent assay, mid-IR transmission measurements on silicon carriers and diffuse-reflectance measurements on gold-coated minitroughs were carried out on dried specimens from 1-μl sample volume each of sera or hemolyzed and centrifuged blood (95). About 300 sera and 90 blood samples were available for calibration. The investigation is, to my knowledge, the most comprehensive study on the feasibility of introducing the IR methodology into the clinical laboratory. For human sera their prediction errors, based on diffuse-reflectance spectra, were as follows (population size between 220 and 306 samples): glucose 9.5 mg/dl, cholesterol 11.3 mg/dl, triglycerides 16.6 mg/dl, total protein 2.4 g/L, and urea 2.0 mg/dl. For hemolyzed blood, the prediction results were significantly worse. Hemoglobin had also been assayed, leading to an average prediction error of 0.78 g/L with a mean concentration value of 13.0 g/L (based on 91 samples).

A further step is to use the blood spectrum of a patient for disease-pattern recognition. For this purpose, dried samples from 1 µl serum deposited on gold-coated minitroughs in disposable substrates were prepared, and spectra were measured by diffuse reflectance (96). There are special changes in the absorption features due to the pathophysiological condition of a diseased person, which can be projected to give a pattern of abnormal substrate concentrations, providing an excellent basis for medical diagnoses. This intermediate step can be overridden, so a fast and efficient diagnostic tool can be developed from the spectral information content by applying sophisticated mathematical methods, searching for typical patterns for different disease categories. This chemometric approach was based on regularized discriminant analysis and binary classification (96). More than 2000 serum spectra from different individuals were analyzed, and binary classifiers were tested, e.g., for "healthy vs. mixture of several diseases," "healthy vs. rheumatoid arthritis," "healthy vs. diabetes," and others, which showed very high levels of discrimination. An expert system can be set up, with the support of minimally invasive techniques taking small blood samples, which could be used for future screening of the status of a patient's health.

### 3.3  Further Biofluids of Clinical Interest

Laboratory examination of biofluids other than blood is frequently carried out. An overview on the analysis of urine and body fluids and their importance in judging the pathophysiological status of the patient was provided by Brunzel (97). To test for renal and metabolic diseases, urine is used, often collected over 24 hours to get representative results, but it can also provide diagnostic clues on infections of the urinary tract. This fluid is easily accessible and available in large quantities. Glucose in urine, for example, is an index component for metabolic disorders, in particular for diabetes. An early mid-IR spectroscopic investigation on the identification of major absorbing components in urine, such as urea, phosphates, sulfates, and glucose, was already mentioned (17). Frequently, multiparameter reagent test strips, which are dipped into urine and visually evaluated, are employed for fast diagnostics. However, many analytes of interest are present at only low concentrations, such as ketones, nitrite, bilirubin, and others, and often small amounts of proteins are excreted (micro-albuminuria). To distinguish between different renal diseases, an assay for protein differentiation, quantitating albumin and various globulins, is required. Despite these challenges, a quantitative IR spectroscopic assay is possible for a number of parameters.

As with blood plasma assays, two options exist for a reagent-free IR spectroscopic multicomponent assay for urine: mid- and near-IR spectroscopy. For glucose, 19 different wavelengths between 1440 and 2350 nm were selected to establish a calibration model using urine near-IR absorbance spectra. However, the applicability of the model was limited, because the glucose concentrations were rather large, lying between 900 mg/dl and 6.3 g/dl (98). In a recent paper, a multicomponent assay was presented for urea, creatinine, and protein based on 123 different urine samples for calibration and a further 50 samples for validation (99). The average prediction errors were 16.6 mmol/L (0.10 g/dl) for urea, 0.79 mmol/L (8.9 mg/dl) for creatinine, and 0.23 g/L for protein. A second calibration model was calculated using urine samples with lower protein concentrations than 1 g/L, which are more frequently found, and the standard error of prediction could be reduced by a factor of 2 compared to the global calibration model. In addition, precision estimates for near-IR predicted concentrations were presented for between-day and within-day measurements, and the effect from spectroscopic noise was also estimated. As expected, there was a decrease in the respective standard deviations. A variance analysis, separating the contributions from the spectral measurement, was also presented.

We recently tested a multicomponent mid-IR assay for urea, creatinine, and uric acid using the micro-Circle cell. Electrolytes such as sulfate and phosphate, which possess significant absorptions in the mid-IR spectral range, were also determined (100). A population of 67 individual urine samples from children and adults was available that had been analyzed with clinical reference methods for the components mentioned earlier and, additionally, for glucose and total protein. However, for the latter two parameters the spread in concentrations was not sufficient to allow for proper calibration modeling. In Fig. 18 some representative spectra from the population of urine samples are shown, which were used for calibration after spectral range selection for calibration optimization. The optimum spectral intervals were 1600–1000 cm$^{-1}$ for urea, creatinine, and uric acid, whereas for glucose and sulfate spectral data from the interval of 1200–950 cm$^{-1}$ were taken. The best-performing calibration models for phosphate were obtained for spectral variables between 1300 and 800 cm$^{-1}$.

As examples, two scatter plots with predicted concentrations versus clinical reference values are shown in Fig. 19 for uric acid, which is quantifiable down to the lowest concentrations, and for phosphate. There is a pH dependency for this inorganic anion, which could possibly be used for the prediction of urine acidity. The characterization of the sample population and calibration results with average prediction errors (RMSEP) are summarized in Table 3.

Another biofluid is saliva, which is easily collected and can also be used

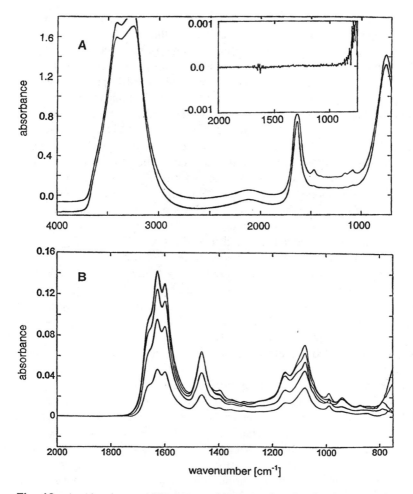

**Fig. 18** A: Absorbance ATR spectra of human urine showing largest and lowest urea concentrations measured by using a micro-Circle cell, respectively; the absorbance noise level as obtained from two consecutive water measurements is shown in the inset. B: Difference spectra of five native urine samples, compensated for water absorbances and with baseline corrections. (Reproduced from Ref. 100.)

as a diagnostic fluid (101). It is produced by different salivary glands, with the effect that the composition of the secreted saliva varies. It has also been analyzed for glucose, but no useful correlation with blood concentrations could be established (see also Ref. 67). The main organic components are several proteins and glycoproteins. Interestingly, inorganic components such as phosphate and thiocy-

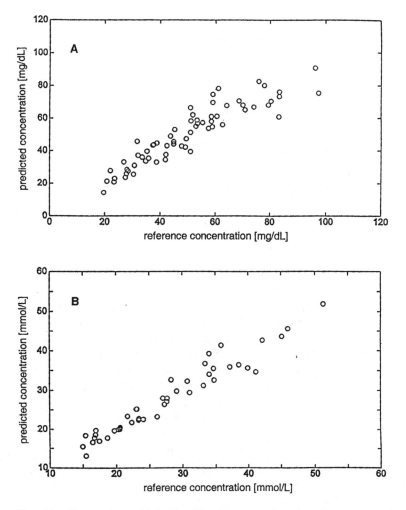

**Fig. 19** Scatter plots with PLS calibration predictions based on processed ATR absorbance spectra versus reference concentration values. A: Results for uric acid and B: for phosphate in native urine samples. (Reproduced from Ref. 100.)

anate are also found; the latter is converted by salivary peroxidases to hypothiocyanite ($OSCN^-$), which shows antibacterial activity. Saliva composition has also been investigated by mid-IR spectroscopy; see, for example, Ref. 52. An IR spectroscopic assay for thiocyanate quantitation has been described that takes advantage of the fact that this anion possesses an undisturbed absorption band at 2058 $cm^{-1}$, a region devoid of bands from other components (102). The integrated

**Table 3**  Concentration Intervals $c_{min}-c_{max}$, Average Values $c_{av}$, and Standard Deviations $\sigma_{pop}$ for Population of 67 Urine Samples from Different Children and Adults, and Average Prediction Errors Obtained with Mid-IR ATR Urine Spectra

| Compound | $c_{min}-c_{max}$ | $C_{av}$ | $\sigma_{pop}$ | RMSEP |
|---|---|---|---|---|
| Urea | 558–2979 mg/dl | 1677 mg/dl | 631 mg/dl | 85.1 mg/dl |
| Creatinine | 24.9–309.9 mg/dl | 92.8 mg/dl | 42.2 mg/dl | 8.7 mg/dl |
| Uric acid | 19.6–129.8 mg/dl | 50.5 mg/dl | 19.0 mg/dl | 7.4 mg/dl |
| Protein | 3.3–31.9 mg/dl | 11.9 mg/dl | 5.4 mg/dl | — |
| Glucose | 0.26–217.2 mg/dl | 4.6 mg/dl | 1.9 mg/dl[a] | — |
| Sulfate | 7.7–31.2 mmol/L | 17.8 mmol/L | 5.8 mmol/L | 0.92 mmol/L |
| Phosphate | 15.0–51.3 mmol/L | 27.7 mmol/L | 9.3 mmol/L | 2.2 mmol/L |

[a] The average concentration and standard deviation were calculated without the single urine sample containing glucose at a concentration of 217.2 mg/dl.

peak area was used for linear regression against thiocyanate concentrations. Small saliva volumes of 5 μl were spread on $BaF_2$ windows and dried to give solid film samples for the spectral measurements. Mean values for saliva, collected in the afternoon, were between 0.83–0.42 mmol/L for 25 different subjects. A circadian dependency of the thiocyanate level could be proven.

Our saliva studies were conducted during an investigation to test whether the blood glucose levels could be followed by noninvasive ATR measurements of the inner lip (52). Under such measurement conditions, saliva films can hinder the tissue spectra analysis. Another problem with ATR probes is certainly the protein adsorption to the crystal material. An example is given in Fig. 20, where spectra of saliva and aqueous glucose are presented with water-absorbance compensation, measured by means of a simple fiber loop and a sample volume of 400 μl. After fiber exposure to the saliva, washing with distilled water, and drying, a spectrum with significant absorption bands from saliva proteins bound as a film on the fiber could be measured.

Novel diagnostic tests for arthritis were recently presented by the group of scientists in Winnipeg, demonstrating a successful analysis of mid-IR spectra of synovial fluid. Synovial fluid is a plasma filtrate formed by filtration across the synovial membranes of joints. When dried the substance contains mainly proteoglycans and the polysaccharide hyaluronic acid (about 80% for the latter), which serve to increase the viscosity of the joint lubrication fluid. In arthritic disorders, the viscosity is reduced, resulting finally in damage to the joint due to lubrication failure. An initial study on dry films of such fluids suggested the existence of $CO_2$ clathrates, showing an absorption band at 2337 $cm^{-1}$ with a half-width of 7 $cm^{-1}$, which appeared to vary with the state of health of the joint, providing possibly an excellent indicator for inflammation diagnosis (103).

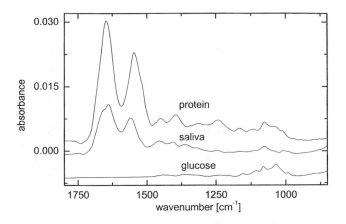

**Fig. 20** Mid-IR ATR spectra obtained with a fiber-optic probe from polycrystalline silver halide material; saliva and aqueous glucose solution (250 mg/dl) were measured with water-absorbance compensation; the protein layer on the fiber surface was obtained after short exposure to two saliva samples and water rinsing. (Reproduced from Ref. 19.)

Further multivariate studies on arthritis diagnosis were based on near-IR spectra of 109 synovial fluid samples. Although the spectra were available for the range between 400 and 2500 nm, exploiting especially the range 2000–2400 nm led to the best results in distinguishing synovial fluid specimens drawn from patients suffering from rheumatoid arthritis, osteoarthritis, and spondyloarthropathy. Linear discriminant analysis (LDA) was applied for the classification (104). The accuracy of this approach was surprising, since it was contrary to the conservative judgment that synovial fluid composition is of limited value in arthritis diagnosis. The protein-level change, as manifested between inflammatory and noninflammatory diseases, plays a role in the correct classification. There were interesting parallels between mid-IR studies of dry films using the CH-stretching modes (3000–2850 cm$^{-1}$) only and the range in the near-IR fluid spectra examined. Here, combination bands that involve the same molecular moieties giving rise to the absorption bands below 3000 cm$^{-1}$ are studied, so similar information content is available in both spectral sections. Further investigations made use of 15 significant spectral subregions between 3500 and 2800 cm$^{-1}$, and classifications were in excellent agreement with clinical diagnosis (96.5% correct results). Infrared spectroscopy can be used for an objective differential diagnosis of arthritis by means of an adequately trained classifier. This methodology is especially attractive, because it requires little operator training, only fluid volumes of 20 µl need to be aspirated, and diagnosis is obtained within minutes (105).

A biofluid of great interest in antenatal diagnostics is amniotic fluid, which is the liquid medium that bathes the fetus throughout its gestation. By guidance through ultrasound imaging, amniocentesis, the collection of this fluid, is now a widespread and safe obstetric procedure. Many tests evaluate the amniotic fluid for fetal pulmonary maturity and fetal distress, required in the clinical management of high-risk pregnancies. In this context, respiratory distress syndrome (RDS) is a common cause of death in preterm-delivered babies, because of insufficient surfactant production within the newborn's lungs, inherently necessary for proper functioning. Three phospholipids [lecithin (L), sphingomyelin (S), and phosphatidylglycerol (P)] are routinely assayed to evaluate fetal lung maturity. The aim of a novel study was to correlate near-IR spectra of amniotic fluid with data obtained from thin-layer chromatography (TLC), in particular the ratio of the first two compounds just mentioned (L/S) is used as a diagnostic indicator (106). The spectral range between 2000 and 2500 nm of the second-derivative spectra of the centrifuged fluids was taken as data input for a PLS calibration. The L/S ratios determined by TLC covered the interval between 0 and 8, and a satisfactory correlation could be achieved by using the optimum PLS calibration model based on near-IR data (correlation coefficient $r = 0.91$).

A larger population of amniotic fluid samples than used in the first investigation was treated for the preparation of dry films on $CaF_2$ substrates and studied by mid-IR spectroscopy, focusing on the C—H stretching region between 3200 and 2800 cm$^{-1}$ (107). A robust linear discriminant analysis method was used to partition samples into normal and abnormal groups based on their spectra. A PLS analysis was again chosen for regression analysis of spectral first-derivative data against TLC-based results. In addition, the accuracy of the phosphatidylglycerol assay by means of IR spectroscopic data was high. The IR-predicted combination of L/S ratios with the phosphatidylglycerol status has certainly the potential to become the established clinical method for screening fetal lung surfactant maturity from the composition of the amniotic fluid. Prediction of respiratory distress syndrome developed in prematurely delivered babies was more successful when based on IR spectroscopically determined L/S ratios than using the TLC results obtained routinely (108).

Further parameters of interest within modern obstetrics are glucose for giving evidence of intra-amniotic fluid infection and lactate concentrations for indicating fetal distress, which have also been determined by the same IR spectrometric methodology (109). Standard error of prediction (RMSEP) for glucose was 0.06 mmol/L (1.1 mg/dl) based on spectral data from 950 to 1180 cm$^{-1}$, as applied in blood analysis (spread between lowest and largest concentrations was significant: 0.5 and 9 mmol/L, respectively). The best concentration predictions for lactate were also achieved with this range (900–1200 cm$^{-1}$), despite the existence of significant and larger absorptivities (e.g., antisymmetric stretching of the carboxylate group of lactate at 1582 cm$^{-1}$) above the high interval limit. Minimum and maximum concentrations were about 5 and 15 mmol/L, and the prediction standard error was 0.18 mmol/L, which is a value double that found for the calibration stage when using a separate training set. This shows that the calibration model had some weaknesses with respect to robustness, indicating the requirement for a training set of larger size and/or better calibration data. Altogether, these achievements can be considered as milestones in IR spectrometry–based clinical assays of biofluids, in particular because the methodology needs only small sample volumes, leaving sufficient material for further clinical tests.

## 3.4 Other Clinical Assays

In this section, some assays are described that differ from those described earlier with respect to a different category of analytes. The study of $CO_2$ in synovial fluid has been already mentioned (103); other condensed-phase studies from the group in the NRCC Institute for Biodiagnostics in Winnipeg followed (110), supporting the view that the source of the endogenous $CO_2$ is metabolic activity. After reading their papers, we inspected our whole-blood spectra recorded by the micro-Circle cell (60) and noticed the absorption band of dissolved $CO_2$ clearly at 2343 $cm^{-1}$. Other fluids studied by the Winnipeg group were amniotic fluid, bronchoalveolar lavage fluid, and human cerebrospinal fluid. In addition, this band is also observed in almost all tissues. We checked our lip mucosa spectra, recorded by diffuse reflectance and connected to the development of a noninvasive blood glucose assay (44), to find out whether a noninvasive monitoring of $CO_2$ could be achieved, but due to intense tissue absorption in this spectral range and the low signal-to-noise ratio, no band could be observed. The assignment of the band at 2343.5 $cm^{-1}$ to dissolved $CO_2$ is supported by the fact that a corresponding band of the $^{13}CO_2$ band exists at 2279 $cm^{-1}$ with the required intensity reduction due to the relative natural abundance.

A different field, which has also been proposed, is enzyme activity assays using IR spectroscopy. There is some analogy to assays for substrates using immobilized enzymes; see, for example, Ref. 111. The determination of α-amylase activity within the clinically relevant range from 50 to 1000 u/L was carried out by mid-IR spectroscopy (112). High values of this enzyme activity in serum have been reported for differentiation of inflammation and hemorrhage of the pancreas and other disorders of the digestive system. The proposed method uses unmodified starch as a substrate, since the enzyme catalyzes the hydrolytic cleavage of the α-1,4 glucan bonds, thus providing sufficient assay information. Difference spectra from a thermostated mixture of the sample and the substrate solution were recorded over two hours, giving evidence on the ongoing enzymatic reaction (optimized incubation time was 20 min). It is advantageous that the unspecific matrix absorption of different serum samples can be eliminated by this procedure.

Other enzyme-activity assays were investigated by Fujii and Miyahara (113), who experimented with lactate dehydrogenase, α-amylase, and creatine kinase using appropriate substrate solutions. The correlation between alkaline phosphatase and spectroscopic data was found to be lower than for the other enzymes mentioned. For their assays, the cryoenrichment technique in combination with ATR measurements was successfully applied (35). The Japanese authors used a PLS regression analysis based on optimum spectral intervals mainly in the fingerprint region, in contrast to the regressions of the previous paper, describing the use of difference spectra and employing absorbance data at just two pairs of wavenumbers.

The quantification of species of extremely low concentration was already mentioned for gas analysis, and special cryotechniques can lower the detection limits to subnanogram levels, as required, for example, in chromatography/FT-IR spectroscopy. A special application is the trace analysis of transition-metal carbonyl-labeled bioligands in the picomole range, which opened the field for a new kind of immunological test, called carbonyl metalloimmunoassay (CMIA) (114). These organometallic labels were selected, since they have very intense absorption bands in a region of the IR spectrum (2200–1800 cm$^{-1}$) that is otherwise devoid of perturbing absorption bands. Immunoassays were developed for the antiepileptic drugs phenobarbital, carbamazepine, and diphenylhydantoin and the steroid hormone cortisone. Most of the assays previously proposed make use of a solvent extraction preparation step for separating the labeled antigen, either free or bound, from the bulk matrix before spectroscopic analysis. To test the working area of multi-CMIA experiments, mixtures of up to three metal–carbonyl complexes were simultaneously quantified with minimum detectable tracer quantities of 10–500 pmol. A further reduction by more than a factor of 30 was reached by using a special light-pipe cell of 20-mm pathlength. The quantitative evaluation of the spectra was carried out by classical least squares and partial least squares (PLS), which provided similar results. Even at the lowest concentrations, a relative error of about 6% was achieved for assay repeatability.

A similar strategy for determining multiple antigens marked by different metal–carbonyl complexes was used by Ismail and Barnett (115). They stressed in particular the simplification and automation of the infrared immunoassay. The tricyclic antidepressant nortriptyline and corresponding polyclonal antibodies were involved in part of these studies. A special solid-phase CMIA test was developed by which the analyte specific antibodies were bound to carboxy-modified polystyrene latex particles, separated in a later step by filtration and studied by transmission IR spectroscopy. The immunoassay involved the competition between the sample and labeled analytes for the limited number of antibody-binding sites. Alternatively, antigens (bovine serum albumin) were adsorbed onto the surface of an ATR crystal, and the binding of antibodies labeled with a metal–carbonyl marker within the solution was monitored by ATR spectroscopy. Competition between the adsorbed antigens and free species in the solution for the antibody-binding sites was the governing principle of such a homogeneous immunoassay. This methodology has several advantages over conventional assays based on radio- or enzyme immunoassays, because the markers are stable and less expensive, as claimed by the developing scientists, and allow for the determination of multiple antigens in a single test. The application of mid-IR fibers for use as simple ATR elements, which includes apparatus miniaturization, may give an upsurge in this methodology in the future.

## 4  ANALYSIS OF NONFLUID SPECIMENS

This area is important for infrared pathology, which embraces the fields of cytology, histology, hematology, microbiology, and clinical chemistry, a list that will be enlarged in the future with the development of novel techniques. One example is in vivo optical imaging over greater distances, based on the relative transparency of biological tissue to SW-NIR radiation, and further non-invasive techniques will be discussed later.

In this section, the investigation of samples from biopsies or postmortem specimens can only be reviewed in part. Early applications of IR microscopic examination of such tissue samples were published in 1989 and a year later (116,117). Since then a tremendous development has taken place, based on powerful IR microscopes with a single detector element or focal-plane detectors, providing spectra with near diffraction-limited spatial resolution. These developments for bioanalytical imaging are described in other chapters of this book. Applications to tissue analysis are nowadays numerous, and several reviews are already available; see, for example, Refs. 118–120. The enormous amount of information contained in spectroscopic images—with spatial coordinates, spectral wavenumbers, and intensities—certainly demands new mathematical tools for interpretation and data reduction as well as for graphical presentation. No extrinsic labeling or staining of the biomaterial is necessary, but interpretation of the results is not always straightforward, when correlating spectral intensity patterns with pathological changes. Other biocompounds and the presence of nonspecific diseases can complicate the diagnosis, and cells at different maturation and cycle stages can confound the interpretation of healthy and diseased tissues or cells (121–124). My part is to demonstrate the versatility of IR spectroscopy for specific clinical applications with a few examples.

Infrared spectroscopy has often been employed for qualitative polymer analysis. A special application of mid-IR microscopic mapping dealing with such materials within a tissue matrix was presented by Ali et al., where they describe the characterization of silicone breast implant biopsies (125). Spectra of polydimethylsiloxane and other polymeric materials, such as polyesters and polyurethanes, were used as reference data for interpreting the spectral maps from biopsy preparations of approximately 1 $mm^2$.

We are engaged in the development of new measurement techniques, and mid-IR fibers based on polycrystalline silver halide material have great potential for use in microprobes for ATR measurements (23). The noise level achieved with a flexible probe containing a sensitive U-turn fiber section (outer diameter 0.7 mm; see also Fig. 2) was good enough to record high-quality spectra from spot sizes of about 100-μm diameter and smaller within 1 minute (signal-to-noise ratio 10,000:1 based on peak-to-peak noise within 1100 and 900 cm$^{-1}$, spectral resolution 4 cm$^{-1}$). Several examples from in vivo skin measurements are illustrated in Fig. 21. Further applications with spectra taken from a cerebella section of a Syrian hamster brain are shown in Fig. 22A. The spectra will not be interpreted, but a special report based on IR microscopy applied to hamster brain sections has been given (126). In part B, spectra from different hair samples from a living male test person and from different moor-mummified corpses are displayed, which give clues for the surface modifications in hair keratin under such preservation conditions.

These new spectroscopic tools are still on their way to clinical acceptance, and there are certainly greater needs for future intensive collaboration between clinical and spectroscopic laboratories. The infrared analysis of urinary calculi is an established method for which extensive component library spectra are available (127,128). The analysis of calculi, on the other hand, is important for the diagnosis, because the composition is essential for understanding their etiology. Early applications made use of the KBr pellet technique, which is an example of the qualitative and quantitative analysis of the principal components. Multivariate techniques were employed for computer-based expert systems, needed for the classification of urinary calculi based on mid-IR spectroscopic analysis of some-

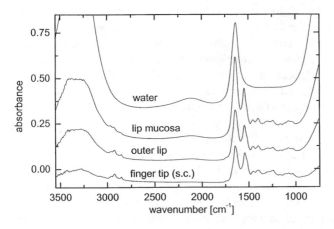

**Fig. 21**   ATR spectra relevant for in vivo spectroscopy of soft biotissues obtained with a miniprobe using a silver halide fiber: water, inner lip mucosa, outer lip skin, and stratum corneum from the fingertip of a single test person. (Reproduced from Ref. 19.)

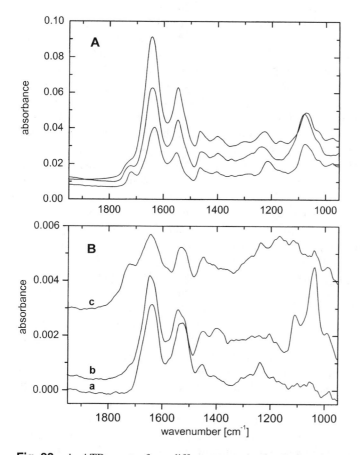

**Fig. 22**   A: ATR spectra from different spots (each area less than about 100 μm²) of the cerebellum section from a Syrian hamster brain obtained with a miniprobe using a silver halide fiber, illustrating the variability within the brain tissue. B: ATR fiber spectra from single hairs: (a) untreated sample from a male person, (b) and (c) samples from moor-mummified corpses (found at Kayhausen and Versen (''roter Franz''), Germany, respectively; the modifications at the hair surfaces still need interpretation).

times-complex mixtures (129). More refined data evaluation using PLS regression for the most frequently occurring compositions reveals the KBr pellet technique as a promising tool for routine quantification in the clinical laboratory (130). Recently, the analysis of special urinary calculi and the incidence and nature of false stones were reported; it was fascinating to read, and it was stated that ''the analysis of each calculus is a special and new case'' (131). Among the 3100 calculi studied, 5% had an unusual composition (101 specimens were false calculi or artifacts, 31 contained drugs or metabolites, and 22 were crystals grown on previously unknown materials). This collection of cases provides clues for the detective work in this field, with encounters of unexpected materials (intentionally factitious stones, by "honest mistake"), which represent jigsaw puzzles to be solved by IR spectroscopy.

A recent investigation on urinary stone layers was carried out, comparing results from IR spectroscopy, Raman spectroscopy, and x-ray powder diffraction (132). While Raman spectroscopy could be used for in situ layer analysis without the use of a microscope attachment, for IR spectroscopy samples were scratched by a scalpel and prepared as KBr pellets for further analysis, which certainly complicated the interpretation of the spectra, because the spatial resolution was different from that achieved by Raman spectroscopy.

In contrast to the latter study on urinary calculi, infrared microspectroscopic mapping was applied to the characterization of human gallstones. Other techniques, such as infrared photoacoustic spectroscopy, usable with minimal sample preparation, and FT-Raman spectroscopy were also employed (133). Gallstones can be grouped into major types due to their composition: cholesterol stones (more than 70% cholesterol by weight), mixed stones (with cholesterol and bilirubinate salts), and brown or black pigment stones with various bilirubinate salts and low cholesterol content. Still to be identified are factors controlling the growth of the different gallstone types. The IR spectra obtained with an IR microscope are certainly affected by the nature of the sample surface, so the diffuse-reflection technique may be applicable. Specular-reflection spectroscopy is to be considered, for example, when a polished surface from cutting a stone in half is exposed to IR radiation. The reflection spectra so obtained can be transformed to normal absorbance spectra by the Kramers–Kronig transformation. We had similar experiences when studying dermis samples from moor-mummified corpses by IR microscopic analysis. Quantitative analysis is certainly adversely affected in the case of a complex mixture of specular and diffuse-reflectance spectra.

A more recent investigation was based on a microscopic analysis using FT-IR and FT-Raman spectroscopy and, in addition, for trace-element analysis, proton-induced x-ray emission (PIXE). The authors' conclusion was that the chemical analysis and categorization of gallstones is optimized by using a combination of techniques, especially since infrared and Raman spectroscopy are complementary in nature, providing an improved information basis for interpretation. However, for black gallstones a strong fluorescence hindered the recording of Raman spectra, so an analysis could be based only on IR data (134).

## 5  NONINVASIVE TECHNIQUES

There is a great demand in clinical chemistry for gentle diagnostic methods, which means that noninvasive spectroscopic techniques are favored. Many medical instruments use the interaction of electromagnetic radiation with matter; for example, magnetic resonance imaging and spectroscopy are used for routine work. Optical techniques, however, offer the potential for the development of small, rugged, and moderately priced instrumentation. Mid-IR probes based on silver halide fibers have already been mentioned, but the ATR technique has only limited opportunities for probing larger tissue depths due to the physics involved. In addition, owing to the large water absorptivities, penetration of mid-IR radiation is not sufficient to establish, for example, transcutaneous measurements of metabolites. A way out of this dilemma is to use near-IR spectroscopy, for which the absorptivities of the biosolvent water are much smaller. In general, due to the optical constants of soft tissues, in describing absorption and scattering, a significant wavelength-dependent penetration depth for near-IR radiation exists. For wavelengths between 600 and 1300 nm ($16,700–7,700$ cm$^{-1}$, the so-called therapeutic window) transmission measurements of body tissues are possible. Other spectral ranges for in vivo measurements are accessible by diffuse-reflectance measurements. In Fig. 23, spectra of lip mucosa, taken as an example of soft tissues, are shown. The diffuse-reflectance spectra, which are dominated by water absorption, were recorded by an optical fiber probe and a rotational ellipsoid–based device, respectively. From these spectra, the influence of the tissue optics on the shape of tissue spectra is evident. The type and collection geometry of the device and the source-detector separation are especially critical factors for in vivo assays. The spectral ranges shown in Fig. 23 are important for the transcutaneous determination of metabolites, and activities for developing noninvasive blood glucose assays will be discussed later, with a particular view back to the in vitro assays for biofluids presented earlier.

For shorter wavelengths, other chromophores dominate the skin spectra. Several of the major chromophores are organized in separate layers, with melanin in the epidermis and hemoglobin in the dermal capillary and venule plexus. Oxygenated and deoxygenated hemoglobins absorb radiation differently and affect the in vivo reflectance spectra accordingly (see also Fig. 24). The absorption bands below 600 nm are assigned to the oxyhemoglobin species; hemoglobin oxygenation found for these measurements was rather high (HbSO$_2$ around 90%). The band maxima can be related to a corresponding blood volume in the skin tissue under investigation. Lip mucosa are well suited for regressing their spectra against changing blood glucose levels due to the large ratio of blood to spectroscopically probed tissue volume. The physiological variances in skin spectroscopy are large because of the inherent role of skin in body temperature and water regulation. Figure 24B shows the effect from hyperemization of the inner lip tissue by application of a fiber sensor head, kept at a temperature of 42°C, which leads to an increase in blood volume and hemoglobin oxygen saturation. Less important for skin oximetry are other chromophores, such as porphyrins, beta-carotene, and bilirubin, which also have absorptions in the visible range.

The monitoring of tissue physiology, including blood and tissue oxygen-

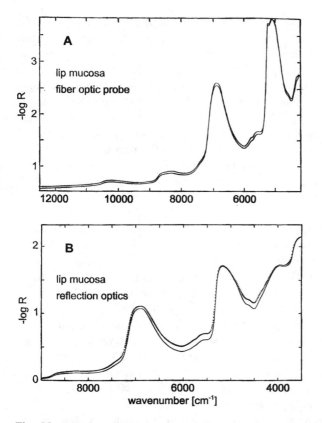

**Fig. 23** Diffuse-reflectance spectra of lip mucosa recorded by using different accessories. A: Lip spectra of two male persons measured against a white reflection standard by using a bifurcated fiber-optic probe. B: Lip mean spectra from a single person and from a multiperson experiment (133 persons; both spectra were offset normalized). (Reproduced from Ref. 15.)

ation, respiratory status, and ischemic damage, is possible by the use of visible and near-IR spectroscopy. Of great interest is the acquisition of knowledge on cerebral hemodynamics as well as on the intracellular reduction-oxidation level of cytochromes, for example, in the brain of human neonates. Functional imaging of the brain is nowadays a research field tackled by many groups. These areas will be described in detail later.

Another large field of application will be optical tomography using time- or frequency-domain–based technology, especially for noninvasive detection of breast cancer (135,136). Both the scattering and absorption properties of tumors

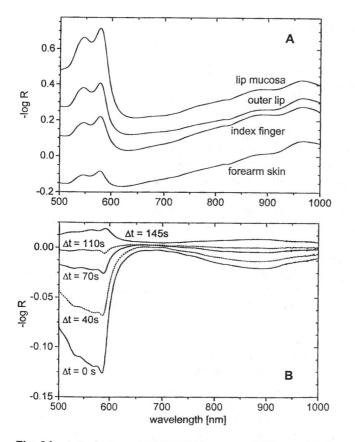

**Fig. 24** A: Reflectance VIS/SW-NIR spectra of different skin tissues of a single person, recorded at normal skin temperature by using a bifurcated fiber-optic probe (for clarity, the individual spectra were offset). (Reproduced from Ref. 28.) B: Difference spectra showing the effect of hyperemization of the inner lip tissue by application of a fiber sensor head thermostated at 42°C (first contact at $\Delta t = 0$ sec, difference spectra were calculated versus a lip spectrum taken after 120 sec). (Reproduced from Ref. 15.)

and the surrounding tissue can be exploited for sensing tissue abnormalities. However, compared to conventional x-ray mammography, the spatial resolution of optical tomography is inferior, due to tissue scattering. For example, the development of a time-domain optical mammograph and first in vivo applications were described by Grosenik et al. (137). At the moment, these activities are certainly at the fringes of conventional clinical applications; however, for screening purposes this technically complex technology needs to be developed further.

## 5.1  Noninvasive Blood Glucose Assays

Self-monitoring of blood glucose is part of the daily routine of diabetic patients, since knowledge of the glucose level is important for its regulation by insulin administrations. Therefore, many research activities tackle the development of a minimally invasive or noninvasive methodology. As shown for in vitro assays, there is potential in the use of near-IR spectroscopic methods, which exploit specific optical glucose characteristics, whether wavelength-dependent absorptivities or refractive indices. Due to the complexity of the integrally probed tissue with the presence of many interfering compounds, only multivariate spectrum measurement and evaluation will be successful. Further complications have their origin in the heterogeneous distribution of glucose in the intravascular, interstitial, and intracellular spaces and corresponding dynamics, whereas the measurement of capillary blood glucose is considered the straightforward ''gold standard'' for an intensified insulin therapy. Many factors have to be considered for an optimization of an in vivo assay, such as skin structure, blood volume, blood flow, and physiological variations (12,15). Until now, the near-IR absorption approach has proven the most effective method for providing in vivo blood glucose–monitoring capabilities. The photoacoustic near-IR measurement technique has also been employed, and promising results were published by Mac-Kenzie and coworkers (37).

The fundamental aspects of multivariate calibrations using near-IR spectroscopy were discussed earlier. One prerequisite for successful application is certainly the reproducible measurement of tissue spectra at very high signal-to-noise ratios. This is necessary because the sample composition is more complex than for biofluids, which can be measured under thermostated and fixed optical pathlength conditions. We developed an accessory that can be thermostated and is especially optimized for diffuse-reflectance measurements of inner lip mucosa (27); see also Fig. 3. Such spectra were the basis for our noninvasive blood glucose calibration experiments. Recent results published from our studies with a single diabetic test person using a spectral variable selection scheme (22,28,44) are shown in Fig. 25. In addition to the two-day blood glucose concentration profiles and the near-IR predicted concentration values (RMSEP = 36 mg/dl, obtained with cross-validation, 26 spectral variables within the spectral interval of 9000–5500 cm$^{-1}$), a scatter plot is also shown, illustrating the deficiencies of the assay especially for the hypo- and normoglycemic concentrations.

Another interesting study was from Blank et al. (138), who used near-IR diffuse-reflectance spectra from forearm skin, studied within the range between 1050 and 2450 nm. Different oral glucose-tolerance tests (OGTT) were performed with nondiabetic test persons on different days to induce blood glucose changes. The consequence of such an experimental design is that the calibration set variance of glucose concentrations was rather low (the standard deviation in

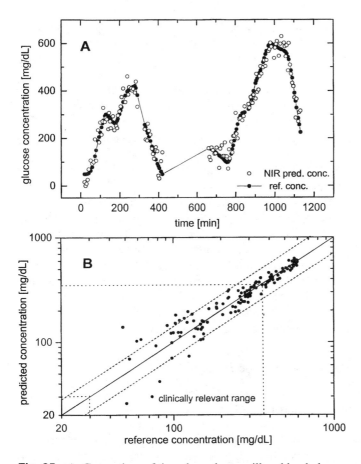

**Fig. 25** A: Comparison of time-dependent capillary blood glucose concentration values (reference data) of a diabetic person and prediction results based on PLS calibration using 26 selected spectral variables from diffuse-reflectance lip spectra (data from nonstandard oral glucose-tolerance testing (OGTT) during two days). B: Scatter plot of predicted concentrations versus reference values (the aspired confidence interval for a 30% relative standard deviation is marked by dashed lines). (Reproduced from Ref. 28.)

reference values around the mean was below 40 mg/dl). Prediction errors based on independent OGTT calibration and test sets were below 20 mg/dl. Some factors, including variations in skin surface roughness, in measurement location, and in skin hydration, the effects of contact pressure, and variations in skin temperature, were mentioned that contribute to changes in the probing of the tissue volume under investigation.

Apart from absorption measurements, a different approach using changes in the scattering of tissue mediated by different glucose concentrations was evaluated by a few groups. There are several effects from dissolving glucose in an aqueous scattering biosystem: the tissue absorption coefficient is influenced by an increase in intrinsic glucose absorption and by water displacement. In addition, further effects are changes in the refractive index and, through the latter, in the scattering coefficient. Utilization of the last effect led to the development of portable instrumentation with two optical sensor heads, each containing four light-emitting diodes of different wavelengths in the visible and SW-NIR range and six photodetectors located at different millimeter distances. The monitoring of tissue glucose concentration was carried out during several glucose clamp experiments (139), and the general variability, i.e., glucose-independent changes in scattering of skin, was investigated. The specificity is the greatest concern here, because many other effects unrelated to glucose can influence the scattering signal. On the other hand, the instrumentation is attractive, due to being simple and affordable, and it would allow continuous monitoring.

There have been other approaches tested; among them are fluorescence spectroscopy in the SW-NIR spectral range (140,141), polarimetry (142,143), and Raman spectroscopy. More details on spectroscopic approaches can be found in reviews on noninvasive glucose measurements (15,144).

## 5.2  Pulsatile Near-Infrared Spectroscopy

Integral tissue probing faces many unknown physiological variables. Further development of noninvasive assay technology can be based on fast subsecond NIR spectroscopic measurements, allowing the probing of parts of the intravascular fluid space through dynamic monitoring of tissue absorption. Photoplethysmography under high time resolution can be used to measure the cardiovascular pulse wave, correlated to changes in blood volume, by spectrometry. At present, the major clinical application is used in pulseoximetry, by which arterial hemoglobin oxygen saturation can be determined (13,145); see also the next paragraph.

Broad near-IR spectra can also be monitored dependent on the cardiac cycle. However, since minor blood volume changes in the arterial blood, which constitutes in itself only a small fraction of the tissue volume, are available, the alternating spectral signal to be evaluated for specific component absorptions is only minor compared to that of the nondynamically measured tissue. Due to limitations in the signal-to-noise ratio, such a measurement principle—although proposed in patents—has not yet been applied for practical metabolite measurements.

First results of time-resolved measurements on human oral mucosa were presented by us using diffuse reflectance spectroscopy (22,44). The first individual lip spectra obtained with fast measurements are shown in Fig. 26A. These are difference spectra smoothed for noise reduction, as calculated versus the first

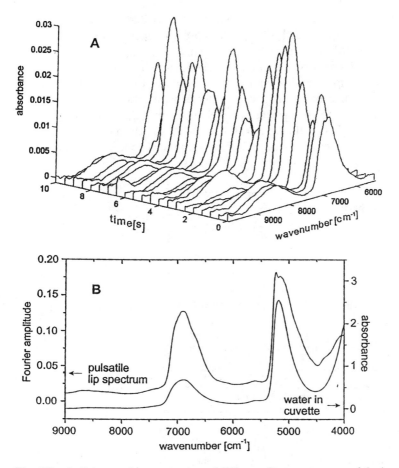

**Fig. 26** A: Subsecond measurements of diffuse-reflectance spectra of the inner lip, presented as differences versus the first measured single-beam spectrum after Savitzky–Golay smoothing. B: Fourier amplitudes (upper trace) illustrate the relative pulsatile spectral components due to cardiac-modulated blood volume variations (for each spectral variable the Fourier coefficients were averaged around the heart beat frequency within an 0.06-Hz interval. For comparison the absorbance spectrum of water with a layer thickness of 0.5 mm is also given. (Reproduced from Ref. 28.)

measured spectrum of the data set, which clearly illustrate the intensity fluctuations caused by changes in the arterial blood compartment associated with the cardiac cycle.

A Fourier analysis of each time-dependent logarithmized wavenumber intensity provided the spectral Fourier amplitudes for frequency components within a certain interval (see also the equivalent data for visible and SW-NIR tissue measurements in Figs. 27 and 28). Slicing up the frequency component of the heartbeat, the pulsatile spectrum can be obtained, as originating from the aqueous, volume-modulated compartment of the skin tissue, which is compared to a water-absorbance spectrum, recorded with a transmission cell of 0.5-mm pathlength. The water-absorbance alterations for the water-band maximum at 6900 cm$^{-1}$ due to the cardiac blood pressure changes are about 20 mA.U., which is equivalent to a water layer of 15-μm thickness (about a factor of 50 smaller than obtained for nondynamic lip measurements). Compared to the cuvette spectrum recorded

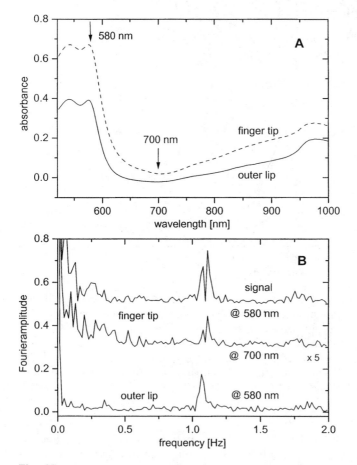

**Fig. 27**  A: Reflectance spectrum of the outer lip and the fingertip of a single person, as averaged from measurements over 1 minute (sampling interval 250 ms). B: Results from a Fourier analysis of the dynamically recorded spectral traces at selected wavelengths.

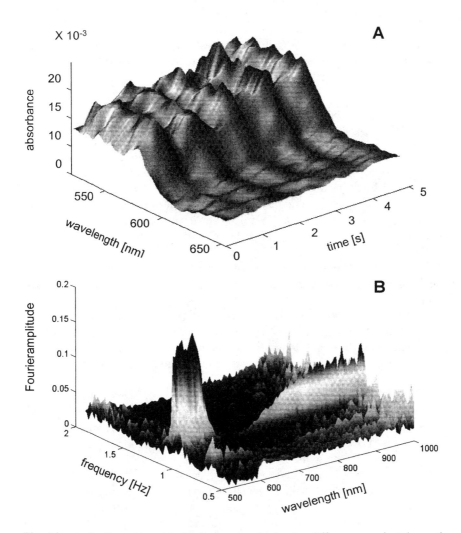

**Fig. 28** A: Section of the pulsatile lip spectra obtained as differences against the section preceding spectrum with a diode-array based PC spectrometer. B: Result from a Fourier analysis of each of the dynamically recorded individual spectral variables, showing clearly the pulsatile spectrum along the heartbeat frequency.

with constant optical pathlength, the difference in the intensity ratio of bands from the pulsatile spectrum, as shown in Fig. 26B, can be explained by the significantly different penetration depths for the near-infrared radiation realized for those wavelengths. Further investigations on improving the signal-to-noise ratio in such pulsatile spectra are required, which can provide the basis for noninvasive blood analysis of substrates, as already accessed by in vitro IR spectrometric assays discussed previously.

### 5.3  Monitoring of Oxygenation and Redox Status of Physiologically Important Compounds

The principles of optical spectroscopy for in vivo studies with respect to hemoglobin oxygenation were developed from laboratory work using in vitro samples. In the early 1930s, considerable efforts were made to develop noninvasive optical techniques for oximetry, providing information about the in vivo state of tissue oxygen supply; see also Ref. 43. With recent developments, the regional and temporal variations in skin tissue oxygenation can be continuously assessed using a near-IR spectroscopy–based imaging system (146). Tissue viability testing, particularly important after plastic surgery, was also tackled by reflectance measurements in the range between 650 and 900 nm (14,147). For presentation, a form of data processing, i.e., fuzzy C-means clustering, was necessary for clustering those spectra with a similar intensity pattern (148). Another compound of interest, similar to hemoglobin in blood, is myoglobin in muscle tissue. A further important chromophore for quantitative tissue spectroscopy is cytochrome $aa_3$ in the cellular mitochondrial membrane, which is the terminal compound in the respiratory chain (149).

Jöbsis (150) published an early paper on the use of near-IR spectroscopy for measurements on intact tissue of centimeter thickness; since then, an upsurge has been noticed in biochemical and medical applications. Spectroscopists extended their measurements to the SW-NIR for noninvasive monitoring of oxygen metabolism, e.g., in the living human brain. Here, hemoglobin still shows significant absorption in both its reduced and oxygenated states. Cerebral monitoring, e.g., for neonates, has been achieved by monitoring a number of additional parameters, such as blood flow, oxygen delivery, and blood volume, using special equipment based on a small number of pulsed laser diodes (typical selections are for 775, 825, 850, and 905 nm; average power is about 2 mW) (151,152).

For newborn infants it is possible to detect photons with head transillumination. However, adult measurements are usually carried out in reflection mode (reflectance oximetry) with different source-detector separations. Mathematical simulations of the photon transport in tissue by the so-called Monte Carlo techniques have shown that the anatomy of the brain-surrounding layers, especially the clear cerebrospinal fluid, affect the distribution of photons (153), which explains in part the large variability of data published by different authors.

A review on cerebral near-IR spectroscopy in neonatal intensive care was published stating that much credible and important data have been accumulated. On the other hand, it has been difficult to take advantage of the continuous and noninvasive nature of the methodology, because the interpretation of the data, especially on redox state measurements of the cytochrome $aa_3$, is still surrounded with doubt (154). Experimental and spectral evaluation algorithm problems, connected to clinical trials, were reported and discussed by Macnab et al. (155). From the same area of neonatology, recent advances in fetal monitoring using near-IR spectroscopy and specially designed optical probes were described by D'Antona et al. (156).

A different setup was used for quantitative measurement of muscle oxygenation by near-IR spectroscopy using two LEDs of 760- and 840-nm peak wavelength and an array of eight photodiodes located at different distances from the IR radiation source. One aim of the study was to test the influences of the subcutaneous fat layer and skin by Monte Carlo simulations of the photon tissue transport and by in vivo tests (157). A correction curve for measurement sensitivity was determined.

Frequency domain multidistance spectroscopy, as applied to a clinical trial for the diagnosis of peripheral vascular disease, was used with a special hemoglobin spectrometer, using 16 emitter fibers, with a two-detector device. Emitter wavelengths were 750 and 830 nm. In addition, the oximeter is capable of imaging and functional imaging (158). Comprehensive literature references on such applications is given by these authors. An example for a measurement device, which can be used for functional imaging, is provided by Yamamoto et al. (159). They developed a 24-channel optical topography system that uses intensity-modulated near-IR spectroscopy at wavelengths of 780 and 830 nm. The system was used to visualize spatial and temporal changes of blood oxygenation in the human brain caused by cortical activity, testing for different language function.

Other developments for a portable, small tissue oximeter based on a two-wavelength LED setup (760 and 840 nm) have been made. Special coefficients for the concentration estimation algorithm were deduced through experiments by varying blood volume and scattering intensity in a tissuelike phantom (160).

Pulse oximetry, which was mentioned earlier, is used widely in hospitals for noninvasive blood gas analysis, monitoring arterial hemoglobin oxygen saturation. It is fundamental for the support of critical care medicine, although some deficiencies still exist. A review on theory and applications, including practical limitations of such a technique, was given by Mendelson (161). Problems mentioned were concerned with low peripheral vascular perfusion, motion artifact, effect from different hemoglobin variants and derivatives (e.g., HbCO), electromagnetic compatibility, and stray light. Pulse oximetry was recommended as a standard care for intraoperative monitoring, and an interesting report on applications and limitations in clinical anesthetic practice was published that was based on the analysis of 2000 incident reports (162).

One of the main problems with pulse oximetry is still measurement error introduced by movement, and special signal processing with nonlinear methodology was proposed for removing such errors (13). Furthermore, the technology appears to be moving toward miniaturization, eliminating motion artifacts, and allowing the use of pulse oximetry outside the critical care area. Another problem investigated was lateral tissue inhomogeneity (163). Near-IR LEDs with wavelengths centered on 820, 1000, 1220, and 1300 nm were employed in a setup largely avoiding ambient-light interference and motion-induced probe-coupling artifacts. The fundamental concern of the study was that the sampled tissue volume, the fingertip with and without applied pressure, is not the same for all wavelengths tested. Use of the 1300-nm and 800-nm wavelengths in a photoplethysmographic device had been described earlier by Schmitt et al. to be used for noninvasive hematocrit measurement (57). An application of near-IR laser diodes, with narrow spectral bandwidth and suitable for high-frequency modulation, was reported for use in pulse oximetry, substituting for the LEDs normally employed (164).

So far, devices were described using at least two or more wavelengths for two- or higher-component systems to be studied. As shown by van Huffel and coworkers, the multivariate determination of hemoglobin, oxyhemoglobin, and cytochrome $aa_3$ can be improved by implementing many wavelengths with wavelength selection, in addition to the use of total least squares, taking different reference data precision into account for regression analysis (165). The use of a multichannel diode array spectrometer for the measurement of hemoglobin derivatives with a limited number of only 38 Si-diode elements was discussed by Mendelson (166).

We also are interested in simple and low-cost instrumentation for multiwavelength measurements in the visible and near-IR spectral range for precise oximetry. The performance of a diode array–based VIS-NIR spectrometer with a special diffuse-reflectance accessory for skin illumination using an integrating sphere and detection through radiation transfer by means of fiber optics was tested for photoplethysmography of skin tissue. The same algorithm for pulsatile spectrum detection was used as described earlier for analytical near-IR pulsatile spectroscopy for metabolites in tissue blood. Spectra recorded were analyzed for mean tissue spectra and for the pulsatile components by Fourier analysis of the original dynamically recorded spectral data with a constant time step of 250 ms. Figure 26A shows the average spectra as obtained for 1 minute as well as the power signal in the frequency domain of special spectral variable-intensity pairs as recorded over one minute. Such duration gives enough resolution for isolating the fundamental heartbeat frequency. Three-dimensional plots are shown in Fig. 28 for part of the dynamically recorded spectra and the data after Fourier analysis of each temporal individual wavenumber absorbance trace. The slice at the heartbeat frequency provides us with the pulsatile spectrum. Aliasing from overtone components can be reduced by faster spectral measurements, but often the separation is good enough to separate the fundamental frequency component from the overtones.

## 6  CONCLUDING REMARKS

Progress in medicine and health care is largely dictated by advances in our ability to collect new information from the analysis of tissues and biofluids. Spectroscopic methodologies spawned a new dimension for clinical applications in the analytical laboratory as well for monitoring in intensive care units, in surgery, and for diabetic patient self-monitoring of blood glucose. The reagent-free multicomponent analysis of biofluids has continuously been improved to become acceptable for the development of routine analyzers with high sampling frequencies. Disposable substrates carrying films formed from physiological fluids may finally be successful. However, further systematic studies must be carried out with respect to the degree of perturbances from, for example, pharmaceutical compounds, which we started to check for blood plasma assays. This will improve the acceptance of the spectroscopic assays as routine analytical tools. Enzyme-activity assays and immunoassays are other promising fields for the future. There are further developments concerned with the user–device interface, but this can be achieved for dedicated instruments with appropriate report presentation. The breakthrough will be reached with mass production of such instrumentation.

Infrared pathology and assays for urinary calculi and gallstones and the analysis of implant materials and their biocompatibility are areas for the implementation of microscopy using infrared and Raman spectroscopy. Intelligent software and comprehensive spectral data banks will be necessary for decision making, speeding up the step from information to diagnosis. Handling the "imaging data cube" obtained in mapping or imaging with efficient classification methodologies is still an ongoing project.

The area of noninvasive technologies is of immense interest. The advantage of IR spectroscopic equipment is that it can be miniaturized with modern electro-optical devices. New types of lasers operated at room temperature are looking promising as radiation sources; IR photometers in combination with fiber optics are other alternatives to the big spectrometers needed in basic research. The design of small spectrometers is profiting from significant advances in integrated electronics, so self-monitoring devices for the patient at home will be possible in the not-too-distant future. Significant interest also exists in the noninvasive equipment for monitoring, e.g., hemodynamics in skin, brain, and other tissues.

More sophisticated mathematical tools developed for information science and chemometrics have to be employed, but the challenges can only be met by tremendous interdisciplinary collaboration. It seems that further research is needed for the understanding of the physiological and histological variances faced by the biospectroscopist and the physician, who need to work hand in hand to achieve health, social, and economic benefits. We look optimistically forward to the scientific offspring created from the marriage of medicine and spectroscopy!

## ACKNOWLEDGMENTS

The author is indebted to Mrs. M. Hillig, Dr. A. Bittner, Dr. R. Marbach, Dipl.-Ing. S. Thomassen, and Dr. L. Küpper for fine collaboration in the past. Further support was granted by Prof. Dr. med. Th. Koschinsky from the Diabetes Research Institute in Düsseldorf, Dr. med. M. Stücker from the Dermatology Department of the Ruhr-Universität in Bochum, Dr. H. Jungmann from the Krebsforschung e.V. in Herdecke. Mrs. J. Kneipp from the Robert-Koch-Institut in Berlin supplied the microtomed hamster brain sample. For several projects, the collaboration with Boehringer Mannheim GmbH (now Roche Diagnostics) in Mannheim and Eppendorf-Netheler-Hinz GmbH in Hamburg is also acknowledged. Further financial support by the Deutsche Forschungsgemeinschaft, the Ministerium für Schule und Weiterbildung, Wissenschaft und Forschung des Landes Nordrhein-Westfalen, and the Bundesministerium für Bildung und Forschung is gratefully acknowledged.

## REFERENCES

1. C Sandorfy, T Theophanides, eds. Spectroscopy of Biological Molecules. Dordrecht, The Netherlands: Reidel, 1984.
2. HH Mantsch, D Chapman, eds. Infrared Spectroscopy of Biomolecules. New York: Wiley-Liss, 1996.
3. J Greve, GJ Puppels, C Otto, eds. Spectroscopy of Biological Molecules: New Directions, 8th European Conference on the Spectroscopy of Biological Molecules. Dordrecht, The Netherlands: Kluwer Academic, 1999.
4. DE Beischer. J Urol 73:653–659, 1955.
5. HM Heise. Lab Med 15:470–476, 1991.
6. P Wilks. Int Lab 5:11A–12A, 1996.
7. B Lacroix, JP Hevenne, M Deveaux. J Chromatog 492:109–136, 1989.
8. KS Kalasinsky, B Levine, ML Smith, J Magluilo Jr, T Schaefer. J Anal Toxicol 17:359–364, 1993.
9. MB Esler, DWT Griffith, SR Wilson, LP Steele. Anal Chem 72:216–221, 2000.
10. WE Heine, HK Berthold, PD Klein. Am J Gastroenterol 90:93–98, 1995.
11. KS Kalasinsky. Cell Mol Biol 44:81–87, 1998.
12. JN Roe, BR Smoller. Crit Rev Therap Drug Carrier Syst 15:199–241, 1998.

13.  MJ Hayes, PR Smith. Appl Optics 37:7437–7446, 1998.
14.  JR Payette, MG Sowa, SL Germscheid, MF Stranc, B Abdulrauf, HH Mantsch. Am Clin Lab 18:4–6, 1999.
15.  HM Heise. In: RA Meyers, ed. Encyclopedia of Analytical Chemistry: Instrumentation and Application. Chichester, UK: Wiley, 2000, in press.
16.  H Günzler, HM Heise. IR-Spektroskopie. 3rd ed. Weinheim, Germany: VCH-Wiley, 1996.
17.  LT Rozelle, LJ Hallgren, JE Bransford, RB Koch. Appl Spectrosc 4:120–124, 1965.
18.  HM Heise. Proc SPIE 553:247–248, 1985.
19.  HM Heise, L Küpper, R Marbach. LEOS Newsletter 13:12–15, 1999.
20.  L Butvina. In: J Sanghera, I Aggarwal, eds. Infrared Fiber Optics. Boca Raton, FL: CRC Press, 1998, pp 209–249.
21.  HM Heise, A Bittner, L Küpper, LN Butvina. J Mol Struct 410/411:521–525, 1997.
22.  HM Heise, A Bittner. In: HH Mantsch, M Jackson, eds. Infrared Spectroscopy: New Tool in Medicine. Proc SPIE 3257:2–12, 1998.
23.  HM Heise, L Küpper, LN Butvina. Sensors Actuators B 51:84–91, 1998.
24.  J Coates. Appl Spectrosc Rev 33:267–425, 1998.
25.  LA Cassis, J Yates, WC Symons, RA Lodder. J Near Infrared Spectrosc 6:A21–A25, 1998.
26.  A Bittner, HM Heise. In: JA de Haseth, ed. Fourier Transform Spectroscopy. 11th International Conference. New York: American Institute of Physics, AIP Conf Proc 430:278–281, 1998.
27.  R Marbach, HM Heise. Appl Optics 34:610–621, 1995.
28.  HM Heise, A Bittner, R Marbach. Clin Chem Lab Med 38:137–145.
29.  HM Heise, A Bittner. J Mol Struct 348:21–24, 1995.
30.  J Lin, CW Brown. Appl Spectrosc 47:62–68, 1993.
31.  J Lin, CW Brown. Anal Chem 65:287–292, 1993.
32.  K Molt, A Niemöller, YJ Cho. J Mol Struct 410/411:565–572, 1997.
33.  K Martin. Appl Spectrosc 52:1001–1007, 1998.
34.  J Wang, M Sowa, HH Mantsch, A Bittner, HM Heise. Trends Anal Chem 15:286–296, 1996.
35.  T Fuji, Y Miyahara, Y Watanabe. Appl Spectrosc 50:1682–1686, 1997.
36.  MG Sowa, HH Mantsch. J Mol Struct 300:239–244, 1993.
37.  HA MacKenzie, HS Ashton, YC Shen, J Lindberg, P Rae, KM Quan, S Spiers. In: EM Sevick-Muraca, JA Izatt, MN Ediger, eds. Biomedical Optical Spectroscopy and Diagnostics/Therapeutic Laser Applications. OSA Trends in Optics and Photonics Series 22:156–159, 1998.
38.  RJ Jakobsen, FM Wasacz, KB Smith. In: JR Durig, ed. Chemical, Biological and Industrial Applications of Infrared Spectroscopy. New York: Wiley, 1985, pp 199–213.
39.  F Cadet, C Robert, B Offmann. Appl Spectrosc 51:369–375, 1997.
40.  FO Libnau, AA Christy, OM Kvalheim. Vib Spectrosc 7:139–148, 1994.
41.  HM Heise, A Bittner. In: JA de Haseth, ed. Fourier Transform Spectroscopy: 11th International Conference. New York: American Institute of Physics, AIP Conf Proc 430:274–277, 1998.
42.  S Berentsen, T Stolz, K Molt. J Mol Struct 410/411:581–585, 1997.

43. J Hoffmann, DW Lübbers, HM Heise. Phys Med Biol 43:3571–3587, 1998.
44. HM Heise, A Bittner, R Marbach. J Near Infrared Spectrosc 6:349–359, 1998.
45. MR Nyden, GP Forney, K Chittur. Appl Spectrosc 42:588–594, 1988.
46. H Martens, T Naes. Multivariate Calibration. New York: Wiley, 1989.
47. R Marbach, HM Heise. Trends Anal Chem 11:270–275, 1992.
48. KU Jagemann, C Fischbacher, K Danzer, UA Müller, B Mertes. Z Phys Chem 191: 179–190, 1995.
49. P Bhandare, Y Mendelson, RA Peura, G Janatsch, JD Kruse-Jarres, R Marbach, HM Heise. Appl Spectrosc 47:1214–1221, 1993.
50. MJ McShane, GL Coté. Appl Spectrosc 52:1073–1078, 1998.
51. MA Arnold, JJ Burmeister, GW Small. Anal Chem 70:1773–1781, 1998.
52. HM Heise, R Marbach. Cell Mol Biol 44:899–912, 1998.
53. MR Riley, MA Arnold, DW Murhammer. Appl Spectrosc 52:1339–1347, 1998.
54. S Kostrewa, HM Heise, W Goemann. Fourier Transform Spectros 12th Int Conf. K Itoh, M Tasumi, eds. Tokyo, Japan: Waseda University Press, 1999, pp 469–470.
55. S Kostrewa, C Paarmann, W Goemann, HM Heise. In: JA de Haseth, ed. Fourier Transform Spectroscopy: 11th International Conference. New York: American Institute of Physics, AIP Conf Proc 430:271–273, 1998.
56. A Roggan, M Friebel, K Dörschel, A Hahn, G Müller. J Biomed Optics 4:36–46, 1999.
57. JM Schmitt, Z Guan-Xiong, J Miller. Proc SPIE 1641:150–161, 1992.
58. G Janatsch, JD Kruse-Jarres, R Marbach, HM Heise. Anal Chem 61:2016–2023, 1989.
59. HM Heise. In: R Macrae, ed. Encyclopedia of Analytical Science. Vol 4. London: Academic Press, 1995, pp 1948–1958.
60. HM Heise, R Marbach, G Janatsch, JD Kruse-Jarres. Anal Chem 61:2009–2015, 1989.
61. HM Heise, R Marbach, Th Koschinsky, FA Gries. Appl Spectrosc 48:85–95, 1994.
62. HM Heise, A Bittner. J Mol Struct 348:127–130, 1995.
63. HM Heise, A Bittner, Th Koschinsky, FA Gries. Fresenius J Anal Chem 359:83–87, 1997.
64. MJ McShane, BD Cameron, GL Coté, CH Spiegelman. Proc SPIE 3599:101–109, 1999.
65. HM Heise, A Bittner. Fresenius J Anal Chem 359:93–99, 1997.
66. HM Heise, A Bittner. Fresenius J Anal Chem 362:141–147, 1998.
67. HM Heise. In: DM Fraser, ed. Biosensors in the Body—Continuous in-vivo Monitoring. Chichester, UK: Wiley, 1997, pp 79–116.
68. R Simhi, Y Gotshal, D Bunimovich, BA Sela, A Katzir. Appl Optics 35:3421–3425, 1996.
69. E Bormashenko, R Pogreb, S Sutovski, I Vaserman, A Katzir. Proc SPIE 3570: 100–106, 1999.
70. R Vonach, J Buschmann, R Falkowski, R Schindler, B Lendl, R Kellner. Appl Spectrosc 52:820–822, 1998.
71. HM Heise, R Marbach, A Bittner, Th Koschinsky. J Near Infrared Spectrosc 6: 361–374, 1998.

72. A Bittner, R Marbach, HM Heise. J Mol Struct 349:341–344, 1995.
73. HM Heise. Mikrochim Acta, 14(suppl):67–77, 1997.
74. KH Hazen, MA Arnold, GW Small. Anal Chim Acta 371:255–267, 1998.
75. JW Hall, A Pollard. Clin Chem 38:1623–1631, 1992.
76. JW Hall, A Pollard. Clin Biochem 26:483–490, 1993.
77. JW Hall, A Pollard. In: GD Batten, PC Flinn, LA Welsh, AB Blakeney, eds. Leaping Ahead with Near Infrared Spectroscopy. Melbourne: Royal Austral Chem Inst, 1995, pp 421–430.
78. KH Norris, JT Kuenstner. In: GD Batten, PC Flinn, LA Welsh, AB Blakeney, eds. Leaping Ahead with Near Infrared Spectroscopy. Melbourne: Royal Austral Chem Inst, 1995, pp. 431–436.
79. MR Gatin, JR Long, PW Schmitt, PJ Galley, JF Price. In: AMC Davies, P Williams, eds. Near Infrared Spectroscopy: The Future Waves. Chichester, UK: NIR Publications, 1996, pp 347–352.
80. G Domján, KJ Kaffka, JM Jákó, IT Vályi-Nagy. J Near Infrared Spectrosc 2:67–78, 1995.
81. RJ Dempsey, DG Davis, RG Buice Jr, RA Lodder. Appl Spectrosc 50:18A–34A, 1996.
82. M Nara, H Kagi, M Okazaki. In: J Greve, GJ Puppels, C Otto, eds. Spectroscopy of Biological Molecules: New Directions. 8th European Conference on the Spectroscopy of Biological Molecules. Dordrecht, The Netherlands: Kluwer Academic, 1999, pp 371–372.
83. K Murayama, K Yamada, R Tsenkova, Y Wang, Y Ozaki. J Near Infrared Spectrosc 6:375–381, 1998.
84. K Murayama, K Yamada, R Tsenkova, Y Wang, Y Ozaki. Fresenius J Anal Chem 362:155–161, 1998.
85. JT Kuenstner, KH Norris, WF McCarthy. Appl Spectrosc 48:484–488, 1994.
86. JT Kuenstner, KH Norris. J Near Infrared Spectrosc 2:59–66, 1994.
87. JT Kuenstner, KH Norris. J Near Infrared Spectrosc 3:11–18, 1995.
88. G Yoon, S Kim, YJ Kim, JW Kim, WK Kim. In: HH Mantsch, M Jackson, eds. Infrared Spectroscopy: New Tool in Medicine. Proc SPIE 3257:126–133, 1998.
89. JL Gonczy, LL Gyarmati. In: I Murray, IA Cowe, eds. Making Light Work: Advances in Near Infrared Spectroscopy. Weinheim, Germany: VCH, 1992, pp 603–609.
90. JW Hall, A Pollard. J Near Infrared Spectrosc 1:127–132, 1993.
91. G Budinová, J Salva, K Volka. Appl Spectrosc 51:631–635, 1997.
92. RA Shaw, S Kotowich, M Leroux, HH Mantsch. Ann Clin Biochem 35:624–632, 1998.
93. RA Shaw, M Leroux, M Paraskevas, FB Guijon, S Kotowich, HH Mantsch. In: HH Mantsch, M Jackson, eds. Infrared Spectroscopy: New Tool in Medicine. Proc SPIE 3257:42–50, 1998.
94. RA Shaw, HH Eysel, KZ Liu, HH Mantsch. Anal Biochem 259:181–186, 1998.
95. GH Werner, D Boecker, HP Haar, HJ Kuhr, R Mischler. In: HH Mantsch, M Jackson, eds. Infrared Spectroscopy: New Tool in Medicine. Proc SPIE 3257:91–100, 1998.
96. GH Werner, J Früh, F Keller, H Greger, R Somorjai, B Dolenko, M Otto, D Böcker.

In: HH Mantsch, M Jackson, eds. Infrared Spectroscopy: New Tool in Medicine. Proc SPIE 3257:35–41, 1998.

97. NA Brunzel. Fundamentals of urine and body fluid analysis. Philadelphia: WB Saunders, 1994.

98. AW van Toorenenbergen. Clin Chem 40:1788, 1994.

99. RA Shaw, S Kotowich, HH Mantsch, M Leroux. Clin Biochem 29:11–19, 1996.

100. HM Heise, G Voigt, S Rudloff, G Werner. Fourier Transform Spectros 12th Int Conf. K Itoh, M Tasumi, eds. Tokyo, Japan: Waseda University Press, 1999, pp 467–468.

101. D Malamud, L Tabak, eds. Saliva as a Diagnostic Fluid. New York: NY Academy of Science, 1993.

102. CP Schultz, MK Ahmed, C Dawes, HH Mantsch. Anal Biochem 240:7–12, 1996.

103. HH Eysel, M Jackson, HH Mantsch, GTD Thomson. Appl Spectrosc 47:1519–1521, 1993.

104. RA Shaw, S Kotowich, HH Eysel, M Jackson, GTD Thomson, HH Mantsch. Rheumatol Int 15:159–165, 1995.

105. HH Eysel, M Jackson, A Nikulin, RL Somorjai, GTD Thomson, HH Mantsch. Biospectroscopy 3:161–167, 1997.

106. KZ Liu, MK Ahmed, TC Dembinsky, HH Mantsch. Int J Gynecol Obstet 57:161–168, 1997.

107. KZ Liu, TC Dembinski, HH Mantsch. Am J Obstet Gynecol 178:234–241, 1998.

108. KZ Liu, TC Dembinski, HH Mantsch. Prenatal Diagnosis 18:1267–1275, 1998.

109. KZ Liu, HH Mantsch. Am J Obstet Gynecol 180:696–702, 1999.

110. CP Schultz, HH Eysel, HH Mantsch, M Jackson. J Phys Chem 100:6845–6848, 1996.

111. B Lendl, R Kellner. Mikrochim Acta 119:73–79, 1995.

112. P Krieg, B Lendl, R Vonach, R Kellner. Fresenius J Anal Chem 356:504–507, 1996.

113. T Fujii, Y Miyahara. Appl Spectrosc 52:128–133, 1998.

114. M Salmain, A Varenne, A Vessières, G Jaouen. Appl Spectrosc 52:1383–1390, 1998.

115. AA Ismail, S Barnett. In: HH Mantsch, M Jackson, eds. Infrared Spectroscopy: New Tool in Medicine. Proc SPIE 3257:101–107, 1998.

116. TJ O'Leary, WF Engler, KM Ventre. Appl Spectrosc 43:1095–1097, 1989.

117. PTT Wong, B Rigas. Appl Spectrosc 44:1715–1718, 1990.

118. M Jackson, HH Mantsch. In: RJH Clark, RE Hester, eds. Biomedical Applications of Spectroscopy. New York: Wiley, 1996, pp 185–215.

119. M Diem, S Boydston-White, L Chiriboga. Appl Spectrosc 53:148A–161A, 1999.

120. R Salzer, G Steiner, HH Mantsch, NE Lewis. Fresenius J Anal Chem 366:712–726, 2000.

121. BR Wood, MA Quinn, B Tait, T Hislop, M Romeo, D McNaughton. Biospectroscopy 4:75–91, 1998.

122. L Chiriboga, P Xie, H Yee, V Vigorita, D Zarou, D Zakim, M Diem. Biospectroscopy 4:55–62, 1998.

123. L Chiriboga, P Xie, V Vigorita, H Yee, D Zarou, D Zakim, M Diem. Biospectroscopy 4:47–54, 1998.

124. M Diem, S Boydston-White, A Pacifico, L Chiriboga. In: J Greve, GJ Puppels, C Otto, eds. Spectroscopy of Biological Molecules: New Directions. 8th European Conference on the Spectroscopy of Biological Molecules. Dordrecht, The Netherlands: Kluwer Academic, 1999, pp 479–482.

125. SR Ali, FB Johnson, JL Luke, VF Kalasinsky. Cell Mol Biol 44:75–80, 1998.

126. J Kneipp, P Lasch, M Beekes, D Naumann. In: J Greve, GJ Puppels, C Otto, eds. Spectroscopy of Biological Molecules: New Directions. 8th European Conference on the Spectroscopy of Biological Molecules. Dordrecht, The Netherlands: Kluwer Academic, 1999, pp. 505–506.

127. A Hesse, G Sanders. Infrarotspektren-Atlas zur Harnsteinanalyse. Stuttgart, Germany: Georg Thieme, 1988.

128. QD Nguyen, M Daudon, eds. Infrared and Raman spectra of Calculi. Paris: Elsevier, 1997.

129. H Hobert, K Meyer. Fresenius J Anal Chem 344:178–185, 1992.

130. M Volmer, A Bolck, BG Wolters, AJ de Ruiter, DA Doombos, W van der Slik. Clin Chem 39:948–954, 1993.

131. JF Sabot, CE Bornet, S Favre, S Sabot-Gueriaux. Clinica Chim Acta 283:151–158, 1999.

132. CG Kontoyannis, NC Bouropoulos, PG Koutsoukos. Appl Spectrosc 51:1205–1209, 1997.

133. E Wentrup-Byrne, L Rintoul, JL Smith, PM Fredericks. Appl Spectrosc 49:1028–1036, 1995.

134. C Paluszkiewicz, WM Kwiatek, M Galka, D Sobieraj, E Wentrup-Byrne. Cell Mol Biol 44:65–73, 1998.

135. JC Hebden, M Tziraki, DT Delpy. Appl Optics 36:3802–3810, 1997.

136. S Fantini, MA Franceschini, G Gaida, E Gratton, H Jess, WM Mantulin, KT Moesta, PM Schlag, M Kaschke. Med Phys 23:149–157, 1996.

137. D Grosenik, H Wabnitz, HH Rinneberg, KT Moesta, PM Schlag. Appl Optics 38:2927–2943, 1999.

138. TB Blank, TL Ruchti, SF Malin, SL Monfre. LEOS Newsletter 13:9–12, 1999.

139. L Heinemann, G Schmelzeisen-Redeker. Diabetologia 41:848–854, 1998.

140. MJ McShane, S Rastegar, GL Coté. Proc SPIE 3599:93–100, 1999.

141. OJ Rolinski, DJS Birch, LJ McCartney, JC Pickup. Proc SPIE 3602:6–14, 1999.

142. BD Cameron, H Gorde, GL Coté. Proc SPIE 3599:43–49, 1999.

143. C Chou, CY Han, WC Kuo, YC Huang, CM Feng, JC Shyu. Appl Optics 37:3553–3557, 1998.

144. OS Khalil. Clin Chem 45:165–177, 1999.

145. JP de Kock, L Tarassenko. Med Biol Eng Comp 31:291–300, 1993.

146. MG Sowa, JR Mansfield, GB Scarth, HH Mantsch. Appl Spectrosc 51:143–152, 1997.

147. MF Stranc, MG Sowa, B Abdulrauf, HH Mantsch. Brit J Plastic Surg 51:210–218, 1998.

148. JR Mansfield, MG Sowa, J Payette, B Abdulrauf, MF Stranc, HH Mantsch. IEEE Trans Med Imaging 6:1011–1018, 1998.

149. CE Cooper, M Cope, V Quaresima, M Ferrari, E Nemoto, R Springett, S Matcher, P Amess, J Penrice, L Tyszczuk, J Wyatt, DT Delpy. In: A Villringer, U Dirnagl,

eds. Optical Imaging of Brain Function and Metabolism II. New York: Plenum Press, 1997, pp 63–73.

150. FF Jöbsis. Science 198:1264–1267, 1977.

151. G Litscher, G Schwarz, eds. Transcranial Cerebral Oximetry. Lengerich, Germany: Pabst Science, 1997.

152. CE Elwell. A Practical Users Guide to Near Infrared Spectroscopy. Joko-cho, Japan: Hamamatsu Photonics, 1995.

153. E Okada, M Firbank, M Schweiger, SR Arridge, M Cope, DT Delpy. Appl Optics 36:21–31, 1997.

154. L Skov, NC Brun, G Greisen. J Biomed Optics 2:7–14, 1997.

155. AJ Macnab, RE Gagnon, FA Gagnon. J Biomed Optics 3:386–390, 1998.

156. D D'Antona, CJ Aldrich, P O'Brien, S Lawrence, DT Delpy, JS Wyatt. J Biomed Optics 2:15–21, 1997.

157. M Niwayama, L Lin, J Shao, T Shiga, N Kudo, K Yamamoto. Proc SPIE 3597: 291–299, 1999.

158. DJ Wallace, B Michener, D Choudhury, M Levi, P Fennelly, DM Hueber, B Barbieri. Proc SPIE 3597:300–316, 1999.

159. T Yamamoto, Y Yamashita, H Yoshizawa, A Maki, M Iwata, E Watanabe, H Koizumi. Proc SPIE 3597:230–237, 1999.

160. T Shiga, K Yamamoto, K Tanabe, Y Nakase, B Chance. J Biomed Optics 2:154–161, 1997.

161. Y Mendelson. Clin Chem 38 1601–1607, 1992.

162. WB Runciman, RK Webb, L Barker, M Currie. Anaesth Intens Care 21:543–550, 1993.

163. LA Sodickson. Clin Chem 45:1687–1689.

164. SM Lopez Silva, R Giannetti, ML Dotor, JR Sendra, JP Silveira, F Briones. Proc SPIE 3570:294–302, 1999.

165. S van Huffel, P Casaer, P van Mele, G Willems. Proc SPIE 2389:743–754, 1995.

166. Y Mendelson. Proc IEEE Engineering in Medicine and Biology, paper no 897, 1996.

# 9
# FT-Infrared and FT-Raman Spectroscopy in Biomedical Research

**Dieter Naumann**
*Robert Koch Institute, Berlin, Germany*

## 1  INTRODUCTION

Recent efforts in the field of biomedical Fourier transform infrared (FT-IR) and FT-Raman spectroscopy of disease states in humans have been published by various groups working on cells, tissues, and body fluids, and a number of excellent papers on the application of biomedical FT-IR and Raman spectroscopy has appeared in the literature (1–37). While FT-IR spectroscopy as a tool for biodiagnostic purposes seems to be established and the first dedicated instrumentation for routine biomedical characterizations has already appeared on the market, FT-Raman spectroscopy, though potentially even more versatile, has not yet reached the stage of development at which routine biomedical applications are possible. This paper will not review details of technological developments and experimental or theoretical progress achieved in biomedical FT-IR and FT-Raman spectroscopy. It will, rather, highlight some of the most important applications and experimental procedures in biomedical vibrational spectroscopy and possible future developments in this field. Emphasis will be put on the interpretation of spectra and the assignment of spectral features frequently observed in cells, tissues, and body fluids and on the discussion of acceptable standards for data acquisition and experimental protocols, the establishment of reproducibility levels, and the problem of data exchange between different laboratories. Only those data will be considered that are concerned with intact cells, tissues, and body fluids. No attempt will be made to comment on the very promising potential of vibrational spectroscopy for the analysis of isolated cell fragments, compounds, macromolecules, and cell metabolites.

## 2 COMPOSITION AND STRUCTURE OF COMPLEX BIOLOGICAL MATERIAL

At the simplest level, all biological systems are composed of water, nucleic acids, proteins, lipids, and carbohydrates. Since these partial structures may specifically assemble into the plethora of highly structured and specialized organisms in nature an understanding of the infrared and Raman spectra of microbial, plant, and animal cells as well as of tissues and body fluids requires a general perception of their composition, major cell types, and chemical structures present. A basic knowledge of cell growth and the differentiation of cells and tissues is required as well. The gross composition of bacterial (prokaryotic), yeast, and mammalian (eukaryotic) cells is given in Table 1.

### 2.1 Microbial Cells

In contrast to mammalian cells, plant and yeast or fungal cells (so-called eukaryotic organisms), bacteria (also called prokaryotes) exist in only a limited number of morphological forms (e.g., rods, cocci, chains, and spirals). Their chemical composition and structures, however, vary considerably. The cytoplasmic structures of bacteria are less organized (compartmentalized), and they are simpler than those of animals, plants, yeasts, or fungi and have complex and very diverse molecular and supramolecular structures outside the plasmic membrane. These include the cell wall, outer membrane, capsules, and, sometimes, specific layers like, e.g., the so-called S-layers. Some bacteria are capable of sporulation or storage material production. Many structural differences providing the possibility of differentiation between bacteria reside in the cell envelope, which is generally

**Table 1** Composition of Prokaryotic and Eukaryotic Cells

|                          | Bacteria[a]        | Yeasts[a]          | Animal cells[a]              |
| ------------------------ | ------------------ | ------------------ | ---------------------------- |
| Radius [μm]              | ~1                 | ~10                | 10–100                       |
| Generation time [hr]     | 0.2–10             | 2–10               | ~20                          |
| DNA base-pairs           | ~4 × 10$^6$        | ~20 × 10$^6$       | 500–5000 × 10$^6$            |
| Number of "genes"        | ~4000              | ~20 000            | >50 000                      |
| DNA % [w/w][b]           | 2–4                | 1–3                | ~5                           |
| RNA % [w/w]              | 5–15               | 3–10               | 10–20                        |
| Proteins % [w/w]         | 40–60              | 40–50              | ~60                          |
| Lipids % [w/w]           | 10–15              | 5–20               | 10–15                        |
| Carbohydrates % [w/w]    | 10–20              | 10–25              | 6–8                          |

From different sources in the literature.
[a] All numbers are rough estimations, which depend on organism/organ, growth conditions, cell division cycle, and other factors.
[b] Content in % of cell dry weight.

defined as the cytoplasmic membrane plus the cell wall (38). Most cell envelopes fall in two categories, the so-called gram-positive bacteria, consisting only of the cytoplasm, the plasma membrane, and the cell wall, and the more complex gram-negative, which, in addition to the cell envelope, contain the so-called outer membrane. Some bacterial species, the mycoplasms, lack any cell wall at all, but express a rather rigid plasma membrane. The bacterial cell wall is a rigid, high-molecular network made up primarily of the peptidoglycan, which has a shape-giving function and protects the cells from osmotic disruption. Its primary structure consists basically of disaccharide-pentapeptide subunits with unusual features such as the occurrence of alternating D- and L-amino acids and a γ-bonded D-glutamic acid residue. Its structural variants are found to be different for various groups of bacteria. Many gram-positive bacteria have an additional polymer, covalently bound to the peptidoglycan, the teichoic and teichuronic acids. The teichoic acids are ribitol- or glycerol-containing macromolecules, built up by a phosphate-carrying backbone with side chains of variable composition. Teichuronic acids or neutral polysaccharides are also sometimes found in the gram-positive cell wall. Gram-negative bacteria exhibit an additional membrane, the so-called outer membrane. The outer membrane is an asymmetric membrane, the inner leaflet of which contains only phospholipids with nearly the same composition as found in the cytoplasmic membrane, while the outer leaflet contains exclusively one particular type of amphiphilic molecules, the lipopolysaccharides (LPS) and the various pore-forming proteins, the porins. The structure of LPS has three basic structural regions, the so-called O-specific side chain (a heterooligosaccharide, responsible for O-antigenicity), the inner and outer core regions, and a lipid anchor that contains the endotoxic principle, the lipid A being responsible for many important immunological reactions of the host (38). The mycobacteria, nocardia, corynebacteria, and some related groups have very unusual cell envelopes, which form thick, waxlike layers around the outside of the cell wall. Major compounds present in this rather impermeable and rigid layer are complex, long-chain fatty acids, the mycolic acids (39). Some bacteria form capsules (sometimes referred to also as "slime layers") surrounding the cell envelope. These are not essential structures and are frequently built up of unusual, negatively charged polysaccharide compounds. Some bacilli exhibit capsules composed of negatively charged homooligopolypeptides such as poly-D-glutamic acids. The expression of such capsules in pathogenic bacteria can inhibit ingestion by phagocytes and may play an important role as "pathogenicity factors" in infectious diseases. Many bacilli and clostridia may form endospores, which are modified cell structures that can survive under unfavorable environmental conditions.

The proteins, lipids, and polysaccharides that make up the membrane, the cell wall, and the capsules of capsulated yeasts have a significant impact on the systematics and phylogeny of yeasts. Only a small number of species of yeasts has been investigated in depth, and even fewer studies have focused on the structural details of these constituents. The cell walls of yeasts are composed of complex polysaccharides and glycoproteins. The major structures of yeast cell walls are β-glucans, in which the glycosyl units are mutually linked by β-(1 → 3), β-(1 → 6), and possibly β-(1 → 2) glycosidic bonds. These structures are found with different molecular weights and branching, and they may form microfibrillar structures of crystalline nature (40). Chitin, a linear β-(1 → 4) polymer of *N*-acetylglucosamine, is a typical constituent of primary septa and budding yeast. Chitosan, a β-(1 → 4)-linked polymer of D-glucosamine, may be considered as a minor yeast polysaccharide, but in some dimorphic fungi belonging to the group of *Zygometes*, it may represent one of the principal wall components. Mannans exist in yeasts as covalent complexes with proteins. The yeast mannoproteins are large molecules (molecular weight up to 500 000) consisting of a covalently linked carbohydrate and a protein. The polysaccharide portion contains up to 150 mannosyl units, being connected via *N*-glycosidically linked polymannose units. A second group of carbohydrate of yeast mannans are short manno-oligosaccharides, O-glycosidically linked to serine and/or threonine residues of the polypeptide (protein) (40). The storage compounds in yeasts have recently been reviewed (41). The principal, readily mobilizable reserve polysaccharide in yeasts is glycogen (see also curve c, Fig. 1). Glycogen, an α-(1 → 6)-linked carbohydrate polymer of glycosyl units, occurs in yeast and mammalian cells in both the cytoplasm and the nucleoplasm either in soluble form or as aggregates of spherical particles having a diameter of 40–50 nm. The glycogen content of yeast cells is highly dependent on the physiological state and may reach up to 20% of the dry weight of the cells. In some yeasts, particularly of the genus *Cryptococcus*, the polymers containing D-glucoronic acid residues are important constituents of extracellular capsules. For example, the capsule of *Cryptococcus laurentii* is made up of an α-(1 → 3)-linked mannose backbone with xylosyl and glucuronosyl residues as side groups (42). Yeast membranes contain a number of lipids and pigments that are not present in prokaryotic cells (bacteria) (38,43). These are sterols, sphingolipids, ergosterins, melanins, and some glycolipids. Culture conditions have a marked influence on the quantitative composition of the cell wall and the total

---

**Fig. 1** (A) FT-IR and (B) FT-Raman spectra of the main macromolecular building blocks present in complex biological samples. (a) Nucleic acids: RNA from yeast. (b) Proteins: ribonuclease A. (c) Carbohydrates: glycogen. (d) Lipids: L-α-dipalmitoylphosphatidylcholine (synthetic). Spectra have been measured from dried films on ZnSe optical plates and as dried pellets by FT-IR (1,4) and FT-Raman spectroscopy, respectively. FT-Raman: Excitation line 1064 nm, approx. 250 mW at the sample, 6 cm$^{-1}$ resolution. Abbreviations: C, U, T, A, G, Phe, and Tyr stand for cytosine, uracil, thymine, adenine, guanine, phenylalanine, and tyrosine, respectively. Boldface numbers in parentheses refer to the assignments given in Tables 2 and 3, respectively. (From Refs. 1, 4.)

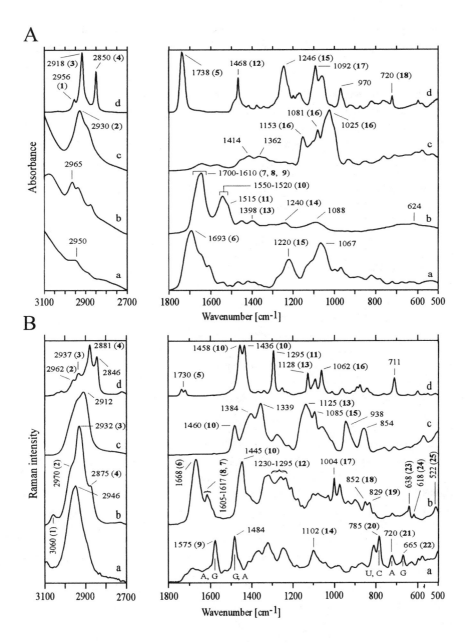

lipid content, and lipid composition of yeasts. Factors controlling, e.g., lipid content and composition are the pH of the medium, the temperature, the time of growth, and the ratio of N- and C-sources. Sterols occur in yeast membranes in the free form and as esters with long-chain fatty acids. Both forms are interconvertible. Free sterols are associated with membrane functions; sterol esters may fulfill a storage, or ''pool,'' function. Common sterol molecules of yeasts are ergosterol, lanosterol, episterol, zymosterol, and fecosterol. Major structures of sphingolipids found in yeasts are the sphingosines, cerebrins (ceramides), sphingomyelines, and cerebrosides. A typical membrane lipid in yeast is ergosterin. Its structure is similar to cholesterol and it belongs to the group of sterines. Further compounds frequently found in the membranes of yeasts are melanins, which are black pigments built up from tyrosine derivatives (38).

## 2.2 Mammalian Cells, Tissues, and Body Fluids

Like microorganisms, cells, tissues, and fluids of the human body are composed primarily of water, different lipids, proteins, carbohydrates, and nucleic acids. These building blocks are organized into about 200 distinct types of cells, which come in a variety of shapes and sizes and which assemble to form the various highly specialized tissues and organs. These cells may have configurations that are spherical or nearly cubic in shape or may be of stratified morphology. Only four basic types of tissues are found in mammalians: epithelial, connective, muscular, and nervous tissue. Epithelial tissues are built up by densely packed, so-called polyhedral cells, which stick to each other very strongly to form extended sheets that cover the inside and outside surfaces of the body, i.e., the skin, inner surface of blood vessels, and the gastrointestinal tract. A major constituent of connective tissues is an extracellular matrix secreted by cells that is composed mainly of protein fibers such as collagen and elastin and a matrix substance made up of various complex mixtures of glycoproteins and proteoglycans. Bone, adipose tissues, and cartilage are specialized forms of connective tissue. Nerve tissues consist of two basic cell types, the neurons (conducting cells) and the glia cells that support and nourish the neurons. The central nervous system (CNS) can visually be divided into gray matter, containing mainly neuronal cell bodies and glial cells, and white matter, which contains mainly the myelinated axons of the neurons. Muscle tissue can be divided into the three main classes—skeletal, cardiac, and smooth muscle—according to different functions, visual appearance, and structural characteristics. For the interpretation of spectral features of cells and tissues, it is important to recognize that in most healthy cells the RNA/DNA ratio is about 5 and that, by dry weight, nucleic acids may account for up to 25% while the proteins account for about 60%, with the rest being from other components, such as carbohydrates, lipids, and low-molecular-weight compounds (see Table 1). It has also been recently recognized that in the cytoplasm of a human cell, the total protein concentration is very high, on the order of 100 mM, and the average concentration of DNA in the cell's nucleus is even higher, indicating that the components in a cell are not really in solution, but rather in a gel-like phase (44). As in microbial cells, the composition of mammalian cells and tissues may vary depending on a number of factors, such as the cell division cycle and nutrition. Spatial and temporal variability of human cells in tissues and also the composition of body fluids are inherent properties of these structures. Temporal changes of cells are also a consequence of alterations within tissues, e.g., as a function of age, or of cyclic processes, as in the female reproductive system.

## 3 THE MAIN BIOLOGICAL BUILDING BLOCKS AND THEIR BAND ASSIGNMENTS

Infrared absorption bands and Raman scattering features observed between approximately 600 and 4000 cm$^{-1}$ arise mainly from the fundamental vibrational modes and—in the group frequency notion—can often be assigned to particular functional groups. At wavenumbers lower than 1400 cm$^{-1}$, vibrational bands tend to result no longer from localized vibrational modes but rather from skeletal and strongly coupled vibrations, which are difficult to describe. For practical purposes, rough band assignments can be obtained from group frequency charts published in several bibliographies (45–48).

Several spectra descriptions and excellent structure–spectra correlations can be obtained from the literature for the most important biological macromolecules (46,48). Most of the efforts to interpret vibrational spectra from biological molecules and to achieve structure–spectra correlations are based mainly on the analysis of known structures, normal coordinate analysis, and isotope exchange experiments. Figure 1 gives the FT-IR (A) and FT-Raman (B) spectra of the most important biological building blocks (nucleic acids, proteins, carbohydrates, and lipids), which, as partial structures, are constantly present in complex biological materials. Some of these bands are numbered, and tentative assignments can be taken from Tables 2 and 3, respectively. Figure 2 shows typical FT-IR and FT-Raman spectra of dehydrated samples of microbial, body fluid, and tissue specimens. Exact assignment to specific structures is certainly too complex a task at present. A number of publications report on this problem while accentuating that improved assignments are in progress (see, e.g., Refs. 3, 6, 20, 21, 25–27, 29, 44). The FT-IR spectra of cells, tissues, and body fluids usually show only broad spectral features. However, some details can be visualized by applying resolution-enhancement techniques (see e.g. upper curve in Fig. 3A). The FT-Raman spectra, in contrast, exhibit a number of rather sharp bands already perceptible without resolution enhancement (Fig. 3B). In general, 50–70 spectral features can be resolved in the infrared and Raman spectra of biological samples and are the

**Table 2** Tentative Assignment of Some Bands Frequently Found in Biological FT-IR Spectra[a]

| Band numbering (cf. Fig. 1A) | Frequency (cm$^{-1}$) | Assignment[b] |
|---|---|---|
| | ~3500 | O—H str of hydroxyl groups |
| | ~3200 | N—H str (amide A) of proteins |
| 1 | ~2955 | C—H str (asym) of −CH$_3$ in fatty acids |
| 2 | ~2930 | C—H str (asym) of >CH$_2$ |
| 3 | ~2918 | C—H str (asym) of >CH$_2$ in fatty acids |
| | ~2898 | C—H str of C—H in methine groups |
| | ~2870 | C—H str (sym) of −CH$_3$ |
| 4 | ~2850 | C—H str (sym) of >CH$_2$ in fatty acids |
| 5 | ~1740 | >C=O str of esters |
| | ~1715 | >C=O str of carbonic acid |
| 6 | ~1680–1715 | >C=O in nucleic acids |
| 7 | ~1695 | Amide I band components |
| | ~1685 | resulting from antiparallel pleated sheets |
| | ~1675 | and β-turns of proteins |
| 8 | ~1655 | Amide I of α-helical structures |
| 9 | ~1637 | Amide I of β-pleated sheet structures |
| 10 | ~1550–1520 | Amide II |
| 11 | ~1515 | "Tyrosine" band |
| 12 | ~1468 | C—H def of >CH$_2$ |
| 13 | ~1400 | C=O str (sym) of COO$^-$ |
| 14 | ~1310–1240 | Amide III band components of proteins |
| 15 | ~1250–1220 | P=O str (asym) of > PO$_2^-$ phosphodiesters |
| 16 | ~1200–900 | C—O, C—C str, C—O—H, C—O—C def of carbohydrates |
| 17 | ~1090–1085 | P=O str (sym) of >PO$_2^-$ |
| 18 | ~720 | C—H rocking of >CH$_2$ |
| | ~900–600 | "Fingerprint region" |

*Source*: Refs. 3, 45, 46.
[a] Peak frequencies have been deduced from the second derivatives and Fourier-deconvoluted spectra.
[b] str = stretching; def = deformation; sym= symmetric; asym = antisymmetric.

basis for assignments to functional groups, known chemical compounds, partial structures, or even different conformational states of a particular macromolecule present. It has frequently been argued in the literature that the sum or a particular combination of these bands provides a spectral fingerprint that can be used to characterize different types of cells or states of cells or even to discriminate between normal and malignant cells or cell aggregates within a given tissue (see, e.g., Refs. 1, 5, 20, 21, 25, and 44). Figures 3A and B show typical FT-IR and

**Table 3** Tentative Assignment of Some Bands Found in FT-Raman Spectra of Biological Specimen

| Band numbering (cf. Fig. 1B) | Frequency (cm⁻¹) | Assignment[a] |
|---|---|---|
| 1 | ~3059 | $(C{=}C{-}H)_{(arom.)}$ str |
| 2 | ~2975 | $CH_3$ str |
| 3 | ~2935 | $CH_3$ and $CH_2$ str |
| 4 | ~2870–2890 | $CH_2$ str |
| 5 | ~1735 | $>C{=}O$ ester str |
| 6 | ~1650–1680 | Amide I |
| 7 | ~1614 | Tyrosine |
| 8 | ~1606 | Phenylalanine |
| 9 | ~1575 | Guanine, adenine (ring stretching) |
| 10 | ~1440–1460 | C—H def |
| 11 | ~1295 | $CH_2$ def |
| 12 | ~1230–1295 | Amide III |
| 13 | ~1129 | C—N and C—C str |
| 14 | ~1102 | $>PO_2^-$ str (sym) |
| 15 | ~1085 | C—O str |
| 16 | ~1061 | C—N and C—C str |
| 17 | ~1004 | Phenylalanine |
| 18 | ~852 | "Buried" tyrosine |
| 19 | ~829 | "Exposed" tyrosine |
| 20 | ~785 | Cytosine, uracil (ring, str) |
| 21 | ~720 | Adenine |
| 22 | ~665 | Guanine |
| 23 | ~640 | Tyrosine (skeletal) |
| 24 | ~620 | Phenylalanine (skeletal) |
| 25 | ~520–540 | S—S str |

[a] str = stretching; def = deformation; sym = symmetric
*Source*: Refs. 26, 45–48.

FT-Raman spectra obtained from intact microbial cells of *Staphylococcus aureus* that have been dehydrated as described (1,28). The FT-Raman spectra of these organisms are characterized by the prominent C–H stretching bands in the range of 2700–3100 cm⁻¹ (see also Fig. 2) and the C—H deformation band around 1450 cm⁻¹, originating most likely from the —CH₃, >CH₂, and C—H functional groups in lipids, amino acid side chains of the proteins, and carbohydrates. In the 1200–1800-cm⁻¹ region, the amide I and amide III bands of the proteins give rise to prominent bands around 1660 and 1250 cm⁻¹, respectively. Additionally, the Raman spectra show a series of well-resolved bands of minor intensity, which

A

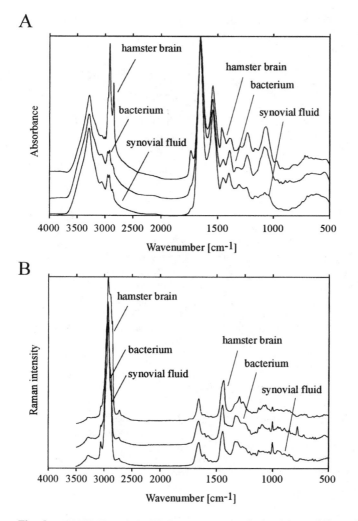

B

**Fig. 2**   (A) FT-IR and (B) FT-Raman spectra obtained from different biological samples. (a) Microbial sample (*Staphylococcus aureus* RKI/WG PS 29). (b) Body fluid (synovial fluid aspirated from a patient suffering from rheumatoid arthritis). (c) Tissue sample (central nervous system material from a Scrapie-infected hamster brain). All samples have been measured as dried films on ZnSe optical plates (FT-IR (1,4)) and as dried pellets (FT-Raman), respectively. FT-Raman: excitation line 1064 nm, approx. 250 mW at the sample, 6 cm$^{-1}$ resolution. (From Refs. 1, 44, 55, and 56.)

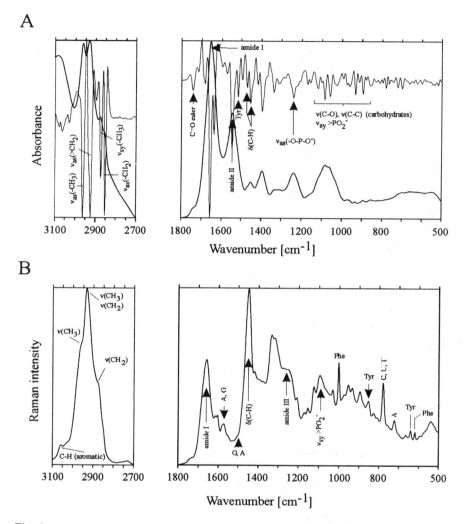

**Fig. 3** Tentative assignments for some major functional group frequencies found in FT-IR and FT-Raman spectra of intact microbial cells (*Staphylococcus aureus* RKI/WG PS 29). (A) *Bottom*: original spectrum of a dried-film sample of *Staphylococcus aureus* cells; *top*: Second-derivative spectrum. (B) FT-Raman spectrum of *Staphylococcus aureus* RKI/WG PS 29. *Abbreviations*: C, U, T, A, G, Phe, and Tyr stand for cytosine, uracil, thymine, adenine, guanine, phenylalanine, and tyrosine, respectively. For assignments, refer also to Tables 2 and 3. For experimental details and abbreviations, see the legend to Fig. 1. (From Refs. 1, 4.) $\nu$ = stretching; $\delta$ = bending; sy = symmetric; as = antisymmetric.

are tentatively assigned to the RNA/DNA nucleotide base-ring vibrations of guanine (G), thymine (T), adenine (A), cytosine (C), uracil (U), and the amino acid side vibrations of tryptophane (Trp), tyrosine (Tyr), and phenylalanine (Phe) of the proteins. The FT-IR spectra of the same strain on the other side reveal only broad superimposed spectral bands, with the amide I and amide II bands usually being the most prominent features in bacterial IR spectra. Different spectral bands can be discriminated only by applying band-narrowing techniques, by, e.g., calculating the second derivatives (as shown in Fig. 3A) or by Fourier self-deconvolution techniques. FT-IR spectroscopy yields much better signal-to-noise (S/N) ratios on biological samples at comparable measurement times than FT-Raman spectroscopy. While FT-IR typically gives S/N ratios better than 2000 at measuring times of less than a minute, about 20–40 minutes are needed in a typical FT-NIR Raman spectroscopic experiment on the same biological sample to obtain an S/N ratio of only 400–800 (28).

Some preliminary conclusions can be drawn with regard to the interpretation of FT-IR spectra of biomedical samples: The region between 4000 and 3100 $cm^{-1}$ is dominated by rather broad spectral features resulting from O—H stretching modes ($\sim$3400 $cm^{-1}$) and from N—H stretching modes (amide A $\sim$3300 $cm^{-1}$ and amide B $\sim$3030 $cm^{-1}$). The region between 3100 and 2800 $cm^{-1}$ exhibits the C—H stretching vibrations of—$CH_3$ and >$CH_2$ functional groups and, hence, is generally dominated by the fatty acids of the various membrane amphiphiles (e.g., phospholipids) and by some amino acid side-chain vibrations. Complementary information can be deduced from the region between 1470 and 1350 $cm^{-1}$, where the various deformation modes of these functional groups are located in the spectrum. In rare cases, a weak band near 3015 $cm^{-1}$ is also observed, resulting from =C—H double-bond stretching modes of unsaturated fatty acid chains. Many of these vibrational modes are known to be sensitive to lipid-phase behavior. The region between 1800 and 1500 $cm^{-1}$ is dominated by the conformation-sensitive amide I and amide II bands, which are the most intensive bands in the spectra of nearly all complex biological systems so far tested. Since infrared spectroscopy is an averaging technique, the amide I and amide II bands cannot provide structure information on a single protein; rather, they indicate the predominance of $\alpha$- or $\beta$-structures present. Useful information can also be obtained from bands near 1740 $cm^{-1}$, essentially resulting from >C=O stretching vibrations of the ester functional groups in phospholipids. Absorptions due to DNA/RNA structures can also be expected in this spectral domain (>C=O, >C=N, >C=C< stretching of the DNA or RNA heterocyclic base structures). A band near 1715 $cm^{-1}$, which is assigned to a C=O stretching vibration, is routinely observed in the spectra of hydrated microbial cells and tissue material, and is known as a sensitive probe of base pairing in nucleic acids. Weak features of nucleic acids between 1600 and 1700 $cm^{-1}$ are generally overlapped by the much stronger protein amide I bands. Weak bands assigned to amino acid side-chain vibrations occur near 1498 $cm^{-1}$ (phenylalanine) and 1516 $cm^{-1}$ (tyrosine) and between 1585 and 1570 cm$^{-1}$ (aspartate and glutamate carboxylate stretching). Complex absorption profiles are observed between 1300 and 1500 cm$^{-1}$, arising predominantly from >CH$_2$ and >CH$_3$ bending modes of lipids, proteins, and ring vibrations of nucleic acids. A characteristic weak feature is often observed around 1400 cm$^{-1}$, which may be attributed to the symmetric stretching vibrations of— COO$^-$ functional groups of amino acid side chains, free fatty acids, or other derivatives. Around 1230 cm$^{-1}$, superimposed bands typical of different >P=O double-bond antisymmetric stretching vibrations of phosphodiester, free phosphate, and monoester phosphate functional groups are observed. In most cases, three to four different weakly pronounced features can be discriminated by resolution enhancement, with the band near 1220 cm$^{-1}$ most probably being due to the phosphodiester functional groups of DNA/RNA polysaccharide backbone structures. The other >P=O double-bond stretching frequencies might be due to head-group vibrations of phospholipids, phosphorylated carbohydrates, or, e.g., phosphorous-containing polysaccharides such as "teichoic acids" and "lipoteichoic acids" (highly charged polymers, found exclusively in gram-positive bacteria), which may be present in substantial amounts. The spectral region between 1200 and 900 cm$^{-1}$ is generally dominated by the symmetric stretching vibration of PO$_2^-$ groups in nucleic acids and a complex sequence of peaks due mainly to strongly coupled C—C, C—O stretching and C—O—H, C—O—C deformation modes of various oligo- and polysaccharides. The region between 900 and 600 cm$^{-1}$ exhibits a variety of weak but extremely characteristic features superimposed on an underlying broad spectral contour. This region may contain weakly expressed bands arising from aromatic ring vibrations of phenylalanine, tyrosine, tryptophane, and the various nucleotides. With the exception of only a few peaks (e.g., a band near 720 cm$^{-1}$, resulting from the >CH$_2$ rocking modes of the fatty-acid chains), valid assignments can hardly be achieved. Therefore, we refer to this spectral domain as to the "true fingerprint region."

Figure 4 shows representative FT-IR and FT-Raman spectra of *Escherichia coli* RKI/A139 cells compared to the FT-Raman spectrum of a typical protein (RNase A). The striking similarity between the bacterial and the pure protein spectra indicates that the major portion of total cell mass is represented by the various cellular and membrane associated proteins, while the contribution of RNA/DNA structures to bacterial IR and Raman spectra is comparably small (see also Table 1). The lipid compounds, although forming less then 10–15% of cell mass, contribute markedly to bacterial Raman spectra, while carbohydrates contribute significantly to both the IR and Raman spectra of biological materials. These experimental findings indicate that all cell constituents can contribute to the observed spectral features according to their proportions within the cells, the vibrational selection rules, and the known physical relation that polar functional groups are more intense in the infrared than in the Raman and, vice versa, the nonpolar functional groups. The changes in the compactness of protein and DNA packing within mammalian cells as a function of the cell cycle may also play a

A

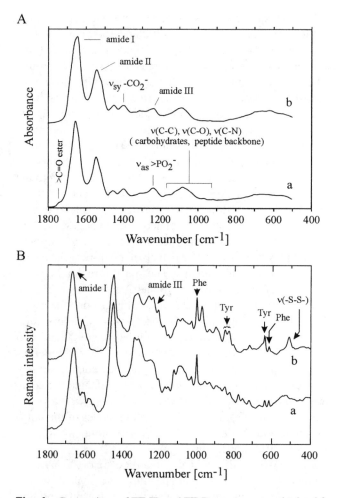

**Fig. 4** Comparison of FT-IR and FT-Raman spectra obtained from a complex biological specimen (*Escherichia coli* RKI A139, curve a) and a typical protein (ribonuclease A, curve b). *Abbreviations*: Phe and Tyr stand for phenylalanine and tyrosine, respectively.

role in the detectability of these structures and will be discussed in the following section (44).

Most authors reporting on biological applications of FT-IR and FT-Raman have recognized that both IR and Raman spectra of complex biological specimen not only provide a number of superimposed IR absorption or Raman scattering bands that in toto describe the composition of each particular ''multicomponent'' sample of interest, but also yield a number of more specific bands that are sensitive to structural changes, various intra- and intermolecular interactions, including H-bonding pattern and lipid-protein interaction, and conformational states such as different secondary structures of proteins or states of order of membranes. It is also a matter of fact that the physical state of the sample (e.g., hydration or aggregation state interaction with ions, etc.) has a severe influence on results and makes it a virtual necessity for the analyst rigorously to standardize sampling, sample preparation, and data-acquisition procedures.

## 4 EXPERIMENTAL METHODOLOGIES

Two major advantages of IR and Raman spectroscopy over other analytical techniques are that nearly any kind of material can be measured and that both techniques are not limited to the physical state of the sample. Samples may be solutions, viscous liquids, suspensions, inhomogeneous solids, or powders. Additionally, there are in principle no restrictions to record spectra of a given sample under very different physicochemical conditions concerning temperature, pressure, state of dispersion, hydration, pH, etc. This is of direct relevance for biomedical analyses, since it is pertinent to test biological specimens under conditions that leave the sample's structures "as they are," preferentially hydrated, unperturbed and nondisintegrated. An important advantage of Raman over infrared spectroscopy for biomedical applications is the virtual transparency of water in the Raman effect, which greatly simplifies the analysis of aqueous or fully hydrated biological specimen. However, with the advent of Fourier transform–infrared spectroscopy (FT-IR) and its inherent physical advantages (49), this particular disadvantage can often be overcome by subtracting water from the raw experimental spectra. This is possible, however, only by using very thin optical pathlengths ($<10$ μm) and by collecting very precise spectra of extremely high signal-to-noise ratio that are free of spectral water vapor contributions (50).

Using laser excitation radiation in the near infrared (NIR), e.g., from a Nd:YAG laser line at 1064 nm, it is possible to obtain high-quality, essentially fluorescence-free Raman spectra on previously intractable biological samples (26,28,46,51–56). Thus, the complementary vibrational spectroscopic information of even colored biological samples is available now by using FT-IR and NIR FT-Raman spectroscopy in tandem.

## 4.1 FT-IR and FT-Raman Techniques Suitable for Biomedical Analysis

In general, a biological sample does not behave ideally in respect to absorption/transmission or scattering. In case of FT-IR, biological specimen can best be analyzed when infrared absorbances of the samples to be compared are not too different, IR bands are not too intensive in order to avoid detector nonlinearities,

Beer's law is at least approximately obeyed, signal-to-noise ratio is sufficiently high, and the problem of varying baseline shifts due to diffuse scattering at the sample surface and/or due to inhomogeneity within the sample itself is minimized. These requirements are still best fulfilled by the traditional absorbance/ transmission (A/T) or the attenuated total reflection (ATR) techniques (49,57). For FT-Raman measurements, penetration depth of the excitation line and the wavelength-dependent absorption of the scattered Raman light (e.g., by bulk water) play a major confining role (46,51). Many different cuvette systems have been designed for FT-IR and FT-Raman measurements of biological samples. Some versatile technical solutions suitable for the analyses of dried samples or aqueous solutions, suspensions, or gel discs are given in Fig. 5.

## 4.2   Sampling of Biological Material and Data Acquisition

These problems concern sampling reproducibility of biological specimen, sample treatment procedures, and the particular spectroscopic techniques and physical parameters used to obtain FT-IR or FT-Raman spectra of a given biological sample. There is certainly no single answer to this crucial point. However, for measurements on cell suspensions, body fluids, and tissue samples a number of reasonable suggestions are already accessible from the literature (1–37). For FT-IR measurements of microbial cells, standardized sampling, sample preparation, data acquisition, and evaluation protocols that are already used in the practice of microbiological laboratories have also been published (1–8,55,56). In the case of microbial cell characterizations, these standardization efforts have been stimulated by the necessity of data exchange between different laboratories in order to construct reference databases for the routine analysis of microorganisms.

FT-IR or FT-Raman whole-cell spectra of microbial samples can be obtained with excellent reproducibility from microorganisms grown either in liquid cultures or on standard solid agar media, provided the microbiological parameters influencing cell growth (composition of growth media, incubation time, temperature of growth, etc.) can be controlled and standardized rigidly. Sample collection, sample preparation, and spectroscopic data-acquisition parameters (spectral resolution, scanning time, etc.) are then of minor importance. Microbial samples suitable for IR measurements can be obtained from liquid cultures or directly from solid nutrient agar plates. These samples can be measured as hydrated pellets or dried films, by FT-IR applying either the A/T or ATR techniques (1–8,55,56). A thin-layer chromatography accessory coupled to a standard FT-IR spectrometer has recently been used as a multisample rack to analyze dried microbial samples in the diffuse-reflection mode (9,10).

A typical, very simple protocol has been reported to run as follows (1–4): Subcultured microbial strains are cultivated on appropriate nutrient solid agar

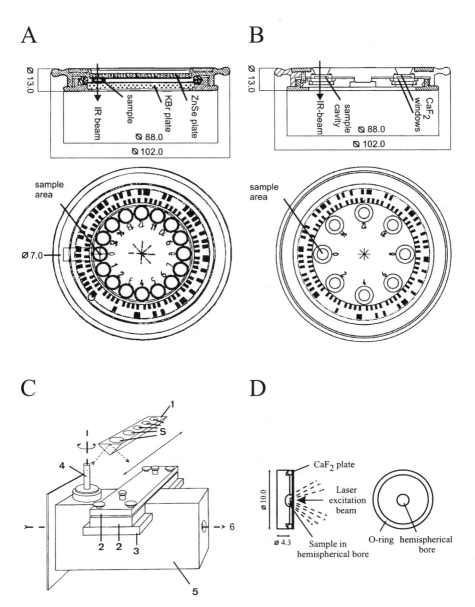

**Fig. 5** Technical drawing of multisample cuvettes useful for measuring dried or fully hydrated biological samples. Layout of two absorbance/transmission cuvettes for (A) dried and (B) fully hydrated samples. (C) Layout of an attenuated total reflectance (ATR) multi-cuvette system with a single-reflection ATR prism. 1: ZnSe prism with five marked sample areas S; 2: demountable, hermetically sealed cartridge (two parts) carrying the ATR prism (1); 3: sliding carriage of the cartridge by which the ZnSe prism may be moved through the IR beam for sample measurements; 4: cartridge drive; 5: housing with reflecting optics; 6: IR beam. (D) Layout of a cuvette for FT-Raman measurements made of stainless steel.

plates. Time and temperature of growth depend on the type of microorganism tested (e.g., 24 h at 37°C for many human pathogens). Microgram amounts of the microbial cells are carefully removed from the agar plate with a standard "calibrated" (1 millimeter in diameter) platinum loop and are suspended in 80 μl of distilled water. Subsequently, 30 μl of the suspensions are transferred to the water-insoluble, infrared-transparent optical plate as small drops covering predefined sample areas of the optical material (cf. Fig. 5A). The drop of microbial suspension is then dehydrated in a desiccator over a drying agent ($P_4O_{10}$ Sicapent from Merck or Silicagel) applying a moderate vacuum ($\approx 25$ torr) to form transparent film discs suitable for A/T measurements. The optical plate is sealed in a gas-tight cuvette cartridge to control humidity and to prevent the instrument from contamination, and is finally transferred to the automatic cuvette holder device of the instrument. Similar procedures have also been applied to body fluids, homogenized tissue specimen, or suspensions of more highly organized cells, such as cells of the immune system (17,18,25,44). Most authors used a nominal physical resolution of 4–8 cm$^{-1}$ and claimed for FT-IR that the signal-to-noise ratio should be sufficiently high to calculate first- or second-derivative spectra as input data for multivariate date analysis (1–8,18,55,56). It has further been advised to take a single-beam reference spectrum through an empty place of the multisample cuvette system directly before the single-beam sample spectrum is obtained to eliminate contributions from impurities on the optical materials and to minimize problems arising from water vapor and $CO_2$ bands due to possible instabilities of dry-air purging of the instruments (1,3,4).

### 4.3 Biological Specimen and the Problem of Reproducibility

The most significant influence on data obtained from independent measurements on biological samples is repeatability and reproducibility. In general, one has to consider different "levels" of reproducibility that define and limit the discriminative and diagnostic power of the IR and Raman techniques. Spectral characteristics of cells, tissues, and body fluids inevitably depend on a number of factors, such as cell cycle, growth conditions, sampling, and sample preparation. These have to be controlled and standardized in order to obtain reproducible results and spectral data that can be communicated to other laboratories.

#### 4.3.1 Body Fluids, and Cells That Can Be Grown in Culture

For measurements on microbial or mammalian cells that can be grown in culture, it is useful to define different reproducibility levels $RL_i$ when repetitive measurements on identical strain samples are to be compared (physical parameters must be kept constant): (a) the reproducibility level $RL_1$ that describes repeatability of measurements on independent IR/Raman preparations obtained from an identical sample according to a strict protocol and (b) the reproducibility level $RL_2$ that defines repeatability of measurements performed on independent IR/Raman preparations, e.g., from cell organisms grown on or in media from different batches over a sufficient period of time. To calculate quantitative numbers for reproducibility, an objective measure for the description and comparison of independent measurements is pertinent. One possibility described in the literature is the crosswise calculation of so-called differentiation indices ($D$) as defined in Refs. 1,3, and 4 between pairs of measured spectra. Typical Poisson-like distributions of calculated individual $D$-values were obtained in practice from FT-IR measurements on microorganisms using first derivatives in the spectral range between 900 and 1200 cm$^{-1}$ of the dried-microbial-film samples, with mean $D$-values of $RL_1$ and $RL_2$ being around $0.4 \pm 0.4$ and $7 \pm 6$, respectively. In general, $RL_1$ was found to be one order of magnitude lower than $RL_2$. Hence, the microbiological parameters (control of cell cycle and growth-state quality of growth medium, sampling of the biomass from the surface of the solid agar plate, etc.) defined the repeatability of FT-IR measurements on independent cultures of microorganisms. Considering that the comparison of strains from different genera yielded $D$-values higher than 300, it was quite obvious that FT-IR spectroscopy, as applied to microorganisms, definitely provided sufficient spectral variance to be used (to) discriminate between different microorganisms. Unfortunately, similarly exact numbers on reproducibility levels of measurements on body fluids or mammalian cells grown in culture have not been published so far.

## 4.3.2 Histological Samples and Tissues

The effect of spatial variability within tissues is self-evident. Quite different spectra will be obtained, depending on the region where the sample spectrum is taken and on the spatial resolution used. This is exemplified by Fig. 6. Curve 1 in Fig. 6 (adapted from Ref. 23) shows the spectrum of a small sample taken from some particular regions of a human breast tumor. For comparison, the spectrum of human type I collagen is shown by spectrum 4. Using an FT-IR microscope and collecting spatially resolved spectra, e.g., with a computer-controlled $x$, $y$-stage from discrete regions of a microtomed tissue (xenografted tumor derived from MT-1 human tumor cells), a complete picture on the spatial distribution of different structures can be obtained. Curves 2 and 3 give spectra of sample areas spatially separated by approximately 20 μm. Absorption bands marked $C$ can be attributed to vibrational features that are diagnostic for collagen. Apparently, these spectra show remarkable differences in collagen content on the microscopic level. Thus, when macroscopic samples of a tissue are analyzed and averaged information is obtained, a statistically significant number of samples of the tissue specimen has to be measured in order to obtain satisfactory information on tissue structure and composition. On the other hand, when micro-sampling techniques are applied, e.g., on microtomed specimens, spectra have to be collected system-

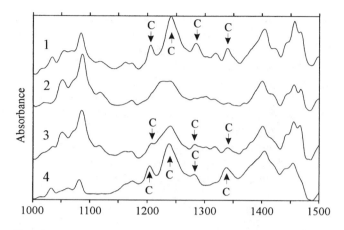

**Fig. 6** FT-IR spectra of a human breast cancer (1) and of human type I collagen (4). For comparison, spectra obtained from two spatially distinct regions of a xenografted tumor derived from MT-1 human breast tumor cells (traces 2 and 3) are given as well. Spectra have been measured with an infrared microscope. Bands marked *C* are typical of collagen. Spectra have been processed by Fourier self-deconvolution to achieve resolution enhancement. (Adapted from Ref. 23.)

atically as a function of spatial coordinates. Additionally, temporal variations in tissue structure and composition as well as differences between the tissues of individuals are also important factors and have to be accounted for. Thus, tissue spectra have to be analyzed from statistically significant numbers of individuals and should ideally be age- and sex-matched to establish reproducibility levels.

## 5   DATA TREATMENT AND EVALUATION TECHNIQUES

When analyzing and comparing hundreds, if not thousands, of spectra collected from various different microorganisms, body fluids, or tissues that all are encoded digitally by a defined number of data points (typically 2000–4000 data points in the mid-infrared region), the use of multivariate statistical techniques is a virtual necessity. The arsenal of modern multivariate statistical techniques provides ample methodologies for the pretreatment, evaluation, and representation of huge and complex data structures. These techniques not only allow the possibility of getting a survey of the data, but also allow the direct analysis and the interpretation of structures and relationships within the "data cloud." While univariate statistical analysis considers only a single property of a given object, multivariate statistics evaluates several properties of the objects at the same time. The advantage of these techniques is that the relationships among the properties can be taken into account as well.

Out of a large number of statistical techniques available for the analyst, four are of particular interest when considering vibrational spectra of complex biological samples as the basis for biomedical diagnosis: factor analysis, hierarchical clustering, linear discriminant analysis, and artificial neural networks (58–63). The differences between these methods are that factor analysis is performed to extract the essential information from large and mixed data sets to achieve data reduction and to facilitate the recognition of patterns, while the hierarchical clustering techniques, so-called unsupervised classification methods, attempt to find intrinsic group structure within the data set without the a priori need of any class assignment or partitioning of the data into training and test data sets. The history of a hierarchical cluster analysis process is generally represented by a minimal spanning tree, also called the ''dendrogram,'' by which the merging process of classes can visually be followed. In contrast to hierarchical clustering, linear discriminant analysis and artificial neural networks are so-called supervised classifiers, which need the class assignment of each individual (sample, spectra, etc.) from the beginning. Partitioning of the whole data into training, validation, and test data sets is generally needed to ensure the reliability of the results.

Particularly interesting applications of cluster, factor analysis, and artificial neural networks to FT-IR microspectroscopic imaging data of thin sections have recently been described (64–67). These methodologies may help to establish a new diagnostic imaging technique that can securely discriminate between different tissue structures in general or healthy and malignant tissue regions in particular without the necessity of tissue staining and with unprecedently high image contrast (see also Fig. 21).

## 6 CHARACTERIZATION OF INTACT CELLS

### 6.1 Particular Components in Whole Cells

This section deals with the detection and identification of particular components in whole cells, which can be identified and analyzed by specific vibrational bands. In this context resolution-enhancement techniques and/or difference spectroscopy turned out to be of particular help to identify compound-specific vibrational bands. The detection of a single component or even a few specific components within a complex mixture of compounds present in cells, tissues, and biofluids is intrinsically problematic. In some cases, however, infrared or Raman spectra were obtained that showed some extra bands or band systems of variable intensities that could not be considered as spectral variation due to experimental conditions or changes of biological parameters. A detailed analysis of these spectral features revealed the presence of particular cell constituents, such as intracellularly accumulated storage materials, cell surface structures, and endospores (5,44,55,56).

### 6.1.1 Storage Material Polyhydroxybutyric Acid (PHB) in Bacteria

Poly-β-hydroxy fatty acids (PHF) are energy and carbon reserves in prokaryotes that are also biotechnologically interesting macromolecules. In many cases, PHF compounds are accumulated under a limitation of nutrients when the supply of energy and carbon is in excess. Under conditions of starvation, PHF can be utilized and degraded by the microorganisms' helping the cells to survive under severe starvation conditions. It is known that the survival rate is related to the amount of PHF, which is intracellularly accumulated as small granules. These granules can easily be detected by light microscopy or by electron microscopy (marked $G$ in Fig. 7A). Poly-β-hydroxybutyrate (PHB), to give an example, is frequently found in bacteria (e.g., *Bacilli, Acetobacter, Pseudomonas*).

In most microbial IR spectra, the ester carbonyl band at 1738 cm$^{-1}$ was described to be only a small, weakly expressed shoulder (1,3–5). This band is caused predominantly by the ester $>$C$=$O-stretching vibration of lipids, e.g., present as phospholipids in cellular membranes. Gram-negative bacteria generally show a stronger ester carbonyl band than the gram-positive organisms due to the presence of an additional membrane layer, the outer membrane, which is built up by an asymmetric bilayer composed of a phospholipid layer at the inner and the so-called lipopolysaccharides at the outer layer leaflet. In spectra of *Legionella* or *Pseudomonas* cells, for instance, a rather prominent ester $>$C$=$O stretching band, accompanied by a number of additional bands between 900 and 1500 cm$^{-1}$, was occasionally observed. Interestingly, these bands turned out not to be permanently present throughout the cell cycle (1,5). In spectra of *Legionella* cells this band reached a maximum after 48 hours, while after 120 hours practically no additional $>$C$=$O ester band could be found. Figure 7B shows the overlaid FT-IR spectra of *L. pneumophila* isolate II8/RKI grown for 48 hours (curve 1) and 120 hours (curve 2), respectively. Inspection of the two spectra reveals a couple of differences, most prominent in the ester carbonyl stretching region around 1740 cm$^{-1}$. Figure 7C gives the difference spectrum as calculated from the two spectra, 1 and 2. This difference spectrum closely resembles the FT-IR spectra recorded for isolated and purified PHB (spectrum not shown). At least 10 bands can be identified and assigned to a typical polyester compound. The quantitative determination of PHB in whole cells as function of time, temperature, and inoculation can be achieved by calculation of the ratio $\alpha = I_1/I_2$ (see Fig. 7B), where $I_1$ is the intensity of the ester carbonyl peak at 1738 cm$^{-1}$ used as a measure marker band of PHB content, and $I_2$ is the intensity of the amide II peak at 1550 cm$^{-1}$ used as an internal standard that measures approximate total cell mass (Fig. 7D). An interesting application in this context is the use of factor analysis that aims at the in situ quantitation of PHB or other intracellularly accumulated storage materials. Figure 8, left panel, shows a factor analysis map ob-

A

C

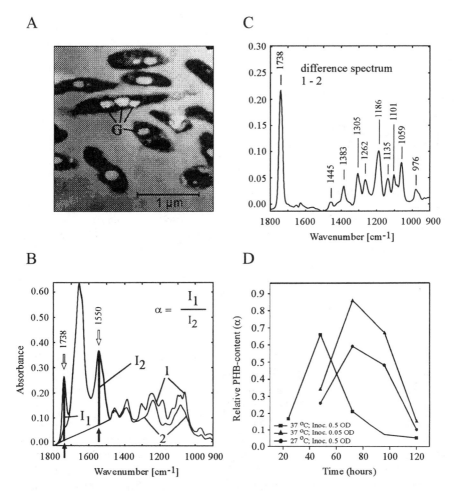

B

D

**Fig. 7** (A) EM micrograph of a thin cross section through cells of *Legionella pneumophila* II8/RKI, grown for 48 hr at 37°C on CYE-agar: up to three polyhydroxybutyric (PHB) granules (*G*) are visible within the cells. Magnification 25,000-fold, bar = 1 μm. Prefixing: *G* in cacodylate buffer; fixing: osmium cacodylate buffer; embedding medium: LR white hard (Science Services GmbH). (B) FT-IR spectra of *Legionella pneumophila* II8/RKI grown on CYE agar plates at 37°C: Spectra obtained after 48 hr (1) and after 120 hr (2). (C) Difference spectrum (1 minus 2). Some absorption bands typical of PHB are annotated. (D) The intensity $I_1$ of the ester carbonyl band at 1738 cm$^{-1}$ and the amide II band $I_2$ near 1550 cm$^{-1}$ can be used to determine the relative PHB amounts α in cells after different times, temperatures, and inoculation of growth in liquid cultures.

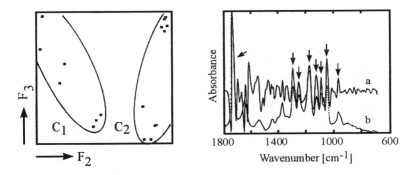

**Fig. 8** Quantification of PHB in bacterial cells by multivariate analysis. *Left panel*: Factor analysis map of factorial coordinates $F_3$ (vertical axis) and $F_2$ (horizontal axis) as obtained for a data set of serologically identical *Legionella pneumophila* strains. $C_1 =$ PHB-positive organisms, $C_2 =$ PHB-negative organisms. *Right panel*: Curve a = factor spectrum 2; curve b = difference spectrum between the means of the original $C_1$ (PHB-positive strains) and $C_2$ (PHB-negative strains) spectra. Arrows mark some typical bands of the polyester compound PHB. Spectra have been pretreated for factor analysis by applying a bandpass filter and then normalized in the spectral window: 700–1800 cm$^{-1}$. (Adapted from Ref. 6 by permission.)

tained from a data set of FT-IR spectra recorded from samples of two different strains of *Legionella pneumophila* that were described to be serologically identical (serogroup 1 strains) and grown for different times in liquid culture. After data compression by principal component analysis, a two-dimensional projection of data along the factorial coordinates (factor loadings) $F_2$ and $F_3$ revealed two main groups independent of strain assignment, which by visual inspection of the class-specific spectra turned out to be the non-PHB containing (left, cluster 1) and the PHB producers (right, cluster 2), respectively. Obviously, the property vector that can separate between the two groups is one of the factor spectra used for projection of data. Indeed, the plotted factor spectrum number 2 (Fig. 8, right panel, curve a) exhibits mainly the spectral characteristics typical of a PHB compound as suggested by comparison with the IR spectrum of pure PHB (Fig. 8B, right panel, curve b). The factorial coordinate $F_2$ scales the "intensity" of this "property vector," i.e., the relative amount of PHB. Thus, after calibration of the data with different known amounts of PHB via an independent technique (multivariate calibration), it is possible to quantitate PHB in unknown samples of intact bacteria in situ.

### 6.1.2 Glycogen in Mammalian Cells

Muscle, liver, and some epithelial cells are able to store the energy-rich molecule glycogen as small intracellular granuals. Glycogen is a glucose polymer (for comparison see FT-IR and FT-Raman spectra of glycogen in Fig. 1 and descriptions in Sec. 2.1) that exhibits a number of strong IR and Raman bands in the "carbohydrate" region between 800 and 1200 $cm^{-1}$, which are due to strongly coupled C—O and C—C stretching and C—O—H deformation modes. Compound diagnostic absorptions in the infrared are observed around 1153, 1081, and 1025 $cm^{-1}$ (see also Fig. 1A). The amount of glycogen within the cell is related to the staging state and cell maturation. The variability of glycogen concentration within the cell depends on a variety of factors and can be a serious obstacle precluding nucleic acid and lipid structure detection within tissues. This has been clearly demonstrated recently (44): Figure 9A shows a photomicrograph of a normal liver biopsy stained by the standard hematoxylin-eosin (HE) technique. The authors have chosen liver tissue due to its homogeneity. Liver cells are terminally differentiated, so any changes observed between pixel spectra could be interpreted in terms of structural, compositional, and /or pathological differences between the various tissue regions tested. Figure 9B shows a photomicrograph of an unstained $2 \times 1$-$mm^2$ section of liver tissue with cirrhosis, and Fig. 9C is an infrared map imaged in $50 \times 50$ $\mu m^2$ pixels. Cirrhosis is described as causing scarring of the liver tissue, which produces regions of high collagen content. Without prior knowledge of the tissue section architecture and the compositional variations to be expected, clustering techniques of the individual spectra were applied to attain grouping according to spectral similarity, and a color code was assigned on the basis of spectral similarities (44). This analysis was performed for two separate spectral regions: the C—H stretching region (2700–3200 $cm^{-1}$) and the spectral region between 800 and 1700 $cm^{-1}$ (see Fig. 9C). Subsequent inspection of the spectra revealed that in some regions the spectral traces contained features of connective tissue devoid of glycogen (see, e.g., traces 1 and 2), while other regions were clearly dominated by spectral features characteristic of glycogen (see traces 3 and 4 in Fig. 9C). Obviously, the main differences evident from these spectra were due to gross variations of different cellular levels of glycogen. The ability of infrared spectroscopy to detect such changes in chemical composition inside cells and tissues without the use of stains and special preparations is a very attractive feature for histopathological analysis.

### 6.1.3 Metabolically Released $CO_2$ in Cells

The detection and quantitation of metabolically released $CO_2$ in cells is another interesting feature of the FT-IR technique. Intracellularly or elsewhere produced $CO_2$, which is usually detected as $CO_2$ hydrates or clathrates in water (17,24), can be determined with extreme sensitivity, since the $CO_2$ antisymmetric stretching band, located near 2343 $cm^{-1}$, is found in a spectral region where the signal-to-noise ratio is optimal and that is usually devoid of overlapping spectral features. Figure 10A shows the spectra of a fully hydrated sample of *Escherichia coli* (1) and of pure water (2) and the difference spectrum of 1 minus 2 (3) (56).

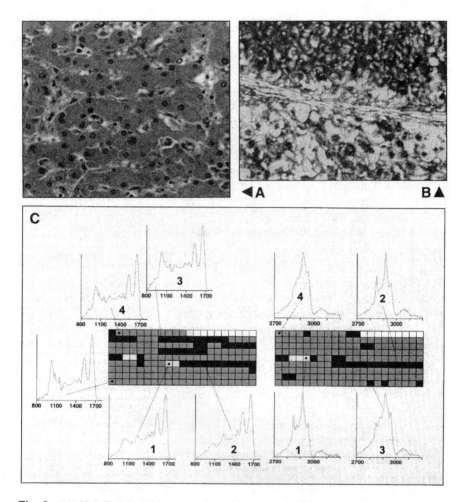

**Fig. 9** (A) H & E stained biopsy section of liver tissue. (B) Unstained liver section with cirrhosis ($2 \times 1.5$ mm²). (C) False color infrared map of liver section shown in Fig 9B. The pixel size is approximately $50 \times 50$ µm². The labeled spectra are typical for regions that contain collagen (1,2) and glycogen (3,4), respectively. (Adapted from Fig. 6 in Ref. 44 by permission.)

The inset gives the expanded spectral region 2400–2300 cm⁻¹, where the anti-symmetric stretching band of $CO_2$ hydrates can be detected at 2343 cm⁻¹. Apparently, the $CO_2$-band intensity can be used to test the level of metabolic activity within the cells. Figure 10B shows the example of $CO_2$ production by a yeast culture supplemented with and without glucose. In this way, the kinetics of $CO_2$

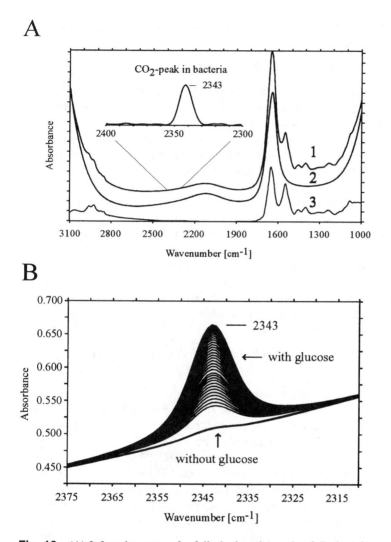

**Fig. 10** (A) Infrared spectra of a fully hydrated sample of *Escherichia coli* (1) and of pure water (2) and the difference spectrum 1 minus 2 (3). The inset shows the expanded region of antisymmetric stretching band of $CO_2$. (B) Quantitation of $CO_2$ after addition of glucose to a yeast culture. Spectra in (A) have been measured using a cell pathlength of 10 μm, spectra of (B) using a pathlength of 250 μm. (Adapted from Ref. 56 by permission.)

release after the addition of various different substrates to a yeast culture can be sensitively monitored.

### 6.1.4   Pigments in Microbial Cells

The FT-Raman spectra of some microbial constituents, like pigments of the carotenoid type, may give rise to additional peaks, which, though being present in only small amounts, gain considerable in intensity due to a preresonance effect. Figure 11A shows two spectra obtained from samples of a *Staphylococcus aureus* and a *Micrococcus roseus* strain that were golden (aureus) and deep rose (roseus) in color, respectively. These spectra show two characteristic Raman bands near 1525 cm$^{-1}$ and 1514 cm$^{-1}$ ($v(C{=}C)$) and 1159 cm$^{-1}$ and 1155 cm$^{-1}$ ($v(C{-}C)$) for the *Staphylococcus aureus* and *Micrococcus roseus* samples, respectively. These bands, being diagnostic for the presence of different carotenoid structures (68), were found to be markedly dependent on the time and temperature of growth. Figure 11B shows, for *Micrococcus roseus* as an example, that the amount of cellular pigment production strongly depends on the time of growth. The relative intensities of the diagnostic $v(C{=}C)$ and $v(C{-}C)$ bands, compared to the amide I band as an internal standard, correlate with the time of growth and thus demonstrate that the expression of this particular bacterial pigment is mainly a growth-dependent process.

### 6.1.5   Structural Changes as a Function of Growth and Cell Cycle

Another example is the in situ Raman spectroscopic detection of relative amounts of protein and/or RNA as a function of the time of growth. Figure 12 gives FT-Raman spectra of *Bacillus subtilis* cultivated for 48 hours (curve a) and 120 hours (curve b), respectively. The comparison of these two spectra strongly suggests a significant decrease in the relative amount of proteins by the reduced relative intensity of the bands at 1679 cm$^{-1}$ (amide I) and 1235 cm$^{-1}$ (amide III), using the band intensity at 1450 cm$^{-1}$ (C—H deformation band) as an internal standard. The difference spectrum obtained by subtraction of spectrum 1 (48 hours time of growth) minus spectrum 2 (120 hours time of growth) (see Fig. 12, curve c) suggests a strong decrease in the cellular amount of RNA, as is indicated by the marker bands at 669 cm$^{-1}$ (guanine), 726 cm$^{-1}$ (adenine), 783 cm$^{-1}$ (uracil and cytosine), 1095 cm$^{-1}$ (symmetric stretching of $>PO_2^-$), and 1576 cm$^{-1}$ (adenine, guanine), respectively (27,69,70). For comparison, curve d gives the FT-Raman spectrum of a purified RNA sample. The observed band-intensity changes may likely be assigned to the transition of the cells from the so-called logarithmic state of growth to the stationary state of growth. It is well known from the literature that the logarithmic state of growth is characterized by a much higher amount of RNA synthesis as compared to resting cells, to guarantee high cell proliferation, i.e., high rates of protein biosynthesis.

A

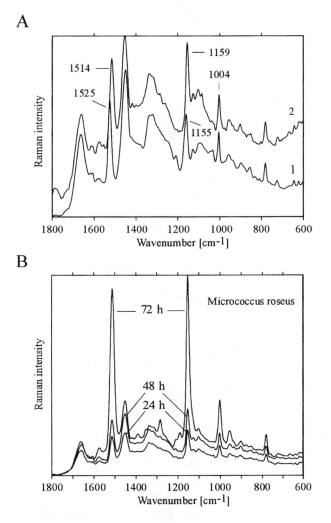

B

**Fig. 11** (A) FT-Raman spectra of bacteria exhibiting preresonance enhanced bands of carotenoid containing pigments *Staphylococcus aureus* DSM 20231 (1), *Micrococcus roseus* DSM 20447 (2). (B) Spectra obtained from *Micrococcus roseus* DSM 20447 after different times of growth. (Adapted from Ref. 55 by permission.)

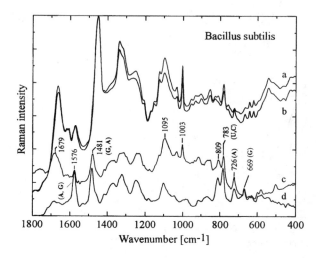

**Fig. 12** Raman spectral changes observed in samples of *Bacillus subtilis* RKI W2GR as a function of time of growth: (a) 48 hours, (b) 120 hours; (c) difference spectrum (a) minus (b); (d) for comparison, the spectrum of RNA. A, C, G, and U stand for adenine, cytosine, guanine, and uracil, respectively. (Adapted from Ref. 55 by permission.)

## 6.2 Differentiation and Identification of Microbial Cells

Cell diversity is always structural and biochemical diversity. FT-IR and FT-Raman spectroscopy of intact microbial cells, for example, provide information on the structure and composition of the whole cells. They are therefore potentially valuable tools for characterizing and differentiating microorganisms at very different levels. However, since vibrational spectra are very complex spectroscopic signals encoding the superposition of hundreds or even thousands of vibrational bands that cannot be resolved by any means, pattern recognition techniques have to be used. These techniques can treat the spectra as fingerprints rather than as a combination of discrete bands with intensities, frequencies, and bandwidths normally taken as characteristics for particular cellular structures. Most classical microbiological techniques give information on cell functions, whereas FT-IR spectroscopy provides spectral traits about cell structures. Both approaches are therefore complementary and can be used in combination.

In recent decades, vibrational spectroscopy has been used by various groups to differentiate and classify microorganisms (1–15,28,29,55,56). In order to appreciate these efforts, some examples of typical bacterial classification trials based on IR and Raman spectra are given that show the flexibility of the technique. Figure 13 shows dendrograms obtained by hierarchical cluster analysis performed on various different microbial FT-IR spectra. Figure 13A gives the

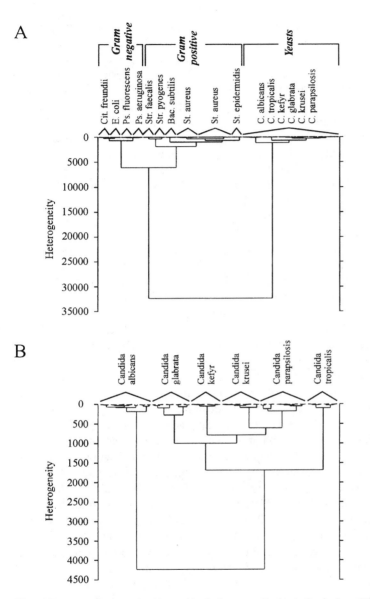

**Fig. 13** Dendrogram of a hierarchical cluster analysis performed on FT-IR spectra of different species and strains of gram-positive and gram-negative bacteria and of yeasts belonging to the genus *Candida*. Cluster analysis was performed using the first derivatives, considering the spectral ranges 2800–3000, 1400–1500, and 900–1200 cm⁻¹ in (A) and 1014–1070, 1099–1171, 1339–1420, and 1622–1670 cm⁻¹ in (B). All spectral ranges used were equally weighted, and Ward's algorithm was applied. For calculating the distance matrix, Pearson's correlation coefficient was used. (Adapted from Ref. 83 by permission.)

dendrogram of a classification trial based on approximately 180 microbial IR spectra obtained from independent measurements on 30 different strains of gram-positive and gram-negative bacteria and different species and strains of yeasts. Three main clusters are recognized, which discriminate perfectly between the gram-positive and gram-negative on the one hand and the yeasts on the other hand. A closer inspection of the fine structure of this dendrogram also reveals a perfect discrimination at the species level, just as is expected from microbiological taxonomy considerations based on molecular genetic techniques. This is also valid for the differentiation of yeasts, as exemplified by the dendrogram of Fig. 13B.

Figure 14 gives the dendrogram of a classification trial on a selection of serologically different *Escherichia coli* strains. The purpose of this analysis was to group the spectra of the isolates according to their O-antigenic structure. O-antigenicity is usually determined by serological and SDS-gel electrophoretic techniques. The O-antigenic epitopes are structurally determined by the O-specific side chains of the lipopolysaccharides, which are hetero-oligosaccharide compounds. Therefore, it can be expected that these structures are observed primarily in the spectral region between 700 and 1200 $cm^{-1}$, where the carbohydrates dominate the spectral features observed. Using Ward's algorithm for cluster analysis, indeed three main clusters are obtained, which perfectly correspond to the O-18, O-25, and O-114 group scheme defined by O-serology.

As for FT-IR, microbial FT-Raman spectra express the total composition and structure of all structures present, which, in turn, may depend on various microbiological parameters, such as time and temperature of growth and nutrient conditions. Thus, FT-Raman spectra of microorganisms were also used in the literature to characterize a number of microorganisms (28,29,53–56). Figures 15A and B show some FT-Raman spectra of very diverse species and strains of bacteria and yeasts. The intensity of the band at 786 $cm^{-1}$ can be taken as a measure of the relative quantity of nucleic acids present, while useful information on the lipid/protein ratio can be calculated from the intensities of the bands at 1450 $cm^{-1}$ ($\delta$(C—H)) and 1660 $cm^{-1}$ (amide I) or 1004 $cm^{-1}$ (phenylalanine)). These relationships may be helpful to quickly estimate gross compositions of microbial cell samples and allow the detection of changes of structural components as a function of, e.g., time, temperature, or growth. As for FT-IR, the total amount of the spectral information provided by Raman spectra can be used to characterize the microbial species and strains under investigation using multivariate pattern recognition techniques. Factor and cluster analysis and the application of data pretreatment techniques, such as the calculation of derivatives and the use of a combination of smaller or larger regions of the spectrum that are most discriminative, may help to achieve optimal classifications. To give an example, Fig. 16A shows the dendrogram of a classification trial on a selection of gram-positive and gram-negative bacteria and of different species and strains of *Candida* yeasts. Like the FT-IR analysis on a similar collection of microorganisms (see Fig. 14A), a dis-

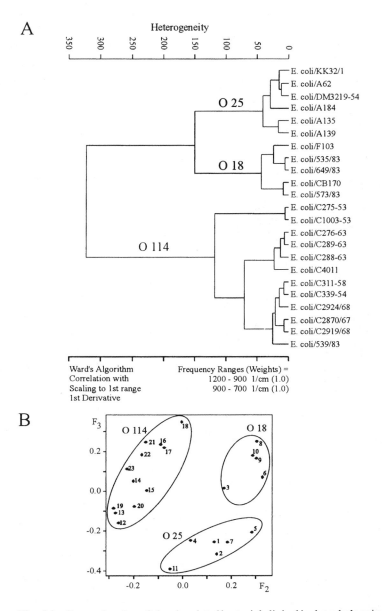

**Fig. 14** Spectral typing of closely related bacterial clinical isolates belonging to the species *Escherichia coli* according to their O-antigenic structure. (A) Hierarchical classification using Ward's algorithm; the distance measure used was Pearson's correlation coefficient as defined in Refs. 1, 3, and 4. (B) Factor analysis map: each point represents the spectrum of an isolate; for projection of data in factor space, the factorial coordinates 2 and 3 were used. In (A) and (B), the first derivatives of the IR spectra considering the spectral ranges 700–900 and 900–1200 cm$^{-1}$ were used. Both spectral ranges were equally weighted.

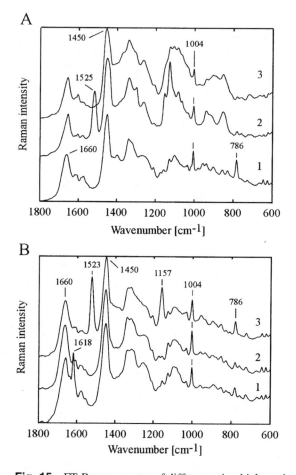

**Fig. 15** FT-Raman spectra of different microbial species and strains. (A) *Bacillus subtilis* RKI W2GR (1), *Mycobacterium smegmatis* HK 144 (2), *Candida albicans*, clinical isolate (3). (B) *Pseudomonas chlororaphis* ATCC 17809 (1), *Escherichia coli* RKI 139(2), *Staphylococcus aureus* DSM 20231 (3). (Adapted from Ref. 55 by permission.)

---

**Fig. 16** (A) Classification trial on a selection of FT-Raman spectra of gram-positive and gram-negative bacteria and yeasts. (B) Dendrogram of a hierarchical cluster analysis performed on the bacterial strains only. The gram-positive, gram-negative, and yeast clusters comprise different strains of *Bacillus subtilis*, *Staphylococcus epidermidis*, *Staphylococcus aureus*, *Escherichia coli*, *Citrobacter freundii*, *Pseudomonas chlororaphis*, and *Pseudomonas aeruginosa* and of *Candida kefyr*, *krusei*, *tropicalis*, *parapsilosis*, and *glabrata*. Hierarchical clustering was applied using Ward's algorithm. The original spectra in the ranges (A) 2800–3000, 3050–2749 and (B) 1140–1020, 1400–1500 cm$^{-1}$ were used. Pearson's correlation coefficient as defined in (1,3,4) was used to calculate interspectral distances. (Adapted from Ref. 56 by permission.)

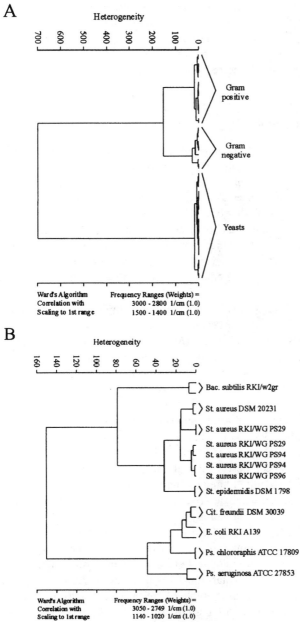

A

Heterogeneity

700 600 500 400 300 200 100 0

Gram positive

Gram negative

Yeasts

Ward's Algorithm
Correlation with
Scaling to 1st range

Frequency Ranges (Weights) =
3000 - 2800 1/cm (1.0)
1500 - 1400 1/cm (1.0)

B

Heterogeneity

160 140 120 100 80 60 40 20 0

Bac. subtilis RKI/w2gr

St. aureus DSM 20231

St. aureus RKI/WG PS29

St. aureus RKI/WG PS29
St. aureus RKI/WG PS94
St. aureus RKI/WG PS94
St. aureus RKI/WG PS96

St. epidermidis DSM 1798

Cit. freundii DSM 30039

E. coli RKI A139

Ps. chlororaphis ATCC 17809

Ps. aeruginosa ATCC 27853

Ward's Algorithm
Correlation with
Scaling to 1st range

Frequency Ranges (Weights) =
3050 - 2749 1/cm (1.0)
1140 - 1020 1/cm (1.0)

tinct clustering of the microbial strains was obtained according to what a microbiologist would expect when applying molecular genetic techniques, i.e., a classification scheme that nicely separates the gram-positive bacteria from the gram-negative ones, with all bacteria quite distant from yeasts. At the same time (see Fig. 16B), all strains belonging to corresponding species and genus groups are nicely grouped together.

## 6.3 Microbial Microcolonies

The characterization of unknown clinical, food-, water-, or airborne microbiological specimens includes the fundamental steps of detection, enumeration, and differentiation of microbial cells present. In practice, rather different microbiological analysis techniques are used to detect, count, and differentiate microorganisms. These techniques include the counting of colony-forming units, the measurement of optical density, the use of cell counters or cell sorters, and the application of the whole arsenal of techniques by which microbial cells can be differentiated, including molecular genetic methodologies. Obviously, the combination of these fundamental steps of microbiological characterization in one single instrument is very attractive.

When applying FT-IR microscopy to microbial samples, three questions have to be addressed: (a) What are the real detection limits for microorganisms? (b) How can nanogram quantities of microorganisms be sampled and reproducibly measured by the spectrometer. (c) Can detection, enumeration, and differentiation of microorganisms be achieved by combining optical, spectroscopic, and computer imaging techniques?

The sensitivity of standard FT-IR microscopic instrumentation was tested and estimated to be sufficiently high to obtain good-quality spectra from microbial microcolonies as small as 30 μm in diameter, corresponding to a few hundred bacterial cells (3,55,56). For sampling microbial cells growing in microcolonies, the following technique has been described in the literature: Aliquots of the microbial cell suspension, sufficiently diluted to guarantee single colony growth on solid agar plates, are plated and incubated over a period of 6–8 hours. After this growing time, colony formation is generally not yet visible by eye. A circular IR-transparent plate made, e.g., of $CaF_2$, $BaF_2$ or ZnSe is then gently pressed to the agar surface. This "microcolony" imprinting technique transfers spatially accurate small amounts of the microcolonies (a few microbial layers) to the plate and provides a replica of dried-microbial-film spots that can be measured by the FT-IR microscope. Figure 17 gives an example of measurements performed on a mixed culture containing three different microorganisms: *Staphylococcus aureus*, *Streptococcus faecalis*, and *Escherichia coli*. Figure 17A, left panel, shows the micrograph from a selected area of the colony replica, which shows three different microcolonies that can already be morphologically differentiated by visual

# A

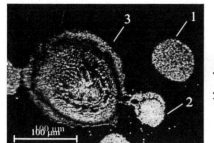

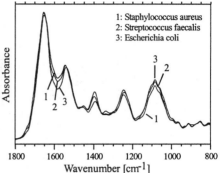

1: Staphylococcus aureus
2: Streptococcus faecalis
3: Escherichia coli

# B

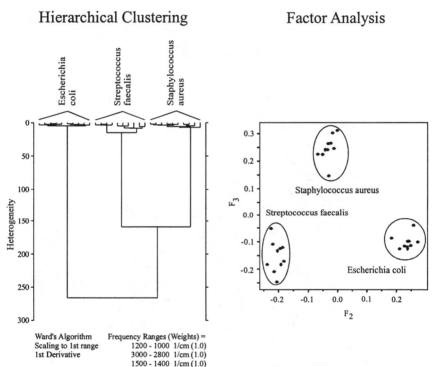

**Fig. 17** Detection, differentiation, and classification of different microbial microcolonies. (A) *Left panel*: Micrograph (magnification approximately 150×) of three different colony spots deposited on BaF$_2$ windows by the described stamping technique (3,55,56). *Right panel*: Infrared spectra obtained from the colony spots shown on the upper left. (B) *Left panel*: Hierarchical cluster analysis performed on the spectral data obtained from approximately 30 different colony spots. *Right panel*: Factor analysis performed on the same spectra. The clusters suggested by both classification techniques are: $C_1 = $ *Staphylococcus aureus*; $C_2 = $ *Streptococcus faecalis*; $C_3 = $ *Escherichia coli*. (Adapted from Ref. 83 by permission.)

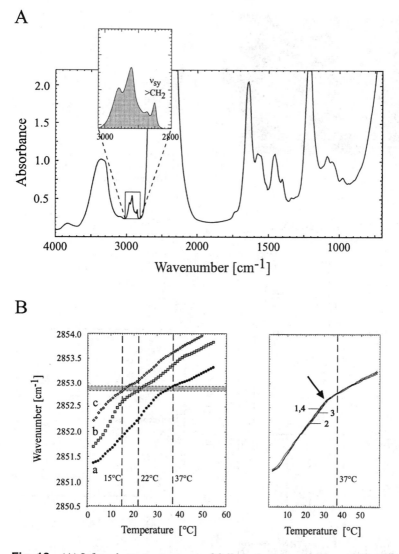

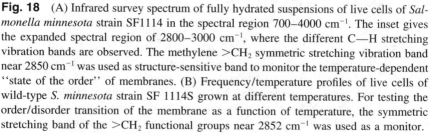

**Fig. 18** (A) Infrared survey spectrum of fully hydrated suspensions of live cells of *Salmonella minnesota* strain SF1114 in the spectral region 700–4000 cm$^{-1}$. The inset gives the expanded spectral region of 2800–3000 cm$^{-1}$, where the different C—H stretching vibration bands are observed. The methylene $>CH_2$ symmetric stretching vibration band near 2850 cm$^{-1}$ was used as structure-sensitive band to monitor the temperature-dependent "state of the order" of membranes. (B) Frequency/temperature profiles of live cells of wild-type *S. minnesota* strain SF 1114S grown at different temperatures. For testing the order/disorder transition of the membrane as a function of temperature, the symmetric stretching band of the $>CH_2$ functional groups near 2852 cm$^{-1}$ was used as a monitor.

inspection using criteria such as different colony size and refractive index. Representative spectra obtained from these three colony spots are shown in the right panel of Fig. 17A. These spectra suggest that the three microcolony imprints have been derived from three different microorganisms. In a second step of analysis, a representative number of FT-IR microscopic spectra of bacterial film spots was subjected to multivariate statistical analysis. Figure 17B shows typical results when using hierarchical clustering (left panel) and factor analysis (right panel), which unequivocally prove that the spectra obtained from the microcolonies are indeed characteristic of three different microorganisms present in mixed culture. It has been anticipated in the literature (3,55,56) that with the aid of a computer-controlled $x$, $y$-stage, spots of microcolonies can be measured automatically or via operator control using video techniques. Additionally, the number of colony spots can be counted, and size classification of these spots should also be possible.

## 6.4 Biological Membranes of Live Bacteria

The pioneers in this spectroscopic area are the members of a group at the National Research Council of Canada in Ottawa (71,72). These authors clearly showed that plasma membranes of *A. laidlawii* can be tested in vivo, when deuterated fatty acids are supplemented to culture media that have been lipid depleted. Cells of *Acholeplasma laidlawii* may directly incorporate externally added fatty acids, since the de novo synthesis of fatty acids can be blocked by avidin. To test the membrane state of order, the frequency value of the symmetric stretching frequency of the $>CD_2$ functional groups of the deuterated fatty acid chains near 2088 cm$^{-1}$ were used as a monitor. A typical experiment was to increase the temperature linearly at approximately 1°C per minute and to collect IR spectra at intervals of 1 degree temperature increase per minute. In this way the frequency values of the $>CD_2$-stretching vibration band could be determined as a function of temperature, and frequency/temperature profiles were constructed to test the

---

*Right panel*: Frequency/temperature profiles obtained from samples of four independent cultivations (1–4) at 37°C were used to test the reproducibility of the technique. *Left panel*: Order/disorder profiles are shown for cells grown at 37°C (a), 22°C (b), and 15°C (c). Shadowed area gives the narrow interval of the $>CH_2$ symmetric stretching band frequencies defined by the different growth temperatures (vertical lines). For FT-IR measurement, pellets of harvested and washed bacterial suspensions (in 1% NaCl/D$_2$O solutions) were prepared as gel films between CaF$_2$ windows equipped with 25-$\mu$m Teflon spacers and were sealed in a gas-tight, temperature-controlled IR cuvette. For temperature profile measurements, the temperature was linearly increased by 0.2°C/min while collecting and storing one averaged and Fourier transformed IR spectrum per 1°C temperature increase. (Adapted from Ref. 75.)

"state of order" of the membranes of intact microbial cells as a function of temperature (71–74).

Since in most cases it is not feasible to selectively incorporate externally added deuterated fatty acids into bacterial membranes, some authors used the $>CH_2$ symmetric stretching vibration near 2850 cm$^{-1}$ as a monitor to test the state of order of the membranes in vivo (75). This methodology, however, is useful only for comparative studies on cells that are not too different in composition or for investigations where only a minimum of changes occur as a function of time of growth, degree of mutation, temperature of growth, nutrient conditions, etc., since the $>CH_2$ band at 2850 cm$^{-1}$ is a composite of many different absorptions. Figure 18A shows a survey spectrum obtained from a fully hydrated cell suspension of *Salmonella minnesota* strain SF1114 between 700 and 4000 wavenumbers (75). The inset gives the marker band (see arrow) that is used to monitor the "state of order" of the membranes. Cultures of *Salmonella minnesota* grown under identical conditions and that have been sampled according to a strict experimental protocol can be measured with high accuracy (see Fig. 18B, right panel). Typically, these samples (in contrast to *A. laidlawii*) yielded only weakly expressed phase-transition features. Interestingly, cells grown at different temperatures gave distinctly different temperature-induced order/disorder profiles (see Fig. 18B, left panel). At the corresponding growth temperatures, however, these profiles indicated a very similar "state of order" of the bacterial membrane at the individual growth temperatures. Consequently, the cells responded to varying growth temperatures by actively adapting the fluidity of the cell's membranes to an optimal value, which, in turn, is a crucial prerequisite for triggering the activity of the membrane-bound proteins. This adaptive behavior of the cells is mediated by changing the composition of the fatty acid and lipid composition of the membrane matrix (75).

## 7  DIAGNOSIS OF DISEASE STATES IN HUMANS

Just as infrared and Raman spectroscopy may provide information on structure and composition of intact microbial cells, they are also applicable to human cells, tissues, and biofluids. A logical extension of this approach is not only to analyze differences between different types of mammalian tissues and cells but also to use IR and Raman spectroscopy to detect changes in structures and compositions induced by the disease process. An overview on the most recent developments in this field can be found in a series of papers and reviews (16–27,30–32,44,52–54,64–67,76–82). They cover a wide range of applications, e.g., the analysis of various biofluids (such as synovial fluids, to diagnose arthritic disorders in joints), blood and serum (clinical diagnosis), the characterization of disease states in tissues, such as a particular type of cancer (skin, breast, or colon cancer), some abnormalities within the central nervous system (multiple sclerosis, Alzheimer's disease), or the analysis of normal lymphocytes or leukemic cells. In the following, examples will be given to outline the possibilities, the great progress achieved to date, and the problems.

## 7.1  Diagnosis of Arthritic Disorders from Synovial Fluids

Arthritis, which can appear in many different forms that lead to severe changes of joint physiology, is still difficult to diagnose. At present, the investigation of synovial fluids (SF) aspirated from the joints of patients suffering from various forms of acute arthritis is the accepted "gold standard" for the differential diagnosis of inflammatory and noninflammatory arthritis. Synovial fluid is produced by filtration of plasma across the synovial membrane. In addition to the normal analytes of plasma, it contains the negatively charged complex carbohydrate hyaluronic acid, which, due to its high viscosity, functions as an effective lubricant. Additional diagnostic techniques are x-ray investigations coupled to the experience of the rheumatologist, imaging techniques such as magnetic resonance imaging and computer tomography, and immunological investigations (e.g., test for "rheumatology factor" in blood, and/or differentiation and analysis of the cells in synovial fluids). The various diagnostic methods, and particularly the combination thereof, make diagnosis labor-intensive and time-consuming. Starting from the hypothesis that the transition from a healthy to an arthritic joint should be accompanied by a change in the composition of the synovial fluid, which should be detectable by IR spectroscopy, a group at the Institute for Biodiagnostics in Winnipeg (National Research Council of Canada) has performed a systematic FT-IR investigation of synovial fluids from healthy persons and patients suffering from various different forms of arthritis, namely, osteoarthritis (OA), rheumatoid arthritis (RA), and spondyloarthropathy (SPA) (17). Initially the authors performed linear discriminant analysis (LDA) on the entire spectral region measured (1000–4000 cm$^{-1}$). Subsequently, they identified some particular spectral subregions with significant discriminatory power using also LDA. One of these regions with the greatest diagnostic potential was the range 2800–3050 cm$^{-1}$, a region of the spectrum populated by C—H stretching vibrations of all compounds present in synovial fluids. Upon application of an "optimal-region-selection-rule" algorithm to the spectral range 2800–3050 cm$^{-1}$, 13 subregions with optimal LDA results were extracted that contributed maximally to the a priori given classification. The results based on these new multivariate analysis methodologies are presented in Table 4. The very good agreement between clinical diagnosis and statistical diagnosis demonstrates that IR spectroscopy could be used as an additional technique for the differential diagnosis of arthritis that requires very little operator training and minute volumes of synovial fluid and permits diagnosis within minutes.

**Table 4**  Results of a Linear Discriminant Analysis Based on the 15 Most Significant
Spectral Subregions Between 2800 and 3050 cm$^{-1}$

| Clinical diagnosis | Classification results | | | | | |
|---|---|---|---|---|---|---|
| | C | RA | OA | SpA | Sensitivity | Specificity |
| Control (C) | 12 | 0 | 0 | 0 | 100 | 98.6 |
| Rheumatoid arthritis (RA) | 0 | 37 | 0 | 1 | 97.4 | 97.9 |
| Osteoarthritis (OA) | 1 | 1 | 23 | 0 | 92.0 | 100 |
| Spondyloarthropathy (SPA) | 0 | 0 | 0 | 11 | 100 | 98.7 |

*Source*: Adapted from Ref. 17.

## 7.2  Disease-Specific Infrared Signals and DNA/RNA Detectability During the Cell Cycle

Infrared spectral variations within the cell cycle were recently investigated in depth by M. Diem and coworkers (44,76,77). Cancerous cells may divide continuously and can be grown in culture ad infinitum. From a culture of myeloid leukemia (ML-1) cells, exponentially growing cells were fractionated according to the cell cycle phases by the process of centrifugal elutriation to achieve separation by size and density. After elutriation, the phase homogeneity was verified by the authors using the cytometry technique of fluorescence-activated cell sorting (FACS). Small volumes from cell suspensions of the elutriated fractions were allowed to dry on ZnSe plates, and approximately 50 single cells were selected from each fraction for FT-IR microscopic investigations and averaged. Trace a of Fig. 19 shows the absorption spectrum of a cell in the G1 phase (gap 1 phase with diploid cells containing two copies of DNA). This spectrum was typical of a cell with a low amount of DNA and resembles the one for "normal" tissues with spectral characteristics between 900 and 1300 cm$^{-1}$ being dominated by RNA spectral features (44). Traces b, c, and d of Fig. 19 show typical spectra obtained from fractions corresponding to early, middle, and late S phase (synthesis phase, the nuclear DNA is actively replicated). Considerable heterogeneity in the S-phase spectra was detected, starting with a gradual increase of the DNA contribution to the spectra between 900 and 1300 cm$^{-1}$ to a very large DNA contribution shown. Trace e was from a cell at the S/G2 junction, whereas trace f was from a cell in the G2 phase (gap 2 phase with quadruploid cells containing four copies of the cell's DNA). The low-frequency region of the spectrum (900–1300 cm$^{-1}$) was found to be very similar to that of the G1 type of spectra, which are also dominated by features characteristic of RNA. M. Diem and coworkers concluded that these results convincingly demonstrated that the different phases through which somatic cells pass during their regular division cycle exhibit very different phase-specific infrared spectra. The small contribution of DNA-specific

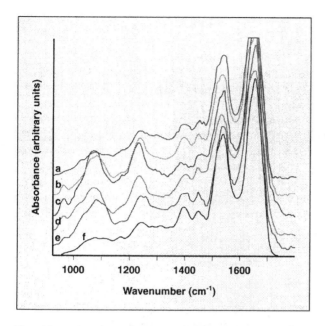

**Fig. 19** Infrared spectra of individual myeloid leukemia (ML-1) cells, selected from fractions of synchronized cells. (a) G1 phase; (b) early, (c) middle, and (d) late S-phase; (e) S/G2 interface; and (f) G2 phase. (Adapted from Ref. 44 by permission.)

features observed in the G1 and G2 phases was explained by the authors in terms of a very interesting hypothesis concerning the DNA detectability during the cell cycle: In the G1 and G2 phases the DNA is not distributed uniformly throughout the nucleosome; rather, it should be densely packed within the histones to form the chromosomes. From molecular weight, packing density, and absorptivity estimations the authors concluded that the DNA's absorptions might have exceeded the optical density of 50 within the nucleosomes of inactive cells (G0, G1, G2 phases) and should consequently appear as "black dots or strings" that cannot contribute to the experimentally observed spectrum of the cells. DNA-specific signals, however, will appear in the spectra within the various stages of the S phase when cells actively replicate their nuclear DNA, a process accompanied by an unwrapping of the DNA/histone complexes and a lowering of DNA packing densities within the nucleosomes. The distinct similarity of spectra from cells in the G1 and G2 phases were taken to support this hypothesis, since G2 cells were in a phase with doubled chromosomes and were quadruploid, with four copies of DNA, while the cells in the G1 phase were still diploid, with only two copies of DNA. The fact that cells in the G2 phase showed basically the same spectra as cells in the G1 phase was interpreted to strongly support the hypothesis that

DNA is not detectable ("black"). The authors further argued that these results should have a very important impact on the interpretation of IR spectra of cancerous cells or tissues. The rationale of the experimental infrared spectroscopic data obtained so far from diseased cells and tissues (particularly cancerous material) have been summarized by Max Diem and coworkers as follows: "Abnormal cells divide more rapidly than normal cells; hence, the overall percentage of time cells spend in the S phase may be increased. Therefore, the spectral features due to DNA are enhanced in cancerous samples" (44). In other words, the IR spectral DNA/RNA signals available from diseased cells and tissues cannot be taken per se as specific monitors for diseased states in general or for cancer in particular. Not even the DNA/RNA signals of different "normal" cells can be expected to be constant. Many cell types (e.g., epithelial or hepatocytes) differ significantly with respect to cell proliferation and transcription rates under normal conditions. Furthermore, structural alterations related to secondary effects of the disease process, such as architectural changes of the tissues and/or infiltration of lymphocytes or leukocytes into the diseased tissue regions as a consequence of the body's immune response, have to be taken into account and, ideally, have to be separated from the spectral changes that are really due to the primary effects of disease development. Thus, though IR spectra of cells and tissues express a significant amount of structural and compositional information, "infrared pathology" as an accepted discipline in histopathology has yet to be established.

## 7.3   Diagnosis of Leukemia Cells

As an illustrative potential application in the clinical environment, recent efforts in the diagnosis of leukemia based on IR analysis of isolated and purified normal lymphocytes and malignant transformed leukemic cells will be discussed (18,81). Leukemia is the most frequent form of cancer of children and of adults below 30 years. While the global differentiation between normal lymphocytes and leukemia cells based on morphological criteria routinely used is easy, the differentiation between the numerous subforms or clones of leukemia is tedious, labor intensive, and accessible only to specialized laboratories. Chronic lymphocytic leukemia (CLL), the most prevalent form of leukemia in Western Europe and North America, is the accumulation of nonproliferating, mature-looking, but functionally immature lymphocytes in the peripheral blood, bone marrow, and lymph nodes. In order to evaluate the diagnostic potential of IR spectroscopy, peripheral mononuclear cells from normal individuals and extensively purified B cells from patients suffering from CLL were examined recently (18). In this study, cell suspensions were deposited on $BaF_2$ windows, dried down as thin films, and subjected to absorbance/transmission measurements. Representative infrared spectra of normal lymphocytes (control samples) and malignant transformed CLL cells are shown in Fig. 20A. For a better visualization of spectral

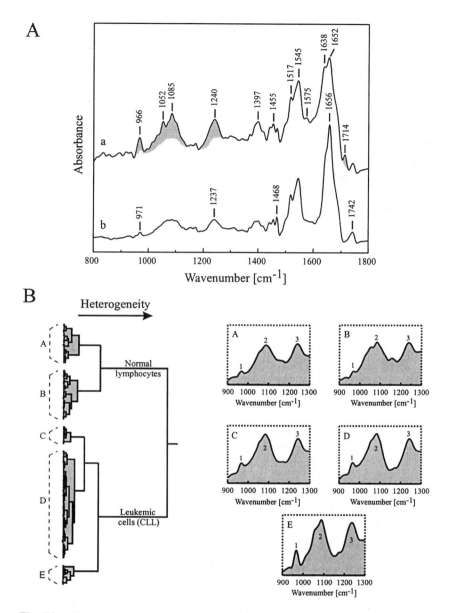

**Fig. 20** Differentiation between normal lymphocytes and chronic lymphocytic leukemic cells. (A) Typical infrared spectra of chronic lymphocytic leukemic cells (a) and normal lymphocytes (b). Spectra are represented as resolution-enhanced spectra using Fourier self-deconvolution techniques. The shaded areas of curve (a) indicate regions related to changes in nucleic acid absorptions. (B) Dendrogram of a hierarchical classification analysis on normal lymphocytes (23 individuals) and chronic lymphocytic leukemic cells (38 patients). Cluster analysis was performed using the first-derivative spectra between 900 and 1300 cm$^{-1}$ after area normalization. The five panels A, B, C, D, and E show the mean spectra of each cluster. (Adapted from Ref. 18.)

differences, the resolution-enhancement methodology of Fourier self-deconvolution was applied on the original spectra. Spectral differences over the entire range between 800 and 1800 cm$^{-1}$ are evident: (a) CLL cells show a more intense band at 1638 cm$^{-1}$ in the protein region, suggesting a higher content of β-sheet structure. (b) The lower intensity of the $CH_2$ deformation band at 1468 cm$^{-1}$ indicates that the lipid content is reduced in CLL cells, which is supported by the reduced intensity of the ester carbonyl band in the spectrum of the leukemic cells. From this the authors concluded that major differences exist between the membranes of the two cell types. (c) The characteristic carboxylate band at 1397 cm$^{-1}$ is increased in the spectra of CLL cells, indicating that the number of molecules carrying carboxylate groups (e.g., corresponding amino acid side chains of proteins) has been increased in these cells. (d) Major spectral differences are observed in the range 900–1300 cm$^{-1}$, where characteristic absorptions of DNA backbone vibrations are primarily observed. The shaded area in Fig. 20A illustrates that the integrated intensity in CLL cells is increased by a factor of almost 2, suggesting that the amount of detectable DNA (see Ref. 44 and Sec. 7.2) is drastically increased. (e) All spectra of leukemic cells reveal an infrared band at 1714 cm$^{-1}$, which is absent or less prominent in the spectra of normal cells. This band is known as a base-pairing marker of DNAs, providing additional evidence for an amount of detectable cell DNA in leukemic cells.

The spectroscopic features for normal and malignant transformed cells were also used by the authors as basis for a hierarchical cluster analysis performed on 177 spectra of normal monomolecular and leukemic cells from 23 normal donors and 38 patients with CLL. The spectral range between 900 and 1300 cm$^{-1}$ was chosen for classification, since it was excepted that this spectral range primarily reflects structural information originating from the DNA sugar and phosphate moieties. The dendrogram obtained in this way is shown in Fig. 20B. Two main clusters discriminating between normal and CLL cells are evident; both clusters are divided into distinct subgroups, labeled A and B (normal cells) and C, D, and E (CLL cells). The mean spectral characteristics of these five subclusters are shown in panels A, B, C, D, and E, respectively, with the majority of CLL cells being allocated to subcluster D. These results clearly demonstrate that IR spectroscopy can differentiate between normal and CLL cells and, interestingly, even between various CLL subclones.

### 7.4 Fourier Transform–Infrared Mapping and Imaging of Malignant Tissues

FT-IR microspectrometry has been described as particularly useful for the examination of small particles in a complex and heterogeneous biological environment (e.g., crystal deposits in tissues, silicone gel in human breast, white deposits in human kidney, and calcium oxalate in bladder tissues) (35–37), and infrared mapping (''chemical mapping'') techniques have been successfully applied to various tissues (33,34,44,76,77,80). However, only a limited number of FT-IR microspectroscopic applications on tissue thin sections have been published that address the problem of how to analyze different histological and diseased structures in tissues systematically by pattern recognition methodologies (64–67). Virtually no FT-Raman microspectroscopic application on the analysis of diseased tissue thin sections has been available from the literature up to now.

Infrared microspectrometry can be used to map a particular functional group across a sample as a function of concentration using the absorbances of its group frequency band. This procedure involves measuring the infrared absorbance in some narrow frequency interval. In this way, chemical information is obtained about the structure for which the functional group stands (33–37). Ratios of absorbances from pairs of functional groups can also be used for mapping experiments (e.g., from amide I and amide II, symmetric and antisymmetric stretching of $>CH_2$). This may particularly be useful when one of the bands can be treated as an internal standard in order to account for unavoidable variations in film thickness across the sample.

The functional-group mapping approach works best when individual spectral signatures, e.g., from small inclusion bodies in tissues, can be identified. The spatial characterization of diseased cells or areas of a tissue in a complex environment of different complex histological structures (which in itself are composed of hundreds of biological macromolecules) is, however, a much more complicated problem. Starting from the generally accepted postulation that disease processes in tissues or cells will induce changes in intrinsic biochemical composition and structure, any disease state should produce unique infrared spectroscopic pattern. However, these signatures, being complex in themselves, are generally superimposed by the various different structures additionally present. Thus, not a single spectral property (e.g., a functional group frequency or absorbance value) but, rather, a multiplicity of defined spectral lines or traits will be necessary to obtain sufficient information for the characterization of ''normal'' and ''diseased'' tissue structures in a given sample. To this end, the necessary amount of IR spectral traits that are diagnostic for the diseased state(s) should only be extractable by pattern recognition and optimal feature extraction techniques. To make this intelligible, an example from the literature on the characterization of human melanoma cancer by multivariate pattern recognition strategies will be given (64,65).

The following brief descriptions of experimental and evaluation strategies are taken from Ref. 65. Microcryotomed tissue specimens of human melanoma were mounted as 8 μm-thin slices on $CaF_2$ windows (2-mm thick microscopic slides). No fixation or embedding or staining procedures were applied. After FT-IR microscopic analysis, the thin sections were stained with hematoxylin eosin (HE) and analyzed light microscopically to confirm differentiation between carci-

1: corium (1a: stratum papillare, 1b: stratum reticulare)
2: melanotic parts of melanoma
3: amelanotic parts
4: tissue embedding medium

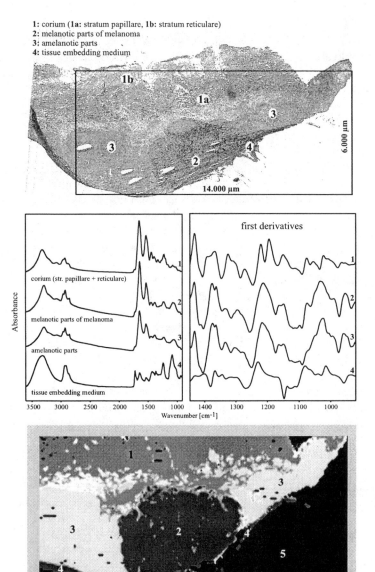

nomatosis and healthy regions of the tissue (see Fig. 21, upper panel). The FT-IR microscope used was equipped with a computer-controlled $x$, $y$-stage. The $4000 \times 5000$-$\mu$m rectangular regions of the samples were mapped with 150-$\mu$m steps in the $x$ and $y$ directions using a 100-$\mu$m aperture and a nominal resolution of 6 cm$^{-1}$ with 50 scans averaged per pixel spectrum. The first step of data evaluation was a "quality test" of raw data. This test included check for water vapor bands, search for spectra that were too low or too high in absorbance (as an indication of preparational artifacts such as microrips due to microtoming and/ or drying of the sample). All spectra that had passed this filter were subsequently baseline corrected and normalized to equal integral absorbance between 1100 and 1750 cm$^{-1}$. From these spectra, functional-group mapping was performed using single band intensities at 1653 cm$^{-1}$ (amide I) or at 2850 cm$^{-1}$ ($>CH_2$ stretching) or combinations of these (lipid/protein ratio). The investigation of various tissue samples by this methodology did, however, not lead to satisfactory results, and a secure differentiation between the healthy and malignant regions of the tissues from different patients generally failed. To improve discrimination capacity and to enhance image contrast, various different pattern recognition techniques, such as principal component analysis and artificial neuronal networks, turned out to be more useful (64,65). Four spectral regions were selected as input data for neural net training in this particular experiment (1372–1392, 1240–1355, 1168–1233, and 1084–1139 cm$^{-1}$). Network training was based on data sets of melanoma biopsy sections from four different patients. Light microscopy was used to confirm regions typical of distinct tissue areas and structures and to define the spectral pattern necessary for network training and simultaneous internal validation. Figure 21, lower panel, shows an IR image based on five spectral patterns that were predefined a priori: corium (1) composed of stratum papillare (papillary dermis, 1a) and stratum reticulare (reticular dermis, 1b), the melanotic parts of the melanoma (2), the amelanotic parts of the melanoma (3), small border regions of the tissue section contaminated by the tissue embedding medium, (4) and "zero pixel spectra," for which the quality test failed (5). For image reconstruction, a feed-forward artificial neural network (ANN) with three neuron layers (one input, one output, and one hidden layer), linked by shortcut connections,

---

**Fig. 21** Infrared imaging of a human carcinoma (melanoma) tissue thin section using IR microspectroscopic maps as a basis of image reconstruction. *Upper panel*: Light microscopic image of a stained thin section from a human melanoma tissue section. *Lower panel*: IR image reconstructed from an IR mapping data set of one patient using an artificial net that has been trained on the base of reference data obtained from four patients. The neural network has been trained to discriminate between the spectral patterns of four different tissue structures. The spectra 1,2,3, and 4 correspond to the means of spectral patterns that were found to be typical for the different tissue structures. (Adapted from Ref. 65.)

was used. The trained ANN was then challenged with an independent data set obtained from a biopsy section from a fifth patient. The output values of the ANN (ranging from 0, no activation = pattern not recognized, to 1, full activation, for one of the four patterns tested) were then used for image reconstruction and to scale image contrast (64,65). In this way, an IR image was obtained that is based on the scaled spatial distribution of spectral patterns, each being characteristic of a particular normal, benign, or malignant tissue structure. The results available from the literature demonstrate that FT-IR microspectrometry may have great potential for histological and pathological analysis in the future. The information contained in mid-infrared spectra is sufficient to distinguish between various tissue structures and pathologies. Functional-group mapping, however, is insufficient for secure tissue differentiation, and various pattern recognition methodologies are definitely required to effectively take advantage of the huge amount of spectral information available. Most of the studies published to date suffer from small numbers of patient materials and sometimes from insufficient feedback from and cooperation with the medical doctors. None of the publications address the problem of interpatient variance versus spectral differences between the various tissue structures quantitatively, and no reference databases necessary to make the new technology applicable in medical practice are available up to now.

## 8  CONCLUSIONS

### 8.1  Characterization of Microbial Cells

The main advantages of FT-IR and FT-Raman spectroscopy over conventional techniques are extreme rapidity compared to conventional techniques, uniform applicability to very diverse microorganisms, and a high specificity that allows differentiation even down to subspecies level. Last but not least, these vibrational spectroscopic techniques require only low amounts of consumables, are computer compatible, and may thus promote results and databases to be exchanged via data nets. While FT-IR spectroscopy seems to be established as a microbiological diagnostic tool and a first dedicated FT-IR instrument for microbial characterizations is already on the market, Raman spectroscopy, though potentially even more versatile, has not yet reached the stage at which routine biomedical applications are already possible. The strength of the FT-IR technique is its ability to conduct rapid epidemiological case studies and large screening experiments. Additional fields of applications are the detection of infection chains and the control of therapy, the maintenance of strain collections, and the differentiation of microorganisms from the environment, for which established systems are not yet available. In the food, water, and pharmaceutical industries, FT-IR and FT-Raman may also help improve microbiological quality control. For the control of biotechnological processes it might also be an alternative or complement to already-existing analytical tools. The prospects of vibrational microspectrometry for microbiological characterizations are very promising. This new technology may help to scale down the number of microbial cells needed to a few hundred, analyze mixed cultures, and detect light-microscopic and spectroscopic features of microorganisms simultaneously. Prospectively, the development of a fully automated system of microbiological analysis that combines detection, enumeration, and identification of microorganisms can be addressed.

## 8.2 Diagnosis of Diseased States from Body Fluids, Cells, and Tissues

Vibrational spectroscopy is proving to be a versatile technique for human cell, tissue, and body fluid characterizations. An increasing number of reports are available from the literature documenting the potentials of FT-IR and FT-Raman spectroscopy to discriminate between different tissue structures and body fluids and to characterize various disease states in humans. These currently include a number of different forms of pathological disorders (such as melanoma, colon carcinoma, cervical and breast cancer, and leukemia), various neurological disorders such as multiple sclerosis and Alzheimer's disease, and some cardiovascular diseases. Given the fact that FT-IR and FT-Raman have been used until now for less than 10 years by only a limited number of groups, the progress so far achieved is very impressive, and new, simple, inexpensive, effective, and very rapid methods for routine biomedical characterizations can be expected in the near future.

## 8.3 Combination of FT-IR and FT-Raman Data Evaluation Methodologies

A particularly attractive future item is the combination of FT-IR and FT-Raman spectroscopy for the analysis of complex biological samples, such as microbial cells, cells of the immune system, body fluids, and tissues. This should make it possible for the first time to establish a combined vibrational spectroscopic analysis of complex biological samples. Especially the full exploitation of the complementary information available by infrared and Raman spectroscopy may open new avenues for biomedical applications in the future. A particularly intriguing challenge is to elaborate computer-based pattern recognition techniques, which provide effective data reduction and optimal classification results from spectral data of biological samples collected in the near-, mid-, and far-infrared or by combining different spectroscopic techniques, such as FT-IR and FT-Raman spectroscopy. Modern neural network analysis strategies and effective optimal-feature extraction algorithms are the possible candidates for addressing this future task of optimal-information extraction from combinations of vibrational spectra.

## ACKNOWLEDGMENTS

The excellent technical assistance of Angelika Brauer in preparing this manuscript is gratefully acknowledged. The assistance of Maren Stämmler and Angelika Brauer in sample preparations, measurements, and evaluations is much appreciated. The technical drawings of Figs. 5A and B have been provided by Bruker Optics GmbH, Ettlingen, Germany. The original spectra of Figs. 6 and 19 have been kindly provided by H. Fabian and S. Keller, respectively. Their discussions in producing these figures are acknowledged. Figures 18, 20, and 21 have been adapted from the literature with direct help of the authors (C.P. Schultz, P. Lasch). Their help is gratefully acknowledged.

## ABBREVIATIONS

| | |
|---|---|
| ANN | artificial neural network |
| A/T | absorbance/transmission |
| ATR | attenuated total reflectance |
| DTGS | deuterated triglycine sulfate |
| FT | Fourier transform |
| FT-IR | Fourier transform–infrared |
| IR | infrared |
| LPS | lipopolysaccharides |
| MIR | mid-infrared |
| NIR | near-infrared |
| PCA | principal component analysis |
| PHB | poly-$\beta$-hydroxybutyric acid |
| PHF | poly-$\beta$-hydroxybutyric fatty acid |

## REFERENCES

1.  D Helm, H Labischinski, G Schallehn, D Naumann. J Gen Microbiol 137:69–79, 1991.
2.  D Naumann, D Helm, H Labischinski. Nature 351:81–82, 1991.
3.  D Naumann, H Labischinski, P Giesbrecht. In: WH Nelson, ed. Instrumental Methods for Rapid Microbiological Analysis. Weinheim, Germany: VCH, 1991, pp 43–96.
4.  D Helm, H Labischinski, D Naumann. J Microbiol Methods 14:127–142, 1991.
5.  D Helm, D Naumann. FEMS Microbiol Lett 126:75–80, 1995.
6.  D Naumann, CP Schultz, D Helm. In: HH Mantsch, D Chapman, eds. Infrared Spectroscopy of Biomolecules. New York: Wiley-Liss, 1996, pp 279–310.
7.  G Seltmann, W Voigt, W Beer. Epidemiol Infect 113:411–424, 1994.
8.  MC Curk, F Peladan, JC Hubert. FEMS Microbiol Lett 123:241–248, 1994.
9.  R Goodacre, ÉM Timmins, PJ Rooney, JJ Rowland, DB Kell. FEMS Microbiol Lett 140:233–239, 1996.

10. MK Winson, R Goodacre, AM Woodward, ÉM Timmins, A Jones, BK Alsberg, JJ Rowland, DB Kell. Anal Chim Acta 348:273–282, 1997.

11. D Lefier, D Hirst, C Holt, AG Williams. FEMS Microbiol Lett 147:45–50, 1997.

12. H Haag, HU Gremlich, R Bergmann, JJ Sanglier. J Microbiol Methods 27:157–163, 1996.

13. C Holt, D Hirst, A Sutherland, F MacDonald. Appl Environm Microbiol 61:377–378, 1995.

14. W Bouhedja, GD Sockalingum, P Pina, P Allouch, C Bloy, R Labia, JM Millot, M. Manfait. FEBS Lett 412:39–42, 1997.

15. GD Sockalingum, W Bouhedja, P Pina, P Allouch, C Mandray, R Labia, JM Millot, M Manfait. Biochem Biophys Res Comm 232:240–246, 1997.

16. DC Malins, NL Polissar, SJ Gunselman. Proc Natl Acad Sci USA 94:259–264, 1997.

17. HH Eysel, M Jackson, A Nikulin, RL Somorjai, GTD Thomson, HH Mantsch. Biospectroscopy 3:161–167, 1997.

18. CP Schultz, KZ Liu, JB Johnston, HH Mantsch. Leukemia Research 20:649–655, 1996.

19. LP Choo, DL Wetzel, WC Halliday, M Jackson, SM LeVine, HH Mantsch. Biophys J 71:1672–1679, 1996.

20. M Jackson, HH Mantsch. In: HH Mantsch, D Chapman, eds. Infrared Spectroscopy of Biomolecules. New York: Wiley-Liss, 1996, pp 311–340.

21. M Jackson, HH Mantsch. In: RJH Clark, RE Hester, eds. Advances in Spectroscopy. Vol. 25: Biomedical Applications of Spectroscopy. Chichester, UK: Wiley, 1996, pp 185–215.

22. LP Choo, JR Mansfield, N Pizzi, RL Somorjai, M Jackson, WC Halliday, HH Mantsch. Biospectroscopy 1:141–148, 1995.

23. H Fabian, M Jackson, L Murphy, PH Watson, I Fichtner, HH Mantsch. Biospectroscopy 1:37–46, 1995.

24. CP Schultz, HH Eysel, HH Mantsch, M Jackson. J Phys Chem 100:6845–6848, 1996.

25. BR Wood, MA Quinn, B Tait, M Ashdown, T Hislop, M Romeo, D McNaughton. Biospectroscopy 4:75–91, 1998.

26. GJ Puppels, J Greve. In:RJH Clark, RE Hester, eds. Advances in Spectroscopy. Vol. 25:Biomedical Applications of Spectroscopy. Chichester, UK: Wiley, 1996, pp 1–47.

27. GJ Puppels, HSP Garritsen, GMJ Segers-Nolten, FFM de Mul, J Greve. Biophys J 60:1046–1056, 1991.

28. D Naumann, S Keller, D Helm, C Schultz, B Schrader. J Mol Struct 347:399–406, 1995.

29. R Manoharan, E Ghiamati, RA Dalterio, KA Britton, WH Nelson, JF Sperry. J Microbiol Meth 11:1–15, 1990.

30. CJ Frank, RL McCreery, DCB Redd. Anal Chem 67:777–783, 1995.

31. DCB Redd, ZC Feng, KT Yue, TS Gansler. Appl Spectrosc 47:787, 1993.

32. SR Hawi, WB Campbell, A Kajdacsy-Balla, R Murphy, F Adar, K Nithipatikom. Cancer Lett 110:35–40, 1996.

33. SM LeVine, DLB Wetzel. Appl Spectrosc Rev 28:385–412, 1993.

34. SM LeVine, DL Wetzel, AJ Eilert. Int J Devl Neuroscience 12:275–288, 1994.

35. L Estepa-Maurice, C Hennequin, C Marfisi, C Bader, B Lacour, M Daudon. Am J Clin Path 105:576–582, 1996.
36. EN Lewis, AM Gorbach, C Marcott, IW Levin. Appl Spectrosc 50:263–269, 1996.
37. LH Kidder, VF Kalasinsky, JL Luke, IW Levin, EN Lewis. Nature Med 3:235–237, 1997.
38. HJ Rogers, HR Perkins, JB Ward. Microbial Cell Walls and Membranes. New York: Chapman and Hall, 1980.
39. PJ Brennan, H Nikaido. Ann Rev Biochem 64:29–63, 1995.
40. FM Klis. Yeast 10:851–869, 1994.
41. V Farkas. In: AH Rose, JS Harrison, eds. The Yeasts. Vol. 3. New York: Academic Press, 1989, pp 317–366.
42. HJ Phaff. In: CP Kurtzman, JW Fell, eds. The Yeasts. A Taxonomic Study. Amsterdam: Elsevier Science, 1998, pp 45–47.
43. C Ratledge, CT Evans. In: AH Rose, JS Harrison, eds. The Yeasts. Vol. 3. New York: Academic Press, 1989, pp 367–455.
44. M Diem, S Boydston-White, L Chiriboga. Appl Spectrosc 53:148A–161A, 1999.
45. D Li-Vien, NB Colthup, WG Fateley, JG Grasselli. The Handbook of Infrared and Raman Characteristic Frequencies of Organic Molecules. Boston: Academic Press, 1991.
46. B Schrader, ed. Infrared and Raman Spectroscopy. Weinheim, Germany: VCH, 1995.
47. NB Colthup, LH Daly, SE Wiberley. Introduction to Infrared and Raman Spectroscopy. New York: Academic Press, 1975.
48. FS Parker. Applications of Infrared, Raman, and Resonance Raman Spectroscopy in Biochemistry. New York: Plenum Press, 1983.
49. PR Griffiths, JA de Haseth. Fourier Transform Infrared Spectrometry. New York: Wiley-Interscience, 1986.
50. H Fabian, C Schultz, D Naumann, O Landt, U Hahn, W Saenger. J Mol Biol 232:967–981, 1993.
51. B Schrader, A Hoffmann, A Simon, J Sawatzki. Vibr Spectrosc 1:239–250, 1991.
52. B Schrader, B Dippel, I Erb, S Keller, T Löchte, H Schulz, E Tatsch, S Wessel. J Mol Struct 480–481:21–32, 1999.
53. B Schrader, B Dippel, S Fendel, S Keller, T Löchte, M Riedl, R Schulte, E Tatsch. J Mol Struct 408–409:23–31, 1997.
54. B Schrader, B Dippel, S Fendel, R Freis, S Keller, T Löchte, M Riedl, E Tatsch, P Hildebrandt. In: HH Mantsch, M Jackson, eds. Infrared Spectroscopy: New Tool in Medicine. Proceedings of SPIE, Vol. 3257. Bellingham, WA: 1998, pp 66–71.
55. D Naumann. In: JA de Haseth, ed. Fourier Transform Spectroscopy: 11th International Conference AIP Conference Proceedings 430. New York: Woodbury, 1998, pp 96–109.
56. D Naumann. In: HH Mantsch, M Jackson, eds. Infrared Spectroscopy: New Tool in Medicine. Proceedings of SPIE, Vol. 3257. Bellingham, WA: 1998, pp 245–257.
57. NJ Harrick. Internal Reflection Spectroscopy. Ossining, NY: Harrick Scientific, 1979.
58. BS Everitt. Cluster Analysis. Toronto: Wiley, 1993.
59. BS Everitt. Statistical Methods for Medical Investigations. Toronto: Wiley, 1994.

60. BFJ Manly. Multivariate Statistical Methods. A Primer. New York: Chapman & Hall, 1996.
61. GJ McLachlan. Discriminant Analysis and Statistical Pattern Recognition. New York: Wiley, 1992.
62. JE Dayhoff. Neuronal Network Architectures. New York: Van Nostrand Reinhold, 1990.
63. J Zupan, J Gasteiger. Neuronal Networks for Chemists. Weinheim, Germany: VCH, 1993.
64. P Lasch, W Wäsche, WJ McCarthy, G Müller, D Naumann. In: HH Mantsch, M Jackson, eds. Infrared Spectroscopy: New Tool in Medicine. Proceedings of SPIE, Vol. 3257. Bellingham, WA: 1998, pp 187–198.
65. P Lasch, D Naumann. Cell Mol Biol 44:189–202, 1998.
66. LM McIntosh, JR Mansfield, AN Crowson, HH Mantsch, M Jackson. Biospectroscopy 5:265–275, 1999.
67. CP Schultz, HH Mantsch. Cell Mol Biol 44:203–210, 1998.
68. F Inagaki, M Tasumi, T Miyazawa. J Raman Spectrosc 3:335–343, 1975.
69. A Cao, J Liquier, E Taillandier. In: B Schrader, ed. Infrared and Raman Spectroscopy of Biomolecules. Weinheim, Germany: Verlag Chemie, 1995, pp 344–372.
70. WL Peticolas, WL Kubasek, GA Thomas, M Tsuboi. In: TG Spiro, ed. Biological Applications of Raman Spectroscopy. New York: Wiley, 1987, pp 81–134.
71. DG Cameron, A Martin, HH Mantsch. Science 219:180–182, 1983.
72. DG Cameron, A Martin, DJ Moffatt, HH Mantsch. Biochemistry 24:4355–4359, 1985.
73. DJ Moore, M Wyrwa, CP Reboulleau, R Mendelsohn. Biochemistry 32:6281–6287, 1993.
74. DJ Moore, R Mendelsohn. Biochemistry 33:4080–4085, 1994.
75. C Schultz, D Naumann. FEBS Lett 294:43–46, 1991.
76. L Chiriboga, P Xie, H Yee, V Vigorita, D Zarou, D Zakim, M Diem. Biospectroscopy 4:47–53, 1998.
77. L Chiriboga, P Xie, H Yee, V Vigorita, D Zarou, D Zakim, M Diem. Biospectroscopy 4:55–59, 1998.
78. PGL Andrus, RD Strickland. Biospectroscopy 4:37–46, 1998.
79. Y Fukuyama, S Yoshida, S Yanagisawa, M Shimizu. Biospectroscopy 5:117–126, 1999.
80. CP Schultz, KZ Liu, PD Kerr, HH Mantsch. Oncol Res 10:277–286, 1998.
81. KZ Liu, CP Schultz, RM Mohammad, AM Al-Katib, JB Johnston, HH Mantsch. Cancer Lett 127:185–193, 1998.
82. GH Werner, J Früh, F Keller, H Greger, RL Somorjai, B Dolenko, M Otto, D Böker. In: HH Mantsch, M Jackson, eds. Infrared Spectroscopy: New Tool in Medicine. Proceedings of SPIE, Vol. 3257. Bellingham, WA: 1998, pp 35–41.
83. D Naumann. In: Encyclopedia of Analytical Chemistry: Instrumentation and Applications. Chichester, UK: Wiley, in press

# COLOR PLATES

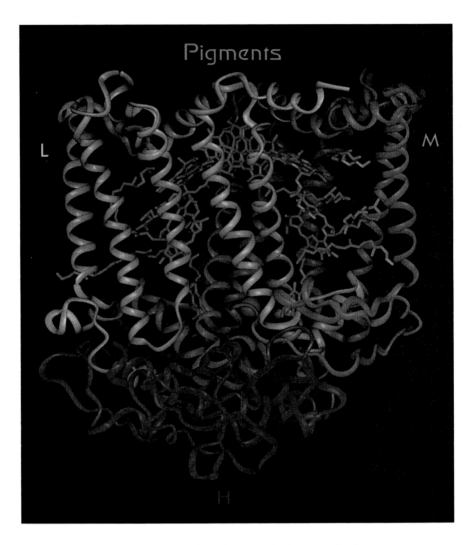

**Fig. 6.23a** Structural model of the bacterial photosynthetic reaction center of *Rhodobacter sphaeroides*.

**Fig. 6.29** Structural model of the H-ras p21 protein with bound GTP.

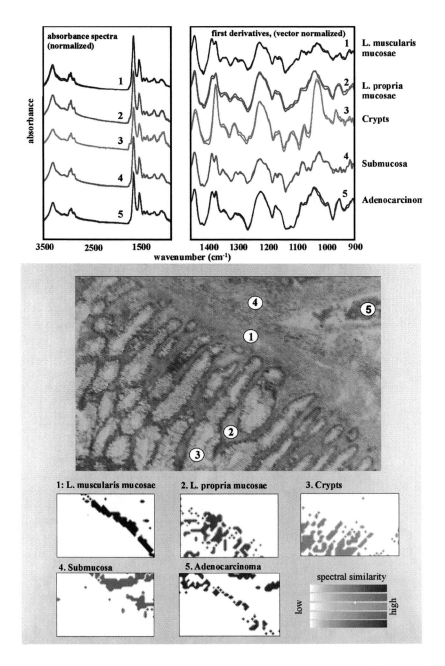

**Fig. 10.9** IR-image reconstruction on the basis of FT-IR spectra applying to artificial neural networks. Upper panel: Absorbance spectra of 5 reference classes (left), first derivatives in the fingerprint region (right). Lower panel: microscopic photomicrograph after HE-staining and below, ANN images after classification.

# 10

# Use of Artificial Neural Networks in Biomedical Diagnosis

**Jürgen Schmitt and Thomas Udelhoven**
*University of Trier, Trier, Germany*

## 1 INTRODUCTION

During recent decades, Fourier transform infrared (FT-IR) spectroscopy has provided a great deal of valuable information on biophysics and biochemistry by investigating isolated biomolecules such as proteins (1), lipids (2), carbohydrates (3), and nucleic acids (4). Within the last five years, attention has begun to focus on the task of transferring the gained knowledge and dedicated instrumentation to studies of complex biological matters as intact microbial or mammalian cells, tissues, and body fluids. In principle, this approach is intended to be taken one step further, where FT-IR and Raman spectroscopy is used for the detection of biochemical abnormalities, associated with or causing a disease. Efforts to develop vibrational spectroscopic techniques, which are suited for biodiagnostics, are now discussed in the research community (5).

Due to their spectral complexity, biological samples require more sophisticated methods for characterization, differentiation, classification, and quantification. This is additionally strengthened by the fact that the transfer of elaborated spectroscopic techniques into routine analysis in microbiology, medicine, and related fields of application strongly depends on its ease of use, the reliability, and the interface to man. Therefore, new methods for data evaluation are required. Self-learning systems such as artificial neural networks (ANNs), with appropriate techniques of feature selection and data pretreatment, are one group of promising techniques. They hold potential to establish large-scale libraries for classification (6), to model nonlinear quantitative relationships, or to reconstruct FT-IR microscopic images, to name just a few applications.

Artificial neural networks are simplified models of the central nervous system and stem from the field of artificial intelligence. They are networks of highly interconnected neural computing elements called "neurons" that have the ability to respond to input stimuli and to learn to adapt to the environment. Aleksander and Morton (7) offered a definition, which is true for most neural networks:

> A neural network is a massively parallel distributed processor that has a natural propensity for storing experiential knowledge and making it available for use. It resembles the brain in two respects. (I) Knowledge is acquired by the network through a learning process. (II) Interneuron connection strengths known as synaptic weights are used to store the knowledge.

The procedure used to perform the learning process is called a *learning algorithm*, the function of which is to modify the synaptic weights of the network. The modification of the synaptic weights provides the traditional method for the design of the neural networks. However, it is also possible for a neural network to modify its own topology, which is motivated by the fact that neurons in the human brain can die and that new synaptic connections can grow.

Artificial neural networks currently used in applied sciences have little in common with their biological counterparts, although research is still being carried out to establish links to neurobiology and artificial intelligence. Although fascinating, a more pragmatic point of view defines ANNs in a first approximation as nonparametric nonlinear regression estimators (8). Nonparametric methods are those methods that are not based on the a priori assumption of a specific model form. And ANNs allow one to estimate relationships between one or several input variables, called *independent variables* or *descriptors*, and one or several output variables, called *dependent variables* or *responses*.

With this concept, ANNs provide a new approach to computing and data evaluation, and they support new, emerging instrumentation and challenging applications of vibrational spectroscopy in medicine, microbiology, and biotechnology. This chapter focuses on biomedical applications of vibrational spectroscopy, where classification and pattern recognition is one of the dominating tasks. In that context, emphasis is put on the application of supervised learning and feed-forward artificial neural networks representing the kernel of ANN methods used.

## 1.1 Benefits of Neural Networks

It is apparent that a neural network derives its computing power through, first, its massively parallel distributed structure and, second, its ability to learn and therefore to generalize. Those two information-processing capabilities make it possible for neural networks to solve complex and large-scale problems that are currently intractable. In practice, however, neural networks cannot provide the solution by working individually. Rather, they need to be integrated into a consistent system engineering approach. Specifically, it is often useful for a complex problem of interest to be broken down into a smaller number of relatively simple tasks.

The use of neural networks offers the following useful properties and capabilities:

*Adaptivity*: Artificial neural networks have a built-in capability to adapt their synaptic weights to changes in the surrounding environment. In particular, a neural network trained to operate in a specific environment can easily be retrained to deal with minor changes in the operating conditions of the environment. In general, the more adaptive we make a system, all the time ensuring that system conditions remain stable, the more robust its performance will be.

*Nonlinearity*: An artificial neuron can be linear or nonlinear. A neural network, made up of an interconnection of nonlinear neurons, is itself nonlinear. Moreover, the nonlinearity is of a special kind, in the sense that it is distributed throughout the network. Nonlinearity is a highly important property, particularly if the underlying physical mechanism responsible for generation of the input signal is inherently nonlinear.

*Input–output mapping*: The paradigm of learning called "supervised learning" involves the modification of the neurons by applying a set of labeled training samples. Each example consists of a unique input signal (e.g., part of a spectrum) and a corresponding desired response. The network is presented with an example picked at random from the set, and the synaptic weights (free parameters) of the network are modified to minimize the difference between the desired response and the actual response of the network produced by the input signal. The training of the network is repeated for many examples in the set until the network reaches a steady state, where there are no further significant changes in the synaptic weights.

*Uniformity of analysis and design*: Basically, neural networks show universality as information processors. The same notation is used in all domains. This feature manifests itself in the way that the neurons in one form or another represent an ingredient to all neural networks. This commonality makes it possible to share theories of learning algorithms in different applications of neural networks. Another feature is that modular networks can be built through a seamless integration of modules.

*Contextual information*: Knowledge is represented by the very structure and activation state of the neural network. Every neuron in the network is potentially affected by the global activity of all neurons in the network. Consequently, contextual information is dealt with naturally by a neural network.

*Fault tolerance*: If a neuron or its connecting links are damaged, the recall of a stored pattern is impaired in quality. However, due to the distributed nature of information stored in the network, the damage has to be extensive before the overall response of the network degrades seriously. Thus, the neural network exhibits a graceful degradation rather than a catastrophic failure. The empirical evidence for robust computation was described by Kerlizin and Vallet (9). This performance is especially valuable when spectral noise, e.g., from monitoring processes, is increased and alters spectral features.

*Implementability*: The massively parallel nature of a neural network makes it potentially fast for the computation of certain tasks and well suited for implementation of very large-scale integrated (VLSI) technology. The particular beneficial virtue is that it provides a means of combining established networks with the measurement device (e.g., the spectrometer) and readout device (computer) to form an integrated and adaptable system for biomedical applications. This advantage leads to the desired potential for highly automatic and rapid processing.

## 1.2  Historical Background

The modern era of neural networks was founded by the pioneering work of McCulloch and Pitts in 1943 (10). McCulloch, a psychiatrist and neuroanatomist, spent some 20 years of research on the representation of an event in the nervous system. Pitts, a mathematician, joined McCulloch in 1942 (11). In their paper, published in 1943, they described a logical calculus of neural networks that united the studies of neurophysiology and mathematical logic, and they gained widespread interest. Their model of a neuron was assumed to follow an ''all-or-none'' law. With a sufficient number of such simple units and synaptic connections, which were set synchronously, they could show that a neural network is able to compute in principle any computable function. It is now generally agreed that this was a milestone in the development of neural networks and artificial intelligence. Networks of these units were then thought to be representative models of the brain.

Another early result was added in 1949 by Hebb (12), who was one of the first to suggest a plausible process for neural learning. With his book *The Organization of Behavior* (12), he presented an explicit statement of a physiological learning rule for synaptic modification. He proposed that the connectivity of the brain is continually changing, as an organism learns differing functional tasks, and that neural assemblies are created by such changes. His ''postulate of learning'' stated, that the effectiveness of a variable synapse between two neurons is increased by repeated activation of one neuron by the other across that synapse. Even today, many of the learning models used by researchers are some forms of ''Hebbian'' learning. Hebb also reasoned that many distributed cell assemblies were used to represent knowledge. This was called ''the connectionist architecture'' a few years later.

Rosenblatt, a physiologist, accomplished the research with his work on perceptrons (13,14). He believed that the brain functioned as a learning associator that computed classifications in response to stimuli. He developed several variations of networks he called *perceptrons* and studied different forms of learning.

The neural network possibilities introduced by perceptrons brought about a flurry of research during the 1960s. Widrow and Hoff introduced the least mean square algorithm and used it to formulate the *adaline* (adaptive linear element) (15). The difference between the adaline and the perceptron lies in the training procedure. One of the first trainable layered neural networks with multiple adaptive elements was the *madaline* (multiadaline) structure proposed by Widrow in 1962 (16). In 1965 Nilsson presented his book *Learning Machines* (17), a classic textbook for pattern classification with neural networks. During this period, it seemed as if neural networks could do anything. In 1969, this atmosphere was seriously damped by Minsky and Papert (18), who published an elegant mathematical analysis of the computational capabilities and limitations of perceptrons. They essentially showed the limits of what logical functions of simple perceptrons could compute. After a phase of intense research and enthusiasm, this was the final blow in killing funding for further research on neural networks. Consequently, most research efforts were reduced or terminated. For the next 10 years, only stalwart researchers continued their work, including James Anderson (19), Teuvo Kohonen (20), Stephen Grossberg (21), Bernard Widrow (22), Christian von der Malsburg (23), John Hopfield (24), S-i Amari (25), and a few others, with substantial contributions. Quiet research was conducted over about 10 years. One of the most important developments of neural network research during that time was the discovery of a learning algorithm to adjust the weights in multilayer feed-forward networks (multilayer perceptrons). This algorithm is known as *back-propagation*, since the weights are adjusted from the output layer backward, layer by layer, to reduce the output errors. This method was discovered at different times by Werbos (1974) (26), Parker (1985) (27), and Rumelhart et al. (1986) (28). The 1986 book *Parallel Distributed Processing: Explorations in the Microstructures of Cognition*, edited by Rumelhart (29), has been a major influence on the use of back-propagation learning, which has emerged as the most popular learning algorithm for the training of multilayer perceptrons. This development opened the way for more general ANN computing by overcoming the limitations suffered by single-layer perceptrons. Perhaps more than any other publication, the 1982 paper by Hopfield (30) and the 1986 two-volume book by Rumelhart and McClelland (29) were the most influential publications responsible for the resurgence of interest in neural networks in the 1980s. Neural networks have certainly come a long way from the early beginning with McCulloch and Pitts and have established themselves as an interdisciplinary subject with deep roots in neuroscience, psychology, mathematics, the physical sciences, and informatics. Needless to say that the beneficial developments in computer science and hardware possibilities contributed much to the recent status and will still help to continue the growth in theory, design, and applications.

## 2   PRINCIPLES OF NEURAL NETWORKS

The multilayer perceptron (MLP) and radial basis function networks (RBFs), whose principles derive from the theory of functional approximation, are two important examples of feed-forward artificial neural networks in spectroscopic applications. Fundamental for the information processing and operation of both types of networks is the neuron. After demonstration of the information flux within a single neuron, the basic concepts of the MLP and of RBF networks will be presented.

### 2.1   Neurons as Basic Elements of Artificial Neural Networks

Although ANN architectures differ in several characteristic ways, a typical neuron is basically an information cell that produces an output if the cumulative effect of the input stimuli exceeds a threshold value. In Fig. 1, the information flux in a neuron is illustrated.

The basic elements can be defined as:

1. *Input signals, synapses, or connecting links*, which are characterized by a weight or strength of their own. An input signal $x_i$ is multiplied by the synaptic weight $w_{ji}$, where the first subscript refers to the neuron in question and the second subscript refers to the input end of the synapse to which the weight refers. The synaptic weight of an artificial neuron may lie in a range that includes negative and positive values. The input $x_i$ values can be real, binary, or bipolar. The weights, which model the synaptic neural connections in biological nets, act to either increase (excitatory input) or decrease (inhibitory input) the input signals of a neuron. The weights are usually assumed to be real (positive for excitatory and negative for inhibitory links).

2. *A propagation function* for summing the input signals, weighted by the respective synapses of the neuron. The operations described here

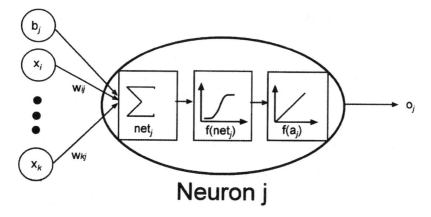

**Fig. 1** Schematic representation of one neuron in a multilayer perceptron (MLP). Abbreviations: see text.

constitute a linear combiner. Quite often, an externally applied bias is used in the model, denoted by $\theta_j$. The bias has the effect of increasing or lowering the net input of the activation function, depending on whether it is positive or negative, respectively.

$$\text{net}_j = \sum_{i=0}^{n} x_i w_{ij} + \theta_j \tag{1}$$

3.  An *activation function* for limiting the amplitude of the output of a neuron. The activation function is also referred to as a *squashing function*, where the permissible amplitude range of the output signal is limited (squashed) to some finite value. The normalized output range is usually written as the closed unit interval [0,1] or alternatively [−1,1].

$$a_j = f(\text{net}_j) \tag{2}$$

The most commonly used activation functions are:

The logistic function:

$$f_{\log}(x) = \frac{1}{1 + e^{-x}} \tag{3}$$

The hyperbolic tangent function:

$$f_{\tanh}(x) = \frac{e^x - e^{-x}}{e^x + e^{-x}} \tag{4}$$

The linear function:

$$f_{\text{lin}}(x) = x \tag{5}$$

4.  The output of the activation function is applied to an output function
    of the neuron $f(a_j)$. In most cases, the *identity function* is used for that
    purpose:

$$o_j = a_j \tag{6}$$

A neuron such as that depicted in Fig. 1 in the output layer of a MLP can
compute a second- or even a higher-class membership grouping of input stimuli
patterns. The components of the input vector $x$, given by $x_1, x_2, x_3, \ldots x_n$ are the
individual stimuli values. In case of a two-layer network (one input and one
output layer), they correspond to the features or attributes of the input (for example, an absorbance value at a distinct wavenumber). In case of three-layer MLPs,
$x$ corresponds to the output vector of the hidden units. In a two-class example, the
output might correspond to class in a diseased/nondiseased classification system.
Computing the membership for inputs, a single neuron behaves like a predicate
that computes class membership by the weight vector value $w$, which defines a
separating hyperplane in $n$-dimensional space. This hyperplane divides the input
values into two regions or classes, class 1 or class 2, depending on whether $f(\text{net}_j)$
is greater or less than zero or another threshold value $y$:

IF $f(\text{net}_j) > y$, then $x$ belongs to class 1.
IF $f(\text{net}) < y$, then $x$ belongs to class 2.

## 2.2  The Multilayer Perceptron (MLP)

In the previous section, we discussed the information flux in one single neuron.
Now we concentrate on the representation capabilities of a complete MLP. Two-
layer MLPs, consisting of an input layer and an output layer, which have only
one layer of trainable weights, have a number of limitations in terms of the range
of functions they can represent. A more general mapping is possible with MLPs
having several layers of adaptive weights. Figure 2 illustrates a typical structure of a fully connected three-layer MLP that can be used for a classification
system. In our earlier example, several classes were coded in only one output
unit. If many classes have to be considered, it is more convenient to use a 1-
of-$n$ coding instead, where each class is represented by one output unit. Exactly one position of the output vector is coded 1 (corresponding to the position
of the class to which the input pattern belongs), whereas all other positions are
coded 0.

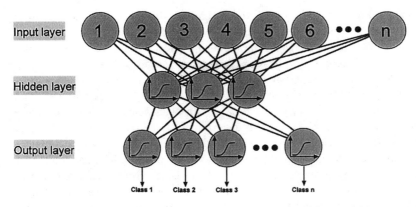

**Fig. 2** Structure of a fully connected multilayer perceptron (MLP) with three layers.

An explicit expression for a fully connected three-layer MLP, consisting of one input, one hidden, and one output layer, is:

$$o_k(\vec{x}) = f\left(\theta_k + \sum_{j=1}^{M} w_{kj} f'\left(\sum_{i=1}^{n} w_{ji} x_i + \theta_j\right) + \sum_{i=1}^{n} w_{ki} x_i\right) \tag{7}$$

$$k = 1, 2, \ldots, K$$

where:

$o_k(\vec{x})$ = activation of output neuron $k$ on the input $x = x_1, x_2, \ldots, x_n$
$w_{ji}$ = a weight in the hidden layer, going from input unit $i$ to hidden unit $j$
$w_{kj}$ = a weight in the output layer, going from hidden unit $j$ to output unit $k$
$w_{ki}$ = a weight in the output layer, going from input unit $i$ to output unit $k$ (shortcut connections)
$\theta_{kj}$ = biases of output and hidden units
$f, f'$ = activation functions for the output and hidden layers
$n, M, K$ = number of input, hidden and output units

A multilayer feed-forward network delineates a function that represents a mapping from a real valued vector domain to a real valued output space. Without hidden layers, an ANN represents a linear mapping of the input variables, which can eventually be transformed by a sigmoidal activation function. The insertion of a hidden layer gives the network the capability to approximate arbitrarily well any functional continuous mapping from one finite-dimensional space with defined bounds to another, provided that nonlinear squashing functions (e.g., sigmoid activation functions) are used and the number of hidden units is sufficiently large. Wu and Massart (31) remark that the term *sufficiently large* is relative in this context, since it strongly depends on the ratio of the number of examples to the number of trainable weights. The higher the ratio of the number of samples to the number of weights, the better is the generalization ability of the network.

If the underlying function is continuous and has defined bounds, a three-layer network is capable of learning this function, but this tells nothing about the generalization capability of the network. In classification problems, such a network provides universal nonlinear discriminant functions, which can be used to model posterior probabilities of class memberships (32). Only a serious violation of a compact domain or nondeterministic functions can hinder any multilayer feed-forward net from learning a function. Otherwise, ANNs are capable of reproducing almost any kind of mapping in the context of multivariate classification or calibration. For three-layer ANNs even violations from the continuity of a function may be tolerable to some degree (33). A higher-order network topology with two or more hidden layers can help to reduce the total number of weights and biases, if the total number of hidden units is held constant. Masters (33) argues that gradient-directed optimization training algorithms are dependent on the degree to which the gradient remains unchanged as the weights in the neural network vary. More than one hidden layer, through which errors must be back-propagated, can result in unstable error gradients and in higher probabilities that the algorithm gets stuck in a local minimum. It was shown that the predictive ability of an ANN decreases proportionally with the number of layers, since errors are accumulated through the network layers (34). Therefore, in the vast majority of spectroscopic applications, feed-forward networks with just one or even no hidden layer should be used.

In spectroscopic applications, the contribution to the response of the input variables is often linear, and nonlinear effects often coincide with small deviations from a linear solution. Such a mixed contribution to the response can be handled with a three-layer feed-forward net with shortcut connections; i.e., the input layer is connected not only to the hidden layer but also to the output layer. Wise et al. (35) call this network type a direct linear feed-through (DLF) neural network. The linear part of the problem is solved by the shortcut connections, whereas the nonlinear part is handled by the hidden layer. It is reported that shortcut connections help to reduce training time when the descriptors are heterogeneous (36).

## 2.3 Commonly Used Learning Algorithms for Multilayer Perceptrons

The adjustable parameters of MLPs, the weights $w$ and biases $b$, are determined by an iterative procedure called *training* or *learning*. They are first given initial random values, when the training starts. In the first step, a forward pass is performed through the ANN with a set of training samples and known and desired output response. At the end of the pass, the magnitude of the error between acquired and predicted output responses is computed and used to adjust all weights of the ANN. This procedure is called a *training cycle* or *epoch*. A new cycle is then initiated with optimized parameters and repeated until adequate performance and error levels are achieved (Fig. 3).

Important for all training algorithms is the scaling of each input variable so that the training process can start within the active range of the nonlinear activation functions. A simple min-max procedure can be applied for that purpose. The scaling parameters must be determined on the training samples. For MLPs having differentiable activation functions, such as a sigmoidal function, there exists a computationally efficient method, called *error back-propagation*, for finding the derivatives of an error function with respect to the weights and biases in the network.

All the training methods described in this study are modifications of standard back-propagation, where the weight adjustment during training is based on the absolute value or the sign of the first derivative of the error with respect to each weight. Most training algorithms can be formulated in a so-called on-line form or, alternatively, in a batch mode. In the *on-line mode*, a weight change is performed after each pattern has been propagated through the network. In the *batch mode*, all patterns are propagated through the net and the individual net errors are summed. At the end, a weight change is performed based on that cumulated error signal.

The most commonly used learning algorithm in MLPs is back-propagation. The numerous variants of this algorithm, of which a few examples will be de-

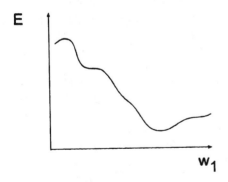

**Fig. 3** Example error curve made by adjusting the value of only one single weight in an MLP. In practice, all weights are adjusted simultaneously during the training process to minimize the overall network error. $E$ = error, $w_1$ = single weight.

scribed shortly, are all based on a steepest-descent minimization of the total error, calculated by the sum of absolute differences or the squared differences between network outputs and the desired outputs for each output unit. The on-line form of back-propagation can be formulated as:

$$\Delta_p w_{ji}(t) = \eta o_{pi} \delta_{pj}[+\alpha \Delta_p w_{ji}(t-1)][-dw_{ji}(t-1)][+q] \tag{8}$$

$$\delta_{pj} = \left\{ \begin{array}{l} f(\text{net}_{pj})(t_{pj} - o_{pj}) \\ f(\text{net}_{pj}) \sum_k \delta_{pk} wk_{jk} \end{array} \right\} \tag{9}$$

where:

$f_t$ = activation function
$\text{net}_{pj}$ = net input of a single unit $j$ from a pattern $p$
$\Delta w_{ji}(t)$ = actual weight change at time $t$
$d$ = weight-decay term
$\alpha$ = momentum term
$t_{pi}$ = target output from unit $i$ by pattern $p$
$o_{pi}$ = output from unit $i$ by pattern $p$
$q$ = flat-spot elimination term
$\eta$ = learning rate

Back-propagation in its elementary form suffers from a very slow convergence rate, and it has a high likelihood of oscillating around the minimum. The optional terms in parentheses can help to circumvent some of these limitations. A weight-decay term, controlled by the parameter $p$, punishes a too-large weight change in the last weight update at $t - 1$. An argument for the consideration of such a term is that large weights lead to steeper and irregular error surfaces. This can result in uncontrolled jumps and oscillations during the learning process. In contrast, an optional momentum term, controlled by parameter $\alpha$, introduces the old weight change as a parameter for the computation of the new weight change momentum term. A larger weight update helps one to travel through flat plateaus and steep valleys of the error curve. Back-propagation with a momentum term is also titled "conjugate gradient descent" (29). Finally, a flat-spot elimination term $q$ is a constant value added to the weight changes. With it, so-called "flat spots" of the error surface can be traversed relative rapidly, since the step size is inversely related to the local derivation of the error surface.

For many problems, the Quickprop algorithm (37) provides faster training as back-propagation. Quickprop assumes that the error surface locally is quadratic. An attempt is made to "jump" in one step from the current position of the error surface directly into the minimum of the parabola, after computing the first gradient with regular back-propagation, by means of

$$\Delta w_{ji}(t) = \frac{\dfrac{\partial E}{\partial w_{ji}}(t)}{\dfrac{\partial E}{\partial w_{ji}}(t-1) - \dfrac{\partial E}{\partial w_{ji}(t)}} \Delta w_{ji}(t-1) \tag{10}$$

where:

$\Delta w_{ji}(t)$ = actual weight change

$\dfrac{\partial E}{\partial w_{ji}(t)}$ = slope of the error function in the direction $w_{ij}$

In extreme situations, the terms $\partial E/\partial w_{ji}(t-1)$ and $\partial E/\partial w_{ji}(t)$ are close to equal. In that case, the formula produces an infinitely wide step; or in even worse situations, a division through zero occurs. Therefore, the maximal step width during the training step is restricted by the maximal growth parameter $\mu$:

$$|\Delta w_{ji}(t)| \leq \mu|\Delta w_{ji}(t-1)| \tag{11}$$

Braun (38) suggests as an additional modification of Quickprop the small weight-decay term $d*w_{ji}(t)$ to inhibit an excessive weight growth.

An alternative to Quickprop for enhancing the speed of the back-propagation algorithm is the introduction of adaptive parameters for the learning rate and the momentum rate of each weight in the MLP. This is the basis of resilient back-propagation (Rprop). Unlike standard back-propagation, Rprop does not consider the harmful influence of the absolute value of the partial derivative for the calculation of the weight changes, but considers only the sign of the derivative to indicate the direction of the weight update (38):

$$\Delta w_{ji}(t) = \begin{cases} -\Delta_{ji}(t), & \text{if } \dfrac{\partial E}{\partial w_{ji}}(t) > 0 \\[2ex] +\Delta_{ji}(t), & \text{if } \dfrac{\partial E}{\partial w_{ji}}(t) < 0 \\[2ex] 0, & \text{otherwise} \end{cases} \tag{12}$$

$$\Delta_{ji}(t) = \begin{cases} \eta^{+}\Delta_{ji}(t-1), & \text{if } \dfrac{\partial E}{\partial w_{ji}}(t-1)*\dfrac{\partial E}{\partial w_{ji}}(t) > 0 \\[2ex] \eta^{-}\Delta_{ji}(t-1), & \text{if } \dfrac{\partial E}{\partial w_{ji}}(t-1)*\dfrac{\partial E}{\partial w_{ji}}(t) < 0 \\[2ex] \Delta_{ji}(t-1), & \text{otherwise} \end{cases} \tag{13}$$

where $0 < \eta^{-} < 1 < \eta^{+}$

The size of the weight change is determined exclusively by the weight-specific update value $\Delta_{ji}(t)$. Every time the partial derivative of the corresponding weight $w_{ji}$ changes its sign, the update value $\Delta_{ji}(t)$ is decreased by the factor $\eta^-$, since the last update was too large and the algorithm jumped over a local minimum. On the other hand, if the derivative retains its sign, the update value is slightly increased by the factor $\eta^+$ in order to accelerate convergence in shallow regions.

All modifications of the gradient descent method provide significantly faster training than the original back-propagation approach (39). Riedmiller and Braun (40,41) and Riedmiller (42) tested standard back-propagation, Quickprop, Super SAB, and resilient back-propagation regarding their sensitivity to the learning parameters. Rprop was most insensitive to the initial value of the learning rate. Braun (38) found in different benchmark experiments (10-5-10 encoder, 6-bit parity, 12-2-12 encoder, spiral) considerably faster training rates for Quickprop and Rprop than for back-propagation. Tuning of the learning parameters of Quickprop was significantly less robust compared to Rprop, a result confirmed in the present study (see later). Quickprop accomplishes fast training when the error surface locally can be approximated by an open parabola and the changes of the weights can be performed independently (39). Since the sigmoid activation function, especially in its saturation range, exhibits properties different from a quadratic polynomial, the error surface locally does not adequately fit a quadratic polynomial function, and the error function may begin to oscillate chaotically. The algorithm has proved to be very sensitive to the parameters $\mu$ and $d$. Zell (39) recommends values for the maximal-growth parameter $\mu$ not larger than 1.75–2.22.

A class of training algorithms that allow even faster training, exists such as with the modifications of the gradient descent method described here. These techniques are based on second-order optimization methods, based on the determination or approximation of the Hessian matrix of partial derivatives of the cost functions. These techniques, which involve the Newton–Raphson method, the Levenberg–Marquardt algorithms, and conjugate gradient optimization, are discussed by Despagne and Massart (36).

## 2.4  Escape from Local Minima

In spite of their relative training speed, none of the foregoing training algorithms guarantees that the global error minimum can be found, since the algorithms can be trapped in local minima on irregular error surfaces. Several ways to escape such local minima exist. With methods such as simulated annealing or genetic algorithms, the weights are adjusted not deterministically but randomly under the assumption of a defined probability distribution (33). Simulated annealing (Fig. 4) is analogous to the annealing of red-hot metal. At high temperatures, the metal atoms have high kinetic energies and they are allowed to build arbitrary atomic configurations. When temperature slowly decreases, the energy of the

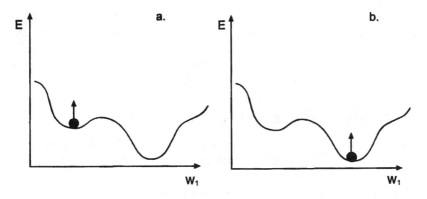

**Fig. 4** Principle of simulated annealing. At a given temperature (indicted by the length of the arrows), the probability of escaping from the first minimum is high (a), but low for the second minimum (b). Temperature is successively lowered during training. $E$ = error, $w_1$ = single weight. (Adapted from Ref. 39.)

atoms declines and finally the system consolidates in a state of minimal energy. If, however, the metal were cooled suddenly, the microstructure would be locked into a random unstable state. At a given temperature, the probability distribution of the energy of the system can be determined by the Boltzmann factor $\exp(-E/(kT))$, where $E$ is the temperature of the system, $k$ is the Boltzmann constant, and $T$ is the absolute temperature (39). Transferred to a neural network, the rough structure of the global state of the network is found at high-temperature terms after initialization of the network. At high temperatures, the tendency to find a state of low energy is very small, but the equilibrium state of the network can be found faster at low temperatures. When the temperature term is lowered during the training, the network reacts on smaller-energy differences and finds smaller minima within the rough structure of the error surface (43).

Simulated annealing can be performed in optimization by randomly perturbing the weights of the neural network and keeping track of the lowest error function value for each randomized set of variables. The temperature is modeled by the standard deviation of the random-number generator. After many tries, the set that produced the best function value is designed to be the center about which perturbations will take place for the next temperature. A more complex method for avoiding local minima are genetic algorithms, which require a number of user-defined parameters to define population size and evolution mode.

Even though random-search and genetic algorithms are guaranteed to find a global minimum, it must be kept in mind that computation time is very high. To overcome this problem, a random search can be combined with classical-learning algorithms. The idea is to start network training with a random-search strategy to optimize the initial set of weights. Conjugate gradient learning or an alternative technique then rapidly finds the next local minima. Once there, one tries to escape from this position to a lower point, again with random search, however, at a lower temperature. This alteration is repeated until an escape from a local minimum is impossible (33).

## 2.5   Radial Basis Function (RBF) Networks

Radial basis function networks are the second type of ANNs often applied in supervised classification problems. In contrast to MLPs, they show a more local modeling behavior. This is an important feature in situations where extrapolations take place. A fully connected RBF network (with shortcut connections between input and output units) is able to represent the following set of approximations:

$$o_k(\vec{x}) = f\left(\theta_k + \sum_{j=1}^{M} w_{kj} h\left(|\vec{x} - \vec{t}_j|, \theta_j\right) + \sum_{i=1}^{n} w_{ki} x_i\right) \qquad k = 1, 2, \ldots, K \quad (14)$$

where:

$|\vec{x} - \vec{t}_j|$ = Euclidean distance between an input vector and the centers of the RBFs

$h$ = basis function

In RBF applications, as in MLPs other types of activation functions are used, the most common being the Gaussian function:

$$f_{\text{gauss}}(x) = e^{x^2/2\sigma^2} \qquad (15)$$

where $\sigma$ denotes a parameter whose value controls the smoothness properties of the interpolating function. The Gaussian function is a localized basis function with the property that $f_{\text{gauss}} \to 0$ as $|x| \to \infty$. The radial basis functions $h$ are applied to the distance, usually taken to be Euclidean, between the input vector and a so-called prototype vector (centers), rather than to a scalar product of input vector and weight vector that is transformed by a monotonic (e.g., sigmoidal) function used in MLPs. The prototype vectors represent the different classes of the training set. The bias of the output units and a direct connection between input and output layers can be used to improve the approximation quality. Shortcut connections, although considered in the last term of Eq. 14, do not, however, improve the learning capability of the network when the Gaussian function is used as a basis function. The biases of the hidden units are used to modify the characteristics of the function $h$. Figure 5 illustrates the principle topology of an RBF network.

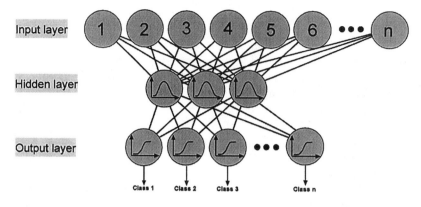

Input layer

Hidden layer

Output layer

Class 1    Class 2    Class 3    Class n

**Fig. 5** Structure of an RBF network with the Gaussian function as a basis function (without shortcut connections).

An RBF network generally consists of a three-layer feed-forward structure. In the hidden layer, localized functions are used as activation functions instead of sigmoid or other nonlinear squashing functions. The output units transform linear combinations of the activations from the hidden layer to generate the outputs.

The RBF network training is performed in several stages. The weights of the links between input and hidden layers, respectively between hidden and output layers, are determined in a different manner from that in an MLP. In the simplest case, the links between the input and hidden layers are assigned to evenly distributed centers $\vec{t}$, which represent input vectors from the training pattern set. All classes to which the training patterns can belong should be characterized by the centers. Suitable centers can alternatively be found by unsupervised training techniques without having to perform a full nonlinear optimization of the network. A commonly used technique is the self-organizing method of Kohonen feature maps (44). Since the determination of the centers is strictly unsupervised, the discriminatory class information cannot be exploited. In other words, there is no way to distinguish relevant from irrelevant inputs when the basis function centers are chosen using the input data alone (45).

Since the topology of such a network generally consists of three layers, the weights of the final layer (the $w_{kj}$ in Eq. 14) and the bias $\theta_k$ can be computed directly without error back-propagation. The weights are found by solving a linear equation system with $M$ equations and $M$ unknown parameters:

$$\sum_{j=1}^{M} w_{kj} h_j(|\vec{x} - \vec{t}_j|) = y_j \qquad (16)$$

The computation requires a reasonable choice of centers $\vec{i}$. Weights of shortcut connections between input and output layers cannot be computed directly, but must be found by network training.

After the determination of the RBF weights, the network error can be further decreased using gradient descent. Several parameter groups can be trained at different learning rates, including centers, links between hidden and output layers, links between input and output layers, and bias of the hidden and output layers.

As shown earlier, the hidden layer representation of MLPs and of RBF networks is different. Whereas MLPs separate class information by using hidden units that form hyperplanes in the input space, RBFs model the class distributions by local kernel functions (Fig. 6). Due to the localized nature of RBFs, both network types react differently in extrapolation situations.

Large Euclidean distances in the hidden layer of RBF networks point to a spectrum that is not well represented by a prototype vector. A local kernel function, such as the Gaussian function, will produce a low activation in this case. Only if the input vector is close to a prototype will the unit have a high activation. The separation of the class distributions by local kernel functions, therefore, allows an identification of outliers. In the hidden-unit representation of MLPs, the hidden units represent hyperplanes in the input space where an extrapolation may take place with unpredictable consequences for the behavior of the MLP. The question of the robustness of MLPs and RBF networks with respect to random errors in the input space is discussed controversially in the literature. Whereas Derks et al. (34) found in a Monte Carlo simulation that RBF networks generally have better robustness properties than MLPs, this finding could

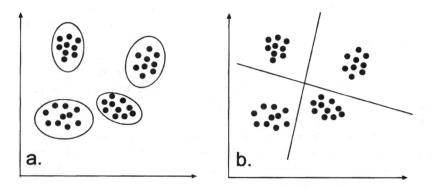

**Fig. 6** Class representation by RBF networks with local kernel functions (a) and by MLPs with hyperplanes (b). (Adapted from Ref. 32.)

not be confirmed by Faber and Kowalski (46) for the modeling of the relation between the physical and mechanical properties of polyethylene terephthalate yarns. In classification problems, Sánchez et al. (47) found comparable results for RBFs and MLPs for small deviations in the input space. However, for higher deviations, the RBF networks detected faster that the input space was changing with respect to the data set used for training the network. For both network types, the samples in the data set should be evenly spread over the expected range of data variability to achieve the best generalization properties. Some useful techniques to design the training sets are discussed by Wu et al. (48) and Wu and Massart (31).

Our own experience with RBF networks in spectroscopic applications is that their differential depth is often lower than for MLPs when many classes have to separated and the number of training samples is limited. On the other hand, the different extrapolation behavior of RBFs may be important in situations where the training samples do not cover the whole possible variance range in the population. This limits extrapolation and is valuable for integration in modular ANNs.

## 3 PRACTICAL ASPECTS OF DATA PREPROCESSING AND ARTIFICIAL NEURAL NETWORK DESIGN

Artificial neural networks represent ideal tools for spectroscopic data elucidation when no a priori indication concerning the nature of the relationship is given or can be deduced. This is especially true in biomedical applications of vibrational spectroscopy, where the spectra are complex and often exhibit little variance between the different objects. The success of an ANN modeling strongly depends on the quality of the spectra, the data preprocessing, and the design of the ANNs. Concerning this, some important aspects are described next.

### 3.1 Spectral Quality

A very important factor controlling the success or failure of ANN models is the data preprocessing aspect. At the very beginning of the measurement, factors such as detector linearity, deviations from the Lambert–Beer law by sample inhomogenities, different particle size, or large particles that lead to scattering effects should be considered. In general, tests of the sampling and measurement procedure are necessary to minimize the sampling variance and should lead to a standardized measurement protocol. For biological investigations and samples from a medical background, this can imply the definition of the storage and conservation temperatures. For mammal sera, we determined significant spectral differences due to a storage temperature of $-20°C$ or $-80°C$, resulting in different classifications. Water vapor during sample handling may also result in bandshape deviations and increasing noise. Problems with scattering, such as encountered in spectroscopic investigations of larger microorganisms like yeasts, actinomycetes, or fungi, are further sources of band distortions and baseline deviations as well as diminishing spectral quality and reproducibility.

Some of these effects can sometimes be corrected using the first or second derivatives of the absorbance spectra. The first derivative especially has been shown to be an appropriate tool to reduce baseline shifts and additionally enhance spectral features compared to absorbance spectra of bacteria (49). The negative aspect in the computation of derivatives is the reduction of the signal-to-noise ratio; thus, thresholds for noise, water vapor content, or further effects should be predefined and integrated in an automatic spectral quality proof procedure. Multiplicative scatter correction (MSC) and standard normal variate (SNV) correction are spectral processing methods used primarily in near-infrared spectroscopy to remove scatter effects, but sometimes nonlinearity is introduced with such methods. And SNV represents such a nonlinear transformation in the wavelength space (50).

## 3.2 Detection of Outliers

The design of a representative ANN model depends on a good representation of the training data set. Before all data are applied for modeling, irregularities or outliers have to be detected by means of numerical and statistical outlier detection methods (36). Although a variety of methods exists, the positive identification of an outlier requires knowledge of the sample and the data-acquisition process. Often, an interaction with the person who is in charge of the measurement or sample preparation is giving valuable further information. It is beyond the scope of this chapter to discuss all methods for outlier detection. Usually, the first choice is the visual inspection of the spectra or, for quantitative measurements, the $X$–$Y$ relationship in a scatter plot. This can be enhanced by the inspection of the score plots of the first principal component, after performing a principal component analysis (51). An alternative approach for the a priori detection is a statistical test such as Grubb's test based on Rao's statistic (52). Centner et al. (52) proposed a procedure based on the development of partial least squares (PLS) and leave-one-out cross-validation models after flagging possible outliers with Grubb's test. Their reason for using the cross-validation is to see if the model is influenced by the outlier. This is certainly a practicable procedure, especially for deterministically constructed models, i.e., discriminant analysis, but it is limited for use with ANNs, as discussed in Sec. 3.6. A final decision about the influence of outliers in ANN models can often be made only on the basis of a comparison of the results of the models built with and without the flagged samples, provided that an exactly identical initialization for the weights is used.

## 3.3   Feature Selection and Data Reduction

It is often difficult to specify an effective architecture for a given problem speci-
fication with an MLP. The appropriate number of input and hidden units must
be found by experimentation.

Adding new features in the input layer can lead, beyond a certain point,
to a reduction in the performance of the classification problem. This is a consequence of the effects of dimensionality coupled with a limited size of the data
set. In spectroscopic applications, where often a limited number of spectra with
high dimensionality is available, this can provide a very poor representation of
the mapping. Reducing the number of input variables can help to improve the
performance for a given data set, although information is being discarded. Too
many features may give extremely poor results. Whereas MLPs can be trained
to ignore irrelevant inputs to some extent and obtain accurate results with a relatively small number of hidden units, this is more difficult for RBF networks,
especially if there are input variables that have significant variance but that play
little role in determining the appropriate output variables. The number of hyper-
cubes (radial basis functions) needed to fill out a compact region of a $d$-dimen-
sional space grows exponentially with $d$ (53). A large number of basis functions,
on the other hand, results in the demand for large training sets to ensure that the
network parameters can be properly determined.

Thus, feature selection and data reduction are crucial preprocessing stages
in ANN development and in pattern recognition in general. In spectroscopy, this
means wavelength selection, which has emerged as an area of intense research
within recent years. Redundant and irrelevant data are cached inside the raw data.
A classifier that uses all features will perform worse than a classifier that relies
only on features that maximize interclass differences and maximize intraclass
differences. Besides, the latter are less complex. Consequently, the ANN model
would have fewer parameters and the dimension of the pattern vector is reduced.
Hence, two motivations for wavelength selection can be mentioned: (a) reduction
of the complexity of the model, (b) improvement of the accuracy of the model.
Reviewing the literature in relation to the topic of wavelength selection in pattern
classification and quantification, it can be observed that there is no such thing
as a general feature selector. In many cases, the selection procedure is dependent
on the application domain and the available data. An appropriate framework for
multidimensional data is given by Devijer and Kittler (54).

In wavelength selection procedures, a clear distinction is made between
the selection criterion, the search strategy, and the stopping condition. Class sepa-
rability is used as the basic selection criterion, and it can be analyzed with statistical methods of error probability, interclass distance, probabilistic distance, probabilistic dependence, and entropy (54). Univariate and multivariate selection

criterions have to be distinguished. One univariate selection criterion, which is robust and easy to compute, is the calculation of an $F$-value, which describes the ratio of the variance within and between the groups. The search strategy in this case is just a ranking of the $F$-values. As stopping condition, a user-defined number of wavelengths with the highest $F$-values can be chosen (6). Alternatively, a user-defined threshold for $F$-values can be defined. An inherent property of any univariate selection criterion is that highly intercorrelated features are not detected. That means the best and second best feature could be linearly dependent but would both be classified as highly discriminant by the algorithm. Thus, nonindependent wavelengths, which tend to form wavelength regions rather than single wavenumbers, are extracted by this algorithm. Single features may be ranked with a low discriminant power, although the multivariate combinations of these wavelengths might prove to be very discriminative. Univariate methods are advisable in certain situations, e.g., if the number of available samples is small and multivariate selection criteria therefore become unstable or if computation time is a limiting factor.

To circumvent the problems of univariate methods and to further reduce the number of features without loosing discriminant power, multivariate approaches, e.g., multiple covariance analysis, can be applied. The principle is to join iteratively all unselected wavelengths (features) with all already-selected variables and to compute a partial $F$-value. That candidate with the highest score is joined to the set of already selected features (the so-called control variables) and retracted from the unselected-features set. The search procedure is repeated until the desired number of selected features is reached or the maximum of all calculated criteria falls below a predefined threshold. The search strategy used here is also referred as *sequential forward selection*. Alternatively, a *sequential backward elimination* can be applied, where at the beginning all features minus 1 are considered as control variables. Variables with the lowest value for the multivariate criterion are then excluded successively from that set (32). It must be mentioned here that the computational costs for both search strategies, used in the context of multivariate criteria, are much higher than in the foregoing method (54).

Alternatively, other recent strategies of wavelength selection comprise the application of genetic algorithms (55). The idea behind this is a search algorithm that abandons the attempt at an exhaustive search and introduces a stochastic search method. They are touted as global optimizers capable of locating the best set of wavelengths for a given large-scale optimization problem. Their application is critically discussed by Brenchley et al. (56). For further reading, the reader is addressed to Jouan-Rimbaud et al. (57,58) Soskic et al. (59), Norinder (60), Kubinyi (61), Bellon-Maurel et al. (62), Wu et al. (63), and Walzak and Massart (64).

## 3.4 Feature Extraction and Compression

When all existing $D$ features are recombined to yield $d$ new features, then we are dealing with feature extraction. Hence, a mapping is performed that transforms any original $D$-dimensional vector into a new $d$-dimensional vector. The mapping conserves or even enhances the discriminatory power while simultaneously reducing the dimension of the vector.

The most popular method for feature extraction and data compression in chemometrics is principal component analysis (PCA). With ANNs, the scores from just a few PCs that describe the variance in the input space sufficiently accurately can be used as input variables for neural networks (60,65). However, one must be aware of some theoretical limitations. The principle components represent sources of successively maximized variance in the data, which lose stability as the number of samples relative to the number of features decreases. This should be kept in mind, since the transformation matrix must be solely derived from the training set, which is eventually biased when it does not cover the whole variance of the population. This argument is also true for other multivariate techniques. In addition, PCA as a linear projection method fails to preserve the structure of a nonlinear data set, and it completely neglects target data information. This implies that the solution from the PCA may be significantly suboptimal to solve the problem (32).

Other methods for feature extraction include Hadamard transforms (67) and wavelet transformation (68). Wavelet transformation is able to describe optimally local information of the spectra, whereas Fourier and PCA decomposition have global and not local character. Information about local structures may be valuable when these segments describe the nonlinearity of the data (64).

## 3.5 Normalization of the Data

For spectral comparability after measurement, normalization is usually applied to the data set. This can be either a vector or a min-max normalization. For the RBF application that uses spherically symmetrical basis functions, it is important to normalize the input variables so that they span equivalent ranges. This is a consequence that the activation of a basis function is determined by a Euclidean distance (32). If autoscaling is used as normalization method, all input variables are centered around a mean of 0 having a unique variance of 1. The effects of autoscaling are a decrease in the distances between the classes, in direction of the largest scattering within the groups, and a spherization of the groups (69). In the MLP applications, there is no need to autoscale the input variables. The biases act as offsets in the model. Artificial neural network training is not based on variance-covariance maximization. Consequently, it is not necessary to scale the input variables to unit variance, since the biases act as offsets in an ANN (66). If the data ranges of the input features are different, PCA or PLS require autoscaling.

## 3.6 Data Splitting and Validation

Overtraining is considered one of the most prominent reasons that neural networks fail their task in some instances. It is a reflection of the training set's not being totally representative of the population. Especially in situations where linear phenomena are modeled by highly flexible ANNs and the number of samples available is a limiting factor, a rapid overfitting of the measurement noise may occur if the topology of the ANN is not carefully designed (66). The danger of overfitting also increases when the training set is small relative to the number of hidden units. A first indication for overfitting is an error increase in the validation set, also often referred as the *monitoring set*. Since the monitoring set is in this way directly involved in deciding when to stop training, it cannot be assumed to be independent of model construction. Rather, a model validation should provide an unbiased indication of the generalization property of a model (70). Therefore, an error from an independent test set is considered a much better estimator of the ANN generalization ability than the validation error (71) (Fig. 7).

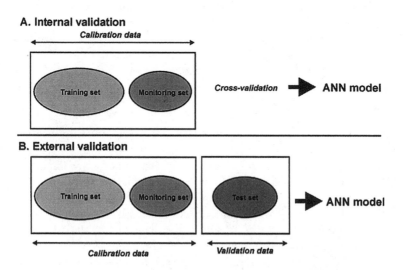

**Fig. 7** Internal and external validation in ANN modeling. A: Internal modeling, estimation of the modeling power, no test of generalization ability possible, risk of overfitting. B: External modeling, estimation of the modeling power, and assessment of the generalization ability. (Adapted from Ref. 36.)

If only a few training samples are available, a leave-$k$-out cross-validation, also called *jackknifing*, is a generally accepted method for estimating the generalization property of a model, provided the adding or deleting of $k$ members from the training set has only a little effect on the model parameters. Discriminant analysis, for example, is deterministically constructed, and for a fixed sample set it will always yield exactly the same model parameters. Since it is based on aggregate information, such as the mean or variance, a small perturbation of the training set will usually have a minor effect on the model. This is distinct for ANNs, since the network error as function of the weights is usually very irregular. The consequence is that different random starting weights can result in completely different results. Even a trivial change in the training set, such as adding or removing one member, may push the weights into an entirely different region, even with identical initializations of the weights (33). A rotation $k$-leave-out procedure can therefore give only a rough indication of the ability of successively trained networks to generalize to new independent observations, but it does not validate a global model (33,36).

## 3.7 Determination of the Number of Hidden Units in Artificial Neural Networks

As was shown in Sec. 2.2, under normal conditions there is no need to use more than one hidden layer in an MLP. Radial basis function networks always consist of only one hidden layer. Another question, however, is how many hidden units should to use in that layer. There exists no universally accepted rule to answer this question (72). Derks et al. (34) found that an RBF network needs significantly more hidden units (base functions) than an MLP to solve a problem with a comparable degree of accuracy. The number of hidden units is an extremely important optimizing problem in all neural network applications. The more adjustable weights an ANN has, the higher its potential information-processing capability (33). Since such a network is able to approximate any nonlinear function, it will also have the tendency to learn insignificant or incorrect aspects of the training set, which are not representative of the whole population. The problem of overfitting grows when the number of patterns in the training set is small relative to the number of adjustable weights in the network. In general, increasing the number of examples increases the overall performance of a neural network, but only up to a certain point (72,73). If the number of input units and output units is fixed, the number of hidden units mainly determines the absolute number of trainable links in the network. If too few hidden nodes are chosen, the building of a correct internal representation is not possible, whereas too many hidden units lead to higher training times and overfitting problems (72).

There are very different approaches in the literature to finding the optimal number of hidden nodes. Masters (33) recommends as a rough guideline to use a geometric pyramid rule for an appropriate number of hidden nodes. Applied to a three-layer ANN with $n$ input neurons and $m$ output neurons, the number of hidden units would be calculated as round(sqrt($mn$)). According to Maren et al. (73), the maximum number of hidden nodes in an ANN with $i$ input and $o$ output units should be of the order $o(i + 1)$. Another possibility for fixing the number of hidden units is to compare different independent models with different numbers of hidden units, starting with a minimal network. The topology of that model is chosen for the final model that minimizes the error in the training set or validation set (74,75). Despagne and Massart (66) summarize some other approaches for determining the number of required hidden units with an initially oversized ANN by an orthogonalization of the hidden layer output matrix with singular value decomposition.

The cascade correlation (CC) technique (76) is able to determine its own neural network topology. For that purpose it uses a specific network architecture, where hidden units are successively added as individual layers to the ANN so that each unit feeds into subsequent units. Beginning with a two-layer fully connected network trained with back-propagation, Quickprop, or Rprop, single hidden units are successively inserted when the network error no longer decreases. They are selected from a pool of so-called "candidate units." The goal is to maximize the covariance $C$ between the activation of the candidate units and the residual error of the net:

$$C = \sum_o \left| \sum_p (v_{po} - \bar{v}_o)(E_{po} - \bar{E}_o) \right| \tag{17}$$

where:

$\bar{v}_o$ = average activation of a candidate unit
$\bar{E}_o$ = average error of an output unit over all patterns $p$
$v_{po}$ = value of a candidate unit
$E_{po}$ = residual error for an output unit

The maximization of $C$ proceeds by training all the links leading to a candidate unit by gradient ascent using:

$$\delta_p = \sum_o \sigma_o \, (E_{po} - \bar{E}_j) f'_p \tag{18}$$

$$\frac{\partial C}{\partial w_i} = \sum_p \delta_p I_{pi} \tag{19}$$

where:

$\sigma_o$ = sign of the correlation between the output of the candidate unit and the residual error at output unit $o$

$I_{pi}$ = value of an input unit or a hidden unit $i$ for the pattern $p$
$f'_p$ = derivative of the activation function

The candidate unit with the highest correlation is added to the net, where its incoming weight is frozen to obtain a new permanent-feature detector. By generating links to the output units, the candidate unit is changed into a hidden unit. Different to the incoming weights, new links between hidden and output units are not fixed and can be adjusted during the next iterations. The procedure is repeated until a sufficiently small error is achieved. The CC algorithm is not only a training algorithm but itself defines network architecture dependent on the properties in the training pattern set (Fig. 8). The behavior of the algorithms depends on some user-defined parameters whose values are chosen empirically.

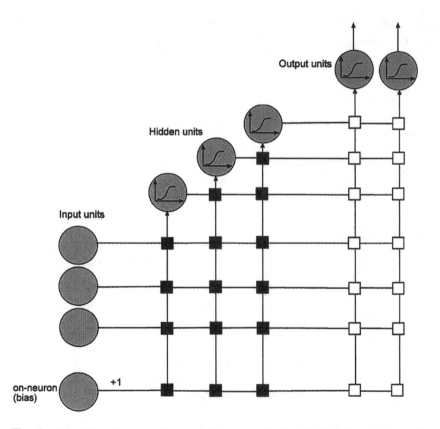

**Fig. 8** Principle of a cascade correlation network. Small black boxes denote fixed weights, white boxes trainable weights. (Adapted from Ref. 39.)

Such CC networks have the capability of incremental learning, which helps to solve the so-called "moving target problem" of back-propagation networks that attempt to adjust all the weights simultaneously (39,77). Here, each hidden unit tries together with all the other hidden units to adjust its weights to become a "useful" detector for a pattern fragment. This task is difficult, since all weights are adjusted simultaneously. In CC networks, the weight of only one hidden neuron at a time is adjusted, when it is added to the network. Its incoming weights are "frozen" so that it becomes a permanent feature for a pattern fragment. With this strategy, it is reported that CC is able to learn many problems significantly faster than all other variants of back-propagation.

### 3.8 Hierarchical System of Neural Networks

Due to the complexity of many classification problems in the field of spectroscopy, the development of a hierarchical and modular ANN, consisting of a prior top level and several following subclassification levels, may be more valuable than one monolithic ANN (6). Hierarchically organized neural networks can be considered a compromise between one multioutput network and many single-output networks. This approach also provides more flexibility in terms of a practical extension to further classes and it is more practical in building up a comprehensive ANN interpretation (78). However, since such a hierarchy provides information on different levels, from general characteristics to more specific features, there is a strong dependence of sublevel networks upon the upper-level network performance. In a recent study (79), multilayer perceptrons (MLPs) and radial basis function networks (RBFs) were used to establish a hierarchical classification system for bacterial FT-IR spectra used for their identification in the laboratory (6,79) (see Sec. 4.2).

### 4 APPLICATIONS

### 4.1 Image Reconstruction from FT-IR Microspectroscopy

Infrared imaging is providing to be an attractive new technique that integrates infrared spectroscopy with high-definition digital imaging. For histopathology, cytology, and microbiology this technique generates spatially resolved spectroscopic information on microstructures of the given sample. The chemical substructures in the field of interest can then be related directly to the microscopic image of the tissue or to microbial colonies (80–83).

The applicability of IR imaging in real-life samples has been demonstrated, e.g., with the in situ detection of β-amyloid protein deposits within a slice of human Alzheimer's diseased brain tissue (84) or the detection of silicon gel compounds in human breast tissue after implant rupture (85). With the continuing development and commercialization of focal-plane array detectors, (FPAs) a further key technology was established and integrated into IR imaging systems (86). Such FPAs are capable of rapidly generating images from unstained biological tissues and cells based on MCT or InSb arrays with 64 × 64 detector elements. Data collection is similar to that for conventional FT-IR studies, except that the entire image planes replace measurements from single detector elements. Each interferogram corresponding to a pixel in the data set is apodized, Fourier transformed, and then divided by averaged air background spectram to yield transmittance images, which are then converted to absorbance. Each complete data set contains 4096 (64 × 64) individual spectra. This would imply that in an average single-scan time of 300 ms at 8 cm$^{-1}$ resolution, about 107–108 bytes have to be read out, transferred, stored, and processed (87). Despite enormous progress in computer hardware, this is still a limiting factor and led to the combination with step-scan technology (88) to deal with that amount of data.

Once the data are collected, due to the multitude of cellular compounds present, only broad and superimposed spectral features are observed, which impede the correct assignment of bands to specific functional groups and structures. This is strengthened by the fact that a comprehensive understanding of the information content of complex biological samples in infrared spectroscopy is still missing and documented in the lack of systematic investigations on tissues in the literature. In this context, a functional-group mapping across a sample using its absorbance and frequency values as input for image construction can be difficult and subjective. An alternative approach to circumvent these difficulties, which is increasingly recognized, is the application of ANNs (81,87). Figure 9 represents a cross-sectional photomicrograph (2000 × 1200 μm) of a human colon carcinoma stained by hemalum eosine (HE). The thin section is made through the colon mucous membrane and the subjacent layers of colon cancer. The tunica muscularis with one outer and one inner circular layer of smooth muscle cells and the submucosa is replaced in this example by the colorectal cancer. The staining was performed after FPA-FT-IR microspectroscopic measurements. Five morphological structures had been identified by light microscopic examinations: (1) *Lamina muscularis* mucosae, (2) *Lamina propria* mucosae, (3) crypts, (4) submucosa, and (5) the adenocarcinoma.

In Fig. 9, upper left panel, five classes of FT-IR absorbance reference spectra were selected, ranging from 3500 to 850 cm$^{-1}$; in the right panel, the corresponding first derivatives of the fingerprint region from 1450 to 900 cm$^{-1}$ are illustrated. Twenty-six different samples of colon carcinoma were investigated, gaining 2356 reference spectra of the five classes. The first derivatives had been vector normalized, averaged over four data points, and a feature-selection method applied to extract 100 input neurons (87). This database was split into a training data set (1180 spectra) and a validation data set (910 spectra). The ANN consisted of a feed-forward network with 100 input neurons, one hidden layer with 40 neurons, and 5 output neurons that were fully connected. For learning, the Rprop

**Table 1** Classification Results of Colon Carcinoma Based on ANN Classification (Validation Data Set)

| Σ spectra | Accuracy | Specifity | Positive predictive value | Negative predictive value | Histology |
|-----------|----------|-----------|---------------------------|---------------------------|-----------|
| 195 | 97.9% | 99.5% | 97.5% | 99.6% | Crypts |
| 123 | 92.7% | 98.5% | 88.4% | 99.1% | *L. muscularis* mucosae |
| 168 | 94.6% | 99.2% | 95.2% | 99.0% | *L. propria* mucosae |
| 126 | 95.2% | 99.1% | 93.0% | 99.4% | Submucosa |
| 298 | 94.0% | 99.4% | 98.2% | 97.9% | Adenocarcinoma |
| 910 | | | | | |

algorithm was applied. Training was stopped after the SSE minimum was reached with 260 training cycles (87). In Table 1, the classification result is summarized. The positive predictive values, compared with the "gold standard" classification of a histopathologist vary between 88.4% for the *L. muscularis* mucosae and 98.2% for the adenocarcinoma. With this database, an ANN spectral library for colon carcinoma was established and all FPA image spectra classified and used for image reconstruction. In Fig. 9, lower panel, the reassembled images of the five classes show a very high correspondence with the histopathological assignment of the HE stained cross section. This shows that ample information is present in mid-infrared spectra of tissue materials, which can be used to define different spectral patterns to complex biological structures of diagnostic value. An essential role in the application of pattern recognition methods and ANNs in imaging is the definition of adequate and representative reference spectra, which is the basis for image construction and contrast enhancement. Research is still needed to find out reliable, patient independent spectral signatures that allow distinguishing not only healthy and diseased states but also transient phases such as precancerous or inflammatory tissue processes. The advantage ANN's can provide to that field of research is their ability to be integrated in large-scale libraries containing these kinds of various reference spectra. Other advantages include their speed in computation during classification of unknown spectra by trained ANNs, accelerating the image processing. The ability of ANNs for on-line processing can be integrated when the user is selecting interactively desired points of interest visually in the microscope image: the corresponding spectra are grouped, automatically pretreated, including feature selection, and trained to form an on-line ANN library. The image is constructed on that preselection. The time needed to include all these processes is no longer the limiting factor with state-of-the-art computer technology. The integration of hardware and new and powerful data-evaluation technologies may lead to the desired automation in image processing.

In combination with the advance of focal-plane array detectors, the standards in biodiagnostics and routine examinations in the clinical lab can probably be met.

## 4.2 Bacteria Identification Based on Modular Artificial Neural Networks

Bacterial identification by FT-IR and ANN (6,88) in this example focuses on a classification problem within a database of FT-IR spectra consisting of four bacterial genera and altogether 126 bacterial species. A hierarchical and modular approach was chosen, consisting of a prior top level and two following subclassification levels (79). We also expected this approach to provide flexibility in terms of practical extension to further spectra (6). In addition, specific questions that may be of therapeutic value, such as an antibiotic resistance, can be implemented in such a modular system in a more dedicated manner. In our example, a top-level network assigns FT-IR spectra into one of the nonexclusive classes of the genera *Pseudomonacae, Staphylococcus, Bacillus*, and *Candida*. Depending upon the activation of the outputs, a more dedicated network was designed to identify individual species within the bacterial genera. Within the class of *Candida*, the hierarchy is extended to one deeper level of refinement, where the species *C. albicans* is classified in respect to their susceptibility against the antibiotic fluconazole. Different network topologies of both types of ANNs, multilayer perceptron (MLP), and radial basis function networks were investigated systematically to provide the best performance.

### 4.2.1 Acquisition of Bacterial Spectra

Bacterial spectra have been recorded on a Bruker IFS 28B and IFS 66 FT-IR spectrometer under cultivation conditions as previously reported (6) and thus included instrument variance. The spectra were recorded with a DTGS detector at 4-cm$^{-1}$ resolution and 64 coadded scans from 4000 to 500 cm$^{-1}$ on a sample holder of ZnSe comprising 15 samples. Each sample and spectrum consisted of an individual cultivation and harvesting at different days, partly in different laboratories to represent the biological variance of growth. Instrumental variance was included by measurements on the IFS 28B and the IFS 66 spectrometer.

### 4.2.2 Artificial Neural Network Design

Two types of ANNs were used in this study, multilayer perceptrons (MLPs) and radial basis functions networks (RBFs). A range of models had been developed and compared due to their different numbers of hidden units. The network that produced the smallest sum of squared errors (SSE) in the validation pattern set was considered as the appropriate one (see upcoming Table 3). The classification results were evaluated with an analysis tool using the functions "winner takes all" (WTA) and "402040." The WTA function accepts an output pattern to be classified if exactly one output node is activated higher than all the others. The "402040" rule (39) requires that within a possible output range from 0 to 1 the activation of exactly one output node is greater then 0.6 (the upper 40% of the output range), whereas all other nodes are activated less than 0.4 (the lower 40% of the output range). Output patterns remain unclassified in all other cases.

### 4.2.3 Data Pretreatment

The FT-IR spectra were checked for quality prior to their use by a microprogram. The program simulated the test of the signal-to-noise ratio, the minimum and maximum absorbance, the water vapor content, and optical fringing. For each parameter, a threshold was set. The next step was testing the spectral compatibility, especially resolution, spectral sampling range, and number of data points. It is a precondition that all spectra within a neural net have identical measurement and data parameters. From absorbance spectra, the first derivative was calculated and afterwards normalized by a vector transformation over the whole spectral range. A Savitzky–Golay (89) smoothing function with nine smoothing points was applied to the first derivatives.

For all wavelengths, a boxcar averaging of three neighboring data points and a univariate variance analysis was computed within predefined windows excluding the nondiagnostic ranges 500–700 $cm^{-1}$, 1800–2800 $cm^{-1}$, and 3000–4000 $cm^{-1}$. Two independent data sets were used for internal calibration of the ANNs, where between 60% and 75% of the available spectra were used for the training and the remaining samples for validation purposes. The number of independent measurements per class varied. The minimum was four, the maximum ten. A third test set was established with 420 spectra, validating the overall performance of the model.

### 4.2.4 Results

The aim of this study was the establishment of a hierarchical classification system, consisting of several ANNs trained for a specific degree for differentiation of bacterial strains (Fig. 10). A top-level network is required for the classification of FT-IR spectra into one of four classes, corresponding to the bacteria genera *Bacillus, Staphylococcus, Pseudomonas*, and the yeast *Candida*. This problem could be solved appropriately with an RBF network. In dependence of the behavior of the top-level network, the subclassification level is entered with a specific ANN. Within the *Candida* strains, a third-level classification could be realized to distinguish fluconazole-sensitive and -insensitive species of *Candida albicans*. The information flux through the connected ANNs is further controlled by specific index tables for each ANN, containing the positions of the wavelengths chosen by the univariate wavelength selection.

The rough top-level classification into one of the five bacteria genera could be accomplished with an RBF network using 50 wavelengths. Here, a classifica-

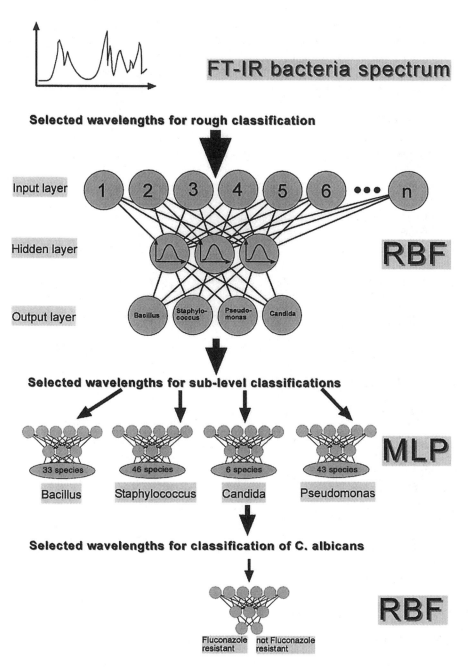

**Fig. 10**   Classification scheme of modular ANNs for bacterial identification.

tion accuracy of 100% was attained for both analyzing functions, "WTA" and "402040." More emphasis was put on the subclassification level, where more classes had to be separated (see upcoming Table 3). This task involved the search for a suitable network type (RBF or MLP), learning algorithms, and network architectures (e.g., number of input and hidden units). The consequence of different numbers of input units (20, 50, 100, 200, and 400), which represent wavelengths of descending univariate $F$-values, is demonstrated for the classification of the *Bacillus* as a typical example. The related MLPs were trained with Rprop and Quickprop. As criterion for the best network, the SSE in the validation pattern set was chosen. The results, summarized in Table 2, indicate a distinct SSE decline with a larger number of wavelengths, up to the maximum number of 400.

The numbers of required learning cycles differed among the models, but, generally, increasing the number of input units reduced the training time of the network. Both learning algorithms, Rprop and Quickprop, required about similar training cycles. The number of hidden units in the ANNs selected to minimize the validation error, varied between 10% and 90% of the number of input neurons. Corresponding results with regard of the number of input neurons were obtained for the classification *Pseudomonas* and *Staphylococcus* strains. The differentiation of *Candida* and the fluconazole susceptibility of *C. albicans* could both be accomplished with 50 wavelengths.

The efficacy of the different learning algorithms on the validation SSEs and the classification accuracy are summarized in Table 3.

The fastest learning and the lowest misclassification rate were obtained by MLPs trained with Rprop and Quickprop. That also includes the CC networks. In each subclassification problem, the resulting MLPs were able to differentiate the bacteria species with a degree of accuracy that varied between 84.8% and 100%. In spite of momentum and flat-spot elimination terms, the back-propagation algorithm achieved only very slow convergence and was, therefore, not able to learn the problems within an acceptable time, so training was terminated manually each after 200,000 training cycles. In the case of the *Bacillus* class, the MLP

**Table 2**  Number of Input Neurons in Relation to Training Cycles and Output Error (SSE) (Bacillus ANN, 65 Validation Spectra)

| Input units | Rprop hidden units | SSE | Cycles | Quickprop hidden units | SSE | Cycles |
|---|---|---|---|---|---|---|
| 20 | 2 | 14.55 | 262 | 12 | 14.23 | 323 |
| 50 | 25 | 15.19 | 161 | 5 | 15.30 | 167 |
| 100 | 20 | 7.52 | 100 | 20 | 7.76 | 139 |
| 200 | 40 | 5.17 | 71 | 20 | 6.13 | 79 |
| 400 | 80 | 3.62 | 85 | 40 | 4.02 | 136 |

**Table 3**  Classification of a Modular ANN

## Toplevel classification

50 wavel., 4 clas.
652 training spectra ;196 validation spectra

| Algorithm | hidden units | SSE | Cycles | "WTA" classification wrong (%) | unknown (%) | correct(%) | "402040" classification wrong (%) | unknown (%) | correct(%) |
|---|---|---|---|---|---|---|---|---|---|
| RBF | 10 | 0.78 | 3000 | 0.0 | 0.0 | 100.0 | 0.0 | 0.0 | 100.0 |

## Sublevel classification (I)

**Bacillus** (400 wavelengths) 36 clas.
144 training spectra; 65 validation spectra

| Algorithm | Hidden units | SSE | Cycles | "WTA" classification wrong (%) | unknown (%) | correct(%) | "402040" classification wrong (%) | unknown (%) | correct(%) |
|---|---|---|---|---|---|---|---|---|---|
| Rprop | 60 | 3.03 | 82 | 3.0 | 9.1 | 87.9 | 3.0 | 12.1 | 84.8 |
| Quickprop | 20 | 3.23 | 90 | 3.0 | 9.1 | 87.9 | 3.0 | 12.1 | 84.8 |
| Backpropagation | 40 | 4.38 | 100000 | 6.1 | 0.0 | 93.9 | 3.0 | 9.1 | 87.9 |
| CC | 1 | 5.07 | 4 | 3.0 | 9.1 | 87.9 | 3.0 | 9.1 | 87.9 |
| RBF | 111 | 21.78 | 50000 | 33.3 | 0.0 | 66.7 | 3.0 | 63.6 | 33.3 |

**Staphyllococcus** (400 wavelengths) 20 clas.
185 training spectra; 46 validation spectra

| Algorithm | Hidden units | SSE | Cycles | "WTA" classification wrong (%) | unknown (%) | correct(%) | "402040" classification wrong (%) | unknown (%) | correct(%) |
|---|---|---|---|---|---|---|---|---|---|
| Rprop | 60 | 0.88 | 200 | 0.0 | 0.0 | 100.0 | 0.0 | 6.1 | 93.9 |
| Quickprop | 20 | 0.54 | 157 | 0.0 | 0.0 | 100.0 | 0.0 | 6.1 | 93.9 |
| Backpropagation | 50 | 4.12 | 200000 | 0.0 | 0.0 | 100.0 | 0.0 | 21.2 | 78.8 |
| CC | 1 | 1.87 | 3 | 0.0 | 0.0 | 100.0 | 0.0 | 12.1 | 87.9 |
| RBF | 129 | 82.67 | 50000 | 18.2 | 0.0 | 81.8 | 3.0 | 87.9 | 9.1 |

**Pseudomonas** (400 wavelengths), 42 classes
243 training spectra ; 83 validation spectra

| Algorithm | Hidden units | SSE | Cycles | "WTA" classification wrong (%) | unknown (%) | correct(%) | "402040" classification wrong (%) | unknown (%) | correct(%) |
|---|---|---|---|---|---|---|---|---|---|
| Rprop | 10 | 2.82 | 162 | 0.0 | 0.0 | 100.0 | 0.0 | 4.8 | 95.2 |
| Quickprop | 10 | 7.07 | 76 | 0.0 | 0.0 | 100.0 | 0.0 | 13.3 | 86.7 |
| Backpropagation | 40 | 66.35 | 200000 | 59.0 | 0.0 | 41.0 | 0.0 | 59.0 | 41.0 |
| CC | 2 | 4.02 | 2 | 0.0 | 0.0 | 100.0 | 0.0 | 7.2 | 92.8 |
| RBF | 144 | 144.83 | 50000 | 44.6 | 0.0 | 55.4 | 3.6 | 80.7 | 15.7 |

**Candida** (50 wavelengths, 6 classes)
80 training spectra; 20 validation spectra

| Algorithm | hidden units | SSE | Cycles | "WTA" classification wrong (%) | unknown (%) | correct(%) | "402040" classification wrong (%) | unknown (%) | correct(%) |
|---|---|---|---|---|---|---|---|---|---|
| Rprop | 0 | 0.02 | 160 | 0.0 | 0.0 | 100.0 | 0.0 | 0.0 | 100.0 |
| Quickprop | 0 | 0.02 | 145 | 0.0 | 0.0 | 100.0 | 0.0 | 0.0 | 100.0 |
| Backpropagation | 15 | 19.62 | 200000 | 85.0 | 0.0 | 15.0 | 0.0 | 100.0 | 0.0 |
| CC | 10 | 0.60 | 10 | 0.0 | 0.0 | 100.0 | 0.0 | 5.0 | 95.0 |
| RBF | 30 | 2.73 | 50000 | 10.0 | 0.0 | 90.0 | 10.0 | 0.0 | 90.0 |

## Sublevel classification (II)

**Candida albicans** (50 wavelengths) 2 clas.
42 training spectra;14 validation spectra

| | hidden units | SSE | Cycles | "WTA" classification wrong (%) | unknown (%) | correct(%) | "402040" classification wrong (%) | unknown (%) | correct(%) |
|---|---|---|---|---|---|---|---|---|---|
| | 0 | 0.00 | 8 | 0.0 | 0.0 | 100.0 | 0.0 | 0.0 | 100.0 |
| Algorithm | 0 | 0.00 | 8 | 0.0 | 0.0 | 100.0 | 0.0 | 0.0 | 100.0 |
| Rprop | 0 | 0.01 | 20 | 0.0 | 0.0 | 100.0 | 0.0 | 0.0 | 100.0 |
| Quickprop | 0 | 0.00 | 0 | 0.0 | 0.0 | 100.0 | 0.0 | 0.0 | 100.0 |
| Backpropagation | 15 | 0.25 | 3000 | 0.0 | 0.0 | 100.0 | 0.0 | 0.0 | 100.0 |
| CC | | | | | | | | | |
| RBF | | | | | | | | | |

trained with back-propagation attained the best classification results: 93.9% of the patterns used for validation could be correctly classified using the "WTA" analyzing function and 87.9% in the case of the "4902040" rule. On the other hand, only 41% ("402040") of the *Pseudomonas* validation spectra could be correctly classified using the related MLP trained with back-propagation and even 0% of the *Candida* spectra.

In contrast to the MLPs, the RBF networks generally led to unsatisfactory classification results in the subclassification level. The application of the "402040" analyzing function on the validation sets resulted in only 33.3% correct classifications for the *Bacillus* class, 15.7% for *Pseudomonas*, and just 9.1% for *Staphylococcus*. More accurate results were obtained for the RBF networks trained for *Candida* with six classes on the species level (90% correct classification) and *C. albicans* with two subclasses (100% correct classification).

The number of hidden units in the MLPs selected to minimize the SSE in the validation data set of the different ANNs varied in subclassification level. No clear dependence between the number of required hidden units and the learning algorithm could be determined. In preliminary examinations, we detected that the number of selected hidden units is more sensitive to the initial random initialization of the weights in the MLP, which results in different combinations of transfer functions in the model.

The CC algorithm generally built MLPs with the lowest number of hidden units, but it should be kept in mind, that in this case the insertion criterion is dependent not on the validation set but on the network error determined by the training set and on some user-defined parameters. The CC network, which classifies the *Pseudomonas* species, contains only two hidden nodes, and no hidden units were selected for the Bacillus and *Staphylococcus* networks. The resulting classification errors in the validation set were, nevertheless, comparable to those of the MLPs trained with Rprop and Quickprop, in spite of slightly higher SSEs.

For the final hierarchical classification system, presented in Fig. 10, each ANN with the lowest classification rate was selected. The single ANNs are logically but not physically connected, since they use different combinations and numbers of wavelengths as input neurons. The information flux from the top-level classification to the subsequent networks is controlled by the "204020" analyzing function, since here many patterns remain "unclassified" that are otherwise eventually classified wrong using the "WTA" function. Only spectra that pass the "204020" function as "classified" are allowed to enter a subsequent network. The overall performance of the modular network was tested using the 420 test spectra, resulting in 1.9% unknown, 3.8% wrong, and 94.3% correctly assigned spectra.

Based on this investigation, it seems to be worth integrating more dedicated ANN libraries to form one seamless data-evaluation module, which can support microbiological investigations based on FT-IR spectroscopy.

## 4.3 Identification of Neurotransmitter by Raman Spectroscopy

Schulze et al. (90) used an ANN model based on Raman spectra of the neurotransmitters acetylcholine, dopamine, epinephrine, norepinephrine, serotonine, histamine, glutamate, glycine, aspartate, and γ-aminobutyric acid in aqueous solution. Raman scattering was excited with the 488-nm line from an Ar+ laser operating at 200 mW and measured with a scanning spectrometer. The spectra were digitized at 1-cm$^{-1}$ intervals, resulting in 1501 data points with an approximate signal-to-noise ratio of 20–50.

For designing the ANNs, fully connected three-layer MLPs with the standard back-propagation algorithm have been used. Feature selection was performed by different spectral smoothing, ranging from 5-point and 10-point moving average, 5-cm$^{-1}$ and 10-cm$^{-1}$ shifted spectra, and the raw uncompressed data.

Their results showed that network accuracy improved with an increase in the number of input neurons. In the shifted data sets, the opposite was found, suggesting that fewer input data may be more useful in a data set where calibration errors are prevalent.

The networks that performed best on the training had a ratio of 0.16 of hidden to input neurons. The effects of different transfer functions on network performance were tested with a 150 (input)–29 (hidden)–10 (output) network configuration. It was concluded that the sine-sigmoid networks, which use a sine function for the hidden layer and a sigmoid function for the output layer, trained 3–5 times faster than a sigmoid-sigmoid transfer function with 4,176 and 15,039 training cycles, respectively. Unfortunately, little is mentioned about the performance of the ANNs based on a validation and test data set.

## 4.4 Classification of Infrared Spectra of Control and Alzheimer's Diseased Tissue

Pizzi et al. (45) described a classification of Alzheimer's diseased tissue with ANNs and a linear discriminant analysis (LDA). They used three different networks: standard back-propagation (E-BPN), back-propagation with fuzzy encoding (FE-BPN), and radial basis functions (RBFs). Their data set consisted of 114 spectra, divided into 49 control tissue and 65 Alzheimer's diseased tissue spectra, applying the leave-one-out method for cross-validation. In the preprocessing cases, they performed principal component analysis (PCA) for feature extraction and used the first $k$ principal components (PCs) that accounted for the cumulative variance as input neurons for both the LDA and the ANNs.

Addressing a two-class problem, they gained accuracy for the ANNs of 100% with the first nine PCs for FE-BPN, 81% for E-BPN, and 88% for RBF. They clearly showed that the ANNs consistently outperformed their LDA counterparts in all cases where PCA was used as a preprocessing technique. In the cases where the original spectra were used, FE-BPNs outperformed E-BPNs and RBFs and had classification results that were only slightly worse than the best results using PCA preprocessing. This highlights the point that PCA is often useful and improves performance results, but there is also a concomitant loss in flexibility in the addition and deletion of data as well as a loss of the ability to analyze relevant features in the original spectra that contributed to the discriminatory power.

## 5   CONCLUSION

This chapter has presented an overview of some applications of ANNs in the field of biodiagnostics based on FT-IR and Raman spectroscopy. There is no doubt that ANNs will find a use in solving many problems in both instrumentation and measurement applications and also in the case of complex and noisy ill-conditioned systems. Artificial neural networks can simplify and accomplish many tasks not feasible using conventional techniques. And ANN software implementation is arriving at a level of maturity in many fields of application. The transfer of ANN knowledge into spectroscopy is certainly promoted by the progress made in instrumentation and by opening the various new application fields of vibrational spectroscopy in the future. It should be noted that many ANN applications are oriented solely toward the use of software solutions, whereas the use of hardware implementation is less common. This is due mainly to the unavailability of on-chip ANNs dedicated to the processing of signals. However, the use of ANNs implemented on-chip could furnish ulterior quick and easy solutions, e.g., for FPA detector and imaging applications.

## ACKNOWLEDGMENTS

The authors would like to thank D. Naumann for providing spectra and a fruitful collaboration over the years. In addition, we are grateful to P. Lasch for his support. Finally, we appreciate the spirit evolved from the ''Forum FTIR Diagnostik'' founded in 1996.

## REFERENCES

1.   M Jackson, HH Mantsch. CRC Crit Rev Biochem Mol Biol 30:95–123, 1995.
2.   HL Casal, HH Mantsch. Biochem Biophys Acta 779:381–394, 1984.
3.   M Mathlouthi, JL König. Adv Carbohydr Chem Biochem 44:7–14, 1986.
4.   E Thaillandier, JL Liquer, A Taboury. In: RJH Clark, RE Hester, eds. Advances in Infrared and Raman Spectroscopy. New York: Wiley, 1985, pp. 65–81.
5.   HH Mantsch, M Jackson, eds. Infrared Spectroscopy: A New Tool in Medicine. Proc SPIE 3257, 1998.

6. J Schmitt, T Udelhoven, D Naumann, HC Flemming. In: HH Mantsch, M Jackson, eds. Infrared Spectroscopy: A New Tool in Medicine. Proc SPIE 3257:237–243, 1998.
7. I Aleksander, H Morton. Introduction to Neural Computing. London: Van Nostrand, 1995.
8. SE Geman, E Bienenstock, R Doursat. Neural Computation 4:1–58, 1992.
9. P Kerlizin, F Vallet. Neural Computation 4:473–482, 1993.
10. WS McCulloch, W Pitts. Bull Mathem Biophys 5:115–133, 1943.
11. W Rall. In: EL Schwartz, ed. Computational Neuroscience, Cambridge, MA: MIT Press, 1999, pp. 3–8.
12. DO Hebb. The Organization of Behavior: A Neurophysiological Theory, New York: Wiley, 1949.
13. F Rosenblatt. Psychol Rev 65:386–408, 1958.
14. F Rosenblatt. Principles of Neurodynamics: Perceptrons and the Theory of Brain Mechanisms. Washington, DC: Spartan Books, 1961.
15. B Widrow, ME Hoff Jr. Adaptive Switching Circuits. New York: IRE WESCON Convention Record, 1960, pp. 96–104.
16. B Widrow. In: MC Yovitz, GT Jacobi, GD Goldstein, eds. Self-Organizing Systems, Washington, DC: Spartan Books, 1962, pp. 435–461.
17. NJ Nilsson. Learning Machines: Foundations of Trainable Pattern-Classifying Systems. New York: McGraw-Hill, 1965.
18. MC Minsky, SA Papert. Perceptrons. Cambridge, MA: MIT Press, 1969.
19. JA Anderson, JW Silverstein, SA Ritz, RS Jones. Psychol Rev 84:413–451, 1977.
20. T Kohonen. IEEE Transactions Computers C-21:353–359, 1972.
21. MA Cohen, S Grossberg. IEEE Transactions Systems, Man Cybernetics 13:815–826, 1983.
22. B Widrow. IEEE Computer 21:25–39, 1988.
23. C von der Malsburg. Kybernetik 14:85–100, 1973.
24. JJ Hopfield. Neurons with Graded Responses Have Collective Computational Properties Like Those of Two-State Neurons. Proc Nat Acad Sci 81:3088–3092.
25. S Amari. IEEE Transactions Electronic Computers, EC-16:299–307, 1967.
26. PJ Werbos. Beyond regression: new tools for predicting and analytics in the behavioral sciences. PhD dissertation, Harvard University, Cambridge, MA, 1974.
27. DB Parker. Ann Math Stat 33:1065–1076, 1985.
28. DE Rumelhart, GE Hinton, RJ Williams. In: DE Rumelhart, JL McClelland, eds. Parallel Distributed Processing: Explorations in the Microstructure of Cognition. Vol 1. Cambridge, MA: MIT Press, 1986.
29. DE Rumelhart, JL McClelland, eds. Parallel Distributed Processing: Explorations in the Microstructure of Cognition. Vols. 1, 2. Cambridge, MA: MIT Press, 1986.
30. JJ Hopfield. Proc Nat Acad Sci 79:2554–2558, 1982.
31. W Wu, DL Massart. Chemom Intell Lab Syst 35:127–135, 1996.
32. CM Bishop. Neural Networks for Pattern Recognition. Oxford, UK: Clarendon Press, 1995.
33. T Masters. Practical Neural Network Recipes in C++. San Diego, CA: Academic Press, 1985.

34. EP Derks, MS Sánchez Paster, LMC Buydens. Chemom Intell Lab Syst 28:49–60, 1995.
35. BM Wise, BR Holt, NB Gallagher, S Lee. Chemom Intell Lab Syst 30:81–89, 1995.
36. F Despagne, DL Massart. Analyst 123:157R–178R, 1998.
37. SE Fahlmann. In: D Touretzky, G Hinton, T Sejnowski, eds. Proceedings of the Connectionists Models Summer School 1988. Pittsburgh, PA: Morgan Kaufmann, 1988, pp. 524–532.
38. H Braun. Neuronale Netze: Optimierung durch Lernen und Evolution. Berlin: Springer, 1997.
39. A Zell. Simulation Neuronaler Netze. Munich: Oldenbourg Verlag, 1997.
40. M Riedmiller, H Braun. RPROP: A fast and robust backpropagation learning strategy. Proceedings of the IEEE International Conference on Neural Networks (ICNN), San Francisco, 1993, pp. 586–591.
41. M Riedmiller, H Braun. A direct adaptive method for faster backpropagation learning: the RPROP algorithm. Proceedings of the IEEE International Conference on Neural Networks (ICNN), San Francisco, 1993, pp. 591–607.
42. M Riedmiller. Computer Standards Interfaces 16:265–278, 1994.
43. DH Ackley, GE Hinton, TJ Sejnowski. Cognitive Sci 9:147–169, 1985.
44. T Kohonen. Self-Organization and Associative Memory. Berlin: Springer Series in Information Sciences, 1984.
45. N Pizzi, LP Choo, J Mansfield, M Jackson, WC Halliday, HH Mantsch, RL Somorjai. Artificial Intell Med 7:67–79, 1995.
46. K Faber, BR Kowalski. Chemom Intell Lab Syst 34:293–297, 1996.
47. MS Sánchez, H Swierenga, LA Sarabia, E Derks, L Buydens. Chemom Intell Lab Syst 33:101–119, 1996.
48. W Wu, B Walzak, DL Massart, S Heuerding, F Erni, IR Last, KA Prebble. Chemom Intell Lab Syst 33:35–46, 1996.
49. D Naumann, H Labischinski, P Giesbrecht. In: WH Nelson, ed. Instrumental Methods for Rapid Microbiological Analysis. New York: VCH, 1991, pp. 43–96.
50. MS Dhanhoa, SJ Lister, R Sanderson, RJ Barnes. J Near Infrared Spectrosc 2:43–49, 1994.
51. S Chatterjee, AS Hadi. Statistic Sci 1:379–385, 1986.
52. V Centner, DL Massart, OE de Noord, Anal Chim Acta 34:330–338, 1996.
53. EJ Hartman, JD Keeler, JM Kowalski. M Neural Computation 2:210–215, 1990.
54. PA Devijer, J. Kittler. Pattern Recognition: A Statistical Approach. London: Prentice Hall, 1982.
55. L Davis, ed. Genetic Algorithms and Simulated Annealing. London: Pittman, 1987.
56. JG Brenchley, U Hörchner, JH Kalivas. Appl Spectrosc 51:689–699, 1997.
57. MS Jouan-Rimbaud, B Walzak, DL Massart, IR Last, K Prebbel. Anal Chim Acta 304:285–293, 1995.
58. MS Jouan-Rimbaud, S Khots, DL Massart, IR Last, K Prebbel. Anal Chim Acta 315:257–265, 1995.
59. M Soskic, D Plasic, N Trinajstic. J Chem Inf Comput Sci 36, 146, 1996.
60. U Norinder. J Chemom 10:95–108, 1996.
61. H Kubinyi. J Chemom 10:119–128, 1996.
62. V Bellon-Maurel, C Vallat, D Goffinet. Appl Spectrosc 49:556–564, 1995.

63. W Wu, S Rutan, A Baldovin, DL Massart. Anal Chim Acta 335:11–22, 1996.
64. B Walzak, DL Massart. Chemom Intell Lab Syst 38:39–50, 1997.
65. T Blank, SD Brown. Anal Chim Acta 277:273–284, 1993.
66. F Despagne, DL Massart. Chemom Intell Lab Syst 40:145–152, 1998.
67. M Dathe, M Otto. Fresenius J Anal Chem 356:17–24, 1996.
68. U Depczynski, K Ketter, K Molt, A Niemöller. Chemom Intell Lab Syst 39:19–27, 1997.
69. M Otto. Chemometrie. Weinheim: VCH, 1997.
70. G Kateman, JRM Smits. Anal Chim Acta 277:179–188, 1993.
71. D Svozil, V Kvasnika, J Pospichal. Chemom Intell Lab Syst 39:43–62, 1997.
72. C Cleva, C Cachet, D Cabrol-Bass, TP Forrest. Anal Chim Acta 348:255–265, 1997.
73. A Maren, C Harston, R Pap. Handbook of Neural Computing Applications. San Diego, CA: Academic Press, 1990.
74. ME Munk, MS Madison, EW Robb. Mikrochim Acta 2:505–514, 1991.
75. MS Sánchez, LA Sarabia. Chemom Intell Lab Syst 28:287–303, 1995.
76. SE Fahlman, C Lebiere. The Cascade Correlation Architecture. Carnegie Mellon University Report CMU-CS-90-100, 1991, pp. 1–13.
77. P Zheng, P Harrington, DM Davis. Chemom Intell Lab Syst 33:121–132, 1996.
78. QC Van Est, PJ Schoenmakers, JR Smits, W Nijssen. Vibr Spectrosc 4:263–272, 1993.
79. T Udelhoven, J Schmitt. Appl Spectrosc, in press, 2000.
80. P Lasch, D Naumann. Cell Mol Biol 44:189–202, 1998.
81. P Lasch, W Wäsche, WJ McCarthy, G Müller, D Naumann. In: HH Mantsch, M Jackson, eds. Infrared Spectroscopy: A New Tool in Medicine. Proc SPIE 3257: 187–198, 1998.
82. M Diem, S Boydston-White, L Chariboga. Appl Spectrosc 53:1–8, 1999.
83. RK Dukor, MN Liebmann, BL Johnson. Cell Mol Biol 43:211–217, 1997.
84. LP Choo, DL Wetzel, WC Halliday, M Jackson, SM LeVine, HH Mantsch. Biophys J 71:1672–1679, 1996.
85. L Kidder, VF Kalasinsky, JL Luke, IW Levon, EN Lewis. Nature Med 3:235–237, 1997.
86. EN Lewis, PJ Treado, RC Reeder, GM Story, AE Dowrey, C Marcott, IW Levin. Anal Chem 67:3377–3381, 1995.
87. P Lasch. Computergestützte Bildrekonstruktion auf Basis FTIR-mikrospektrometrischer Daten humaner Tumoren. PhD dissertation, Freie University, Berlin, 1999.
88. EN Goodacre, EM Timmins, PJ Rooney, JJ Rowland, DB Kell. Fems Microbiol Lett 140:233–239, 1996.
89. A Savitzky, MJ Golay. Anal Chem 36:1627–1638, 1964.
90. HG Schulze, MW Blades, AV Bree, BB Gorzalka, LS Grek, RF Turner. Appl Spectrosc 48:50–57, 1994.

# 11

# Biological Applications of Raman Spectroscopy

**Elizabeth A. Carter**
*University of Sydney, Camperdown, Australia*

**Howell G. M. Edwards**
*University of Bradford, Bradford, West Yorkshire, United Kingdom*

## 1 INTRODUCTION

Vibrational spectroscopic studies of biomolecular polymers have demonstrated a number of advantages over other spectroscopic methods, including minimal or no damage to the sample and ease of sampling arrangements. Spectral sensitivity to a range of parameters that affect biomolecular structures is an added advantage for studies of environmental change, in particular, tissue degradation resulting from changes in pH, temperature, degree of hydration, or bacterial or drug attack. Prior to the advent of Fourier transform (FT) Raman spectroscopy using near-infrared excitation, classical Raman spectroscopic studies of molecules and materials of biological interest were fraught with difficulties arising principally from the generation of fluorescence by visible radiation and the real possibility of sample degradation from the use of the relatively high laser powers necessary for sample illumination. Hence, the major advantages of Raman spectroscopy over infrared spectroscopy, namely, the weak Raman scattering of water and of glass cells, could not be exploited fully. For the particular studies reported here, the power and potential of the Raman spectroscopic technique for the analysis of biological tissues is now realized.

### 1.1 Origin of Protein Vibrations

Proteins can be described as linear polymers, which are constructed from a combination of the 20 amino acids that are available for biosynthesis. The first stage of biosynthesis is the assembly and linkage of the amino acids to form polypeptide chains. The most basic covalent linkage found in proteins is the peptide bond. This is formed between the carbonyl and amino groups of amino acids as a result of a condensation reaction. A number of these condensation reactions will ultimately produce a polypeptide chain; see Fig. 1 (1). The sequence and chemical structure of amino acids within a peptide, polypeptide, or protein is known as the *primary* structure. Raman spectroscopy cannot be used to establish the sequence of these amino acids, but it can be used to detect the presence of certain types of amino acids, for example, the aromatic amino acids tyrosine, tryptophan, phenylalanine, and histidine.

The *secondary* structure of a protein is defined by the spatial arrangement of the amino acids in polypeptide chains of the protein. This secondary structure defines the function and properties of the protein. There are a number of types of secondary structures, including: the α-helix, $3_1$-helix, β-sheet, β-turn, and amorphous form, also known as the *random coil*. Raman spectroscopy can be used to determine the type of secondary structures present within a protein, the geometry of certain bonds, and the environment of some side chains.

The peptide group (—CONH—) is the most characteristic functionality bond within a protein. The group is considered to be nearly planar, and this is attributed to bond resonance stabilization. The vibrations of the atoms give rise to a number of distinctive vibrational features, namely, the amide I, II, III, IV, V, VI, VII, A, and B bands. Due to different selection rules, the amide I, II, III,

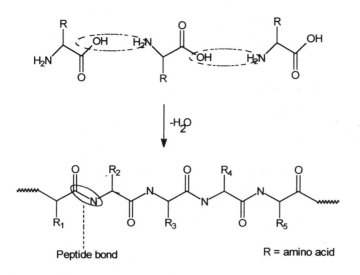

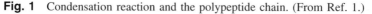

**Fig. 1** Condensation reaction and the polypeptide chain. (From Ref. 1.)

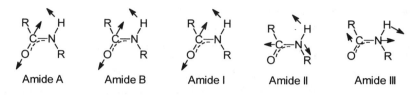

**Fig. 2** Vibrational modes of the peptide bond.

A, and B bands are observed in the infrared spectra of proteins, whereas only the amide I, III, A, and B bands are observed in Raman spectra. The vibrational modes of these bands are depicted in Fig. 2 (2–5).

The amide I band, arises mainly from the C=O stretching vibration of the peptide group with a small contribution from N—H in-plane bending. The amide II band, observed only weakly if at all in Raman spectra, is due primarily to N—H bending, with a minor contribution from C—N stretching vibrations. The amide III band arises from a combination of N—H bending and C—N stretching of the peptide group. Both the amide A and B bands originate from Fermi resonance, a result of the interaction of the first excited state of the N—H stretching vibration and the overtone of the amide II vibration (6–7).

The amide I, II, and III bands are conformationally sensitive and can be used to identify the type of secondary structure present in a protein (see Table 1). The positions of the amide bands are characteristic for various secondary structures and can be used in conjunction with other vibrations, such as the C—C skeletal backbone vibrations (2,5). For identification purposes it is recommended that a combination of two bands be used to provide a more reliable structural confirmation.

Vibrations arising from amino acids can be classified into two groups: those that originate from the $CH_2/CH_3$ groups of amino acids and the vibrations that originate from the aromatic rings of phenylalanine, tyrosine, and tryptophan. The C—H vibrations of amino acids are observed in both IR (weak) and Raman spec-

**Table 1** Characteristic Vibrations of Various Secondary Structures

|  | α-Helix | β-Pleated sheet | Random coil |
|---|---|---|---|
| Amide I | 1660–1645 | 1680–1665 | 1670–1660 |
| Amide III | 1310–1260 | 1240–1225 | 1260–1240 |
| Skeletal C—C | 950–885 | 1010–1000 | 960–950 |

*Source*: Refs. 4–7.

tra (strong). The aromatic amino acids in particular generate very distinctive Raman bands and are weak or not observed in IR because of low content (8).

Cystine is a very important amino acid, oxidation of the thiol groups results in disulphide bond formation, and this ultimately leads to stabilization of the protein structure. The bond is homonuclear and is therefore Raman active and IR inactive. The disulphide bonds form crosslinks between adjacent protein chains or different parts of the same protein. The frequency of this vibration is dependent upon the arrangement of the atoms, $-C_\alpha-C_\beta-S-S-C_{\beta'}-C_{\alpha'}-$, involved in these crosslinkages. Raman bands observed in the region 550–510 cm$^{-1}$ have been assigned to various conformations of S$-$S stretching vibrations and are depicted in Fig. 3 (4).

## 1.2  Sample Preparation and Presentation

### 1.2.1  Sample Preparation

A major advantage of the application of Raman spectroscopy or microspectroscopy to the analysis of biological materials is the minimal sample preparation required for presentation of the specimen to the spectrometer. Unlike most other analytical techniques, no chemical or mechanical pretreatment is necessary; of particular relevance to biological and biomedical studies is the ability to obtain Raman spectra from specimens in their state of natural hydration. The weak Raman scattering of hydroxyl groups and silica means that water and glass will not strongly affect the observation of Raman spectra. In other cases, specimens that have been prepared for optical microscopy and protected using standard glass cuvettes or slide-covered slips can also be studied using a Raman microscope without any changes being effected.

### 1.2.2  Sample Presentation

The signal-to-noise (S/N) ratio of a Raman spectrum is related directly to the applied laser power and signal collection time. An increase in laser power will improve the S/N ratio in a given time interval, but may also increase the risk of sample photodegradation. Therefore, it is necessary to investigate the stability of the sample in relation to the intensity of the laser source and to assess the likelihood of sample damage. A method often used to improve spectral quality is to increase the sample density. This is often achieved by designing a sample holder appropriate for the sample to be analyzed. This is the most efficient means of improving the S/N ratio, limiting the possibility of photodegradation. Schrader has demonstrated how sample holders with highly polished conical indentations, or backing mirrors, cause multiple reflections of the exciting and scattered Raman radiation, considerably enhancing the intensity of the Raman signal (9).

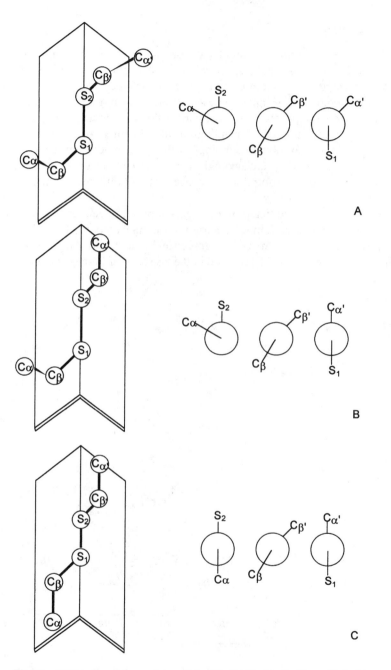

**Fig. 3** Different conformations of the disulphide bond. (From Ref. 4.)

### 1.2.3 Fluorescence

The swamping of weaker Raman spectra by strong fluorescence emission in the visible region of the electromagnetic spectrum is an occupational hazard of Raman spectroscopy using laser excitation, particularly in the blue and green regions (400–520 nm). In some cases sample fluorescence arises from impurities in the specimen, and possible sample preparation or purification may be necessary. This may be a trivial or a difficult task, depending on the particular specimen being studied. For example, the swabbing of biological tissue to remove surface contaminants is not effective when archaeological biomaterials are being studied, which may have involved the absorption and concentration of fluorescent materials over a long period of time.

It has often been reported that prolonged exposure of a specimen to a laser beam will produce an enhanced Raman spectrum or a certain amount of the little-understood physical phenomenon of ''fluorescence burnout''; by this means, good-quality Raman spectra can be recorded in the presence of fluorescent mate-

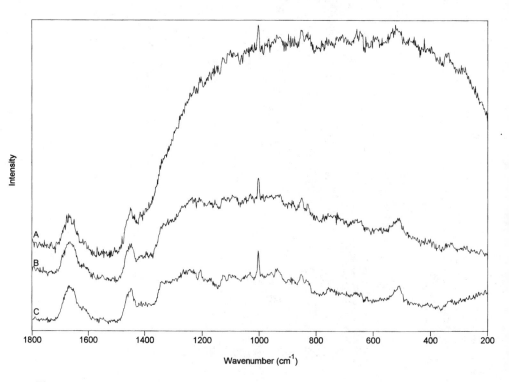

**Fig. 4** Dissipation of fluorescence in wool on extended exposure to laser radiation (633-nm excitation, 5-sec acquisition time, untreated wool fiber). A: 5 sec; B: 30 sec; C: 210 sec. (From Ref. 10.)

rial; see Fig. 4 (10). However, the removal of the fluorescence emission is only temporary. Some biomaterials, such as skin and wool, exhibit strong fluorescence in the blue region of the spectrum, and their Raman spectra generally cannot be recorded using laser radiation around 4880 nm.

The choice of exciting-line wavelength is of critical importance for the observation of Raman spectra of biomaterials without attendant fluorescence emission. A particular advantage of the move toward the red excitation (800 nm) and particularly the near-infrared (NIR) (1064 nm) is the minimization of fluorescence from organs and tissues, such as skin. Other methods of combating fluorescence have involved the use of pulsed laser excitation and dc chopping or synchronization of the scattered radiation to electronically filter out the background emission.

## 2  RAMAN SPECTROSCOPY OF BIOLOGICAL TISSUES

Raman spectra of proteins are dominated by numerous C—H and aromatic ring vibrations and are characterized by the presence of a number of amide bands generated from the peptide group. Figures 5 and 6 present a stack plot of the

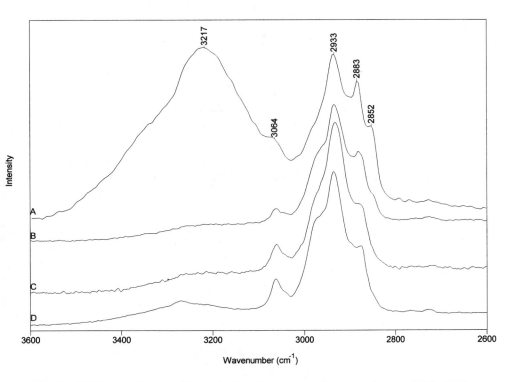

**Fig. 5**  FT-Raman spectra of keratotic biopolymers over the spectral range 3600–2600 cm$^{-1}$. A: stratum corneum; B: quill; C: horn; D: beak.

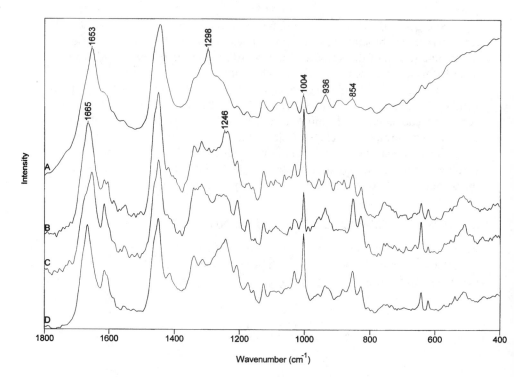

**Fig. 6** FT-Raman spectra of keratotic biopolymers over the spectral range 1800–400 cm$^{-1}$. A: stratum corneum; B: quill; C: horn; D: beak.

Raman spectra of a number of keratin proteins; a table of assignments for a number of proteins discussed in this chapter can be found in Table 2.

## 2.1 Keratin Proteins

Keratin proteins are characterized by their high cystine content and can be classified according to the amount of sulphur present in the protein. Sulphur may be found either in the form of cystine or its reduction and oxidation products, cysteine and cysteic acid. These sulphur-containing amino acids are collectively known as "half-cystine" residues. Oxidation of cystine thiol groups results in the formation of disulphide bonds, which ultimately lead to stabilization of the protein structure. These bonds can form crosslinks between different protein chains (intermolecular) or between different parts of the same chain (intramolecular).

The disulphide bonds produce a three-dimensional network, the stability of which increases with the number of crosslinks. These bonds confer a high

**Table 2** Vibrational Modes and Assignments of Various Keratin Proteins

| Approximate assignment of vibrational mode | Wool (16,18) | Hair (29,35) | Nail (29) | Stratum corneum (29) |
|---|---|---|---|---|
| Amide A ν(NH) | 3308 | | — | — |
| Amide B δ(CNH) | 3057 | | 3060 | 3070 |
| ν(CH₃) asymmetric, ν(CH) | 2971 | | — | — |
| ν(CH₂) asymmetric | 2931 | | 2931 | 2935 |
| ν(CH₂) symmetric | 2877 | | 2874 | 2882 |
| ν(CH₃) symmetric | 2848 | | — | 2854 |
| ν(CH) | 2732 | | 2733 | 2727 |
| ν(C=O) lipid | — | — | — | 1733 |
| ν(C=O) Amide I; β-pleated sheet | 1670 | | 1666* | — |
| ν(C=O) Amide I; α-helix | 1655 | 1654 | 1654 | 1651 |
| Tyr, Trp ν(C=C) | 1614 | 1615 | — | 1617 |
| Phe ν(C=C) in phase | 1604 | 1604 | — | 1606 |
| Phe ν(C=C) symmetric | 1585 | 1586 | 1585 | 1584 |
| Trp | 1553 | 1556 | — | 1558 |
| δ(CH₂)(CH₃) | 1448 | 1451 | 1450 | 1443 |
| Trp, ν(NH) | 1420 | 1423* | — | — |
| CH₂, Trp | 1338 | — | 1341 | 1338 |
| ν(Cα-H) | 1318 | 1314 | 1310 | 1316 |
| Amide III ν(CN) δ(NH) α-helix | 1266 | 1271 | — | 1272 |
| Amide III β-pleated sheet | 1238 | 1239 | — | 1253? |
| Tyr, Phe ν(C-C6H5) | 1207 | 1209 | — | 1208 |
| Tyr ν(CH) in phase | 1176 | 1177* | — | 1177 |
| ν(CN) | 1155 | 1158* | — | 1158 |
| ν(CN) | 1125 | 1128* | 1127* | 1128 |
| ν(CN) | 1096 | 1102* | — | — |
| ν(CN) | 1080 | 1081* | 1087* | 1080 |
| ν(CC) | | 1061 | — | 1062 |
| Phe ν(CH) in phase | 1032 | 1033 | — | 1031 |
| Phe ν(C=C) symmetric | 1001 | 1003 | 1003 | 1003 |
| (CH₂) | 952 | 953* | — | 959 |
| ν(CC) skeletal α-helix | 934 | 928* | 935 | 937 |
| ν(CC) skeletal α-helix | 898 | 893* | — | 895 |
| Trp skeletal and ν(NH) pyrrole | 881 | 876* | — | — |
| Tyr (ring breathing) | 851 | 853 | 855 | 855 |
| Tyr (out of plane ring breathing) | 828 | 825 | — | 828 |
| Trp | 757 | 751 | — | 744? |
| ν(C—S) cystine | 661 | — | — | — |
| Tyr (CC) ring twist | 642 | 643 | 643 | 642 |
| Phe (CC) ring twist | 618 | 627 | 622 | 619 |
| ν(S—S) cystine trans-trans | 545 | — | — | — |
| ν(S—S) cystine gauche-trans | 532 | 529 | — | — |
| ν(S—S) cystine gauche-trans | 519 | | — | — |
| ν(S—S) cystine gauche-gauche | 512 | 506 | 520 | — |

* Authors' assignments differ from those listed in the table.

degree of physical and chemical stability to keratin proteins. Structures such as wool, hair, hooves, horns, claws, beaks, and feathers are classified as "hard" keratins, because the sulphur concentration in these proteins is greater than 3%. Keratin proteins containing less than 3% sulphur, such as the outermost layer of skin, are classified as "soft" keratins (11,12).

Keratin proteins can also be classified on the basis of one of two x-ray diffraction patterns. The first pattern was denoted as being that of an α-keratin and was observed in structures such as wool and hair fibers. From the x-ray diffraction pattern it was deduced that the polypeptide chains of an α-keratin protein were arranged in a coiled helical configuration; they are referred to as the α-helix. The second pattern was denoted as β-keratin and was observed in structures such as feather keratin and stretched mammalian keratin. This pattern suggested that the polypeptide chains existed in an extended form arranged in sheets; they are referred to as a β-pleated sheet conformation (13). See Fig. 7.

## 2.1.1 Wool

### 2.1.1.1 Morphological Structure of Wool

Wool is a natural biopolymer composed of two morphologically and chemically different layers known as the *cortex* (inner layer) and the *cuticle* (outer layer). Clean wool is essentially pure protein, 82% of which is classified as keratinous based on the fact that this type of protein contains more than one half-cystine residue for every 33 residues. And 17% is classified as nonkeratinous; that is, the protein contains less than one half-cystine residue for every 33 residues (11). The remaining 1% of the fiber weight is attributed to the wool lipids. Figure 8 depicts the organization of the various wool fiber components.

2.1.1.1.1 CORTEX. The cortex is the major structural component of the wool fiber and represents approximately 90% of the total fiber weight. The cortical cells are composed of crystalline protein, approximately 7 nm in diameter, called *microfibrils*, which are formed from low-sulphur proteins that favor the formation of α-helices. The microfibrils are embedded in an amorphous protein matrix composed of high-sulphur protein, high-glycine/high-tyrosine proteins, and water. The microfibrils are then arranged into large cylindrical units called *macrofibrils*, which are approximately 0.3 μm in diameter and range in length from 10 to 95 μm (11,13–15).

Other cortical cell components are nuclear remnants and intermacrofibilar material derived from the nucleus and cytoplasm of once-living cells, which represent about 13% of the nonkeratinous protein present in the cortex. Electron microscopy has been used to classify three types of cortical cells based on the composition and arrangement of the microfibril/matrix system within the macrofibrils of the cortical cells, and they are known as the *orthocortex, mesocortex,* and *paracortex* (11,12,14).

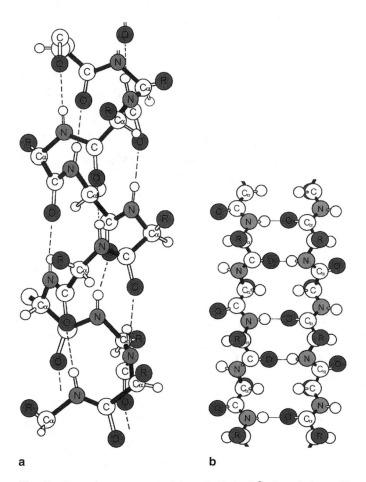

**Fig. 7** Secondary structure of the α-helix and β-pleated sheet. (From Ref. 1.)

2.1.1.1.2 CUTICLE. The single layer of overlapping cells, which encapsulate the cortex, is known as the *cuticle*. These cells constitute approximately 10% of the mass of the whole fiber, are rectangular in shape (20 × 30 × 0.5 μm), and are scalelike in appearance. The tips of the cuticle cells point away from the root of the fiber, giving a serrated edge (14). The cuticle cells are composed of three subcomponents: the *epicuticle*, *exocuticle*, and *endocuticle*; these are depicted in Fig. 9.

The cuticle cells overlap and are arranged longitudinally and circumferentially, with the thickness of the cells decreasing from the distal, or protruding,

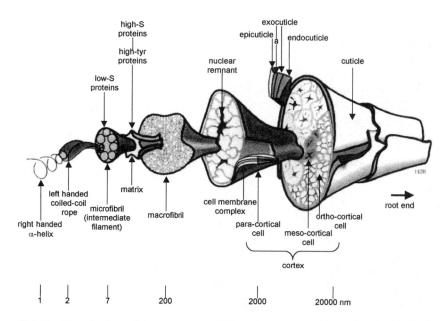

**Fig. 8** Organization of the various wool fiber components. (Reproduced with the kind permission of the C.S.I.R.O. Division of Wool Technology.)

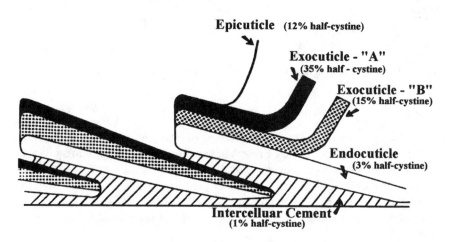

**Fig. 9** Substructure of the cuticle. (From Ref. 11.)

edge to the attached proximal end. The cuticle thickness, for fine wool, is greater on the intrados of the crimp (the paracortical side) than on the extrados (the orthocortical side). The epicuticle is the outermost layer of the wool fiber. It is a hydrophobic, chemically resistant, semipermeable membrane, with a thickness of 30–60 Å, which is chemically different from the bulk of the fiber and accounts for around 0.1% of the fiber mass. The exocuticle is a keratinous layer that lies underneath the epicuticle. It is approximately 0.3 μm thick and represents 60% of the total cuticle mass. It is particularly rich in cystine and averages one cross-link per five amino acid residues in the polypeptide chains, which is twice the amount present in the bulk of the wool fiber. Due to its high density of cross-linkages, the exocuticle is extremely resistant to enzyme attack. The exocuticle is composed of two sublayers. The dense "A layer" is about 0.1 μm thick and lies adjacent to the epicuticle; it is believed to contain a higher level of cystine than the "B layer" (11). The endocuticle lies between the exocuticle and the cell membrane complex and is approximately 0.2 μm thick. The endocuticle is approximately 40% of the mass of the cuticle and has a relatively low crosslink density, with one half-cystine residue for every 33 amino acid residues, and is therefore classified as a nonkeratinous protein. Due to its low cystine content, the endocuticle is mechanically weak and more susceptible than the exocuticle to chemical attack.

### 2.1.1.2   Raman Spectroscopy of Wool

The first report of a Raman spectrum of wool was by Lin and Koenig and was provided without experimental details in a review by Frushour and Koenig (16). The first FT-Raman spectrum of wool was published independently by Hogg et al. and Carter et al. in 1994 (17,18).

The predominant secondary structure present in wool is the α-helix; however, there are a number of other protein conformations present, including the random coil, or amorphous form, and the β-pleated sheet (19–23). The position of certain bands, such as amide bands, are characteristic of these various secondary structures; these are summarized in Table 1. An FT-Raman spectrum of wool is presented in Fig. 10. The positions of the amide I band, at 1654 cm$^{-1}$, and the C—C stretching vibration of the skeletal backbone, at 931 cm$^{-1}$, are indicative of an α-helical conformation. In accordance with characteristic literature values, the position of an amide III band of an α-helix would be expected within the region of 1310–1260 cm$^{-1}$. A number of weak features can be observed in the amide III region, and the feature at 1266 cm$^{-1}$ is assigned to the α-helix structure.

As previously mentioned, the cuticle and cortex are both chemically and morphologically different. Cuticle cells have a higher proportion of the amino acids cystine, proline, serine, and valine compared with cortical cells. These resi-

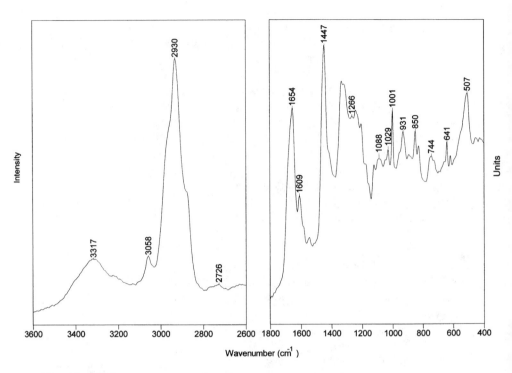

**Fig. 10**   FT-Raman spectrum of wool.

dues are generally considered to be nonhelical forming, and the proteins of the cuticle cells are thought to have a more amorphous structure that the cortical cell proteins. Transmission electron microscopy results confirmed the presence of α-helical–rich filaments in cortical cells but not the cuticle cells. The FT-Raman spectra of cuticle and cortical cells were found to have significant differences in the band position, intensity, and lineshape of the amide I band. The structure of the amide I band was analyzed by a curve-fitting technique (24).

The amide I vibration is a very complex region for most proteins, for a number of reasons. The functional groups of the aromatic amino acids phenylalanine, tyrosine, and tryptophan are observed within this region. The broadness and general asymmetry of the band can be attributed to the presence of a number of secondary structures within the protein. For example, within one type of secondary structure the dihedral angles of the peptide backbone chain vary over a wide range, and this leads to band broadening. The secondary structures also have similar frequencies, which overlap—again resulting in band broadening. Techniques such as deconvolution, curve fitting, and derivative spectroscopy can be used to assist in resolving overlapping bands.

Wool fibers are reported to undergo conformational changes during stretching in water from an α-helix to a β-pleated sheet structure. Within the literature it is reported that extension of a wool fiber in air does not lead to formation of the β-pleated sheet, but partially destroys the α-helix (25). In a recent study, Raman spectroscopy was used to study wool fibers extended to various degrees in air (26). Analysis of the amide I band, using curve-fitting techniques, together with analysis of other spectral regions revealed that wool in its natural state (i.e., with crimp) had a significantly higher α-helical content than wool extended 22%. This suggested that a significant destruction of the α-helix and a substantial increase in the β-sheet content had occurred.

### 2.1.2   Nail

#### 2.1.2.1   *Morphological Structure of Nail*

Nail is a specialized epidermal structure that has evolved with physical properties differing widely from those of normal stratum corneum (27). The nail unit is composed of four different keratinized epithelial structures known as the *proximal nail fold*, the *matrix*, the *nail bed*, and the *hyponychium*. The *nail plate*, more commonly referred to as the nail, is a flat roughly rectangular, transparent/translucent horny structure. The nail is surrounded on three sides by nail folds, and the cuticle surrounds the fourth side. The cuticle is dead tissue that is an extension of the skin of the finger; its function is to help protect the matrix from infection and injury. The keratin protein of the nail consists of a crystalline phase, composed of α-helical proteins, and an amorphous protein matrix phase. Stabilization of the matrix phase is attributed to the crosslinking disulphide bonds of cysteine residues, together with van der Waal's forces, hydrogen bonds, and ionic interactions (28). The hydration state is believed to be an important factor that influences the physical and mechanical properties of the nail.

#### 2.1.2.2   *Raman Spectroscopy of Nail*

Structural characterization of nail using FT-Raman spectroscopy was first undertaken by Williams et al. in 1994 (29). The potential of this technique was not further exploited until Gniadecka investigated the molecular conformation and interactions of water and nail protein (28,30).

An FT-Raman spectrum of nail is presented in Fig. 11; it is very similar to other keratin proteins, as illustrated in the stack plot in Figs. 5 and 6. The position of the amide I, III, and C—C backbone vibrations indicate that the predominant secondary structure of the nail protein is that of an α-helix; see Table 1.

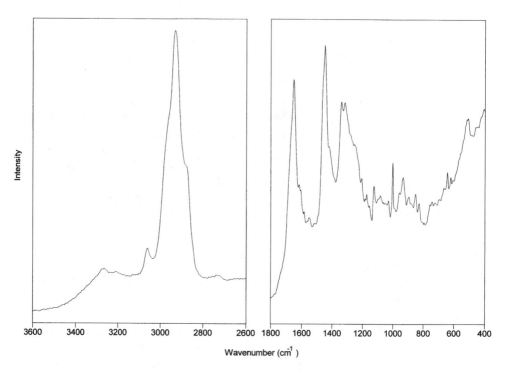

**Fig. 11**  FT-Raman spectrum of nail.

This is expected, for most keratin proteins exist in the α-helical form. Another point of interest to note is the relative intensity of the ν(CS) and ν(SS) modes, which are attributed to the high sulphur content present in nail—a "hard" keratin.

Raman spectroscopy has been used to investigate the molecular conformation of proteins and the interaction of water in human nails. In biological systems, water is hydrogen bonded either to biomolecules, such as proteins and glycoaminoglycans, or to other water molecules. *Free water* is the term used to describe water molecules hydrogen bonded to other water molecules that produce a vibration at approximately 180 cm$^{-1}$, assigned to a ν(OH) mode. *Bound water* is the term given to describe water molecules that form heterogeneous structures with other molecules, such as proteins, via hydrogen bonds (30).

It is difficult to study the low-frequency region of the Raman spectrum where, for example, the ν(O· ·O) of water is observed, due to the presence of an intense Rayleigh line. To eliminate the effect of the Rayleigh line, Eq. (1) (referred to as the R(ν) representation) has been used (31,32):

$$R(\nu) = \nu \left[ 1 - \exp\left( \frac{-hc\nu}{kT} \right) \right] I(\nu) \tag{1}$$

where $I(\nu)$ is the intensity in the Raman spectrum at the Raman shift on $\nu$ cm$^{-1}$, $c$ is the velocity of light, $T$ is the absolute temperature, $h$ is Planck's constant, and $k$ is Boltzmann's constant (31,32).

To estimate the water content of nail, a ratio of the intensities of the $\nu(CH)$ mode, at 2930 cm$^{-1}$, and the $\nu(OH)$ mode at 3250 cm$^{-1}$ was calculated. This ratio is said to be related to hydrogen bonding, or hydration, between the water and protein in the nail. From an analysis of the $R(\nu)$ representation of the Raman spectra, Gniadecka et al. proposed that over 90% of the total water molecules present in dry nail exist in the bound form (30).

The physical and mechanical properties of nail are believed to be influenced by the level of hydration. Fourier transform Raman spectroscopy was used to monitor water uptake and establish if structural changes were occurring during nail hydration (28). Water uptake was monitored by calculating the intensity ratio of the $\nu(OH)/\nu(CH_2)$ bands. (Note that this is the inverse of the previously described ratio.) Again the $R(\nu)$ representation of the low-frequency range of the Raman spectra was used for spectral interpretation. It was noted that mainly bound water was found in both dry and wet nails, which implied a water–protein interaction. The intensities of the $\nu(CC)$ skeletal backbone vibration at 932 cm$^{-1}$ and the $\nu(SS)$ vibration at 510 cm$^{-1}$ were observed to increase with water uptake. It was suggested that hydration may separate the proteins in the amorphous matrix and crystalline phases of the nail, resulting in the observed spectral changes.

### 2.1.3 Hair

#### 2.1.3.1 Morphological Structure of Hair

Hair fibers have three distinct cellular regions: the *cuticle*, the *cortex*, and a porous *medulla*. The gross structure of a hair fiber is similar to that of wool; however, there are a number of subtle variations in the shape and surface architecture of the cuticle and in the internal arrangement of different cell types (12).

The major structural component of hair is the cortex, which is composed of α-helical keratin proteins and represents 65–95% of the total fiber weight. The cortex is composed of three types of cells that are classified as ortho-, para-, and mesocortical based on the composition and arrangement of the microfibril/matrix system within the macrofibrils. The cortex is surrounded by the multilayered scales of the cuticle, which are 0.2–0.4 μm thick and are overlaid from root to tip of the fiber. The medulla, the innermost component of the hair fibers contains an irregular, girderlike framework of cortical material. Other constituents include water, lipids, pigments, and trace elements. Lipids contribute from 1 to

9% of hair matter and are derived from sebum, consisting primarily of 56% free fatty acids and 44% neutral acids (esters, glyceryl wax, hydrocarbons, and alcohols).

### 2.1.3.2 Raman Spectroscopy of Hair

Application of Raman spectroscopy to hair analysis has centered on studies that characterize the molecular and conformational nature of the sample and on those that assess the technique for applications such as biomedical diagnosis or for the development and formulation of cosmetic products for hair.

Fourier transform Raman spectra of a number of human keratotic biopolymers (stratum corneum, callus, hair, and nail) were obtained and characterized by Williams et al. in 1994 (29). It was noted that the spectra of these samples were very similar, and the principal structural dissimilarities were attributed to differences in the sulphur content of the "hard" and "soft" keratin proteins. Hair is categorised as a "hard" keratin protein, and the relative intensity of the $\nu(SS)$ modes, around 530–510 cm$^{-1}$, are stronger when compared with those of "soft" keratin proteins.

Melanin pigments are granules that provide the coloration in the hair fiber. These granules are commonly found in the cortex but may also be found in the cuticle and medulla (33). The melanin granules produce large amounts of fluorescence, and hair fibers have been reported to scorch almost immediately on exposure to a laser beam (34). In order to overcome this problem, workers have used near-infrared excitation for analysis or have limited their investigation to only naturally grey or white hair samples (34,35).

Confocal Raman microscopy is a recent technological development that provides a method of obtaining two- or three-dimensional images of small samples, which may or may not be embedded within strongly scattering substrates. This can be achieved by means of a spatial filtering system of optically conjugated pinholes (36,37). Basically, in a confocal Raman microscope, only light from the focal plane is allowed to enter the detector. Light from planes above and below the focal plane is partially attenuated. Because only the light from a localized spatial volume at the focal point is detected, it allows for discrimination between areas, at different depths, within a sample (38).

The molecular orientation and chemical composition of the constituent proteins of the cuticle and cortex of human hair fibers have been investigated using confocal Raman microscopy. Only naturally grey and white hair fibers were investigated due to fluorescence, because black, brown, and blonde hair samples scorched on exposure to an argon ion laser (514.5 nm). The authors reported observing small, but significant, differences between spectra obtained at the surface and 10 μm below the hair fiber (34).

The effects of hair bleaching, permanent waving, and photodamage have been investigated by several workers using FT-Raman spectroscopy (35,39,40).

Bleaching is a process by which strong oxidizers are used to bleach the melanin pigment. The oxidant has to diffuse into the cortex and can damage the hair fiber during the process. Comparison of the spectra of bleached and untreated hair showed the development of a vibration at approximately $1045-1040$ cm$^{-1}$ and a decrease in the relative intensity of the $\nu$(SS) mode of the disulphide group. These spectral changes were attributed to the formation of cysteic acid, an oxidation product of cystine, which formed as a result of the bleaching process.

A permanent waving treatment involves application of a reducing agent to the hair, which is then wound onto rollers, rinsed, and neutralized. The reduction and oxidation of the disulphide bond and the production of thiol groups can be monitored using Raman spectroscopy. Incomplete reoxidation of the disulphide bond is considered to be the main cause of damage to the physical and mechanical properties of the fiber. A number of spectroscopic techniques, including Raman spectroscopy, were used to investigate structural changes that human hair fibers undergo during a permanent waving treatment. X-ray diffraction and NMR data revealed a partial disruption of the $\alpha$-helical structure in the microfibrils; this occurred predominately during the reduction process. Raman and FTIR spectroscopy were employed to establish whether the observed structural changes were due to the formation of a random-coil or $\beta$-sheet structure. Spectra revealed that the $\alpha$-helix underwent a conformational change to a random-coil structure during reduction followed by oxidation, rather than the $\beta$-sheet structure (41).

### 2.1.4 Other Keratotic Biopolymers

The wide range of keratinous materials found in the animal, reptilian, and avian kingdoms illustrates the versatility and sophistication of this tissue: exoskeletal materials such as horns, hooves, hair, fur, feathers, claws, quill, beaks, and turtle shell. From primary fibrous keratin, such as rhinoceros horn, highly complex adaptions of different keratinous compositions are often found in the same species that fulfill different purposes; e.g., the soft, flexible hinges of squamate scales composed of $\alpha$-helical keratin are in contrast to the hard $\beta$-sheet conformation found in the scales themselves. Other animal horns have evolved from hair, and their protein conformations are quite similar.

The potential of FT-Raman spectroscopic characterization of keratinous materials of this sort has been realized recently, and novel information is now emerging from the application of this technique to the study of archaeological materials (42,43). This has become of special relevance to keratotic materials such as horn, tortoiseshell, and hooves, which have been fashioned into items of jewelery and adornment and can now be identified nondestructively using vibrational spectroscopy (42). The scope of FT-Raman spectroscopy, with its demonstrable spectral quality from these fluorescent materials, is an exciting possibility for future identifications in this area.

## 2.2  Skin

### 2.2.1  Structure of Skin

Skin is the largest organ of the body. Its primary function is to act as a barrier to protect the body from physical, chemical, and microbial attack, and it also acts as a sensory organ for monitoring of the external environment. Skin protects itself by producing a number of keratinized structures, such as nail, hair, and the stratum corneum (the outermost layer of skin). Human skin consists essentially of three integrated layers—the epidermis, the dermis, and the subcutaneous fatty layer; see Fig. 12 (44–45.)

The thickness of the epidermis varies over different parts of the body. It is thickest on the load-bearing parts of the body, such as the palms of the hands and the soles of the feet. The outermost layer of the epidermis, which completely covers the body, is known as the *stratum corneum* and typically consists of 10–15 layers of flattened anucleated, keratinized cells embedded in a lipoidal matrix, and for much of the skin surface it is only approximately 10 μm thick when dry.

The *dermis* is composed of two layers and is considered to be the primary structure of skin. It consists of thick, coarse interlacing collagen fibers as well as reticular and elastin fibers that are embedded in a glycosaminoglycan matrix.

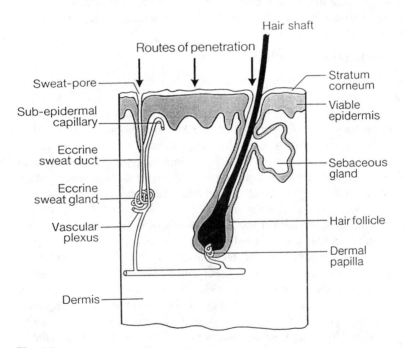

**Fig. 12**  Cross section of human skin tissue, showing the dermis and epidermis. (From Ref. 58.)

The dry weight of skin is 40–50% collagen, the bulk of which is composed of the fibrous types I and III (46). Collagen and elastin are found in a number of structural tissues, including tendons, blood vessel walls, skin, and bone. Collagen fibers comprise a number of smaller units known as *tropocollagen*. Tropocollagen is formed by three separate polypeptide chains twisting into a three-stranded coil with a right-handed screw sense. This forms a $3_1\alpha$ left-handed helix that is stabilized by proline and hydroxyproline residues. Elastin is also a fibrous protein, but previous Raman investigations have revealed that this protein exists in a disordered conformation (47). Collagen and elastin are both proteins that permit skin to stretch easily yet retain its shape; they have very distinctive secondary structures that make Raman spectroscopy an ideal technique for the molecular characterization and investigation of skin and its individual layers.

## 2.2.2 Raman Spectroscopy of Skin

Fourier transform Raman spectroscopy was first used to characterize human stratum corneum in 1992 (48). The barrier nature of the stratum corneum is derived from the unique morphology of its constituents, which are typically 75–80% protein, 5–15% lipids, and 5–10% unidentified material on a dry weight basis; see Fig. 13 (44). The predominant secondary structure of the corneocytes protein

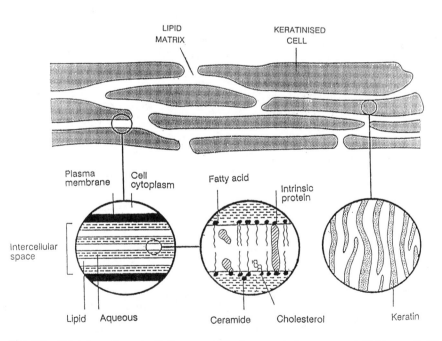

**Fig. 13** ''Bricks and mortar'' diagram of the human stratum corneum (thickness 6–30 μM) showing lipid matrix and keratinized cell construction. (From Ref. 58.)

fraction is that of an α-helix (approx. 70%), with some β-keratin (approx. 10%) and cell wall envelopes (approx. 5%) (45). The high percentage of α-helical content is reflected in the position of the conformationally sensitive amide and C—C skeletal backbone vibrations in the FT-Raman spectrum of human stratum corneum; see Fig. 14.

Fundamental to the regulation of the skin barrier function is the multilamellar lipid continuum, within which the corneocytes are embedded. The major lipids present in the stratum corneum are ceramides, cholesterol, and free fatty acids (49,50). The continuous lipid domain is thought to constitute the major pathway for most drugs penetrating across the membrane. Stratum corneum lipids exist in crystalline, gel, and liquid crystalline forms at physiological temperature. The gel phase is the most predominate form, and the lipid backbone (C—C) is arranged such that the alkyl chains are maximally extended in an all-trans structure. Conformational changes can occur during thermal or chemical perturbation and have been investigated using FT-Raman spectroscopy.

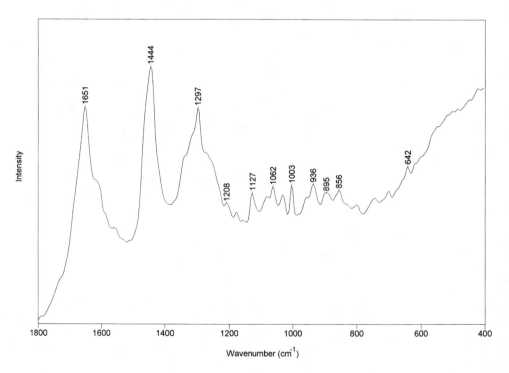

**Fig. 14** FT-Raman spectrum of human stratum corneum.

Penetration enhancers are a group of chemicals that temporarily and reversibly diminish the barrier function of the stratum corneum, allowing for increased or improved percutaneous penetration of drugs either for local therapeutic effect or for systemic therapy. Anigbogu et al. used FT-Raman spectroscopy to investigate the mechanism by which the penetration enhancer dimethyl sulfoxide (DMSO) disrupted lipid organisation (51). Curve fitting of the amide I region illustrated that the stratum corneum keratin underwent a conformational change from an $\alpha$-helix to a $\beta$-sheet structure. Furthermore, at concentrations that enhance drug flux, DMSO was found to cause conformational changes in the lipid domains from an all-trans gel phase to a trans-gauche liquid crystalline phase.

As demonstrated by Anigbogu et al., one advantage of using Raman spectroscopy is the ability to monitor conformational changes in the lipid hydrocarbon backbone (51). Lawson et al. used an FT-Raman spectrometer coupled to a heating chamber to investigate thermally induced molecular rearrangements of the model lipid system, dipalmitoylphosphatidyl choline (DPPC), and stratum corneum (52). The 3100–2700-$cm^{-1}$ spectral region is a very complex region comprising vibrations that originate from both the proteinaceous and lipid components of human stratum corneum. However, it was established that thermally induced conformational lipid changes could be observed in this region, for it represented the total hydrocarbon component and provided information on structural changes that occurred on a gross level. In this study the assignments for the $\nu(CH)$ region were modified after examination of deconvoluted spectra of exhaustively lipid-extracted and normal stratum corneum.

A novel technique developed for spectral analysis of the complex $\nu(CH)$ region was developed by combining the change in the wavenumber position of the $\nu(CH_2)$ asymmetric and symmetric stretching modes and plotting this band separation and rate of band separation as a function of temperature. This technique was found to provide a clearer description of the molecular order–disorder phenomena in lipid systems and avoided the use of intensity measurements, which can require comprehensive deconvolution routines that introduce large errors and overcomplicate measurements.

One novel approach used to characterize the function and structure of stratum corneum lipids has been to examine mixtures of isolated lipid components. The interactions and thermotropic properties of ceramides and mixtures of ceramides with cholesterol and fatty acids were investigated using FT-Raman spectroscopy and differential scanning calorimetry (DSC). Raman spectroscopy was used to measure the degree of order of ceramide alkyl chains, monitor bilayer reorganization as a function of temperature, and examine the effect of hydration on the phase transition temperature (53–57).

Recently, there has been much interest in evaluating the potential of Raman spectroscopy as a dermatological diagnostic tool for the analysis of normal and pathological skin samples (58–61). A study comparing normal, hyperkeratotic (callus), and psoriatic human stratum corneum found that the principle spectroscopic differences between normal and diseased skin were attributable to delipidization of the stratum corneum. Interpretation of Raman spectra indicated that biochemically the hyperkeratotic activity near the dermis–epidermis interface resulted in a weakening of the "bricks and mortar" structure of the stratum corneum, particularly for the psoriatic diseased tissue. Both psoriatic and callus tissue were found to have very similar Raman spectra, indicating that both conditions arose from a delipidization process that produced a predominantly keratotic material (58).

Verruca, also known as plantar warts, is another skin condition that results from hyperkeratotic activity in the epidermis. Verrucae are contagious, benign tumors that result from a viral infection and appear as small papules or nodules on the skin or mucous membrane. The most commonly used agent in preparations prescribed for verruca treatment is salicylic acid. Raman spectroscopy was used to investigate verrucae treated with a local application of a salicylic acid paint in order to elucidate the molecular basis for the therapeutic action of this preparation. Spectroscopic evidence implied that the salicylic acid selectively bonded with the human papillomavirus-containing verruca tissue, which was found only in the center of each wart (59).

One of the greatest advantages that Raman spectroscopy offers is noninvasive, nondestructive in situ analysis of biological systems. The first in vivo Raman spectroscopic study assessed intra- and intersample variation, the effect of sample hydration and the stability of human stratum corneum to laser exposure under in vitro conditions and compared spectra obtained from human stratum corneum in vitro and in vivo (62). Minor intersample variations were observed in the intensity of some vibrations; however, the peak wavenumber positions were consistent. These variations were attributed to differences in the lipid content between donors. Tissue hydration had little effect on the Raman spectrum, and the human stratum corneum samples were found to be stable to laser excitation. Significant spectral differences were observed between the in vitro and in vivo human stratum corneum spectra, and it must be noted that the spectra were from different body sites (in vitro from the abdomen and in vivo from the dorsal forefinger surface). Vibrations between 1130 and 1030 cm$^{-1}$ were more intense in the in vivo spectrum, as compared to in vitro, and were associated with the skeletal backbone of lipids. The major spectral difference was a broad feature centered around 3230 cm$^{-1}$ that was not observed in the in vitro spectrum; it was assigned to a $\nu$(NH) vibration. The origin of this vibration was proposed to be due to the presence of enzymes, sweat, bacteria, or site-to-site variation.

Recent studies comparing in vivo and in vitro Raman spectra of human skin have also reported a number of significant spectral differences (63–65). The differences have been attributed to variations in: molecular composition and structure of different skin layers, the anatomical region of analysis, lipid content, and constituents of the natural moisturizing factor (NMF) of the stratum corneum. The NMF is an efficient humectant found exclusively in the stratum corneum, and it is composed of a highly hygroscopic mixture of free amino acids, amino acid derivatives, and salts (63,64).

Figure 15 presents the FT-Raman spectra of in vivo full-thickness human skin together with stratum corneum, dermis, and collagen obtained in vitro. This figure illustrates that, in this particular instance, the spectrum obtain in vivo is predominately that of the dermis, which has a spectrum very similar to that of collagen. This would indicate that the laser is probing the dermis layer of skin, and the contribution from the epidermal membrane is minimal (65). The individual layers of skin have been characterized by both Raman and FT-Raman spectroscopy, and characteristic vibrations of these layers are found in Table 3 (65–68). Interpretation of in vivo Raman spectra must therefore be approached with caution, and it is recommended that the analyst have some understanding of the chemical and morphological structure of the sample under investigation.

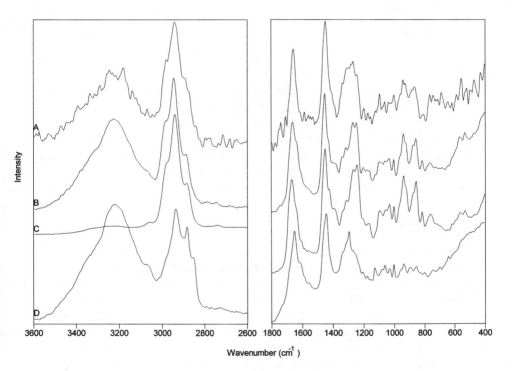

**Fig. 15**   FT-Raman spectra of (A) full-thickness skin (in vivo), (B) dermis, (C) collagen, and (D) stratum corneum.

**Table 3** Characteristic Vibrational Markers of Individual Skin Layers

| Assignment | Stratum corneum | Dermis |
|---|---|---|
| $\nu_a(CH_3)$, $\nu(CH)$ | 2975 | 2985 |
| $\nu_a(CH_2)$ | 2930 | 2940 |
| $\nu_s(CH_3)$ | 2881 | 2887 |
| $\nu_s(CH_2)$ lipids | 2854 | — |
| $\nu(C{=}O)$ helix | 1649 | 1660 |
| $\delta(CH_2/CH_3)$ | 1442 | 1450 |
| $\delta(CH_2)$ lipid | 1297 | — |
| Amide III | — | 1271 |
| Amide III | 1266 | 1247 |
| $\nu(C{-}C)$ skeletal helix | ? | 937 |
| $\nu(C{-}C)$ proline ring | — | 920 |
| $\nu(C{-}C)$ skeletal backbone | 903 | — |
| $\nu(C{-}C)$ hydroxyproline | — | 874 |
| $\nu(C{-}C)$ proline | — | 855 |

*Source*: Ref. 68.

Caspers et al. have demonstrated that Raman spectra of individual skin layers can be obtained in vivo (63). This was achieved by using a custom-built instrument with confocal capabilities, which had a microscope objective capable of translating vertically, thereby altering the depth at which the laser beam is focused into the skin sample. In vivo Raman spectroscopy is a technique likely to develop into a powerful method for future applications, such as pathologic, pharmaceutical, and cosmetic research.

## 2.3 Calcified Tissues

Bone, ivory, and teeth all belong to a group collectively referred to as *calcified tissues* (69). Collagen is the principle organic component of calcified tissues. The major inorganic component of these tissues are mineral apatites, in particular, calcium hydroxyapatite $Ca_{10}(PO_4)_6(OH)_2$. Calcified tissues, such as bone and teeth, can undergo compositional and structural changes as a result of aging, living conditions, use, and disease. A greater understanding of the microstructure and chemical constituents of these tissues can be obtained using Raman microspectroscopy. Application of techniques such as Raman microspectroscopy allow for two-dimensional functional-group mapping which illustrates the distribution of various tissue components.

## 2.3.1 Raman Spectroscopy of Teeth

The major human dental organ is the tooth, which comprises four distinct tissues: enamel, dental pulp, dentine, and cementum. *Dentine*, the major component of a tooth, is composed of about 70% inorganic mineral, crystalline hydroxyapatite $Ca_{10}(PO_4)_6(OH)_2$ and about 30% organic matter, mainly collagen. The *enamel* layer is the hardest tissue in the body, containing approximately 98% hydroxyapatite and 1% protein, and protectively surrounds the exposed crown of *dentine*. A layer of *cementum*, compositionally but not structurally similar to dentine, surrounds the root dentine.

The spatial distribution of the organic and inorganic components of human tooth enamel, dentine, and in particular the enamel–dentine junction have been analyzed using FT-Raman microscopic mapping techniques by Wentrup-Byrne et al. (70). The distribution of organic components was examined using the $v(CH)$ bands of collagen between 2880 and 2700 cm$^{-1}$. Phosphate and carbonate distribution was determined using the symmetric stretching vibrations at 961 cm$^{-1}$ ($v(PO)$) and 1070 cm$^{-1}$ ($v(CO)$). A map of the organic-to-inorganic ratio, at a spatial resolution of 10 μm, revealed a sharp increase in the organic constituents across the enamel–dentine junction, which decreased and leveled off toward the dentine center. The concentration of the carbonate ions was found to increase from the outside of the enamel toward the enamel–dentine junction. Previously, destructive sampling methods have hindered the study of this junction; the techniques used in this study illustrate that spectral information can easily be obtained and related to the physical properties and biological functions of complex biological tissues (70).

Lee et al. (71) used Raman microscopy, in combination with energy dispersive spectroscopy (EDS), to establish the identification and location of various mineral phases present in an extensively mineralized chiton tooth (Mollusca: Polyplacophora). The major lateral teeth of all chitons comprise a central tooth core composed of either apatitic calcium phosphate or iron(III) phosphate, depending on the individual species. An iron oxide cap overlies the softer core and extends down the posterior cutting surface. In the chiton species, *Acanthopleura hirtosa*, four iron minerals are known to be present: ferrihydrite ($5Fe_2O_3 \cdot 9H_2O$), goethite ($\alpha$-FeOOH), lepidocrocite ($\gamma$-FeOOH), and magnetite ($Fe_3O_4$). Previous studies have suggested the presence of lepidocrocite but could not establish its precise location. In order to identify and localize lepidocrocite, both EDS and Raman spectroscopy were used to map a line scan from the cutting edge of the posterior cusp to the anterior of a mature mineralized tooth. The Raman spectrum obtained at the cutting edge of the posterior cusp was found to be that of magnetite. The spectrum measured inside the boundary separating the two outer regions was characteristic of lepidocrocite, and the central region of the tooth contained both lepidocrocite and apatite. The anterior surface was found to be composed of a significant amount of lepidocrocite and no apatite. The spatial distribution of various mineralized components of teeth is often difficult to elucidate due to the complexity of the biomineralization process. Raman microscopy is a powerful analytical technique that has been successfully used to identify and localize mineral phases present in an extensively mineralized chiton tooth.

Modern dental amalgams are composed of composite resins, the retention of which depends on an adhesive bond forming at the tooth/material interface. Currently, there is no general treatment available to provide effective bonding at the dentine/adhesive interface. To identify the factors contributing to the failure of this interface requires an understanding of the chemical reactions that occur at these surfaces. Characterization of the interface using a number of analytical techniques, including FTIR and Raman microspectroscopy, has indicated the presence of exposed protein, which has resulted due to incomplete adhesive penetration. The lack of adhesion provides a conduit for the penetration of bacteria and fluids, the exchange of which can lead to recurrent decay, postoperative sensitivity, and breakdown of the bond (72).

### 2.3.2   Raman Spectroscopy of Bone

Bone is a complex tissue composed of an organic matrix, predominately collagen, which supports an inorganic phase comprising nonstoichiometric amounts of various phosphate and carbonate species (73). Other constituents include water, glycoaminoglycans, glycoproteins, lipids, peptides, and various ions. Traditional analytical techniques used for structural and compositional analysis of bone include light and electron microscopy, x-ray diffraction, and chemical analysis. The sample preparation required for these techniques can alter bone structure and composition. The advantages of using vibrational spectroscopy for analysis include: minimal tissue preparation, nondestructive data acquisition, the ability to obtain information at the molecular level, and the ability to analyze samples under physiologic conditions (74).

Previously, the scope of Raman spectroscopic studies of bone has been limited due to fluorescence, which was believed to originate from within the organic matrix. To overcome this problem, samples were deproteinated to remove the bulk of the collagen and consequently isolate the mineral phase of bone. With the advent of FT-Raman spectroscopy, fluorescence is no longer a problem and deproteination of bone samples prior to analysis is no longer necessary. However, FT-Raman spectroscopic studies of deproteinated bone samples are still of interest, because significant differences have been observed in the spectra of deproteinated bone and synthetic hydroxyapatite, the only common vibration being a symmetric stretching mode of the phosphate group observed at 952 cm$^{-1}$ (74,75).

Synthetic hydroxyapatites are widely used for medical and dental applications, such as prostheses and bone implants. Raman spectroscopic investigations in this area have included; monitoring the mineralization and composition of synthetic hydroxyapatite-coated implants and characterization of pre- and postimplanted dental screws (76,77).

### 2.3.3 Raman Spectroscopy of Ivory

It is now of paramount importance to environmental bodies and law enforcement agencies to be able to identify both ivory and ivory artefacts. Although it is a relatively straightforward matter to identify ivory as tusk or tusk sections, the characterization of ivory that has been fashioned into jewelery (e.g., netsukes) or that has been inlaid in other materials, such as wood and metal, is more difficult. A nondestructive means of identifying suspected illegal ivory at ports of entry is required.

Elephant and mammoth tusk ivory comes from the modification of upper incisor teeth. Most commercial ivory is obtained from the African elephant, since the tusks of Asian elephants are usually small and occur only in males of the species. However, in recent years, quality ivory from extinct *Proboscidea*, the woolly mammoths, common to the Pleistocene epoch, has been found to provide a further source of elephant ivory material. It is estimated that approximately one-quarter of commercial ivory is obtained from freshly killed elephants, and the remainder derives from carcasses of indeterminate age.

Currently, the identification of ivory and ivory substitutes is based upon macroscopic and microscopic physical characteristics in combination with a simple test of the materials using a hand-held ultraviolet light. Fourier transform Raman spectroscopic studies of ivory and similar material have concentrated on the identification of genuine ivory specimens as opposed to fakes; where genuine ivory is composed of collagen protein in an inorganic phosphatic matrix, the latter are often polymer composites based on polymethylmethacrylate and polystyrene or polyurethane resins to which may be added pulverized calcite to simulate the texture and density of natural ivory. The conclusions of those experiments are very encouraging, in that the viability of the FT-Raman technique for the nondestructive study of genuine and fake ivories was established for a variety of specimens; the latter consisted of carved objects of unusual shape, which provided a challenge for practical spectroscopy.

Of critical importance in the advancement of this work to the characterization of ivory specimens from animals other than elephants is the ability to monitor the changes in the observed Raman spectra arising from different compositions of organic and inorganic components in the ivories. Ivory is composed of osteons, which contain lamellae of collagen embedded in an inert inorganic matrix of carbonated hydroxyapatite, generally formulated as $Ca_{10}(PO_4)_6(CO_3) \cdot H_2O$; the former confers elasticity and tensility and the latter hardness and rigidity on ivory samples. Whereas true ivory is applied strictly only to the tusks of *Loxodonta africana* and *Elephans maximus*, the African and Asian elephant, respectively, other species of animals are ivory producers, such as narwhal, sperm whale, hippopotamus, walrus, and wart hog.

Vibrational spectroscopic assignments for mammalian ivories from their Raman spectra are shown in Table 4 (78). The similarities between the spectra

**Table 4** FT-Raman Frequencies (cm$^{-1}$) and Vibrational Assignments of Elephant Ivory, Woolly Mammoth Ivory, Adult Human Tooth, and Collagen

| Approximate assignment of vibrational mode | Collagen | Mammoth | Asian elephant | African elephant | Tooth dentine | Tooth enamel |
|---|---|---|---|---|---|---|
| v(C=CH) | 3059 w | 3060 w | 3060 w | 3063 w | 3063 w | — |
| v(CH$_3$) symmetric | 2978 m, sh | 2978 m, sh | 2978 m, sh | 2978 m, sh | 2979 m, sh | 2980 m, sh |
| v(CH$_2$) symmetric | 2940 ms | 2939 ms | 2940 ms | 2940 ms | 2940 ms | 2940 ms |
| v(CH$_2$) asymmetric | 2884 m | 2884 m | 2885 m | 2884 m | 2883 m | 2883 m |
| v(CH$_3$) symmetric | — | — | — | — | 2861 w | — |
| v(CH$_2$) | — | — | — | — | 2840 vw | — |
| v(C=O) citrate | — | — | — | 1721 vw | 1735 vw | — |
| v(C=O) amide I; α-helix | 1667 m | 1665 m | 1665 m | 1666 m | 1667 m | 1667 w |
| v(C=C) | 1638 mw, sh | 1638 mw, sh | 1638 mw, sh | 1638 mw, sh | 1637 w | — |
| v(CCH) aromatic ring quadrant | 1611 w | 1605 w | 1604 w | 1607 w | 1605 vw | — |
| v(CCH) aromatic ring quadrant | 1584 wv | 1588 vw | 1588 vw | 1588 vw | 1586 vw | — |
|  | 1561 vw | — | 1566 vw | 1565 vw | — | — |
|  | — | — | — | 1504 vw | — | — |
| δ(CH$_2$) | — | — | — | — | 1460 w | — |
| v(NH) | 1451 ms | 1450 ms | 1451 ms | 1451 ms | 1451 ms | — |
| δ(CH$_2$) | 1425 m, sh | 1427 m, sh | 1427 m, sh | 1427 m, sh | 1425 w | — |
| $\nu_3$(CO$_3^{2-}$) | — | — | — | — | — | 1415 w |
| δ(CH$_2$) | 1391 w, br | 1382 w | 1380 w | 1383 w | 1380 w | — |
| δ(CH$_3$) | 1342 w | 1342 w | 1343 w | 1342 w | 1340 w | — |
| γ(CH$_2$) | 1316 w | 1315 w | 1317 w | 1317 w | 1317 w | — |
| δ(NH$_2$) amide III | 1270 m | 1272 m | 1270 m | 1271 m | 1270 m | — |
| δ(NH$_2$) disordered phase | 1245 m | 1245 m | 1245 m | 1244 m | 1245 m | — |
| ω(CH$_2$) in phase | 1205 vw | 1205 vw | 1205 vw | 1205 vw | 1202 vw | — |
|  | — | 1177 w | 1177 w | 1177 w | — | — |
| v(CCC) out of phase | 1171 br | 1166 w | 1166 w | 1166 w | 1174 w | — |

| Assignment | | | | | | |
|---|---|---|---|---|---|---|
| | 1124 ww | 1124 ww | 1124 ww | 1125 ww | 1101 vw | 1072 ms |
| $\nu$(PO) out of phase; $\nu$(CO$_3^{2-}$) | 1098 mw | 1100 w | 1101 w | 1100 w | 1071 ms | — |
| $\nu_3$(PO) out of phase; $\nu$(CO$_3^{2-}$) | 1057 w | 1071 s | 1067 ms | 1069 ms | — | — |
| $\nu_2$(PO) | — | 1044 m, br | 1046 w | 1047 w | 1045 ms | 1045 m |
| | 1033 mw | 1033 mw | 1032 mw | 1033 mw | 1032 w | 1032 w |
| $\nu$(CC) aromatic ring, phenylalanine | 1004 mw | 1005 mw | 1003 mw | 1002 mw | 1003 mw | — |
| $\nu$(PO) symmetric | 939 m | 960 vs | 960 vs | 960 vs | 960 vs | 960 vs |
| $\omega$(CH$_2$) out of phase | 921 mw | 921 mw | 921 mw | 921 mw | 921 w | — |
| $\nu$(CC), proline | 873 mw | 875 mw | 875 mw | 875 mw | 875 mw | — |
| $\delta$(CCH) aromatic | 855 m | 855 m | 856 m | 855 m | 855 mw | — |
| $\nu$(COC); $\nu$(CC) | 815 mw | 816 mw | 816 mw | 816 mw | 815 mw | — |
| $\nu$(CCO) citrate | — | — | — | — | 781 vw | — |
| $\nu_4$(CO$_3^{2-}$); $\nu$(CCO) | 760 mw, br | 765 mw, br | 770 mw, br | 765 mw, br | 760 w | 758 w |
| $\rho$(CCH); $\nu$(CCO) | 725 w | 722 vw | 720 w | — | — | — |
| $\delta$(PO) | — | 671 vw | 673 vw | 673 w | — | — |
| $\delta$(PO) | 591 w | 610 w | 610 w | 609 w | 609 w | 609 w |
| $\delta$(CCC) | — | 585 m | 582 m | 583 m | 590 m | 590 m |
| | — | — | — | — | 581 | 581 |
| | 565 | — | — | — | — | — |
| $\delta$(CCC); $\delta$(PO) | 533 w | 532 w | 534 w | 536 w | 535 vw | — |
| | — | 448 vw | 453 vw | — | 448 w | — |
| $\nu_2$(PO); $\nu$(SiO) | 418 w, sh | 431 m | 428 m | 429 m | 430 m | 430 m |
| $\delta$(CCC) | 391 mw | 390 mw | 392 mw | 393 mw | 387 w | — |
| | 299 mw | 296 w | 296 mw | 297 mw | — | — |
| | 255 vw | — | — | — | — | — |

$\nu$: stretching; $\delta$: deformation; s: strong; m: medium; w: weak; sh: shoulder; br: broad.
*Source*: Ref. 78.

from the different species should be noted, and this highlights one of the major problems associated with the adoption of the Raman spectroscopic technique for ivory identification. The spectral differences may be summarized in terms of the change in relative intensity of the bands, particularly composite features in the 700–300-cm$^{-1}$ region, associated with the $\delta$(PO) modes of the carbonated hydroxyapatite matrix, and near 1060 cm$^{-1}$, associated with collagen and $\nu(CO_3^{2-})$ modes of the matrix.

## 2.4   Gallstones/Kidney Stones

The function of the kidneys is to process blood and remove both metabolic waste products and foreign chemicals. Water and inorganic electrolytes are excreted that regulate the volume, osmolarity, mineral composition, and acidity of the body (79). Precipitation of urinary constituents, often due to biochemical and metabolic disorders, results in the formation of kidney stones, also known as *urinary calculi*. The presence of these calculi can cause severe pain, obstruction, and infection of the urinary tract and complete renal damage (80,81).

Gallstones are complex biomineralized deposits formed in the gallbladder and develop as a result of a disease known as *cholelithiasis*. Gallstones represent the gradual collection of precipitated bile components and can cause obstruction or infection of the gallbladder or the common bile duct. Symptoms of cholelithiasis include abdominal pain, jaundice, and fever, and treatment is either surgical (cholecystectomy) or medical (bile salts taken orally to dissolve gallstones) (82,83).

The classification, diagnosis, and correct treatment of kidney stones are dependent upon a knowledge of the chemical composition of the stone (84,85). In order to develop preventative medical treatments for gallstones it is important to understand the mechanism of nucleation and growth (86–88). Structural and compositional methods of identification of these stones have included x-ray diffraction, light and electron microscopy, and wet chemical techniques that can be time-consuming and destructive (89,90). Vibrational spectroscopic techniques have been applied for the nondestructive analysis of both types of stones with great success and shall be examined in later sections.

### 2.4.1   Chemical Composition of Gallstones

The chemical constituents of gallstones include: cholesterol, bile pigments, glycoprotein, protein, calcium, and other metal salts. Gallstones can be categorized into four groups based on these constituents: cholesterol (stones containing greater than 70%), mixed (varying amounts of cholesterol and bilirubinate salts), and two kinds of pigment (black and brown); see Table 5. Brown stones and calcium bilirubinate are associated with stasis and bacterial infection and occur

**Table 5**  Gallstone Categories

| Group | Category | Chemical constituents |
|-------|----------|----------------------|
| 1 | Cholesterol | >70% Cholesterol |
| 2 | Mixtures of cholesterol and bilirubinate | Varying amounts of each component |
| 3 | Pigment stones: black | Calcium phosphate and calcium carbonate |
| 4 | Pigment stones: brown | Up to 25% of calcium palmitate |

*Source*: Ref. 91.

throughout the biliary tract. Black stones are associated with hemolysis, cirrhosis, and old age and form predominantly in the gallbladder.

### 2.4.2  Raman Spectroscopy of Gallstones

There is no single vibrational spectroscopic technique capable of providing all the spectral information required for gallstone characterization. An extensive study conducted by Wentrup-Byrne et al. (91) used a combination of IR microscopy, microspectroscopic mapping, photoacoustic spectroscopy (PAS), and FT-Raman spectroscopy to examine the chemical composition and microstructure of different categories of human gallstones. A total of 50 gallstones were investigated using FT-Raman spectroscopy. Spectral variation between stones and within a stone was reported, and it was necessary to obtain spectra from several points in a stone to establish all the chemical constituents present. Of the categories studied, only Raman spectra of the black pigment stones were dominated by a fluorescent background and were featureless; this was found to be an inherent characteristic of these stones.

Infrared microscopy and application of mapping techniques were reportedly excellent methods for obtaining spectral information from stones, particularly those with microstructural characteristics. Although sample preparation involved cutting the stone in half, it was less destructive than present methods of gallstone identification, which involve crushing, extraction, and analysis of constituents using wet chemical techniques. Photoacoustic spectroscopy proved to be the most suitable vibrational spectroscopic technique for gallstone characterisation, for data about the whole stone could be obtained with minimal sample preparation (91).

Thalassemia is a group of genetically inherited diseases in which the impaired synthesis of hemoglobin leads to chronic anemia. A consequence of chronic anemia is an iron overload that accumulates in tissues as nanoscale iron(III) oxyhydroxide particles. Pathological biomineralization deposits, such as gallstones, are a major cause of morbidity and hospitalization in thalassemia patients. Clinically, it is of interest to characterize these gallstones in order to obtain a better understanding of the thalassemic stone formation and occurrence.

Mössbauer, FTIR, and FT-Raman spectroscopic techniques were employed to study the chemical composition of pigment gallstones obtained from patients suffering from β-thalassemia (87). Visual classification of the stones categorized all stones as black. The FTIR spectra obtained from all the samples were very similar to the spectrum of the control black pigment stone, with one exception. Analysis of the FTIR spectrum of this one stone revealed that this sample had a low cholesterol content, and distinctive vibrations attributed to $CaCO_3$ were also observed, although it could not be ascertained in what form the $CaCO_3$ existed. Vibrations in the FT-Raman spectrum of this sample indicated that the $CaCO_3$ was present in the calcite form.

Spectroscopic results indicated that the thalassemic gallstones investigated were not identical in composition and that further subcategorises existed. The presence of iron(III) and high $CaCO_3$/low cholesterol content in one sample could not be explained. A more extensive study was planned to establish whether low cholesterol could be linked with the concomitant presence of high $CaCO_3$ and high iron content in certain thalassemic stones. The heterogeneous nature of most gallstones precludes characterization by any one technique. The findings of this study highlight the benefits of utilizing several spectroscopic techniques for analysis.

In China, pigment gallstone formation commonly occurs in the intrahepatic ducts of the liver. The surgical procedure to remove these gallstones is difficult and the pathogenesis for pigment stones is not well understood. A study by Zhou et al. (92) investigated a number of pigment gallstones using a suite of spectroscopic techniques. Elemental analysis using atomic emission spectroscopy (AES), induced coupled plasma–AES, and atomic absorption spectroscopy (AAS) revealed the complexity of pigment gallstones, with up to 20 kinds of ions detected. Polyacrylamide gel electrophoresis (PAGE) studies identified the presence of pigments, polysaccharides, proteins, and glycoproteins. Smaller molecules such as cholesterol, bilirubin and bile salts were also detected.

Three further subcategories of gallstones (carbonate, apitate, and the salt of fatty acids) were observed using FTIR spectroscopy. Brown pigment stones were observed to have a strong band at 300 cm$^{-1}$, attributed to calcium bilirubinate complexes producing Ca–O and Ca–N vibrations. A very broad band in the spectra of black pigment stones was said to be indicative of multiple metal ions (Ca, Cu, Mg, Fe) present in high concentrations. No characteristic metal-ligand vibrations were observed in cholesterol gallstones, and this supported elemental analysis data that indicated these stones had a very low metal content.

The main gallstone categories were easily distinguished using FT-Raman spectroscopy, and characteristic Raman vibrations were tabulated; see Table 6. As in previous studies, brown pigment stones were found to produce good-quality

**Table 6** Assignment of the FT-Raman Spectra of Gallstones

| $cm^{-1}$ | Assignment | Gallstone type |
|---|---|---|
| 2964s, 2935s, 2904s | $v_{as}(CH_2, CH_3)$ | Cholesterol |
| 2886s, 2850s, 2807s | $v_s(CH_2, CH_3)$ | |
| 1671 m | $v(C=C, C-O)$ | |
| 1461m, 1438m, 1328m | $\delta(CH_2, CH_3)$ | |
| 1618s | $v(C=C, C=O, C-N)$ | Brown gallstones |
| 1577m, 1501w | $v(C=C, COO)$ | |
| 1456m, 1341m | $\delta(CH_2, CH_3)$ | |
| 1624w | $v(C=C, COO)$ | Black gallstones |

*Source*: Ref. 92.

spectra, whereas fluorescence and broad ill-defined bands dominated the spectra of black stones. The FT-Raman spectrum of copper bilirubinate was found to be very similar to that of black pigment gallstones, and it is believed that these stones contain a particularly high content of polymeric copper bilirubinate complexes.

### 2.4.3 Chemical Composition of Kidney Stones

The chemical structure of a kidney stone is dependent upon a number of factors, such as diet, age, sex, environment, occupation, and metabolism. The typical analytical techniques used for analysis include x-ray diffraction (XRD), scanning electron microscopy (SEM), polarization microscopy, and wet chemical procedures, which can be destructive and time-consuming. Reliable and efficient treatment of kidney stones relies on a knowledge of the chemical composition of the stone, which in turn enables classification of the disease. Vibrational spectroscopic techniques are ideal for kidney stone analysis, because they provide rapid, sensitive, and accurate information about the major chemical components of the stone.

### 2.4.4 Raman Spectroscopy of Kidney Stones

Kidney stones have been categorized into six groups, depending on their phase composition using vibrational spectroscopy, the results of which are presented in Table 7 (84). Proton-induced x-ray emission (PIXE) was used for determination of elemental composition, and atomic emission spectrometry (AES) confirmed that lead accumulated in high concentrations in calcium-containing stones. Previous studies have indicated that trace elements are significant in the formation and growth of stone particles. Therefore, it is necessary to use a combination of analytical techniques to better identify the composition and location of major and

**Table 7**   Renal Stone Categories

| Group | Chemical name | Chemical formula | Mineralogical name |
|-------|---------------|------------------|--------------------|
| 1 | Magnesium ammonium phosphate hexahydrate | $MgNH_4PO_4 \cdot 6H_2O$ | Struvite |
| 2 | Calcium phosphates | $Ca_5(PO_4)_3(OH),$ $Ca_3(PO_4)_2$ | Hydroxyapatite, whitlockite |
| 3 | Mixtures of phosphates and oxalates | $CaC_2O_4 \cdot H_2O$ | |
| 4 | Calcium oxalate monohydrate and/or dihydrate | $CaC_2O_4 \cdot H_2O$ | Whewellite |
| | | $CaC_2O_4 \cdot 2H_2O$ | Wedellite |
| 5 | Mixtures of oxalates and uric acid | | |
| 6 | Uric acid | $C_5N_4O_3H_4$ | |

*Source*: Ref. 84.

minor components within kidney stones, for such information may indicate a specific process of stone nucleation or growth.

It is estimated that 75% of all kidney stones are composed mainly of pure or mixed forms of hydrated calcium oxalate (86). Calcium oxalate monohydrate (whewellite) and dihydrate (weddellite) are two forms of calcium oxalate that can easily be distinguished using vibrational spectroscopy. Characteristic Raman vibrations of the monohydrate include a doublet at 1493/1468 cm$^{-1}$ assigned to a $v_s(COO^-)$ and a $v(CC)$ mode at 898 cm$^{-1}$. In the Raman spectrum of the dihydrate the $v_s(COO^-)$ is a sharp singlet at 1477 cm$^{-1}$ and the $v(CC)$ is observed at 915 cm$^{-1}$. If synthetic mixtures are used to calibrate the peak heights of the $v_s(COO^-)$ and $v(CC)$ bands, then the proportions of the mono- and dihydrate forms can be found (88).

Application of vibrational spectroscopy to the analysis of kidney stones has been extensively reviewed by Carmona et al. (88). This review examines the characteristic vibrations of a number of kidney stones, including phosphates, uric acid, urates, and cystine, and is also extended to literature regarding stones from canine, feline, and equine animal species. The advantages in application of Fourier deconvolution techniques to resolve overlapping peaks of these multicomponent systems and accurate spectral subtraction are pointed out for both Raman and infrared spectra. Because of the inherent fluorescence that arises from the organic matrices of urinary calculi when irradiated with visible laser radiation, the near-infrared (1064 nm) excitation employed in FT-Raman studies of kidney stones has a demonstrable advantage in spectral quality, but it has not yet been applied extensively in this area.

The state of hydration of the calcium oxalate component in renal stones is readily distinguished in Raman spectroscopy; there is still some conjecture about the mechanism of deposition of the calcium oxalate, especially in mixed stones. On one hand, it is believed that the monohydrate is deposited initially and that this is then partially converted to dihydrate. An alternative hypothesis is that the metastable dihydrate, formed first, is then converted to monohydrate—a phenomenon that has been identified in synthetic specimens. This is an area for future Raman studies.

## 2.5 Eyes

Physiological and pathological changes in ocular tissues are normally studied using biochemical and morphological techniques, both of which require some destructive sampling or biopsy of the tissues concerned. During this sampling process it is well known that postsampling biochemical changes can occur, for example, artificial oxidation of the sulfhydryl groups of lens proteins (93). Raman spectroscopy has been used to probe the physiological condition of eye tissue in situ and nondestructively, particularly for cataract development.

Information about eye lens proteins ($\alpha$-, $\beta$-, and $\gamma$-crystallins), to the extent of 33% by weight of the eye lens, and lens water can be obtained using Raman spectroscopy, and the presence of key indicator molecules in cataract formation, such as cysteine, tyrosine, and tryptophan, have been targeted; lens dehydration is known to occur in the normal aging processes, while lens hydration occurs in the early stages of lens opacification and cataract development. Implication of the formation of S—S linkages and of high-molecular-weight protein aggregates in cataractogenesis can be monitored by Raman studies of the diminution in intensity of the $\nu$(SH) band at 2580 cm$^{-1}$ with time; the involvement of tyrosine and tryptophan residues in these aggregation processes has also been studied using characteristic bands in the 900–700-cm$^{-1}$ region of the Raman spectra, from which microenvironmental changes in the tyrosine and tryptophan residues involved in lens protein aggregation are studied (94,95).

Imaging experiments on cataractous spots in the human eye lens using a Raman probe molecule, filipin, for cholesterol with unesterified 3-$\beta$-OH groups have demonstrated the presence of these cholesterols in increased concentration inside the cataractous spots (96). Filipin accumulates significantly more in the radial shades than in the surrounding healthy lens tissues, and this has been related to the presence of larger lipid concentrations. Also it has been shown that the cataractous spots are deficient in phenylalanine and that disulphide bridge of specific trans-gauche-trans geometry were present (97).

Applications of SERS techniques to eye lens pigments (98) and lens extracts (99) have also been undertaken and were shown to be effective in the identification of selected material at high sensitivity. Although naturally involving destructive Raman analysis, the methods provide novel information about pigments and amino acids in eye lens extracts.

## 2.6 Neoplastic Tissues

### 2.6.1 Breast

Laser-based diagnostic probes using Raman spectroscopy have recently been used to address two major areas in the characterization of human breast tissues, namely, breast cancer diagnosis and the effects of silicon breast implants on the degeneration of encapsulated tissue. Breast cancer accounted for nearly 20% of all cancer-related deaths in the United States alone in 1996. The early-warning ramifications of identification of the presence of breast cancers in their early stages are of supreme importance. The normal mammographic detection methods require the removal of relatively large amounts of tissue; alternative methods, such as aspiration cytology and needle biopsy are less invasive of breast tissue but require skilled postsurgical sample treatment for maximum diagnostic results. McCreery et al. have pioneered the use of Raman spectroscopy in the differentiation between normal, malignant (ductal carcinoma) and benign (fibrocystic change) human breast tissue (100–102). The Raman spectra of malignant and benign tissues both showed an increase in protein content and decrease in lipid content when compared with normal breast tissue. However, because of the spectral similarities between tissue types, it was not possible to use the Raman spectra to distinguish between benign and malignant tissues. A major problem was encountered in the use of conventional Raman spectroscopy in these studies, since relatively large numbers of breast cells were sampled spectroscopically and only a small proportion of them were cancerous in the early stages.

Recent experiments in this area have addressed the problem using an imaging apparatus in which spatially resolved Raman spectra are analyzed on a molecular basis from samples or "footprints" that are effectively only several hundred nanometers in size (103). The complexity of breast tissue structure itself poses problems in the characterization of diseased specimens, since a relatively large variation in fat, water, and muscular tissue components can be encountered. Again, chemometric approaches are now being used to resolve the spectral identification problems. It is clear, however, that strong signal from heme-type proteins are generated in the cancerous tissues; when used with lipid and β-carotene changes from fatty and muscular tissue, a complex spectral picture is yielding to analysis.

In the second area of investigation, the detection of silicone features in lymph node biopsy specimens from women with ruptured breast implants has been successfully accomplished by Raman spectroscopy (104). In post mastectomy patients who have undergone surgical breast reconstruction following the removal of diseased tissue, the condition is worsened by enlarged lymph nodes. Raman spectroscopy has been used to detect polydimethylsiloxane gel in lymph node dissection related to neoplastic flow or for repetitive breast reconstruction surgery. Clear Raman features at 487 and 705 cm$^{-1}$ were identified in the biopsy tissue from silicone gel infiltration (silicone lymphadenopathy). Adaptation to fiber-optic sampling and in vivo probe work is an obvious target for the future. In a related study, the decomposition or breakup of polyester and polyurethane patches securing the implants in the reconstructed breast have been characterized by Raman spectroscopy (105). In these studies, talc ($MgCO_3$) was identified in 42 out of 86 cases, intracellular within the macrophages.

### 2.6.2 Artery

In the first observation of the Raman spectra of cardiovascular tissue, Clarke et al. reported strong Raman features from calcified deposits within a coronary artery and on an aortic value; the phosphate mode at 960 cm$^{-1}$ was absent from the surrounding healthy tissue sites (106). This work was extended to in vitro human coronary samples displaying atherosclerosis lesions (fatty plaques) (107). Despite extensive dehydration of the samples due to prolonged visible laser radiation and fluorescence, Raman spectra showed the presence of carotenoids, which are believed to contribute to the yellow coloration of the atherosclerotic fatty plaques.

Using longer-wavelength radiation at 810 nm, the fluorescence emission from human arterial tissue was sufficiently reduced to allow the rapid observation of the Raman spectra (108). Raman spectra obtained in 1-second exposures were of sufficient quality to permit the identification of calcified and healthy aortal tissue; a 5-minute exposure using 20 mW of incident power with the charged-couple detector (CCD) system and 810-nm excitation compared favorably with a 35-minute exposure using 500 mW at 1064 nm with an FT-NIR system (109). However, fluorescence emissions are still large enough at 810 nm to warrant the use of special spectral subtraction techniques.

### 2.6.3 Brain

Previous applications of Raman spectroscopy to the characterization and analysis of brain tissue were fraught with problems relating to florescence of the specimen and decomposition of sensitive material by the visible laser wavelengths used. Recent application of NIR-FT-Raman spectroscopy, however, has provided some good-quality information from human brain tissue and tumors. The Raman spectra from normal but edematous human grey and white matter were similar; the Raman spectra of glioma grades II and III, acoustic neurinoma, and central neurocytoma were essentially the same as that of grey matter. One of the neurinoma contained a carotenoid that exhibited strong Raman bands at 1157 and 1524 cm$^{-1}$, whereas another neurinoma did not contain any carotenoid. The central neurocytoma of the choroid plexus exhibited a strong Raman band at 960 cm$^{-1}$, indicative of phosphate, and suggested calcification had taken place (110,111).

## 3   ARCHAEOLOGICAL APPLICATIONS

A major focus for the interest in Raman spectroscopic techniques and their application to the determination of molecular information from biological samples has arisen because of the realization that the method is nondestructive, requiring minimal sample preparation in the form of chemical or mechanical pretreatment. When this is coupled with the low Raman scattering intensity from water and the fact that specimens need not be subjected to desiccation prior to their analysis, the application of Raman spectroscopy to archaeological biomaterials is now assuming importance (112).

### 3.1   Skin

The processes by which biomaterials such as skin, hair, nail, bones, teeth, ivories, resins, and plant and animal fibers are degraded or preserved are of historical and scientific interest, since they reflect the geological, environmental, and archaeological record. The presence of exogenous materials in ancient tissue samples, for example, can indicate a mummification or preparation ritual and provide an insight into ancient processes and technologies. Mummification can represent either a natural desiccation process, which has been observed for mummies found in hot or cold desert environments, or an artificial process in which the body has been treated with chemicals and preservatives to promote preservation. The epitome of artificial mummification was achieved during the Middle Kingdom dynasties in ancient Egypt (ca. 4000 y BP), where lengthy and involved ritualistic preservation processes were undertaken. Perhaps surprisingly, it is not always certain whether archaeologically excavated human and animal tissues have been preserved artificially, and this is often a contentious issue in anthropological and archaeological research. For the first time, recently, FT-Raman spectroscopy of ancient human skin has revealed the presence of balsamic lipids and waxes in several mummified species from the Chinchorro culture (ca. 1000 y BP) in the Atacama desert (113); it was very significant that other mummified human remains from the same excavation did not show signs of this artificial mummification, and it was deduced that only one in four mummies from the cache had been so treated. This was a surprising result for archaeologists, and the reasons for this are not understood; perhaps, mummification was in an experimental stage, or there may have been a hierarchical reason.

A very important result that can be forthcoming from Raman spectroscopic studies of ancient biomaterials is novel information about the trade routes and sources of supply for ancient cultures and the elucidation of the chemical damage that may have occurred in a burial or storage environment (114); Fig. 16 gives a flow diagram of the role of the Raman technique in the analysis of archaeological materials. A very graphic demonstration of this is provided by the nondestructive analysis of traces of resinous materials adhering to funerary pottery artefacts; for

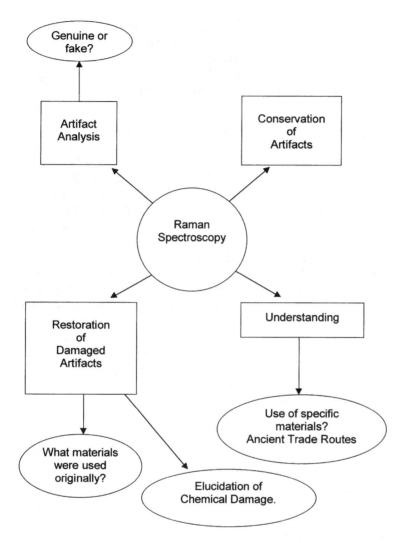

**Fig. 16** Raman spectroscopy and the conservation of artifacts.

several thousand years, resins have been used as decorative finishes (e.g., shellac, dammar), coloring agents (dragon's blood), and adornment (amber) and as practical sealants and adhesives (colophony, rosin) in many cultures.

Often, there is little material available for investigation from archaeological excavations, and for the conservation, storage, and restoration of historical items, the characterization of these ancient biomaterials is a fundamental necessity.

Also, the influence of UV radiation, humidity, and temperature on conserved specimens and the chemical processes of degradation that can occur under storage conditions are especially important. Fourier transform Raman spectroscopic studies of resins from several excavations have been undertaken to demonstrate the novel application of the technique to archaeological science:

1.  Analysis of organic residues from burial urns from the Hau Xa culture, Iron Age, in Central Vietnam dating from 2300 BP (115).
2.  Sealants from Paiute, Apache, and Washo American Indian pottery artifacts dating from 3000 BP from the Mojave, Sonoran, and Colorado Desert regions of the western United States (116); in Fig. 17, a wavenumber expansion of the 1725–1475 cm$^{-1}$ region of several contemporary resins are compared with the spectrum of a Paiute Indian "tus." From this comparison, the similarity between the Paiute resin and the local *pinyon* pine resin (I) is clearly seen. Other possibilities are very different in composition and chemical content.
3.  "Dragon's blood" resins from several geographical sources; "dragon's blood" resin has been used for over 2000 years as a decorative medium on pottery because of its rich, deep red color; it was highly

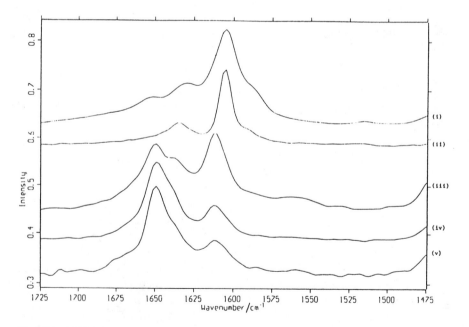

**Fig. 17** FT-Raman spectra of (i) *Pinus monophylla* pinyon resin, (ii) Paiute water bottle resin, (iii) colophony resin, (iv) rosin, and (v) Pinacae pine resin (2000 scans).

prized and became the source of much mysticism in medieval times. The Raman spectroscopic study of dragon's blood resin from several contemporary sources has been accomplished successfully. The spectra from alternative materials (including *Daemonorops draco, Draceana draco* Liliacae) have been used (117) to characterize and differentiate the Socotran resin (*Dracaene cinnibari* L.), which is currently in only limited supply but which, it is believed, provided the true origin of "dragon's blood" resin in antiquity, from several dozen synthetic and natural possibilities available today.

The first Raman spectroscopic study of ancient human skin tissue was accomplished following the discovery in September 1991 of a late Neolithic man in a glacial field between Austria and Italy that offered a uniquely preserved archaeological sample. Commonly known as Iceman (or Ötzi, having been found in the Tyrolean Ötztaler Alps), the body is the oldest to be retrieved from an alpine glacier and is one of the best preserved mummified humans ever discovered; see Fig. 18. Initially thought to date from the Early Bronze Age, Iceman dates from the Copper Age (Chalcolithic), as verified by chemical analysis of the axe he carried at the time of death. The remarkable preservation of the body is attributable to his location in a hollow some 3240 m above sealevel and the prevailing atmospheric conditions at the time of death; it is likely that Iceman froze to death before being dehydrated via a process akin to freeze-drying. By interment in a hollow under a glacier, biological degradation was minimized when compared with bodies previously found in peat bogs. Indeed, even for a glacially interred body, Iceman is remarkably preserved, in that the corpse had not putrefied and, unusually, was intact on discovery and had not been subjected to predatory attacks or mutilation. Of critical importance to the Raman spectroscopic study is the state of preservation that has arisen without the use of exogenous chemicals, as is usually the situation in dynastic Egyptian mummified tissues, which have been subjected to harsh treatment with bitumen, natron, and mixtures of hot waxes and resins.

The corpse is that of a male who was between 25 and 40 years of age when he died of natural causes, it is believed, 5200 years ago, some 2000 years before the boy Pharaoh, Tutankhamen. The circumstances surrounding the death of Iceman have been the subject of considerable speculation. Initially, consequent to the belief that the cadaver was that of an accident victim, the importance of the find was not appreciated after walkers in the area discovered the corpse. Samples of skin and bone from the corpse were sent for radiocarbon dating at Oxford and Zurich, producing calibrated figures of 5100 and 5300 years before present. From a small sample (16.2 mg) of Iceman skin radiocarbon dated at Oxford, the molecular characterization of the tissue using FT-Raman spectroscopy was determined, and the nature of Iceman skin was compared with that of contemporary skin and

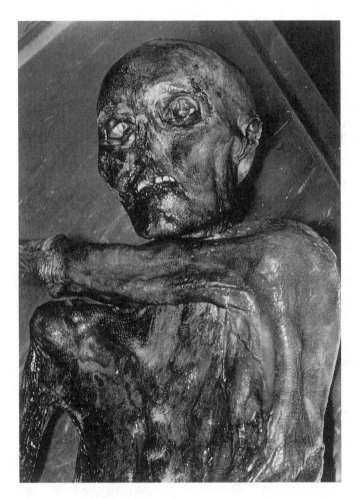

**Fig. 18**   Photograph of Iceman.

with that of freeze-dried skin, used in an attempt to mimic approximately the conditions under which the archaeological sample was preserved. During the Raman spectroscopic analysis, two distinct types of spectra were obtained from the stratum corneum and the dermis parts of the Iceman skin. By comparison with the spectra obtained from contemporary stratum corneum and dermal tissue it is possible to analyze the extent of biodeteroration of the skin tissue; it appears that the structure of Iceman skin is essentially intact following 5200 years of glacial interment but that the keratotic proteinaceous content of the stratum corneum has suffered some degradation. The lipids, though showing some evidence of C=C

**Table 8** Major FT-Raman Frequencies ($cm^{-1}$) and Approximate Descriptions of Vibrational Modes for Skin of Iceman and Contemporary Man

| Assignment | Contemporary stratum corneum | Freeze-dried contemporary stratum corneum | Iceman stratum corneum | Contemporary dermis | Freeze-dried contemporary dermis | Iceman dermis |
|---|---|---|---|---|---|---|
| $\nu(CH)$ olefinic | 3060 w | 3062 w | — | | 3059 w | 3059 w |
| $\nu(CH_3)$ symmetric | 2931 s | 2930 s | 2928 m | 2977 mw, sh | 2978 m, sh | 2977 mw, sh |
| $\nu(CH_3)$ symmetric | | | | 2938 s | 2935 s | 2935 s |
| $\nu(CH_2)$ asymmetric | 2883 ms | 2883 m, br | 2895 m, br | 2881 m | 2880 m | 2880 m |
| $\nu(CH_2)$ symmetric | 2852 m | 2853 m | 2852 m | — | — | — |
| $\nu(CH)$ aliphatic | 2723 w | 2725 w | — | — | — | — |
| $\nu(C=O)$ amide I | 1652 s | 1654 s | 1655 vw | 1669 mw | 1668 ms | 1666 vw |
| $\nu(C=C)$ | 1602 w | 1604 mw | 1596 m, br | — | — | 1595 w, br |
| $\delta(CH_2)$ scissoring | 1438 s | 1442 s | 1438 ms | 1450 s | 1450 s | 1445 ms |
| $\delta(CH_2)$ | 1296 ms | 1296 ms | 1299 ms | 1316 m, br | 1315 ms, br | 1312 m, vbr |
| $\nu(CN)$ and $\delta(NH_2)$ amide III | 1274 mw, sh | 1274 mw, sh | — | 1270 m | 1267 m | 1282 m, vbr |
| $\nu(CC)$ skeletal, trans conformation | 1126 mw | 1123 mw | 1120 vw | 1125 w | — | — |
| $\nu(CC)$ skeletal, random conformation | 1082 mw | 1081 mw | 1081 vw | 1094 w | 1097 mw | — |
| $\nu(CC)$ skeletal, trans conformation | 1062 mw | 1062 mw | 1061 mw | 1061 mw | 1061 mw | — |
| $\nu(CC)$ skeletal, cis conformation | 1031 mw | 1031 vw | 1031 vw | 1034 mw | 1032 mw | 1029 mw |
| $\nu(CC)$ aromatic ring | 1003 m | 1003 w | 1003 w | 1003 m | 1003 m | 1003 w |
| $\nu(CS)$; amide IV | 644 w | — | — | — | — | — |
| $\nu(CS)$ | 623 w | 627 w | — | — | 621 w | — |
| $\nu(CS)$ | 526 w | — | — | — | 532 w | — |

$\nu$: stretching; $\delta$: deformation; s: strong; m: medium; w: weak; sh: shoulder; br: broad.

oxidation, are still essentially intact. The dermal tissue of the ancient skin is identical with that of contemporary freeze-dried human skin. Details of the vibrational Raman assignments are given in Table 8, and sample spectra that demonstrate the keratotic degradation are shown in Figs. 19 (the $\nu(CH)$ stretching region) and 20 (the amide I stretching region) (118,119).

Following this study, a similar FT-Raman investigation of the ice-mummy cache from Qilakitsoq in Greenland (ca. 500 y BP) (120) revealed a similar conclusion, namely, that protein degradation had occurred with the human skin but to a lesser extent than that exhibited by the alpine Iceman.

The discovery of an ice-mummified penguin 110 miles from open water on Alexander Island, Antarctica, provided a novel ancient sample for comparison with the human ice-mummies; Fig. 21 shows a stack-plot of the $\nu(CH)$ stretching region Raman spectrum of ice-mummified penguin skin with those of contemporary and ice-mummified human skin for comparison. Clearly, a different process has been operational here, and the penguin sample still retains it keratotic protein-

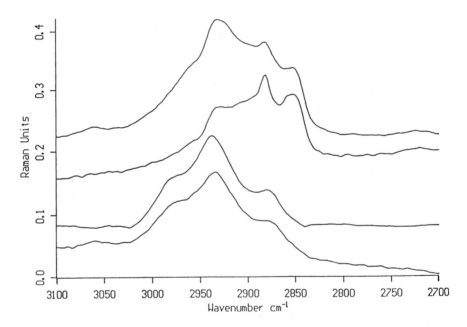

**Fig. 19**  FT-Raman spectra of the C—H stretching modes of (from top to bottom) contemporary human stratum corneum, Iceman stratum corneum, contemporary dermis, and Iceman dermis. The spectra clearly show differences between the molecular structure of stratum corneum and of dermal tissue and that Iceman has retained these distinct tissue layers.

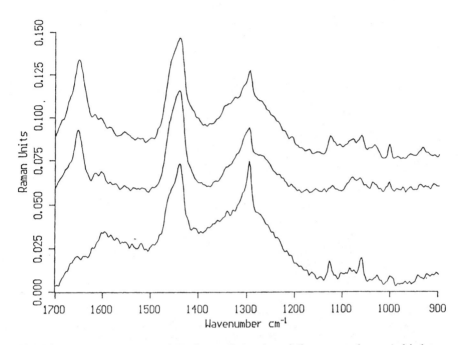

**Fig. 20** FT-Raman spectra of the fingerprint region of (from top to bottom) dried contemporary stratum corneum, freeze-dried contemporary stratum corneum, and Iceman stratum corneum. The spectra demonstrate that the protein component of the Iceman skin has degraded during its approx. 5200-year interment in an alpine glacier, whereas the lipoidal moiety of the tissue is largely intact, although some oxidation of olefinic moieties is suggested. No evidence of modern external chemical contamination can be seen in the spectrum from Iceman skin.

aceous integrity, although obviously losing lipid intensity in the 2900–2800-cm$^{-1}$ region (121).

Another unique Raman spectroscopic study of ancient mummified material has involved the analysis of a cat mummy from 3800 y BP. There has been little effort expended in the study of mummified animals, which often accompanied human burials as votive offerings to the deities. Also, it has been reported that large numbers of such mummies were used as combustible material or for medicinal purposes in the past, so their survival has not been considered very important. Prior to this Raman spectroscopic study, two mummified cats (*Felis silvestris libyca*) from Beni Hassan dating from the Middle Kingdom (ca. 1800 BC) had undergone examination in Manchester Museum; an interesting and novel discovery of artificial eye beads in the eye sockets of the animals was made. It was believed that the material of these beads was amber (122).

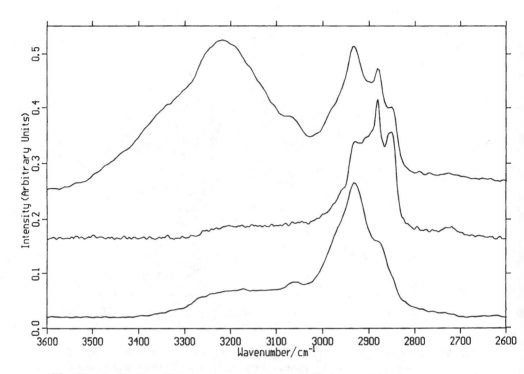

**Fig. 21** Stack-plotted FT-Raman spectra of freeze-dried human skin (upper spectrum), skin of the mummified alpine Iceman (middle spectrum), and skin of the mummified penguin from Antarctica (lower spectrum). Wavenumber range 3600–2600 cm$^{-1}$.

The FT-Raman spectra of the artificial eyes from both mummified cats' heads were identical. The spectra clearly indicate that the material of the artificial eyes is organic; a prominent $\nu$(CH) stretching feature near 3000 cm$^{-1}$ indicates aliphatic CH content, but the mid-wavenumber region is dominated by medium-strong features at 1668, 1447, and 1240 cm$^{-1}$ (broad). These are characteristic of proteinaceous material, being assigned to $\nu$(CO) amide I, $\delta$(NH), and $\delta$(CH$_2$) modes. Comparison of the Raman spectra of the artificial eye beads with that of fossilized amber resin reveals that the beads are clearly not amber, as was first suspected. The Raman spectrum of amber has a medium-strong band at 1646 cm$^{-1}$ assigned to the $\nu$(C=C) stretching, which is not present in the artificial eyes. Also, the sharp aromatic rings stretching band $\nu$(CC) at 1003 cm$^{-1}$, which is present in the spectrum of the artificial eyes, is not present in amber. However, the spectra of the artificial eye beads and those obtained from a contemporary cat's claw show remarkable similarities, with only small differences, which we can ascribe to the effects of processing and molding treatment in the bead production, such as the diminution in $\nu(SS)$ band intensity of the keratins around 500 $cm^{-1}$.

## 3.2 Hair

Other studies of archaeological samples that have undergone degradation have centered on human hair; human hair does not normally undergo biogenic alternation postkeratinization—it exhibits degradation resulting from cosmetic practices. This process continues postmortem under microbial or environmental exposure conditions. The study of ancient hair specimens in archaeology or forensic science is useful for dietary reconstruction, and the interpretation of structural changes in human hair with burial or storage environment is of fundamental importance.

Fourteen samples of human hair from disparate microenvironments, ranging from medieval to modern times, gave Raman spectra that showed evidence not only of degradation, particularly in the amide I $\nu(CONH)$ modes, but also of the $\nu(SS)$ cystine bond breakage to form cysteic acid residues. An intense feature here was the influx of contaminants from the burial matrix into the degraded samples as a result of their increased porosity. Hence, samples from lead-lined coffins and limestone rock–cave burials showed evidence of the presence of lead and calcium carbonates, respectively, in the hair samples.

All of the human hair specimens showed evidence of degradation and proteinaceous breakdown, illustrated by $\alpha$-helical to $\beta$-sheet structural changes in the amide I $\nu(C{=}O)$ band near 1650 $cm^{-1}$ and a loss in intensity of the NH deformation mode near 1451 $cm^{-1}$. Comparison of the Raman spectroscopic results with recent infrared studies on Egyptian mummy hair reveals that a diminution in band intensity of the $\nu(C{=}O)$ at 1739 $cm^{-1}$ had also occurred (123).

## 3.3 Bone and Teeth

Finally, the Raman spectroscopic analysis of archaeological hard-tissue specimens, viz. bones and teeth, have also yielded novel information about degradation processes. Spectra of the fossilized teeth of dinosaurs (70M y BP) show the incorporation of carbonate and silica in the mineralization process, with no detectable collagenic protein being exhibited. Ice-preserved material such as mammoth tusks, dating from about 12,000 y BP, still show their proteinaceous content, but with accumulation of carbonated content. More recent examples of Roman-Britain and medieval human teeth have yielded information about postburial decay processes, including bacterial attack. Archaeological interest in the temperatures to which human bone has been subject in excavated cremations sites, especially Anglo-Saxon and Viking tombs dating from about 1000 y BP, is also receiving attention, giving rising to the Raman spectroscopic study of inorganic and organic components of heated human bone, a study of some complexity and importance that relates the disappearance of the organic collagen material to the changes in the inorganic hydroxyapatite matrix.

## 4   FUTURE TRENDS

The application of Raman spectroscopy to biomedical diagnostics has provided several examples of the provision of novel information from diseased or abnormal human tissue. However, the future expansion of studies in this area will depend critically on the acceptability of the technique by the medical profession and by patient care organizations. Of fundamental importance is the optimization of the instrumentation to enable the collection of Raman spectra under clinical conditions in day care centers, in monitoring surveys, and in operation theater environments. Good-quality Raman spectra with high signal-to-noise ratios are essential for the observation of small differences between healthy and diseased tissue specimens, with the requirement of minimal exposure times and maximum patient comfort for in vivo studies.

Technical advances in reliable long-wavelength laser excitation ($\sim$780 nm or higher) with improved detector response times is an essential prerequisite for the acceptability of the technique in clinical conditions with live patient subjects. For this to be effective from the points of view of the patient and the surgical team, Raman spectra of the highest possible quality recorded in significantly less than 1 minute seems to be the future target.

The interpretation of the Raman spectra of diseased and healthy tissue depends critically on the provision of a suitable database and the identification of key molecular indicators of tissue degradation. The comparison of spectral database information and in-field raw spectra requires considerable effort and assessment of the effects of external agencies on the spectra and band appearances, for example, what effect humidity has on the observed spectra, skin temperature, applied chemicals such as skin creams, alcohol from skin swabs, etc. A major problem that needs to be addressed is the effect of pigmented and hirsute skin on the quality of the spectra. The analysis of tissue spectra form laboratories worldwide will require the application of chemometric techniques and possible curve-deconvolution algorithms; it should be borne in mind, however, that any data treatment that requires extensive and expert laboratory systems (and time!) will be self-defeating with respect to the acceptance of the technique by nonspecialists.

Several vibrational spectroscopic conferences have been called to address these matters in full or in part, and the activity generated therein is to welcomed—for example, BIOS, San Jose, CA, 1998 and 1999; Spectroscopy Against Disease, the Millennium conference, Winnipeg, Canada, 2000.

## REFERENCES

1. L Stryer. Biochemistry. 3rd ed. New York: Freeman, 1987.
2. J Twardowski, P Anzenbacher. Raman and IR Spectroscopy in Biology and Biochemistry. New York: Ellis Horwood, 1994.
3. T Miyazawa, ER Blout. J Amer Chem Soc 83:712–719, 1960.
4. AT Tu. Raman Spectroscopy in Biology. New York: Wiley, 1982.
5. AT Tu. In: RJH Clark, RE Hester, eds. Spectroscopy of Biological Systems. London: Heyden, 1986.
6. LG Tenomeyer, EW Kauffman II. In: HA Havel, ed. Spectroscopic Methods for Determining Protein Structure in Solution. New York: VCH Publishers, 1996, pp. 69–95.
7. I Harada, H Takeuchi. In: RJH Clark, RE Hester, eds. Spectroscopy of Biological Systems. London: Heyden, 1986.
8. EG Bendit. Symposium on Fibrous Proteins, 1976, pp. 386–396.
9. B Schrader, ed. Infrared and Raman Spectroscopy: Methods and Applications. New York: VCH, 1994.
10. RH Brody, EA Carter, HGM Edwards, AM Pollard. In: JC Linder, GE Trarter, and JL Holmes, eds. Encyclopedia of Spectroscopy and Spectrometry. New York: Academic Press, 1999, pp. 649–657.
11. JA Rippon. In: DM Lewis, ed. Wool Dyeing. England: Society of Dyers and Colourists, 1992, pp. 649–657.
12. JA Swift. In: RS Asquith, ed. Chemistry of Natural Protein Fibers. New York: Plenum Press, 1977.
13. P Alexander, RF Hudson. Wool: Its Chemistry and Physics. London: Chapman and Hall, 1953.
14. JA Maclaren, B Milligan. Wool Science: The Chemical Reactivity of the Wool Fiber. Sydney, Australia: Science Press, 1981.
15. DFG Orwin, JL Woods, SL Randford. Aust J Biol Sci 37:237–255, 1984.
16. BG Frushour, JL Koenig. In: RJH Clark, RE Hester, eds. Advances in Infrared and Raman Spectroscopy. Vol. 1. London: Heyden, 1975, pp. 35–97.
17. LJ Hogg, HGM Edwards, DW Farwell, AT Peters. J Soc Dyers Colour 110:196–199, 1994.
18. EA Carter, PM Fredericks, JS Church, RJ Denning. Spectrochim Acta 50:1927–1936, 1994.
19. Anon. Wool Sci Rev 25:1–16, 1965.
20. Anon. Wool Sci Rev 4:35–49, 1949.
21. Anon. Wool Sci Rev 22:1–16, 1635.
22. P Alexander, C Earland. Nature 166:396–397, 1950.
23. F Happey. Nature 166:397–398, 1950.
24. JS Church, GL Corino, AL Woodhead. Biopolymers 42:7–17, 1997.
25. R Postle, GA Carnaby, S de Jong. The Mechanics of Wool Structure. New York: Ellis Horwood, 1988, pp. 37.
26. EA Carter. Vibrational spectroscopic studies of wool. PhD dissertation, Queensland University of Technology, Brisbane, Australia, 1997.
27. N Zaias. The Nail in Health and Disease. New York: Spectrum, 1980, pp. 1–18.

28. M Gniadecka, Private communication, 1999.

29. AC Williams, HGM Edwards, BW Barry. J Raman Spectrosc 25:95–98, 1994.

30. M Gniadecka, O Nielsen, DH Christensen, HC Wulf. J Invest Dermatol 110:393–398, 1998.

31. O Faurskov Nielsen. Annu Rep Prog Chem Sec C Phys Chem 90:3–44, 1993.

32. O Faurskov Nielsen. Annu Rep Prog Chem Sec C Phys Chem 93:57–99, 1997.

33. ML Ryder. Hair. London: Edward Arnold, 1973, pp. 1–26.

34. LE Jurdana, KP Ghiggino, KW Nugent, IH Leaver. Textile Res J 65:593–600, 1995.

35. CM Pande. J Soc Cosmet Chem 45:257–268, 1994.

36. HGM Edwards. In: Encyclopedia of Applied Physics. Vol. 16. Weinheim, Germany: VCH, 1999, pp. 1–43.

37. G Turrell, M Delhaye, P Dhamelincourt. In: G Turrell, J Corset, eds. Raman Microscopy Developments and Applications. London: Academic Press, 1996, pp. 27–50.

38. P Dhamelincourt, J Barbillat, M Delhaye. Spectrosc Europe 5:16–26, 1993.

39. S Tanaka, H Iimura, T Sugiyama. J Soc Cosmet Chem Japan 25:232–239, 1992.

40. W Akhtar, HGM Edwards, DW Farwell, M Nutbrown. Spectrochim Acta 53:1021–1031, 1997.

41. N Nishikawa, Y Tanizawa, S Tanaka, Y Horiguchi, T Asakura. Polymer 39:3835–3840, 1998.

42. HGM Edwards, DE Hunt, MG Sibley. Spectrochim Acta 54A:745–757, 1998.

43. W Akhtar, HGM Edwards. Spectrochim Acta 53A:81–90, 1997.

44. GL Wilkes, IA Brown, RH Widnauer. CRC Crit Rev Bioeng pp 453–458, 1973.

45. PM Elias. J Invest Dermatol 80:44–49, 1983.

46. ND Light. In: Methods in Skin Research. Chichester, U.K.: Wiley, 1985, pp. 560.

47. BG Frushour, JL Koenig. Biopolymers 14:379–391, 1975.

48. BW Barry, HGM Edwards, AC Williams. J Raman Spectrosc 23:641–645, 1992.

49. NJ Schurer, PM Elias. Adv Lipid Res 24:27–56, 1992.

50. PW Wertz, DT Downing. In: LA Goldsmith, ed. Physiology, Biochemistry and Molecular Biology of the Skin. 2nd ed. Oxford: Oxford University Press, 1991, pp. 205–236.

51. ANC Anigbogu, AC Williams, BW Barry, HGM Edwards. Int J Pharm 125:265–282, 1995.

52. EE Lawson, ANC Anigbogu, AC Williams, BW Barry, HGM Edwards. Spectrochim Acta 54A:543–558, 1998.

53. M Wegener, R Neubert, W Rettig, S Wartewig. Int J Pharm 128:230–213, 1996.

54. R Neubert, W Rettig, S Wartewig, M Wegener, A Wienhold. Chemistry Physics Lipids 89:3–14, 1997.

55. M Wegener, R Neubert, W Rettig, S Wartewig. Chemistry Physics Lipids 88:73–82, 1997.

56. S Wartewig, R Neubert, W Rettig, K Hesse Chemistry Physics Lipids 91:145–152, 1998.

57. S Wartewig, R Neubert, W Rettig, M Wegener. Mikrochimica Acta S14:263–264, 1997.

58. HGM Edwards, AC Williams, BW Barry. J Mol Struct 347:379–388, 1995.

59. EE Lawson, HGM Edwards, BW Barry, AC Williams. J Drug Targeting 5:343–351, 1998.

60. M Gniadecka, HC Wulf, N Nymark Mortensen, O Faurskov Nielsen, DH Christensen. J Raman Spectrosc 28:125–129, 1997.

61. B Schrader, B Dippel, S Fendel, R Freis, S Keller, T Lochte, M Riedl, E Tatsch, P Hildebrandt. Proceedings of Infrared Spectroscopy: New Tool in Medicine, San Jose, CA, 1998, pp. 66–71.

62. AC Williams, BW Barry, HGM Edwards, DW Farwell. Pharm Res 10:1642–1647, 1993.

63. PJ Caspers, GW Lucassen, R Wolthuis, HA Bruining, GJ Puppels. Biospectroscopy 4:S31–S39, 1998.

64. PJ Caspers, GW Lucassen, R Wolthuis, HA Bruining, GJ Puppels. Proceedings of Biomedical Applications of Raman Spectroscopy, San Jose, CA, 1999, pp. 99–102.

65. EA Carter, AC Williams, BW Barry, HGM Edwards. Proceedings of Biomedical Applications of Raman Spectroscopy, San Jose, CA, 1999, pp. 103–111.

66. EA Carter, AC Williams, BW Barry, HGM Edwards. Proceedings of Infrared Spectroscopy: New Tool in Medicine, San Jose, CA, 1998, pp. 72–77.

67. GW Lucassen, PJ Caspers, GJ Puppels. Proceedings of Infrared Spectroscopy: New Tool in Medicine, San Jose, CA, 1998, pp. 52–61.

68. EA Carter, AC Williams, BW Barry, HGM Edwards. Unpublished Results.

69. A MacGregor. Bone, Antler, Ivory & Horn: The Technology of Skeletal Materials Since the Roman Period. London: Croom Helm, 1985.

70. E Wentrup-Byrne, CA Armstrong, RS Armstrong, BM Collins. J Raman Spec 28:151–158, 1997.

71. AP Lee, J Webb, DJ Macey, W van Bronswikj, AR Savarese, G Charmaine de Witt. J Biological Inorganic Chem 3:614–619, 1998.

72. RM Lemor, MB Kruger, DM Wieliczka, JR Swafford, P Spencer. Proceedings of Biomedical Applications of Raman Spectroscopy, San Jose, CA, 1999, pp. 73–79.

73. A Carden, JA Timlin, CM Edwards, MD Morris, CE Hoffler, K Kozloff, SA Goldstein. Proceedings of Biomedical Applications of Raman Spectroscopy, San Jose, CA, 1999, pp. 132–138.

74. I Rehman, R Smith, LL Hench, W Bonfield. J Biomed Mater Res 29:1287–1294, 1995.

75. R Smith, I Rehman. J Biomed Mater Res 5:775–778, 1995.

76. B Dippel, BT Mueller, A Pingsman, B Schrader. Biospectroscopy 4:403–412, 1998.

77. AM Tudor, CD Melia, MC Davies, D Anderson, G Hastings, S Morrey, J Domingos-Sandos, M Barbosa. Spectrochim Acta 49A:675–680, 1993.

78. HGM Edwards, DW Farwell, JM Holder, EE Lawson. J Mol Struct 435:49–58, 1997.

79. AJ Vander. Renal Physiology. 5th ed. New York: McGraw-Hill, 1995, pp. 2–15.

80. MS Dunnill. Pathological Basis of Renal Disease. 2nd ed. London: Bailliere Tindall, 1984, pp. 490–500.

81. VR Kodati, GE Tomasi, JL Turumin, AT Tu. Appl Spectros 45:581–583, 1991.

82. R Thompson. Lecture Notes on the Liver. Oxford: Blackwell Scientific, 1985, pp. 156–163.

83. BH Ruebner, CK Montgomery. Pathology of the Liver and Biliary Tract. New York: Wiley, 1982, pp. 316–317.

84. C Paluszkiewicz, M Galka, W Kwiatek, A Parczewski, S Walas. Biospectroscopy 3:403–407, 1997.

85. VR Kodati, AT Tu, R Nath, JL Turumin. Appl Spectrosc 47:334–337, 1993.

86. VR Kodati, GE Tomas, JL Toormina, AT Tu. Appl Spectrosc 44:1408–1411, 1990.

87. E Wentrup-Byrne, W Chua-Anusorn, TG St Pierre, J Webb, A Ramsay, L Rintoul. Biospectroscopy 3:409–416, 1997.

88. P Carmona, J Bellanato, E Escolar. Biospectroscopy 3:331–346, 1997.

89. JS Wei, HM Juang, WC Shyu, CS Wu. Clinical Chemistry 35:22472249, 1989.

90. VR Kodati, AT Tu, JL Turumin. Appl Spectros 44:1134–1136, 1990.

91. E Wentrup-Byrne, L Rintoul, JL Smith, PM Fredericks. Appl Spectros 49:1028–1036, 1995.

92. XS Zhou, GR Shen, JG Wu, WH Li, ZY Xu, SF Weng, RD Soloway, XB Fu, W Tian, Z Xu, T Shen, GX Xu, E Wentrup-Byrne. Biospectroscopy 3:371–380, 1997.

93. Y Ozaki, A Mizuio. Laser Applications in Life Sciences: Lasers in Biophysics and Biomedicine. Moscow: 1990, pp. 710–719.

94. Y Ozaki, K Itola, K Iriyana. J Biol Chem 262:15545–15551, 1987.

95. NT Yu, DC De Najel, JFR Kuck. In: TG Spiro, ed. Biological Applications of Raman Spectroscopy. Vol 1. New York: Wiley, 1987, p. 47.

96. C Otto, CJ deGrauw, JJ Duindam, NM Sijtsema, J Greve. J Raman Spectrosc 28:143–150, 1997.

97. JJ Duindam, GFJM Vrensen, C Otto, J Greve. Invest Opthalm Vis Sci 39:94–103, 1998.

98. S Nie, CG Castillo, KL Bergbauer, JFR Kuck, IR Nabiev, NT Yu. Appl Spectrosc 44:571–575, 1990.

99. KV Sokolov, SV Lutsenko, IR Nabiev, S Nie, NT Yu. Appl Spectrosc 45:1143–1148, 1991.

100. CJ Frank, RL McCreery, DB Reed. Anal Chem 67:777–783, 1995.

101. CJ Frank, DB Redd, TS Gansler, RL McCreery. Anal Chem 66:319–326, 1994.

102. DC Redd, ZC Feng, TY Kwok, TS Gansler. Appl Spectrosc 47:787–791, 1993.

103. NJ Kline, PJ Treado. J Raman Spectrosc 28:119–126, 1997.

104. CJ Frank, RL McCreery, DCB Redd, TS Gansler. Appl Spectrosc 47:387–390, 1993.

105. JL Luke, VF Kalasinsky, RP Turnicky, JA Centeno, FB Johnson, FG Mullick. Plastic Reconstructive Surg 6:1558–1656, 1997.

106. RH Clarke, EB Hanlon, JM Isner, H Brody. Appl Opt 26:3175–3176, 1987

107. RH Clarke, Q Wang, JM Isner. Appl Opt 27:4799–4800, 1988.

108. JJ Baraga, MS Feld, RP Rava. Appl Spectros 46:187–190, 1992.

109. JJ Baraga, MS Feld, RP Rava. Proc Natl Acad Sci USA 89:3473–3477, 1992.

110. S Keller, B Schrader, A Hoffmann, W Schrader, K Metz, A Rehlaender, J Pahnke, M Ruwe, W Budach. J Raman Spectrosc 25:663–671, 1994.

111. A Mizuno, H Kitajima, K Kawauchi, S Muraishi, Y Ozaki. J Raman Spectrosc 25:25–29, 1994.

112. HGM Edwards. Applications of Raman Spectroscopy to Archaeology. Proceedings of the XVth International Conference on Raman Spectroscopy, Capetown, South Africa, 1998, pp. 81–84.

113. M Gniadecka, HGM Edwards, JP Hart Hansen, O Faurskov Nielsen, DH Christensen, SE Guillen, HC Wulf. J Raman Spectrosc 30:147–153, 1999.

114. HGM Edwards, DW Farwell, L Daffner. Spectrochim Acta 52A:1639–1648, 1996.

115. HGM Edwards, MG Sibley. Spectrochim Acta 53A:2373–2382, 1997.

116. HGM Edwards, MJ Falk. J. Raman Spectrosc 28:211–218, 1997.

117. HGM Edwards, DW Farwell, A Quye. J Raman Spectrosc 28:243–249, 1997.

118. AC Williams, HGM Edwards, BW Barry. Biochim Biophys Acta 1246:98–105, 1995.

119. HGM Edwards, DW Farwell, AC Williams, BW Barry, F Rull. J Chem Soc Faraday Trans 91:3883, 1995.

120. M Gniadecka, HC Wulf, O Faurskov Nielsen, DH Christensen, JP Hart Hansen. J Raman Spectrosc 28:179, 1997.

121. HGM Edwards, DW Farwell, DD Wynn-Williams. Spectrochim Acta, in press, 1999.

122. HGM Edwards, DW Farwell, CP Heron, H Croft, AR David. J Raman Spectrosc 30:139–146, 1999.

123. G Lubec, G Nauer, K Siefert, E Strouhal, H Portedo, J Szilvassy, M Teschler. J Arch Sci 14:113, 1987.

# 12
# Vibrational Studies of Enzymatic Catalysis

## Hua Deng and Robert Callender
*Albert Einstein College of Medicine, Bronx, New York*

## 1 INTRODUCTION

Infrared and Raman spectroscopies measure the vibrational frequencies of a group of bonded atoms in a molecule. The frequencies are determined by the masses of the atoms, the force constants, and the geometry of the molecule. Thus, it is possible, at least in principle, to determine structural properties of a molecular group, such as bond orders, bond length/angle, and ionization state of ionizable moieties, from vibrational frequencies. Furthermore, the interactions that take place between molecular groups when a small molecule binds to a protein, such as hydrogen bonding and other bond-polarizing electrostatic interactions, geometry distortion from steric clashes, and the formation of new bonds, affect vibrational frequencies and, hence, are reported by the changes in frequencies. The accuracy of determining these parameters from vibrational frequencies is very high, generally much better than probes of protein structure such as x-ray crystallographic and multidimensional NMR studies. On the other hand, while vibrational spectroscopy is a very fine-scale probe, it is not (currently) feasible to determine the full structure of a protein from its vibrational spectrum. There is thus a high degree of synergism among these experimental probes in the information they provide.

Enzymes bind their reacting substrates and distort them with the goal of accelerating the reaction between them. Often, this means lowering the transition-state energy found in the corresponding solution reaction, although a particular enzyme may find an entirely new reaction pathway. In any case, the bond distortions imposed on bound substrates are generally small compared to the bond strength itself, because noncovalent interactions are typically involved. However,

**477**

it is precisely these interactions that bring about the large rate enhancements associated with enzymatic catalysis. Hence, measurement of the vibrational spectra of bound substrates, with its high resolution, is generally very valuable in understanding enzymatic catalysis on a molecular level. In fact, it may be said that an understanding of how a particular enzyme binds substrates and inhibitors and brings about rate accelerations in the chemistry of reacting molecules will often require this approach. It is a daunting technical problem, however, to determine the vibrational spectrum of a small molecule embedded in a much (by orders of magnitude) larger one. Some early studies were able to exploit resonance Raman spectroscopy in the study of chromophoric substrates (1,2). Recent developments in FT-IR and Raman difference techniques make it possible to measure the vibrational spectra of *any* small molecule when it is bound to a protein. In such an experiment, a protein spectrum is measured, as is that of a protein complexed with the molecule. Subtraction of the two yields the spectrum of the bound ligand. Alternatively, an atom within a bond of interest is labeled with a stable isotope ($^2$H, $^{13}$C, $^{15}$N, $^{18}$O, etc.), and this shifts the frequency of the modes that involve the motion of the labeled atom. Subtraction of labeled and unlabeled protein spectra yields an "isotopically edited" difference spectrum whereby only those modes involving the labeled atom show up in the spectrum. These general approaches have stimulated rapid progress in studies of enzymes using vibrational spectroscopy.

Here, we will discuss the general approaches used on enzymic systems and try to answer how the normally complex spectra of proteins can be understood and assigned using the new difference experimental techniques and how the vibrational data are related to structural and functional issues that are of importance to enzymes. Several specific enzymes are discussed in some detail. Many other systems are cited (we have tried to be reasonably thorough but have inevitably left some out), and the reader is referred to the references for more details on them. Techniques for the Raman/FT-IR difference spectroscopies and how these experiments are performed is given in the first section of this chapter. Since more and more vibrational data have become available for small molecules bound to proteins, interpretational tools that can relate the vibrational frequencies to biologically interesting quantities, such as bond order, bond length, interaction energy, and geometry of the molecular groups, have become more and more important. Several tools that have been developed empirically will be discussed, as will interpretations of the vibrational spectra based on ab initio vibrational normal-mode calculations.

## 2  SPECTROSCOPIC TECHNIQUES

### 2.1  Raman Difference Spectroscopy

A typical Raman difference setup uses a conventional Raman spectrometer employing a optical multichannel analyzer (OMA) system, as depicted in Fig. 1

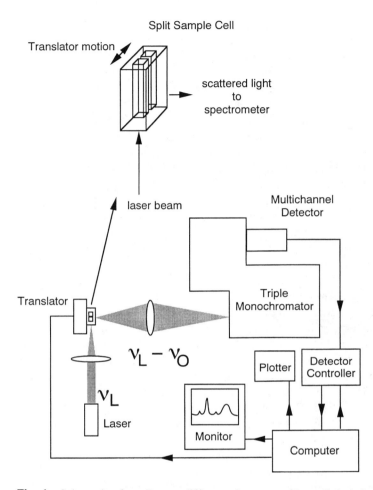

**Fig. 1** Schematic of our Raman difference instrument. Laser light is incident on a specially fabricated split cell from the bottom. The Raman scattered light at $\nu_L$ is frequency shifted from the incoming laser light by $\nu_0$, the frequency of a vibrational mode. Raman spectroscopy is an especially useful form of vibrational spectroscopy for bio-macromolecules because the water in protein solutions yields a negligible Raman background. The split cell has been designed so as to make the light paths in both halves as equal as possible in order to reduce subtraction artifacts. The Raman scattered light is collected by a monochromator and multichannel detector combination. Scattered light is collected from one side of the split cell for about 30 minutes, and this spectrum is stored in the computer. The cell is translated and scattered light is collected from the second side. The two spectra are then subtracted in the computer to form the difference spectrum.

(3–5). The parent spectra that make up the difference spectrum are obtained using a split-cell cuvette, which has been specially made for this purpose. The split-cell cuvette is mounted on a translator stage-stepping motor combination. Typically, 100 mW of visible light from either an argon ion or krypton ion laser is used to excite Raman scattering, and an experiment takes 1–2 hours. About 1200 cm$^{-1}$ of the Raman spectrum can be detected simultaneously. Each side of the split cell has an inside dimension of 3 × 2.5 mm and is loaded with 30 μl of sample in the setup depicted. Excellent spectra can be obtained using sample concentrations of 1 mM. This means that about 1.2 mg of a 40-kDa protein is required. We believe that advances in spectrometer technology, in terms of light collection and detector efficiency as well as sample cell configuration, involving no new principles, can be developed to reduce the amount of protein needed by a factor of 100. In fact, Peticolas and his coworkers have employed a Raman microscope system that requires only 2.5 μl of sample volume (6), and Carey and coworkers have demonstrated the feasibility of measurement at a sample concentration of ca. 100 μM (5,7). Several precautions may be taken to minimize systematic errors in forming the difference spectrum. The entire spectrometer system, including the exciting laser, is mounted on a vibration-free table, and the room temperature and humidity are maintained at a constant level.

The Raman difference spectroscopic system in our lab can discern the differences between two spectra at the level of 1 part in 1000 or better (8). A recent example is given by studies on liver alcohol dehydrogenase (LADH), a protein of 40 kDa per binding site, complexed with coenzyme nicotinamide adenine dinucleotide (NADH) and bound inhibitor (9). The *n*-cyclohexylformamide (CXF) and LADH (or LADH/NADH/labeled-CXF) samples were loaded into a split Raman cell and the Raman spectra were taken alternately from the two samples (8). Up to 10 runs were taken from each sample, and the spectra were then summed and averaged. The averaged spectrum so obtained for the LADH/NADH/CXF complex is shown in Fig. 2a, and Fig. 2b shows that of LADH. As a control experiment, the odd-numbered runs and even-numbered runs on the

---

**Fig. 2** (a) Raman spectrum of LADH/NADH/N-cyclohexylformamide (CXF) at 4°C in 100 mM pyrophosphate buffer, pH 9.6 (pH meter reading) in $D_2O$. (b) Raman spectrum of LADH at 4°C in 100 mM pyrophosphate buffer, pH 9.6 (pH meter reading) in $D_2O$. (c) Difference spectrum formed between even-numbered LADH runs minus odd-numbered LADH runs, the result multiplied by a factor of 20. (d) Difference spectrum between (a) and (b), the result multiplied by a factor of 5. (e) Raman difference spectrum between LADH/NADH/[$^{13}$C]CXF and LADH, the result multiplied by a factor of 5. (f) Raman difference spectra between LADH/NADH/N-cyclohexylformamide (CXF) and LADH/NADH[$^{13}$C]CXF at 4°C in 100 mM pyrophosphate buffer, pH 9.6 (pH meter reading) in $D_2O$. The Raman bands marked with an asterisk are due to bound NADH. (From Ref. 9.)

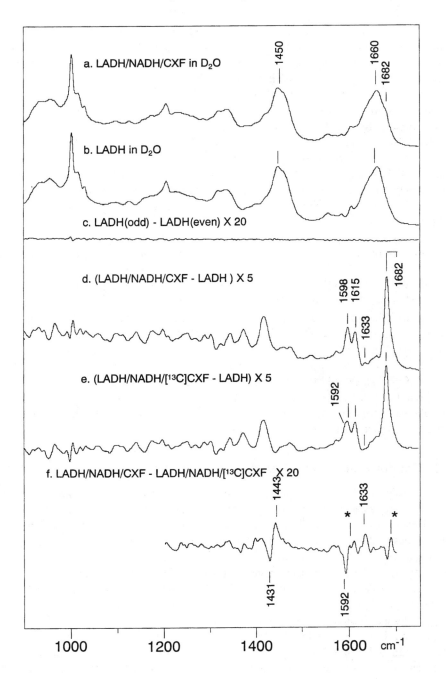

LADH sample were averaged, their difference spectrum is shown in Fig. 2c (enhanced by a factor of 20). This difference spectrum should null completely. In fact, the noise in this spectrum, all of which is simple shot noise and is free from systematic subtraction artifacts, is 0.2% of the amide I band of LADH. In the subtraction process, two Raman bands were used as internal references, the amide I band at 1660 cm$^{-1}$ and the $\delta$CH band at 1450 cm$^{-1}$. A multiplication factor, typically ranging from 1.03 to 0.97, was applied to one of the spectra and adjusted so that these two bands were no longer visible in the final difference spectrum. Since the peak-to-peak intensity in the spectrum of the control (Fig. 2c) is about 0.2% of the 1655-cm$^{-1}$ band intensity of the original spectrum, the noise level in complex minus labeled-complex difference spectra is expected to be the same. Figure 2d shows the LADH/NADH/CXF minus LADH difference spectrum, with both the 1450 cm$^{-1}$ and 1660 cm$^{-1}$ bands subtracted properly; Fig. 2e shows the difference spectrum between LADH/NADH/$^{13}$C labeled CXF minus LADH. In these difference spectra, all the major peaks are from the dihydronicotinamide moiety of NADH in the ternary complex (cf. 10,11). The intensity of the most prominent band at 1682 cm$^{-1}$ is about 30% of the major protein band at 1660 cm$^{-1}$. All other visible peaks in the difference spectra have intensities of about 1% of the 1660-cm$^{-1}$ protein peak or higher, well above the noise level. The Raman intensity of the CXF bands are very weak compared to the Raman bands of NADH. For example, the small peak at 1633 cm$^{-1}$ in Fig. 2d is the C=O stretch of CXF. This assignment is supported by its disappearance in $^{13}$C-substituted CXF (Fig. 2e) and the concomitant appearance of a shoulder to the 1598-cm$^{-1}$ band at 1592 cm$^{-1}$, suggesting that the $^{13}$C=O stretch lies at 1592 cm$^{-1}$ in the labeled ternary complex. These peaks becomes more prominent in the isotope edited difference spectrum shown in Fig. 2f.

## 2.2  Fourier Transform Infrared Difference Spectroscopy

The primary tool for vibrational studies of enzymes has been Raman difference spectroscopy, as just described. However, there are times when certain vibrational bands are only weakly allowed or are forbidden in the Raman spectrum. In most cases, these bands will be IR allowed. For example, antisymmetric stretch modes of phosphate are only weakly allowed in the Raman spectrum but are quite strong in the IR spectrum. Thus, Raman and FT-IR spectroscopy complement each other. Several FT-IR difference techniques have been developed to study protein–ligand or enzyme/substrate interaction. For example, a number of protein–ligand interaction problems, especially those involved with a C=O bond, have been studied by FT-IR difference technique using isotope editing (12–17). In addition, FT-IR difference spectroscopy using photolysis techniques and ''caged'' phosphate compounds (18) has been applied to GDP binding proteins (G-protein).

Here, we describe preliminary studies using isotopic editing in FT-IR difference spectroscopy in our lab to observe IR-allowed bands of substrate bound to proteins. Two IR cells, made as identical as possible, are mounted on a shuttle translation stage. One cell is moved into the IR beam, and the IR absorption spectrum is taken. A reference spectrum of the residual water vapor is taken (which is quite low, on the order of 0.002 mOD), and this is subtracted. The cell holder is translated to bring the second cell into the IR beam, and the sequence is repeated. The major difficulty in forming the difference spectrum is the strong water IR band that may overlap with the frequency region of interest, such as the C=O stretch mode. This can be avoided by using $D_2O$. Studies of protein solutions in $D_2O$ move the strong water band from ca. $1630 \, cm^{-1}$ to $1200 \, cm^{-1}$. In a separate run, the water (or $D_2O$) background is measured, and this background is subtracted away. Figure 3 shows the IR spectra of liver alcohol dehydrogenase (LADH) with bound inhibitor, $N$-cyclohexylformamide (CXF) (9; see earlier). The top spectrum (a) is that of LADH complexed with $^{12}C$-substituted, while the bottom spectrum (b) is with $^{13}C$-substituted CXF. The $D_2O$ spectrum in each protein complex has been subtracted, with the relatively weak $D_2O$ spectra around $1800 \, cm^{-1}$ being nulled. There is a remnant of $D_2O$ at $1200 \, cm^{-1}$ in the two difference spectra, showing up as a negative band, because the $D_2O$ OD is around 1.5 OD units and there is a small nonlinear response of the spectrometer. The bands at 1639 and $1457 \, cm^{-1}$ are the protein amide I′ and amide II′ absorption bands, respectively (the ′ denotes deuterated protein solutions). The panel on

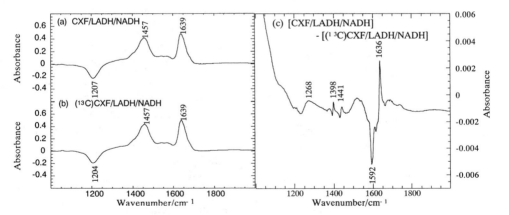

**Fig. 3**   IR isotope edited difference spectrum of CXF complexed with LADH. Top panel (left) is the spectrum of CXF/LADH complex with $^{12}C$-substituted CXF; the bottom panel, with $^{13}C$ substituted. The panel on the right is the difference between these two spectra. IR cell spacing was 15 μm; protein concentration was 1.2 mM; total sample volume was 15 μl.

the right (c) shows the difference spectrum between these two spectra. The two primary spectra do not have exactly the same OD, since the two IR cells have slightly different path lengths. Hence, the difference spectrum has been formed between these two by multiplying one of the spectra by a factor near 1 and adjusting this factor until the amide protein peaks disappear. The peak at 1636 cm$^{-1}$ is the $^{12}$C═O stretch of bound CXF, while that at 1592 cm$^{-1}$ is the $^{13}$C═O stretch. Comparing Fig. 3 with Fig. 2, it becomes obvious that the intensity of the C═O stretch mode of CXF is stronger relative to the major protein bands in the IR (ca. 1%, Fig. 3) than in Raman (ca. 0.3%, Fig. 2). On the other hand, the intensity of the C–N stretch mode in the Raman spectrum (band near 1431/1443 cm$^{-1}$, Fig. 2f) is somewhat stronger than in the IR difference spectrum.

## 3 INTERPRETATIONAL TOOLS OF THE VIBRATIONAL SPECTRA

### 3.1 Interpretations Based on Comparison with Model Compounds in Solution

A long-used method for the interpretation of vibrational spectra has been based on comparisons of the spectra of complex molecules to those of simpler molecules. This has often been combined with the notion of group modes. That is, the normal-mode motions of atoms often are localized to a specific chemical group, and the frequencies carry over when the chemical group is a constituent of a complex molecule. Identification of a group mode is normally based on its characteristic frequency and can be definitively assigned by isotopic labeling of the atoms that contribute to this mode. This is often the method of choice in many cases, since it may provide the most definitive conclusion.

Comparisons between the spectra of bound substrates and the solution spectra of models are also of value when a new covalent bond is formed between the ligand and the protein. This new bond is characterized by a specific normal mode involving motions of the newly bonded atoms, and a new peak in the vibrational spectrum of the complex will likely appear. Assignment of this peak can be achieved by studying the solution model compound and by isotopic labeling of the atom on either side of the bond. For example, a Schiff base (─C═N─) linkage forms between the ε-amino group of active-site lysine-258 of *E. coli* aspartate aminotransferase (AATase) and the carbonyl moiety of enzyme bound cofactor, pyridoxal 5′-phosphate (PLP). In Raman difference studies (19,20), the spectrum of the low- and high-pH forms of this complex were studied and Raman bands characteristic of PLP protonated Schiff base and unprotonated Schiff base were determined, respectively (Figs. 4a and 4b). An inhibitor, α-methyl-L-aspartate (αMeAsp), complex with enzyme-PLP was also studied, because it is believed to be a Michaels complex of the enzyme. Four forms of the complex are possible, since the inhibitor may, in principle, form a Schiff base with lysine-258 (so-called internal Schiff base) or with PLP (the "external" Schiff base), and may be protonated or unprotonated. As shown in Fig. 4c, the Raman spectrum of αMeAsp bound to AATase complex was found to contain a 1630-cm$^{-1}$ band, which was assigned to the unprotonated internal Schiff base C$=$N stretch mode based on a downshift of 18 cm$^{-1}$ with $^{15}$N labeling of the Lys258 ε nitrogen (Fig. 4e). This study also showed that the protonated external Schiff base form coexisted in this complex, a conclusion supported by the insensitivity of the observed $v_{C=NH+}$ of the αMeAsp–enzyme complex at 1655 cm$^{-1}$ to lysine $^{15}$N substitution, but shifts down to 1640 cm$^{-1}$ upon sample deuteration (Fig. 4d). No significant amount of the other two possible Schiff base forms, namely, the protonated internal or unprotonated external forms, were found. Interestingly, the Raman study, which of course was performed on protein solutions, showed that the solution structure of the *E. coli* AATase–αMeAsp complex differs significantly from the reported crystal structures in terms of the equilibrium concentrations found between the four forms (19).

The direct spectral comparison method has also been applied in a number of studies on other enzyme systems, including that to determine the hydration of the N1$=$C6 bond of the purine ring in adenosine deaminase (see Sec. 4.1), the protonation state of the N5 of dihydrofolate in dihydrofolate reductase/NADP$^+$ complex (see Sec. 4.2), the conformation of NADH bound to lactate dehydrogenase (see Sec. 4.5), and the conformation of the thioester in acyl proteases (21).

## 3.2 Entropic Effects

It is believed that an important factor in enzymatic catalysis has to do with bringing the reactants close enough together and with the proper orientation so that they can react by providing appropriate binding sites. Such specific binding patterns can bring about rate enhancements of some 10$^8$-fold by themselves (22–24), although rate enhancements from this factor more likely give rise to enhancements of around 10$^4$-fold for bimolecular reactions. These entropic effects may play a role in enzymatic catalysis by lowering the $T\Delta S$ component of $\Delta G$ between the ground state and the transition state along the reaction coordinate for the enzyme-catalyzed reaction relative to the reaction pathway in solution. The bandwidths of a vibrational spectrum are a composite of its homogeneous bandwidth, which is an intrinsic parameter of the molecule, and its heterogeneous width, which comes about from different molecules in the ensemble of molecules experiencing (perhaps slightly) different environments. It is generally observed that bandwidths decrease substantially for ligands bound to proteins; this is a reflection of the more uniform binding environment at the protein binding active site

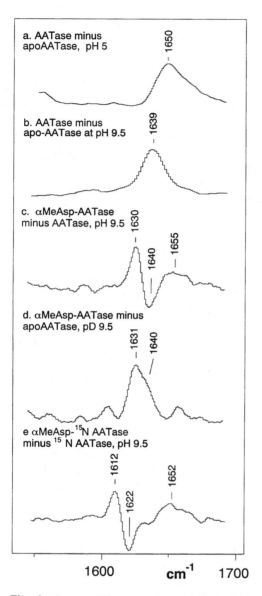

**Fig. 4** Raman difference spectra of *Escherichia coli* aspartate aminotransferase (AA-Tase) complexes with pyridoxal 5′-phosphate (PLP) and α-methyl-L-aspartate (αMeAsp). The structural assignments in the figure are discussed in the text. The α and ε indicate the nitrogen substituent of αMeAsp and the ε-N atom of Lys-258, respectively. Enzyme concentrations were 2–3 mM, $T = 4°C$. Buffer solutions were adjusted to the indicated pH values by the addition of NaOH or NaOD. (a) The AATase-minus-apoAATase Raman

compared to solution. A crude measure then of how tightly substrates or inhibitors are bound to protein binding sites compared to solution is then given by the inhomogeneous bandwidth of the respective vibrational bands. This can be related to a decrease in entropy by $S = R \ln \Omega$, where $\Omega$ is the number of (heterogeneous) states available to the normal coordinate and $R$ is the gas constant (ca. 2 cal/mol). A lower limit to $T \Delta S$ may be estimated by substituting the ratio of the heterogeneous bandwidths of the solution bandwidth to the bound bandwidth for coordinates that are part of the reaction coordinate. On the basis of this, we found that of the 4.2 kcal/mol decrease of the transition-state barrier for hydride transfer caused by the loop closure in lactate dehydrogenase (effectively tightly holding the reactants together), it could be estimated from an analysis of bandwidth changes in key modes involved in the reaction coordinate that at least 1.4 kcal/mol arises from the elimination of various nonproductive conformations of bound NADH and pyruvate (25,26).

## 3.3 Empirical Correlations

In a number of systems, the frequency changes of specific modes that occur when a ligand binds to a protein can be readily detected by Raman and/or FT-IR difference spectroscopy, but tools are required to relate the vibrational frequencies with the structural information of the molecular group and its interactions with the binding pocket. Some of the tools that have been used extensively in vibrational studies of enzymes are summarized next.

### 3.3.1 Bond Length/Order Versus Frequency Correlations

Early studies of simple di- and triatomic molecules suggest that the internuclear distances of polyatomic molecules can be predicted with considerable accuracy from vibrational data (27). One such example is the linear correlation that relates the $C{=}O$ bond length with its stretch frequency in various compounds (28):

$$L_{C=O} = 2.00988 - 0.0004563\nu_{C=O} \tag{1}$$

---

difference spectrum at pH 5.5 in 0.2 M sodium acetate. (b) The AATase-minus-apoAATase Raman difference spectrum at pH 9.5 in 0.2 M CHES. (c) The $\alpha$MeAsp-AATase-minus-AATase Raman difference spectrum at pH 9.5 in 0.2 M (CHES) (Sigma Chemical Co.). The concentration of $\alpha$Me-$DL$-Asp (Sigma Chemical Co.) was 100–150 mM; thus, the fraction of AATase bound to $\alpha$Me-L-Asp was over 90%; the D isomer is not recognized by the enzyme. (d) The Raman difference spectrum of $\alpha$MeAsp-AATase-minus-apoAATase at pD 9.5 in 0.2 M CHES in $\geq$ 98% $^2H_2O$. (e) Same as spectrum (b) with the $\epsilon$-nitrogen atom of active-site Lys-258 and other atoms quantitatively enriched with $^{15}N$. 100-mW 568.2-nm laser line from a krypton laser was used, and spectral resolution was from 6 to 7 cm$^{-1}$. (From Ref. 19.)

In addition, structural information about bonding in more complex molecular moieties, such as metal and nonmetal oxides, often can be obtained from vibrational spectroscopy by using two types of empirical relationships. One is the bond-length/bond-strength correlation pioneered by Brown and coworkers (29,30); the other is the bond strength–vibrational frequency correlation formulated by Hardcastle and Wachs (31) and developed further by us (32). In the Brown and Wu relationship, a network paradigm is used to define bond order, $s$, in terms of the average number of electron pairs per bond so that $\Sigma\ s$ for any atom in a crystalline compound is equal to the formal valence of that atom. When defined in this way, bond strength can be referred to as valence bond strength and is expressed in terms of valence units, vu (33). For the nonbridging P(V)$\cdot$ $\cdot$O bonds of phosphates and vanadates, there exist very accurate empirically derived relationships between bond length/bond order and bond order/vibrational frequency. The equations for phosphates are (32):

$$s_{PO} = (r_{PO}/1.620)^{-4.29} \tag{2}$$

where $s$ is the bond order of the PO bond (in valence units), $r_{PO}$ is the length of the PO bond (in angstroms), and

$$s_{PO} = [0.175\ \ln(224{,}500/u)]^{-4.29} \tag{3}$$

where $u$ is the frequency of the phosphate stretch. The following equations hold for vanadates (30–32):

$$s_{VO} = (r_{VO}/1.791)^{-5.1} \tag{4}$$

$$s_{VO} = [0.2912\ \ln(21{,}349)/v)]^{-5.1} \tag{5}$$

These equations hold for any ionization state of the phosphate (vanadate) moiety. The symmetric stretch frequency can sometimes be used in $s/u$ correlations in phosphates (33). However, in general, and particularly in situations where the changes in the angle of the O$\cdot$ $\cdot$P$\cdot$ $\cdot$O bonds are important, a suitable average or "fundamental" frequency, which is the mean root square of all symmetric and antisymmetric stretch frequencies, i.e., $u = [(u_s^2 + (n - 1)u_a^2)/v]^{1/2}$ ($n$ is the number of nonbridging P$\cdot$ $\cdot$O bonds, 3 for dianionic phosphate esters and 2 for monoanionic phosphate) must be used (32). Hence, bond lengths and bond orders can be determined if the stretch frequencies are known. Moreover, from the symmetric and antisymmetric stretch frequencies, changes of the O$\cdot$ $\cdot$ P$\cdot$ $\cdot$O bond angle can be estimated (see Sec. 3.3.2). The error in these relationships is estimated to be about $\pm0.04$ vu and $\pm0.004$ Å for bond orders and bond lengths, respectively, and better at estimating changes in the parameters for a given molecule as it goes from one environment to another (32). Hence, the accuracy found only in small-molecule diffraction studies heretofore is now available for studies of phosphates and vanadates in solution as well as when bound to proteins using vibrational spectroscopy. In principal, this methodology can be applied to other polyatomic moieties. An example for the applications of Eqs. (4) and (5) can be found in Sec. 4.6.

### 3.3.2 Bond Angle Versus Frequency Correlations

For molecular groups with $C_{2v}$ or $C_{3v}$ symmetry, such as $PO_2^-$, $CO_2^-$ or $PO_3^{2-}$, $VO_3^{2-}$, respectively, it is possible to determine the bond angle based on the observed symmetric and antisymmetric stretch frequencies. For molecular groups with $C_{2v}$ symmetry, the following equations define the relations between frequencies and bound angle (32):

$$u_s^2 = \left(\frac{1}{\mu} + \frac{\cos\theta}{M_p}\right)(f_s + C_{ss}) \tag{6}$$

$$u_a^2 = \left(\frac{1}{\mu} - \frac{\cos\theta}{M_p}\right)(f_s - C_{ss}) \tag{7}$$

Here, $f_s$ is the stretching force constant for the P⋅⋅O bond, $\mu$ is the reduced mass of the P⋅⋅O bond, $M_p$ is the mass of the phosphorus, $C_{ss}$ is the coupling constant for the two stretching modes, and the bond angle for the O⋅⋅P⋅⋅O group is $\theta$. For molecular groups with $C_{3v}$ symmetry, such as the $PO_3^{2-}$ group, the following equations define the relations between frequencies and bond angle:

$$u_s^2 = \left(\frac{1}{\mu} + 2\frac{\cos\theta}{M_p}\right)(f_s + 2C_{ss}) \tag{8}$$

$$u_a^2 = \left(\frac{1}{\mu} - \frac{\cos\theta}{M_p}\right)(f_s - C_{ss}) \tag{9}$$

The use of these equations in selected enzymic systems to determine the O⋅⋅P⋅⋅O angles of phosphates and angle changes upon binding is discussed in the last section of this chapter.

A very accurate relationship between bond angle and frequency has to do with the indole ring mode of tryptophan (W3) that has an intense band at 1542–1560 cm$^{-1}$, which has a major contribution from the $C_\gamma$–$C_{\delta1}$ stretch (34). Besides being affected by the polarity of the solvent environment, studies of a series of crystalline tryptophan derivatives have shown that the W3 frequency increases sigmoidally with $\chi^{2,1}$, the dihedral angle about the bond connecting the indole ring with the $C_\beta$ atom of the tryptophan side chain (35,36), and an empirical equation was found as follows:

$$v(W3) = 1542 + 6.7[\cos(3|\chi^{2,1}|) + 1]^{1.2} \tag{10}$$

where $v(W3)$ is the W3 frequency in $cm^{-1}$ and $\chi^{2,1}$ is expressed in degrees.

### 3.3.3  Badger–Bauer Relationships

Badger and Bauer (27) suggested quite some time ago that the enthalpy of formation of a hydrogen bond, $\Delta H$, is linearly related to the vibrational frequency shift,
$\Delta v$, of the O—H stretch frequency of an alcohol. A stronger hydrogen bond
weakens (polarizes) the O—H bond and the stretch force constant; this gives rise
to a significant decrease in the observed stretching frequency. Such thermodynamic correlations have been investigated by many workers, and sometimes the
relationship is very direct and simple (37). The general methodology of this body
of work is to determine the association constants, enthalpies, binding entropies,
and vibrational frequencies of a number of donor–acceptor pairs in nonpolar
organic solvents (reviewed in Refs. 37–39), and Badger–Bauer relationships
have been found for a number of O—H, C=O, N—H, and other groups in various
molecules (see Table 1). The exact relationship depends on the nature of the
group and nearby chemical constituents. It also needs to be stressed that the
interaction energy that is measured from the polarization of a polar group must
be given as an *effective* interaction energy. It may very well be true (more often
than not for proteins) that there is more than one proton donor to a C=O group,
for example, and even long-range electrostatic interactions that bring about the
bond's polarization.

In general, the empirically derived relationships from the thermodynamic
method extend over only a fairly small range in $\Delta H$ (<10 kcal/mole) and $\Delta v$
(<15 $cm^{-1}$) and for a relatively limited number of molecules. The reasons are
that the molecule of interest must be soluble in a hydrophobic solvent, such as
carbon tetrachloride, so that the solvent does not compete for hydrogen bonding
formation of the donor–acceptor pairs, and also must contain only isolated polarizable groups so that the hydrogen bonding on one polarizable group will not
affect the vibrational frequencies of another polarizable group. Thus, the correlation between $\Delta H$ and $\Delta v$ for primary and secondary amides, for example, cannot
be obtained by this method.

Recently, an alternative method was developed to overcome some of the
limitations so that the correlation between $\Delta H$ and $\Delta v$ for secondary amides can
be determined empirically over a somewhat larger range than that by the direct
thermodynamic method. This method is based on the assumption that the linear
relationship between $\Delta H$ and $\Delta v$ exists for the C=O bond of the amide and based
on the concept of (electron) "acceptor numbers" (AN) for various solvents,
which are empirically determined (40). A linear relationship between the acceptor
numbers of a series of solvents and vibrational frequencies of a given molecule

**Table 1** Badger–Bauer Linear Relationships Between Change in Normal-mode Frequency, $\Delta v$, and Hydrogen Bond Enthalpy, $\Delta H$, Between the Bond of the Compound Listed and Various Hydrogen Bond Donors or Acceptors

| Mode | Compound | Slope ($a$) | Intercept ($b$) |
|---|---|---|---|
| C=O stretch | Acetone[a] | 0.47 | −0.11 |
| | Acetophenone[a] | 0.31 | −0.506 |
| | Benzophenone[a] | 0.36 | −1.05 |
| | Methylacetate[b] | 0.40 | −5.488 |
| | Retinal | 0.27 | |
| | 4-Methylbenzoyl S-ethyl thioester[c] | 0.30 | −2.1 |
| | 5-Methylthienylacryloyl methyl ester[d] | 0.33 | −3.9 |
| | N-Cyclohexylformamide[e] | 0.26 | |
| O—H stretch[f] | Phenol | 0.0105 | 3.0 |
| | p-Fluorophenol | 0.0103 | 3.1 |
| | t-Butanol | 0.0106 | 1.7 |
| | Perfluoro-t-butanol | 0.0106 | 3.9 |
| | 2,2,2-Trifluoro-ethanol | 0.0121 | 2.7 |
| | 1,1,1,3,3,3-Hexafluoro-isopropanol | 0.0115 | 3.6 |
| N—H stretch | Pyrrole[g] | 0.0123 | 1.8 |
| | Indole[h] | 0.0116 | 0.8 |
| | N-Methylaniline[i] | 0.0394 | 2.1 |
| Tyrosine ring mode($v_8$) | Tyrosine[j] | 0.8 | −0.64 |

The linear relationship is defined by $-\Delta H = a\,\Delta v + b$, where $H$ is in kcal/mol and $v$ in cm$^{-1}$.
[a] Ref. 81; [b] Ref. 82; [c] Ref. 47; [d] Ref. 83; [e] Ref. 9; [f] reviewed in Ref. 84; [g] Ref. 85; [h] Ref. 37; [i] Ref. 86; [j] Ref. 87.

in those solvents can often be found and is a measure of the polarizability of a given bond. This approach was taken in analyzing the spectra from the studies of the LADH/CXF interaction by the Raman difference technique (see earlier; Ref. 9). However, a correction to the observed frequencies was necessary, since ab initio calculations suggested that only about 70% of C=O stretch frequency shift of CXF is due to the hydrogen bonding changes near the C=O bond, while the rest is due to the hydrogen bond change near N—H (see Sec. 3.4). Altogether, the conversion yields $\Delta\Delta H_{CXF} = 0.26\,\Delta v$ for the CXF C=O stretch mode. The correlation coefficient of the relationship, 0.26, is very similar to that determined for N,N-dimethylpropionamide using the thermodynamic method, which is 0.24 (unpublished data).

Keto (aldehyde) groups inside enzymes have received much attention in vibrational studies because they are important in many enzymic systems and partly for technical reasons. The C=O stretch sometimes lies in an otherwise spectroscopically silent region (higher in frequency than the amide I protein

**Table 2** Shifts in Frequency in Wavenumbers (cm$^{-1}$) of C=O Vibrational Bands When a Substrate Binds to Its Enzyme and the Deduced Effective Binding Energy Between the C=O and the Protein as Derived from Badger–Bauer Relationships

| Group | $\Delta v$(in situ minus solution) | $-\Delta H$(kcal/mol) |
|---|---|---|
| C=O of bound substrates | | |
| DABA in LADH[a] | −94 | −14 |
| Pyruvate adduct in LDH[b] | −35 | −17 |
| CXF in LADH[c] | −21* | −5.5 |
| 5-Methylthienylacryloyl in acyl enzyme[d#] | −54 | −14 |
| C=O of APAD's COCH$_3$ group bound to | | |
| LDH[e] | −10 | −2.8 |
| sMDH[f] | −9 | −2.5 |
| mMDH[f] | −9 | −2.5 |
| DHFR[g] | +5 | +1.4 |

The derived binding $\Delta H$ is expressed with respect to water. See text and the listed references for details.
[a] Ref. 46; [b] Ref. 45; [c] Ref. 9; [d] Ref. 83; [e] Ref. 88; [f] Ref. 89; [g] Refs. 10, 90.
* The frequency reported here is after correcting the effect of the hydrogen bonding change on the amide N—H bond; see text for details.
# The frequency shift is not relative to the solution value; see Ref. 83.

band near 1660 cm$^{-1}$). Significant bond polarization of C=O groups has been found in recent studies of acyl-proteases (6,12,13,15,41,42), yeast aldolase (43), in phopholipase A$_2$ (14), citrate synthase (44), lactate dehydrogenase (26,45), liver alcohol dehydrogenase (9,46), 4-chlorobenzoyl-CoA dehalogenase (47), and acyl cysteine proteases (48). Table 2 summarizes some of these results where a Badger–Bauer analysis has been performed. A number of examples of applications of these relationships in the studies of enzyme catalysis are given in Sec. 4.

### 3.3.4 Structure/Reactivity Relationship

With the sensitivity of vibrational spectroscopy, it is feasible to determine systematically changes in structure in a series of enzyme–substrate complexes and relate them to enzyme-induced rate enhancements. These studies are, in principle, very powerful at elucidating how an enzyme works. Let us take, for example, the interaction between a substrate containing a C=O bond where it is believed that the transition state contains substantially more $^+$C—O$^-$ character than the ground state. In this case, a stabilizing electrostatic interaction (like hydrogen bond donation) would be stronger in the transition state than the less polar ground state. Use of reaction rate theory then suggests that an Arrhenius plot of $k_{cat}$ versus the shift in carbonyl frequency should yield a straight line, assuming a linear (Badger–Bauer-like) relationship between the energy of the transition state and the frequency of C=O stretch, as is found for the ground state. We then could write:

$$k_{cat} \sim \exp(-\Delta E^{\neq}/RT) \sim \exp(x\Delta v_{C=O}/RT) \tag{11}$$

where $x$ is an unknown numerical factor that depends on the relationship between the transition-state $C^{+} \cdot {}^{\cdot}O^{-}$ stretch and its enthalpic interaction with the H-bond donor(s). Thus, a linear relationship between the C=O frequency shift ($\Delta vC=O$) of the substrate upon binding to enzyme and $\log(k_{cat})$ (or $\Delta\Delta G$) is expected, provided certain conditions are met (26). Carey and Tonge and their colleagues (49) reached the same conclusion with a somewhat different argument. This approach has been of enormous help in elucidating how lactate dehydrogenase and the serine proteases bring about their catalytic-rate enhancements and are discussed later (see Sec. 4).

## 3.4   Ab Initio Normal Mode Calculations

Vibrational analyses based on quantum mechanical ab initio normal-mode calculations have the potential to bring vibrational spectroscopy to the level of a general analytical tool. In such analyses, the conclusions are made based on the comparison of calculated frequencies of a model system (and Raman intensities and depolarization ratios) to the experimentally observed values. The validity, accuracy, and usefulness of ab initio calculations for vibrational analysis have been established in the studies of several enzyme/substrate complexes where definitive conclusions can be achieved independently by other means, such as the method discussed in Sec. 3.1. For example, it has been applied successfully to confirm the conformational changes of a molecule imposed by binding to a protein (50,51) and a specific ionization state of a molecular group in an enzyme (9,52–56). More importantly, in the studies of protein–ligand interactions, such an analysis can also be used for systematic assessment on how particular interactions between a molecule and its environment affect vibrational frequencies. This permits modeling of protein binding sites and a ranking of the importance of various active site interactions (9,52,54,56–58).

We present two examples where ab initio calculations have been invaluable. The first example is a modeling of the active site of the myosin S1 subfragment, specifically the S1 · MgADP · Vi complex, which is believed to be a transition-state complex mimic for the hydrolysis of ATP to ADP and phosphate. In this study (57), the goal was to reproduce the experimentally observed spectra of the bound vanadate moiety by ab initio calculations and, hence, to understand what in the myosin active site is important in bringing about the distortions on the vanadate moiety. Cartesian force constants and the Raman polarization tensor of an isolated dianionic vanadate or methylvanadate were calculated at the HF/3-21g* level. The Cartesian force field was then converted to an internal force field and subsequently "scaled" (by a set of scaling factors) so that the calculated vibrational frequencies and isotopic shifts of the vanadate derivatives measured in solution were in good agreement with those observed ones. This scaling procedure is required to overcome the problem related with the well-known, systematic overestimation of the vibrational frequencies calculated by ab initio methods. The Raman intensities (and their dependencies on incident-light polarization) of the new vibrational modes were also calculated. These quantities were then compared with the experimental spectra to evaluate the scaling factors. Once a set of satisfactory scaling factors was obtained, calculations on a series of models of the active site simulating the S1 · MgADP · Vi complex were then performed, with the geometrical arrangement as determined by diffraction studies of this complex serving as a structural starting point. The resulting calculated force fields are then appropriately scaled by the predetermined set of scaling factors to produce vibrational modes, Raman intensities, and polarization dependencies and then compared with the measured quantities for the bound vanadate in S1 · MgADP · Vi complex.

The results of the calculations simulating methylvanadate dianion in solution are presented in Fig. 5A. Geometry optimizations indicate that the bridging V—O bond is not collinear with the bridging O—C(H) bond at equilibrium. One of the nonbridging V·⁻·O bonds is longer as compared to the other two, so the $C_{3v}$ symmetry expected for the three nonbridging V·⁻·O bonds is slightly broken. Three nonbridging V·⁻·O stretch modes, at 869, 865, and 851 cm⁻¹, respectively, are predicted after application of the scaling procedure described earlier (Fig. 5A). This is so because the calculations do not take into account fast rotation of the methyl group about the bridging bond, which would result is an average equal environment for each of the three nonbridging V·⁻·O bonds. Nevertheless, the 869-cm⁻¹ band corresponds to an approximate symmetric stretch mode, with the ratio of perpendicular-to-parallel polarization intensity very close

---

**Fig. 5** (A) Calculated Raman spectrum of dianionic methylvanadate from ab initio calculations as described in the text. The top curve is the simulated parallel polarized spectrum, and the bottom curve is the perpendicular polarized spectrum. The individual Raman bands are simulated by a Gaussian function, and the band width is set to 22 cm⁻¹. The curves identified at 867 cm⁻¹ (top) and at 864 cm⁻¹ (bottom) are the envelope (observable) spectra arising from the three bands. (B) Calculated Raman spectrum of dianionic methylvanadate complexed with a water molecule at a water oxygen-vanadium distance of 2.5 Å. The top curve is the simulated parallel polarized spectrum, and the bottom curve is the perpendicular polarized spectrum. The Raman bands are simulated by a Gaussian function, and the band width is set to 10 cm⁻¹. The curves identified at 822 cm⁻¹ (top) and 869 cm⁻¹ (bottom) are the envelope (observable) spectra arising from the three bands. (From Ref. 57.)

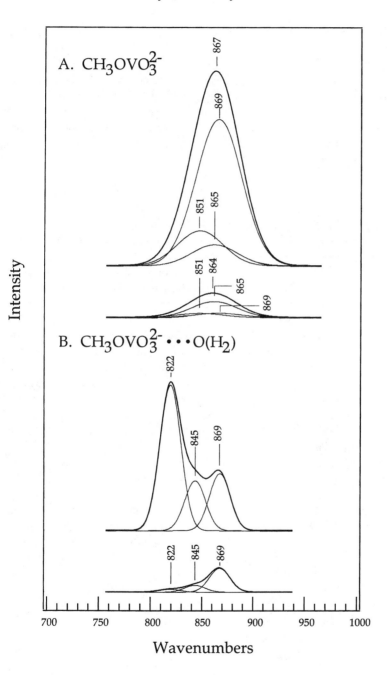

Wavenumbers

to zero (Fig. 5A). The character of the 865- and 851-cm$^{-1}$ modes are quite close to an asymmetric geometry and ratios of perpendicular to parallel polarization intensity of asymmetric modes. Calculations were also performed on methyl vanadate, where the nonbridging oxygens were replaced by $^{18}O$. The 869-cm$^{-1}$ band moves to 816 cm$^{-1}$, the 856-cm$^{-1}$ band to 835 cm$^{-1}$, and the 851-cm$^{-1}$ band to 830 cm$^{-1}$. For presentation, all bands were assigned a bandwidth that matches the experimentally determined bandwidth; the bandwidth does not come out of the calculations. It can be seen that the agreement between the measured (see upcoming Fig. 10A) and calculated Raman spectra (Fig. 5A) is reasonable, as are the $^{18}O$ induced shifts. This then sets the values of the scaling factors used in subsequent Raman spectrum calculations of vanadate protein complexes.

To simulate the vanadate moiety in S1 · MgADP · Vi complex, a water molecule opposite of the bridging V—O bond of methylvanadate dianion was placed at various distances. The results of calculations on several model complexes with V—O(H$_2$) distance ranging from 2.25 Å to 4.0 Å (and at infinity, e.g., no interaction) were obtained. It was found that for a water oxygen-vanadium distance of about 2.5 Å, the calculated Raman spectrum of the nonbridging V$\cdot\cdot$O bonds shows a pattern quite similar to that observed in the Raman spectrum of S1 · MgADP · Vi complex. The Raman spectra calculated from this 2.5-Å model complex are presented in Fig. 5B. The 822-cm$^{-1}$ band moves to 780 cm$^{-1}$ upon $^{18}O$ replacement of the nonbridging V$\cdot\cdot$O oxygens, the 869-cm$^{-1}$ band to 839 cm$^{-1}$, and the 845-cm$^{-1}$ band to 816 cm$^{-1}$. The calculated pattern of bands and their intensity ratios between parallel and perpendicular polarizations, and shifts upon $^{18}O$ substitution, are in rather good agreement with the observed spectra (compare upcoming Fig. 10B and Fig. 5B). The crystallographic study of the S1 · MgADP · Vi complex places the V—O(H$_2$) distance at 2.27 Å (59), in very good agreement with the present results.

The second example has to do with understanding the coupling of C=O stretch with N—H bend in an amide so that Badger–Bauer-like relationships could be developed. The ab initio calculations on various model complexes of CXF with hydrogen bonds to C=O and/or N—H (9) showed that the observed frequency shift of the C=O stretch mode, $\Delta v_{C=O}$ can be treated as the sum of two terms: one is due to the hydrogen bonding change on the local C=O bond, $\Delta v_{C=O}^{local}$, and the other is due to the change near the remote N—H bond, $\Delta v_{C=O}^{remote}$. Similarly, the observed frequency shift of the N—H bending mode, $\Delta v_{N-H}$, can also be treated as the sum of the two terms, $\Delta v_{N-H}^{local}$ and $\Delta v_{N-H}^{remote}$, due to the local change near the N—H bond and due to the remote change near the C=O bond, respectively. Quantitatively, it was found that:

$$\Delta v_{C=O} = \Delta v_{C=O}^{local} - 0.3\Delta v_{N-H}^{local} \tag{12}$$

$$\Delta v_{N-H} = \Delta v_{N-H}^{remote} - 0.4\Delta v_{C=O}^{local} \tag{13}$$

Thus, the frequency shifts of the C=O stretch and N—H bending modes that are due to the local interaction only can be calculated from the observed frequency shifts of these two modes based on the foregoing equations.

## 4  EXAMPLES

### 4.1  Adenosine Deaminase

Adenosine deaminase (ADA) catalyzes the hydrolysis of adenosine or deoxyadenosine to the respective inosine product and ammonia. Purine ribonucleoside is a competitive inhibitor of adenosine deaminase, with an apparent inhibition constant of $2.8 \times 10^{-6}$ M (60). Recent Raman difference studies have shown that the differences between the Raman spectra of purine ribonucleoside in solution and purine ribonucleoside bound in ADA are simply too drastic to be explained by the vibrational-mode shifts of the same molecule in a different environment (Fig. 6) (54). However, the bound purine ribonucleoside spectrum is very similar to that of a model compound, 6-methoxyl-1,6-dihydropurine ribonucleoside, in solution and in enzyme. Thus, the Raman results suggested that in its complex with adenosine deaminase, purine ribonucleoside is hydrated at the N1 and C6 positions to yield 6-hydroxyl-1,6-dihydropurine. This conclusion is consistent with carbon-13 NMR measurements (61) and x-ray crystallographic studies (62). Since the deamination reaction pathway is expected to have a transition state (or intermediate) in which N1 is protonated and C6 becomes tetrahedral, the ADA/6-hydroxyl-1,6-dihydropurine complex is believed to be a very good transition-state (or intermediate) analog.

  A detailed vibrational analysis of the data also suggest that the (unobserved) N1—H bending frequency of the ADA-bound inhibitors shifts up by 50–100 $cm^{-1}$ compared to its frequency in aqueous solution (54). An estimate can be made that such a frequency shift corresponds to about a 4–10 kcal/mole of the hydrogen bonding energy in the N1—H bond upon forming the enzyme complex over that in aqueous solution (54). In the proposed reaction mechanism of ADA-catalyzed deamination of adenosine (62,63), the COOH group of Glu217 interacts with N1 of adenosine in the ground state; the COOH group subsequently donates the proton to N1, followed by the attack of a hydroxide to the C6 of adenosine to form the transition state (or reaction intermediate). (Simultaneous proton and hydroxide additions are also possible.) Since the purine ring is protonated at N1 and its C6 carbon is tetrahedral in the ADA/inhibitor complex, it is believed to be an analog to the transition state (or the intermediate). Assuming that the interaction energy between COOH of Glu217 and N1 of adenosine in the ground state of ADA/adenosine complex is similar to the interaction between N1—H of 6-methoxyl-1,6-dihydropurine ribonucleoside and a water molecule in solution, and if the interaction between N1—H of the inhibitor and COO⁻ of Glu217 in the

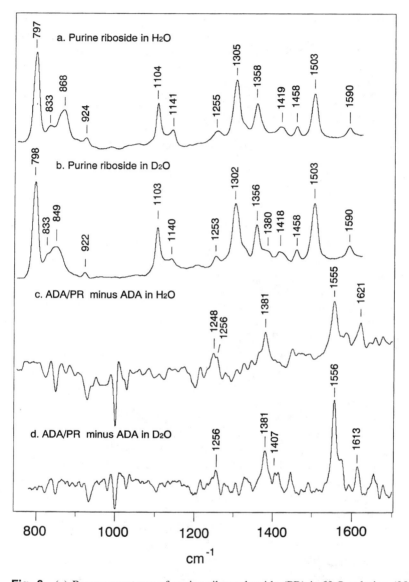

**Fig. 6**  (a) Raman spectrum of purine ribonucleoside (PR) in $H_2O$ solution (200 mM) at 4°C and pH 7.5. (b) Same as in (a) except the sample was in $D_2O$. (c) Difference Raman spectrum between adenosine deaminase (ADA) complex ADA/PR and ADA alone ([ADA]/[PR] = 4 mM/3.5 mM) at 4°C in 20 mM tris buffer, pH 7.5. (d) Same as in (c) except the sample was prepared in $D_2O$. (From Ref. 54.)

ADA/inhibitor complex is comparable to that between adenosine and enzyme in the transition state, the Raman results then suggest that the interaction between $COO^-$ of Glu217 and N1—H of hydrated adenosine can reduce the transition-state energy by up to 4–10 kcal/mol if entropic effects are ignored. This estimation of the transition-state stabilization contribution of the interaction between Glu217 of ADA and N1—H of substrate is in reasonable agreement with previous studies of a number of Glu217 mutants (64). These studies have shown that when Glu217 is mutated to either Asp, Ala, or Arg, the value of $K_m$ changes only slightly but $k_{cat}$ is decreased substantially. For example, in the Glu217Asp mutant, $K_m$ does not change within experimental error but $k_{cat}$ is reduced by 700-fold. The 700-fold reduction of $k_{cat}$ corresponds to a 3.9-kcal/mol destabilization of the transition state, just under the value estimated from our current Raman studies.

## 4.2 Dihydrofolate Reductase

Dihydrofolate reductase (DHFR) catalyzes the reduction of 7,8-dihydrofolate ($H_2$ folate) to 5,6,7,8-tetrahydrofolate ($H_4$ folate) by facilitating the addition of a proton to N5 of $H_2$folate and the transfer of a hydride ion from NADPH to C6. And DHFR is an important enzyme because it is required for the production of purines, thymidylate, and a few amino acids and is the target for both antitumor and antimicrobial drugs. Hence, its catalytic mechanism has been under intense scrutiny for some time.

A key issue has to do with whether or not N5 is protonated in the ground state of the DHFR/NADPH/$H_2$folate complex and hence precedes hydride transfer or occurs later in the reaction pathway. A recent study using difference Raman spectroscopy of the DHFR/$NADP^+$/$H_2$folate complex (65), believed to be an accurate mimic of the productive DHFR/NADPH/$H_2$folate complex, identified two N5=C6 stretch "marker" bands, indicating unprotonated ($1650$ $cm^{-1}$) or protonated ($1675$ $cm^{-1}$) N5 (Fig. 7a). The assignments were based on $^{13}$C6 isotope labeling (Fig. 7c), deuterium shift of these bands (Fig. 7b), comparisons to the spectra of solution models, and extensive ab initio calculations (56). Based on these assignments, N5 in the binary DHFR/$H_2$folate complex was found to be unprotonated at near-neutral pH values. In the ternary DHFR/$NADP^+$/$H_2$folate complex, however, the "protonated" band at $1675$ $cm^{-1}$ was observed. A titration study, using the 1650- and protonated 1675-$cm^{-1}$ marker bands as indicators for unprotonated and protonated species, respectively, showed that the pKa of N5 is raised from 2.6 in solution (66) to 6.5 in this complex (65). Hence, N5 is protonated at physiological pH values.

From a mechanistic point of view, that the protein environment raises the pKa of N5 by 4 units to 6.5 in the productive ternary complex means that there is 4 orders of magnitude more productive complex in the physiological pH range, provided the transition-state energy barrier of hydride transfer is lower in proton-

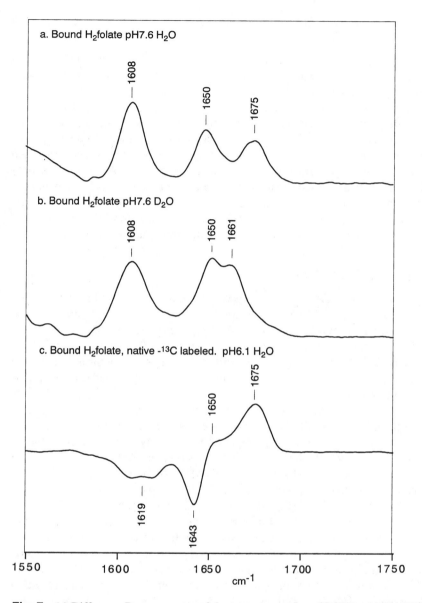

**Fig. 7** (a) Difference Raman spectra of the ternary complex of DHFR with NADP$^+$ and H$_2$folate ([DHFR/NADP$^+$]/[H$_2$folate] = 3.4/2.5) minus the spectrum of DHFR/NADP$^+$ at 4°C in 20 mM tris buffer containing 0.5 M KCl, pH 7.6. The difference bands are about 10–30% of the protein amide I band and have been assigned to bound H$_2$folate; (b) the same as (a) except the samples have been suspended in D$_2$O so that all the exchangeable protons of the sample have been deuterated. (c) $^{13}$C6 isotope edited Raman difference spectrum of DHFR/NADP$^+$/H$_2$folate at pH 6.1, that is, the difference spectrum formed by subtracting the spectrum of DHFR/NADP$^+$/[$^{13}$C6]H$_2$folate from that of DHFR/NADP$^+$/ [$^{12}$C6]H$_2$folate; only modes whose motions involves C6 show up in the $^{13}$C6 isotope edited difference spectrum. (From Ref. 56.)

ated H$_2$folate compared to unprotonated H$_2$folate. This is almost certainly true, since this makes C6 more positive (see, e.g., Ref. 56) and so can more readily accept the negatively charged hydride ion. Assuming that the transition-state barrier for protonated H$_2$folate is lower than that for unprotonated substrate, additional factors that make up the rate acceleration brought about by DHFR have to do with the rate at which protons arrive to N5 in the DHFR/NADPH/H$_2$folate ternary complex and by the rate, which may be dynamically controlled at least partly, at which the hydrogen on C4 of NADPH comes close enough to C6 of bound H$_2$folate to permit hydride transfer.

Although much is known about the structure of DHFR and its complexes with cofactor and substrates, it is not clear just what structural factors are responsible for raising the pKa of N5 by 4 units. These is no obvious proton acceptor at the active site that is close enough to stabilized protonated N5 by forming a hydrogen bond. The Raman results even seem to rule out a hydrogen-bonded network, which places a structural water molecule close enough to hydrogen bond to N5—H (56) since it appears that the immediate environment of N5 is quite hydrophobic in the DHFR/NADP$^+$/H$_2$folate complex. One possible explanation is that the negative charge of the carboxylate of the active site Asp27 would stabilize a protonated substrate even though this group is on the other side of bound substrate relative to N5. It is known that Asp27 is crucial to raising the pKa of N5, but additional protein structural factors must also be involved. For example, it has been shown in Raman difference studies that DHFR undergoes a conformational changes with a pKa of 6.5 (67), and it seems likely that this conformational changes affects the electrostatic environment of N5, stabilizing its protonation. Clearly, however, this story is still quite incomplete.

## 4.3 Energy Landscape of Liver Alcohol Dehydrogenase

A comparison between the CFX spectra in solution and the spectra of CXF bound in the LADH/NADH/CXF complex, obtained by Raman/FTIR difference spectroscopy, indicate that the local interaction with the C=O bond is stronger in the protein complex than that in aqueous solution, while that with the N—H bond is weaker (see Figs. 2 and 3 and Ref. 9). It is also observed that the C=O stretch shifts by $\Delta v_{C=O} = -48$ cm$^{-1}$ and the N—H bending shifts by $\Delta v_{N-H} = +36$ cm$^{-1}$ in the LADH/NADH/CXF complex relative to their respective frequencies for CXF in methylene chloride, which yields $\Delta v_{C=O}^{local} = -43$ cm$^{-1}$ and $\Delta v_{N-H}^{local} = +19$ cm$^{-1}$ using Eqs. (12) and (13). Therefore, the local interactions with both the C=O and N—H bond are stronger in the protein complex than that in methylene chloride. Comparing CXF in water relative to their respective frequencies for CXF in methylene chloride ($\Delta v_{C=O} = -32$ cm$^{-1}$ and $\Delta v_{N-H} = +41$ cm$^{-1}$, Table 1), $\Delta v_{C=O}^{local} = -23$ cm$^{-1}$ and $\Delta v_{N-H}^{local} = +32$ cm$^{-1}$. Our results indicate that the hydrogen-bonding strength pattern for the C=O moiety is

LADH/NADH/CXF complex > water > methylene chloride. On the other hand, the pattern for the N—H group is water > LADH/NADH/CXF complex > methylene chloride.

By using the Badger–Bauer relationship between $\Delta H$ and $\Delta v$ for CXF (as discussed earlier and in Table 1), it can be determined that the change in the effective *local* interaction energy between the C=O bond of the CXF and the enzyme resulting from a $-21$-cm$^{-1}$ frequency shift is about 5.5 kcal/mol more stable in the LADH/NADH/CXF complex compared to that in aqueous solution. Although a direct quantitative relationship between the hydrogen-bonding enthalpies and the O—H or N—H bending frequencies has not been firmly established, we have obtained an approximate correlation between the bending frequency shift and interaction energy change: every 10-cm$^{-1}$ shift of the bending frequency corresponds to about 0.8–1.0 kcal/mole change of the energy (54). Thus, the $-18$-cm$^{-1}$ shift of the N—H bending caused by the local interaction on the N—H bond corresponds to about 1.5 kcal/mol change in enthalpy in favor of the aqueous solution over that in the LADH ternary complex. On the basis of these observations, we conclude that the amide moiety of CXF contributes a net of about 4.0 kcal/mol of enthalpic energy stabilizing inhibitor binding to the protein.

Thus, the energy landscape for the binding of CXF to LADH is mapped out. Such results are of great importance in understanding the mechanism of LADH-catalyzed reactions. A major discovery here is that the C=O bond of the bound inhibitor has been polarized at the active site relative to its value in aqueous solution and even more so relative to a hydrophobic environment. The aldehyde bond is polarized by favorable, essentially electrostatic, interactions between the C=O moiety and the protein, presumably due mostly to the active site zinc with a contribution from the hydrogen bond to the hydroxyl group of Ser-48. The total of these interactions is estimated in results to be equivalent to a hydrogen bond whose enthalpy is 5.5 kcal/mol stronger than it is for CXF in water or 11 kcal/mol stronger than found in a hydrophobic (methylene chloride) environment. It is generally believed that the transition state for hydride transfer in the dehydrogenases contains considerable polar +C—O$^{-}$ bond character. Hence, stabilization of such polar resonance structures would lower the height of the transition-state barrier for hydride transfer. The current results show directly that the effect of binding to the active site of LADH is to stabilize such structures, as reported in the measurement of the ground-state, Michaelis-like, complex. By performing additional studies such as those discussed in Sec. 3.3.4, it is also possible to estimate how much the transition state may be lowered when substrate similar to CXF binds to LADH/NADH.

The downward shift in the amide N—H bend of CXF when it binds to the enzyme/NADH complex from water is indicative of the loss of hydrogen bond interaction energy (a loss of about 1.5 kcal/mol). However, there is still apparent hydrogen bonding between the N—H moiety and the protein, since the N—H bend frequency still lies at a higher frequency than found in the hydrophobic nonhydrogen-bonding environment of methylene chloride. In fact, it is estimated that there is an effective stabilizing interaction energy of the N—H group of bound CXF of about 1.5 kcal/mol relative to a hydrophobic environment. In the crystallographic results of the LADH/NADH/CXF complex (68), there is no usual proton acceptor for the N—H moiety. However, the position of binding of the CXF to the enzyme/NADH complex suggested that the N—H of the amide could form a cation–π interaction (69) with the benzene ring of Phe-93, since the distance between the N and CE1 of Phe-93 is 3.0 Å (68).

Finally, the measured inhibition constant of CXF in aqueous solution suggests that the total binding Gibbs free energy ($\Delta G$) on the CXF in the LADH/NADH/CXF complex is about 7 kcal/mol more favorable than that in aqueous solution. Since this made up of both enthalpic and entropic components, and the entropic $T \Delta S$ term of the Gibbs free energy is almost certainly unfavorable, the binding enthalpy is greater than 7 kcal/mol. Our results show that net of about 4 kcal/mol, the enthalpic binding energy is from the amide moiety and the rest apparently is from the hydrophobic interaction of the cyclohexyl ring with the protein residues. The binding of the carbonyl group to the catalytic zinc and the hydrogen bonding of the oxygen to the hydroxyl group of Ser-48 would account for about 5.5 kcal/mole to the enthalpic term. The binding of the amide N—H moiety is destabilizing. However, it clearly would be even more destabilizing if not for the cation–π interaction at the binding site.

## 4.4  Lactate Dehydrogenase

Lactate dehydrogenase accelerates the hydride transfer rate between NADH and pyruvate to form lactate and $NAD^+$ by a factor of $10^{11}$ or more over the rate found in solution (70–72). A most significant interaction between pyruvate and LDH is the exceptionally strong and favorable electrostatic interaction between pyruvate's carbonyl oxygen and His195 and other groups in the enzyme's active site. Assuming that the essential structure of the reacting components is not affected by the mutation except in regard to the H-bonding interaction, a linear relationship between $\Delta \nu_{C=O}$ in a series of mutants $\log(k_{cat})$ of the mutants should exist, as outlined earlier (Sec. 3.3.4).

Figure 8 shows a plot of the hydride transfer step rate as a function of shift in carbonyl frequency for the wild-type protein and several mutants of LDH (see Table 1) and the solution rate of hydride transfer (26). It can be seen that reasonable agreement is obtained except for the R109Q mutant. In the R109Q mutant, the hydride transfer rate is substantially below what would be predicted from the observed polarization of the C=O bond. This is not unreasonable. It is known that binding of substrates to LDH triggers the closing of a ''loop'' of residues over the active site and bound substrate in the competent form of the enzyme.

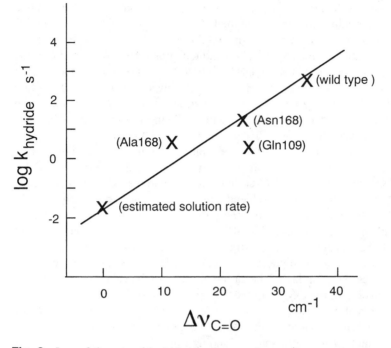

**Fig. 8**  Log of the rate of hydride transfer step (in sec$^{-1}$) versus the observed change in C=O stretch frequency of pyruvate upon binding. The solution hydride transfer rate, which has not been measured, is estimated from the ratio of enolization rate of pyruvate in solution and in LDH and taking the transfer rate in LDH to be 750 sec$^{-1}$. (From Ref. 26.)

The replacement of arginine by glutamine almost certainly affects loop closure, resulting in an active site structure that is substantially altered (25,73). The broadness of the C=O stretch band in the R109Q mutant data (26) also suggest that this is true. This study shows that, of the $10^{11}$-fold rate enhancement brought about by LDH, about $10^6$-fold involves stabilizing (H-bonding) interactions that work on the key C=O group. Other factors bringing about rate enhancement must also include reactant immobilization (see Sec. 3.2). Key vibrational bands associated with the reaction coordinate between pyruvate and lactate (the C=O stretch of pyruvate and the C4-H stretch of NADH) are observed to narrow substantially when the LDH/NADH/pyruvate ternary complex is formed. Interestingly, this is reproduced in molecular dynamics calculations on LDH and substrate complexes. Atomic motions in the productive ternary complex are considerably smaller than in the binary complex (74).

The reaction catalyzed by LDH is extremely stereospecific. The transfer of hydride to and from the *re*-face of the nicotinamide ring of $NAD^+$ and NADH exceeds transfer from the *si*-face by 1 part in $10^7$. This requires a discrimination of over 10.4 kcal/mol favoring hydride transfer from the *re*-face compared to the *si*-face. Two different mechanisms can produce this specificity. One is that either the *re*- or *si*-face points toward substrate with roughly equal probability, but the reaction occurs only from the *re*-face. Alternatively, catalysis would not distinguish between *re*- and *si*-face transfer, but rather geometrical considerations would dictate, so the cofactor would be bound in only one conformation. It is difficult to see how the protein could arrange for such a substantial difference in chemical reactivity between the two faces, but certainly not impossible. However, Raman difference spectroscopy of NADH and $NAD^+$ binding to LDH from pig heart (Deng et al., 32) provides evidence for strong interactions between the carboxamide group of NADH and the enzyme. Using Badger–Bauer relationships, the shifts in the stretching frequency of the carboxamide carbonyl of $NAD^+$ and the rocking motion of the carboxamide $NH_2$ group of NADH upon binding LDH, it has been estimated that the enthalpy of interaction between the carboxamide group of the coenzyme and LDH changes to 9–11 kcal/mol, in very good agreement with the required 10.4 kcal/mol (see Table 2 and Ref. 39). A recent electrostatics calculation showed that there is little steric differences between when the *re* and *si* geometries of the nicotinamide group end up with the nicotinamide C4 hydrogen pointing toward substrate, but that there are a number of strong and not so strong but numerous hydrogen bond interactions between the carboxamide group in the *re* geometry and not in the *si* geometry (74). The favorable electrostatic interactions between the carboxamide group of the NADH coenzyme and protein residues of the active site of LDH can account for much if not all of the stereospecificity of the LDH-catalyzed reaction.

## 4.5 Serine Proteases

Another example of structure/reactivity relationship is found in the beautiful studies of acylserine proteases. Extensive work on acyl intermediates of a series of serine proteases and functional mutants has shown that the hydrogen-bonding enthalpy, the deacylation rate, and the $C\!=\!O$ bond length are likewise correlated (49). For the series of acyl enzymes, the 54-$cm^{-1}$ shift of the $C\!=\!O$ stretch mode correlates to a change of the hydrogen-bonding strength of $-27$ kJ/mol, which is very close to the change in the measured activation energy of $-24$ kJ/mol (Fig. 9). The deacylation rate is directly related to the $C\!=\!O$ stretch frequency of the acyl enzyme in the ground state. A shift of the $C\!=\!O$ stretch mode of the acyl enzyme to higher frequencies is linearly correlated with the logarithm of the decrease of the deacylation rate, $\log(k_3)$. As the $C\!=\!O$ frequency shifts upward by 54 $cm^{-1}$, the deacylation rate decreases by 16,300-fold. An empirical relation-

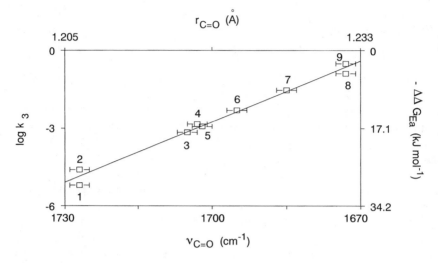

**Fig. 9** Correlation between the carbonyl stretching frequency, $v_{C=O}$ and log $k_3$, where $k_3$ is the maximal deacylation rate at pH 10 for the series of acyl-serine protease. Each point represents an acyl-enzyme. $v_{C=O}$ and log $k_3$ are also recast in the forms of $r_{C=O}$ (carbonyl bond length) and $\Delta\Delta G$ (free energy of activation for deacylation), respectively. (From Ref. 42.)

ship between C=O stretch frequency and bond length yielded that the C=O stretch frequency difference of 54 cm$^{-1}$ corresponds to a difference of the C=O bond length of 0.025 Å (Fig. 9), about 10% of the change from a C=O double bond to a C—O single bond. Such relationships established that a tetrahedral intermediate is formed at the transition state, since a polarization of the C=O bond, which lowers its stretch frequency, in the enzymic acyl intermediates reflects a ground-state distortion toward the transition state, where the C=O double bond character is substantially reduced (49).

## 4.6 Protein–Nucleotide Interactions and Enzymic Phosphate Hydrolysis

There have been a number of vibrational studies probing the structures of proteins that bind nucleotides and catalyze phosphate hydrolysis. A number of enzymes have been studied. The p21 ras and the elongation factor Tu (EF-Tu) proteins bind GDP and GTP (18,55,75–77). EF-Tu is responsible for transporting and coupling aminoacyl tRNA (aa-tRNA) to the bacterial ribosome, where the amino acid is added to the nascent peptide chain. Ras proteins are associated with cell division or proliferation. Both these so-called ''G-proteins'' are biologically active in the protein/GTP state and inactive upon enzymic-catalyzed hydrolysis of the γ-phosphate of GTP to form the protein/GDP complex. Another study was on the muscle protein myosin (57). The binding of ATP to myosin reduces the affinity of myosin for actin, and the subsequent hydrolysis of ATP yields the energy needed to drive the sequence of steps leading to the myosin/actin sliding motion. A fourth system concerned ribonuclease A, which catalyzes hydrolysis of the RNA phosphodiester linkage in two steps (78). A quite through characterization of the hydrogen bond interaction energies between protein and the guanine moiety of GDP/GTP and EF-Tu and ras was performed (75,76).

Most of the studies have focused specifically on phosphate hydrolysis mechanisms, which have been probed in two ways. One approach has been to examine the structure of bound vanadate analogs that replace the phosphate moiety that is undergoing the hydrolysis reaction. Vanadate typically (but not always) replaces phosphate and binds quite tightly to phosphate-binding proteins. It is often believed to be bound as a transition-state analog for the enzymic-catalyzed hydrolysis reaction. Hence its structure is close to or resembles the structure of the transition state. The second approach is to examine perturbations of the ground state for the phosphate substrates. The transition-state paradigm has it that enzymes bind transition states selectively and tightly. Thus, a substrate should be distorted toward the transition state when it binds. These distortions may be very small, particularly if the electronic nature of the ground state does not resemble that of the transition state. Nevertheless, the great accuracy provided by vibrational spectroscopy is often able to discern even the smallest of distortions. For example, the published x-ray structure (79) of the Rnase/vanadate transition-state complex is not consistent with the structure of the expected (from other studies) transition state of a $S_N^2$-like reaction, because the reported lengths of the nonbridging V$\cdot$$\cdot$O bonds are substantially longer than those in a normal vanadate diester. However, the Raman results (78) reporting the vanadate bond lengths (using $^{18}$O—$^{16}$O Raman difference spectroscopy and relating frequencies to bond lengths from empirical relationships; see earlier) show that the V$\cdot$$\cdot$O bond length (1.650 Å) of the transition-state complex increases only 0.012 Å from the solution value, about 0.1 Å lower than the crystallographic value. It should be noted that the vibrational study is accurate to about ±0.004 Å (see Sec. 3.3.1). This value for the V$\cdot$$\cdot$O bond length obtained in the Raman study is consistent with a $S_N^2$-like reaction. Moreover, the Raman study and a recent study based on the $^{18}$O isotope effects of the RNase-catalyzed reaction (80) both yielded the same results on how much the bond orders change from the ground state to the transition state, with both studies suggesting that about 0.11–0.20 vu is lost to the nonbridging oxygen atoms in the transition state of the chain cleavage step.

Figure 10 shows the polarized $^{18}$O–$^{16}$O Raman difference spectra of the symmetric and antisymmetric vibrations of the nonbridging V$\cdot$$\cdot$O bonds of vanadate in solution (Fig. 10A) and in a complex formed by myosin subfragment 1

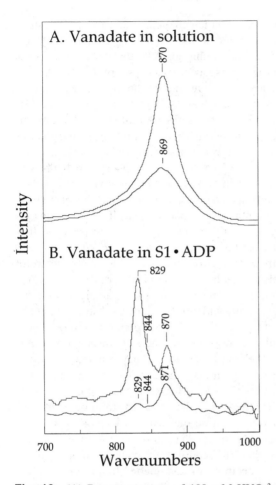

**Fig. 10** (A) Raman spectrum of 100 mM HVO$_4^{2-}$ at pH 10.5. The top curve is the parallel polarized spectrum with the Raman excitation laser beam parallel to the spectrometer's entrance slit plane; the bottom graph is the perpendicular spectrum (excitation laser beam perpendicular to the spectrometer's entrance slit plane). (B) Raman difference spectrum between myosin S1 · MgADP · Vi complex and myosin S1 · MgADP · AlF$_4^-$ complex, pH 8. The top curve is the parallel polarized spectrum with the Raman excitation laser beam parallel to the spectrometer's entrance slit plane; the bottom graph is the perpendicular spectrum (excitation laser beam perpendicular to the spectrometer's entrance slit plane). (From Ref. 57.)

(S1), MgADP, and vanadate to form the transition-state complex S1 · MgADP-vanadate (Fig. 10B; Ref. 57). The first obvious feature of the spectra is that the normal $C_{3v}$ symmetry found for dianionic vanadate in solution is slightly broken in the protein, where two nondegenerate antisymmetric modes are observed compared to the single doubly degenerate antisymmetric band observed in solution. The "fundamental" frequency of the vanadate solution model is calculated to be 869.3 cm$^{-1}$ from the measured values of $v_s = 870$ cm$^{-1}$ and $v_a = 869$ cm$^{-1}$ (Fig. 10A), which yields a bond order of 1.430 vu and a bond length of 1.669 Å for each of the three nonbridging V$\cdot\cdot$O bonds using Eqs. (4) and (5). Using the data of Fig. 10B, the three frequencies at 829, 844, and 870 cm$^{-1}$ yield bond orders of 1.327, 1.365, and 1.433 vu and bond lengths of 1.694, 1.685, and 1.669 Å, respectively. The total bond order of the three nonbridging V$\cdot\cdot$O bonds sums to 4.29 vu for the vanadate model in solution and 4.13 for the bonded vanadate as part of the S1 ·MgADP · Vi complex. This means that, since the bond orders sum to 5 in both cases, the total remaining bond order for all other bonds has increased by 0.16 vu when vanadate binds to subfragment 1.

The increase of the angle $\vartheta$ between the nonbridging O$\cdot$·P·$\cdot$O bonds upon vanadate binding to S1 · MgADP · Vi complex can be estimated quantitatively using the analytical expressions for $v_s$ and $v_a$ of dianionic vanadate (Eqs. 8 and 9). For dianionic vanadate in solution, the angle $\vartheta$ can be estimated by ab initio calculations ($\vartheta = 111°$), fs can be determined directly from the fundamental stretch frequency, as pointed out earlier, and fs/Css can be determined from the difference between $v_s$ and $v_a$ (fs/Css = 10.5). Since the fs/Css ratio does not change much according to ab initio calculations, $\vartheta$ can be calculated exactly for a VO$_3^{2-}$ moiety having $C_{3v}$ symmetry. Since the symmetry has been broken in the protein complex, the analytical formulas cannot be applied strictly. As an approximation, assuming the $C_{3v}$ symmetry and taking $v_s = 870$ cm$^{-1}$ and $v_a = 829$ cm$^{-1}$ in the S1 · MgADP · Vi complex (see Fig. 10), it is calculated that $\vartheta = 119°$ (a value of 120° is a planar VO$_3^{2-}$ moiety). Since the other V$\cdot$·O stretch mode is observed at 844 cm$^{-1}$, we expect the average angle $\vartheta$ between pairs of V$\cdot$·O bonds to be somewhat smaller. Nevertheless, $\vartheta$ has "flattened" considerably from its value in solution; the VO$_3^{2-}$ moiety has become 5–7° more planar.

How does the protein bring about these structural changes? The changes in frequencies of the symmetric and asymmetric stretch modes that occur upon binding are very sensitive to specific interactions; both the size of the interaction as well as geometry are important because the symmetry of the two mode classes is different. As discussed in Sec. 3.4, ab initio calculations could reproduce very well the experimental spectra by placing a water molecule about 2.5 Å from the vanadium atom (compare Fig. 5B with Fig. 10B). Even the lifting of the degeneracy of the two asymmetric modes is reproduced. Figure 11 shows a representation of the active site. This water molecule depicted in Fig. 11 is believed to be the attacking nucleophile in the hydrolysis reaction.

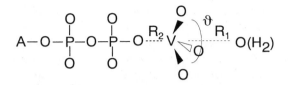

**Fig. 11** Schematic of ADP•∙Vi•∙O(H$_2$) at the S1 binding site, which defines the molecular coordinates used in the text to describe the reaction coordinate. For the trigonal bipyramidal transition-state structure believed to occur in phosphotransfer reactions (91), the nonbridging P•∙O bonds are planar. The following conventions have been used to categorize the structure of the transition state: the bond order of R1 + R2 at the transition state is much smaller than the reactant states for a dissociative transition state and substantially larger than the bond order of the ground state bridging P•∙O bond for an associative transition state. The bond order of the two axial bonds remains the same as that of the original PO bond for a synchronous reaction ($S_N^2$-like). The bond orders, as defined by its valence bond strength (see text), of the ground states for various bridging RO—PO$_3^{2-}$ bonds are close to 1.

Thus, the Raman spectral changes of the vanadate upon forming the S1 · MgADP · Vi complex can be explained by the following physical changes of the VO$_3^{2-}$ moiety: an increase of the angle between any two V∙∙O bonds (hence VO$_3^{2-}$ becomes more planar) with a concomitant small decrease in the average V∙∙O stretch force constants fs (hence, a small increase of the V∙∙O bond lengths). Although there is a small change in the nonbridging V•∙O bonds, most of the bond order associated with the vanadate–water linkage is "taken" from the bond of the leaving group, ADP-Vi, linkage. The value of R1 + R2 in the transition-state complex is close to that measured for RO—VO$_3^{2-}$ solution models. A very similar story is given for studies of ras/GTP (53) except that the changes are smaller, as expected for a ground-state species compared to a transition state–like species. It is found that the attacking water molecule found at the active site makes a very weak bond with the γ-phosphorous of bound GTP, the γ-phosphorous O•∙P•∙O bonds "flatten" some (about 1–2°), and most (ca. 75%) of the small amount of bond order (0.03 vu) formed between GDPPO$_3^{2-}$ ∙∙O(H$_2$) is taken from the GDP-Pi bond. It was additionally found that the GDP leaving group was quite stabilized by binding to ras. Hence the reaction mechanism in both proteins appears to be close to a concerted Sn2-like reaction where the total bond order of the new and old axial bonds remains roughly constant throughout the hydrolysis reaction. An important structural attribute of the binding domain in both proteins appears to be one that can hold the attacking water tightly and in a geometry that is needed for nucleophilic attack.

## ACKNOWLEDGMENT

We wish to acknowledge the support of the National Institutes of Health from the Institute of General Medicine (grant GM35183), whose financial support has made much of the research discussed here possible.

## REFERENCES

1. PR Carey. Biochemical Applications of Raman and Resonance Raman Spectroscopy. New York: Academic Press, 1982, pp. 1–262.
2. PR Carey. J Raman Spectrosc 29:7–14, 1998.
3. KT Yue, H Deng, R Callender. J Raman Spec 20:541–546, 1989.
4. R Callender, H Deng, D Sloan, J Burgner, TK Yue. In: Proccedings of International Society for Optical Engineering, Los Angeles, 1989, pp. 154–160.
5. J Dong, D D, PR Carey. Appl Spectrosc 52:1117–1122, 1998.
6. WL Peticolas, K Bajdor, TW Patapoff, KJ Wilson. In: J Stepanek, P Anzenbacher, B Sedlacek, eds. New Methods of Studying Enzyme–Substrate Interactions Using Ultraviolet Resonance Raman and Microscopic Raman Difference Technique. Amsterdam: Elsevier, 1987, pp. 249–270.
7. J Dong, D D, PR Carey. Appl Spectrosc 52:1117–1122, 1998.
8. R Callender, H Deng. Annu. Rev. Biophys. Biomol Struct 23:215–245, 1994.
9. H Deng, JF Schindler, KB Berst, BV Plapp, R Callender. Biochemistry 37:14267–14278, 1998.
10. D Chen, KT Yue, C Martin, KW Rhee, D Sloan, R Callender. Biochemistry 26:4776–4784, 1987.
11. KT Yue, CL Martin, D Chen, P Nelson, DL Sloan, R Callender. Biochemistry 25:4941–4947, 1986.
12. PJ Tonge, M Pusztai, AJ White, CW Wharton, PR Carey. Biochemistry 30:4790–4795, 1991.
13. PJ Tonge, PR Carey. In: RJH Clarke, RE Hester, eds. Raman, Resonance Raman and FTIR Spectroscopic Studies of Enzyme–Substrate Complexes. New York: Wiley, 1993, pp. 129–161.
14. PK Slaich, WU Primrose, D Robinson, K Drabble, CW Wharton, AJ White, GCK Robert. Biochem J 288:167–173, 1992.
15. AJ White, CW Wharton. Biochem J 270:627–637, 1990.
16. AJ White, K Drabble, S Ward, CW Wharton. Biochem J 287:317–323, 1992.
17. JE Baenziger, KW Miller, KJ Rothschild. Biophys J 61:983–992, 1992.
18. V Cepus, RS Goody, K Gerwert. Biochemistry 37:10263–10271, 1998.
19. H Deng, JM Goldberg, JF Kirsch, R Callender. J Am Chem Soc 115:8869–8870, 1993.
20. JM Goldberg, J Zheng, H Deng, YQ Chen, R Callender, JF Kirsch. Biochem 32:8092–8097, 1993.
21. PJ Tonge, PR Carey, RH Callender, H Deng, I Ekiel, R Muhandiram. J Am Chem Soc 115:8757, 1993.

22.  MI Page, WP Jencks. Proc Nat Acad Sci USA 68:1678–1683, 1971.
23.  WP Jencks. In: F Chapeville, A-L Haenni, eds. What Everyone Wanted to Know About Tight Binding and Enzymatic Catalysis but Never Thought of Asking. New York: Springer Verlag, 1980, pp. 3–25.
24.  A Fersht. Enzyme Structure and Mechanism. 2nd ed. New York: Freeman, 1985, pp. 1–475.
25.  H Deng, J Burgner, R Callender. J Am Chem Soc 114:7997–8003, 1992.
26.  H Deng, J Zheng, A Clarke, JJ Holbrook, R Callender, JW Burgner. Biochemistry 33:2297–2305, 1994.
27.  RM Badger, SH Bauer. J Chem Physics 5:839–855, 1937.
28.  G Horvath, J Illenyi, L Pusztay, K Simon. Acta Chim Hung 124:819–822, 1987.
29.  ID Brown. Acta Crystallogr B48:553–572, 1992.
30.  ID Brown, KK Wu. Acta Crystallogr B32:1957–1959, 1976.
31.  FD Hardcastle, IE Wachs. J Phys Chem 95:5031–5041, 1991.
32.  H Deng, J Wang, WJ Ray, R Callender. J Phys Chem B 102:3617–3623, 1998.
33.  JWJ Ray, IJW Burgner, H Deng, R Callender. Biochemistry 32:12977–12983, 1993.
34.  H Takeuchi, I Harada. Spectrochim Acta A 42:1069–1078, 1986.
35.  T Miura, H Takeuchi, I Harada. J Raman Spectrosc 20:667–671, 1989.
36.  T Maruyama, H Takeuchi. J Raman Spectrosc 26:319–324, 1995.
37.  M Joesten, LJ Schaad. Hydrogen Bonding. New York: Marcel Dekker, 1974, pp. 1–622.
38.  SN Vinogradov, RH Linnel. Hydrogen Bonding. New York: Van Nostrand Reinhold, 1971, pp. 1–319.
39.  H Deng, R Callender. Comments Mol Cell Biophys 8:137–154, 1993.
40.  V Gutmann. The Donor–Acceptor Approach to Molecular Interactions. New York: Plenum Press, 1978, pp. 1–279.
41.  PR Carey, PJ Tonge. Chem Soc Rev 19:293–316, 1990.
42.  PJ Tonge, PR Carey. Biochemistry 29:10723–10727, 1990.
43.  JG Belasco, JR Knowles. Biochemistry 22:122–129, 1983.
44.  LC Kurz, GR Drysdale. Biochemistry 26:2623–2627, 1987.
45.  H Deng, J Zheng, J Burgner, R Callender. Proc Nat Acad Sci USA 86:4484–4488, 1989.
46.  R Callender, D Chen, J Lugtenburg, C Martin, KW Ree, D Sloan, R VanderSteen, KT Yue. Biochemistry 27:3672–3681, 1988.
47.  J Clarkson, PJ Tonge, KL Taylor, D Dunaway-Mariano, PR Carey. Biochemistry 36:10192–10199, 1997.
48.  JD Doran, PR Carey. Biochem 35:12495–12502, 1996.
49.  PR Carey, PJ Tonge. Acc Chem Res 28:8–15, 1995.
50.  PJ Tonge, PR Carey, R Fausto. J Chem Soc Faraday Trans 93:3619–3624, 1997.
51.  H Deng, L Huang, M Groesbeek, J Lugtenburg, RH Callender. J Phys Chem 98:4776–4779, 1994.
52.  H Deng, AY Chan, CK Bagdassarian, B Estupinan, B Ganem, RH Callender, VL Schramm. Biochemistry 35:6037–6047, 1996.
53.  JH Wang, DG Xiao, H Deng, MR Webb, R Callender. Biochemistry 37:11106–11116, 1998.
54.  H Deng, L Kurz, F Rudolph, R Callender. Biochemistry 37:4968–4976, 1998.

55. JH Wang, DG Xiao, H Deng, M Webb, RH Callender. Biochemistry 37:11106–11116, 1998.
56. H Deng, R Callender. J Am Chem Soc 120:7730–7737, 1998.
57. H Deng, J Wang, RH Callender, JC Grammer, RG Yount. Biochemistry 37:10972–10979, 1998.
58. JH Wang, DG Xiao, H Deng, R Callender. Biospectroscopy 4:219–228, 1998.
59. C Smith, I Rayment. Biochemistry 35:5404–5417, 1996.
60. C Frieden, LC Kurz, HR Gilbert. Biochemistry 19:5303, 1980.
61. LC Kurz, C Frieden. Biochemistry 26:8450–8457, 1987.
62. DK Wilson, FB Rudolph, FA Quiocho. Science 252:1278–1284, 1991.
63. DK Wilson, FA Quiocho. Biochemistry 32:1689–1694, 1993.
64. KA Mohamedali, LC Kurz, FB Rudolph. Biochemistry 35:1672–1680, 1996.
65. Y-Q Chen, J Kraut, RL Blakley, R Callender. Biochemistry 33:7021–7026, 1994.
66. G Maharaj, BS Selinsky, JR Appleman, M Perlman, RE London, RL Blakley. Biochemistry 29:4554–4560, 1990.
67. Y-Q Chen, J Kraut, R Callender. Biophys J 72:936–941, 1997.
68. S Ramaswamy, M Scholze, BV Plapp. Biochemistry 36:3522–3527, 1997.
69. JC Ma, DA Dougherty. Chem Rev 97:1303–1324, 1997.
70. JW Burgner II, WJ Ray. Biochemistry 23:3636–3648, 1984.
71. JW Burgner II, WJ Ray. Biochemistry 23:3626–3635, 1984.
72. JW Burgner II, WJ Ray. Biochemistry 23:3620–3626, 1984.
73. AR Clarke, DB Wigley, WN Chia, JJ Holbrook. Nature 324:699–702, 1986.
74. J van Beek, R Callender, M Gunner. Biophysical J 72:619–626, 1997.
75. D Manor, G Weng, H Deng, S Cosloy, CX Chen, V Balogh-Nair, K Delaria, F Jurnak, RH Callender. Biochemistry 30:10914–10920, 1991.
76. G Weng, D Manor, Z Chen, V Balogh-Nair, R Callender. Protein Sci 3:22–29, 1993.
77. V Cepus, C Ulbrich, C Allin, A Troullier, K Gerwert. FTIR Photolysis Studies of Caged Compounds. Methods in Enzymology 291:223, 1998.
78. H Deng, J Burgner, R Callender. J Am Chem Soc 120:4717–4722, 1998.
79. A Wlodawer, R Bott, L Sjolin. J Biol Chem 257:1325–1332, 1982.
80. GA Sowa, AC Hengge, WW Cleland. J Am Chem Soc 119:2319–2320, 1997.
81. R Thijs, T Zeegers-Huyskens. Spectrochemica Acta 40A:307–313, 1984.
82. L Vanderheyden, J Th Zeegers-Huyskens. J Mol Liquids 25:1, 1983.
83. PJ Tonge, R Fausto, PR Carey. J Mol Struct 379:135–142, 1996.
84. CNR Rao, PC Dwivedi, H Ratajczak, WJ Orville-Thomas. J Chem Soc Faraday Trans 71:955–966, 1975.
85. MS Nozari, RS Drago. J Amer Chem Soc 92:7086–7092, 1970.
86. JH Lady, KB Whetsel. J Phys Chem 71:1421–1429, 1967.
87. PG Hildebrandt, RA Copeland, T Spiro, GJ Otlewski, MJ Laskowski, FG Prendergast. Biochem 27:5426–5433, 1988.
88. H Deng, J Zheng, J Burgner, R Callender. J Phys Chem 93:4710–4713, 1989.
89. H Deng, J Burgner, R Callender. Biochemistry 30:8804–8811, 1991.
90. J Zheng, YQ Chen, R Callender. Eur J Biochem 215:9–16, 1993.
91. AS Mildvan. Proteins: Structure, Function, Genetics 29:401–416, 1997.

# 13

# Infrared and Raman Spectroscopy and Chemometrics of Biological Materials

**Yukihiro Ozaki**
*Kwansei-Gakuin University, Nishinomiya, Japan*

**Koichi Murayama**
*Kobe University, Kobe, Japan*

## 1  INTRODUCTION

The aim of this chapter is to present an up-to-date view of applications of chemometrics to the analysis of infrared and Raman spectra of biological materials. Infrared (1–6) and Raman (5–10) spectra, in general, contain a wealth of valuable information about the physical and chemical properties of molecules. However, it is not always easy to extract such information from the spectra. This is particularly true for complicated biological molecules and biological materials consisting of many components. In order to select useful information from rather intricate infrared and Raman spectra, various spectroscopic analytical methods, such as second-derivative, difference spectra, deconvolution, and curve-fitting, have been employed. These are very useful, but all of them have disadvantages as well as advantages. Recently, two powerful methods for spectral analysis have been introduced to the fields of infrared and Raman spectroscopy. One is two-dimensional (2D) correlation spectroscopy, proposed by Noda (11,12) and another is chemometrics (13–17).

*Chemometrics* can be defined as the use of statistical and mathematical methods for the design or optimization of chemical experiments and for the efficient extraction of information from chemical data (13–17). Chemometrics has been used extensively in near-infrared (NIR) spectroscopy for the years since Norris et al. demonstrated its potential in quantitative analysis of foods in 1968 (18). However, it is only relatively recently that it became a routine technique for spectral analysis in infrared and Raman spectroscopy, probably for several reasons. The major reason is that both infrared and Raman spectroscopy were long recognized as useful tools for the identification and qualitative analysis of materials and studies of molecular structure, which do not always need the aid of chemometrics. In addition, Raman spectroscopy had three major difficulties that hindered the application of chemometrics. For example, Raman spectra were often interfered by background luminescence from fluorescent or phosphorescent processes, or both. Moreover, the high photon flux of the laser beam produced unwanted photochemical effects on the samples. An intensity change in Raman scattering with the fluctuation of the laser power and/or slight variations in the optics near a sample made the reproducibility of Raman scattering intensity poor.

Of course, infrared spectroscopy was also utilized for quantitative analysis 30 or 40 years ago. However, in those days, only univariate calibration was employed. Therefore, it was difficult for infrared spectroscopy to be used for the quantitative analysis of complex mixtures. It was in the early 1980s that classical least squares (CLS) calibration was often applied to infrared data (19–21). This might be a starting point for today's wealth of multivariate statistical methods in infrared spectroscopy. During the 1980s, Fourier transform infrared (FT-IR) spectrometers made remarkable progress, resulting in a dramatic revitalization and expansion in quantitative infrared spectroscopy. At the end of the '80s, a number of pioneering studies that demonstrated the potential of infrared and Raman spectroscopy in league with multivariate statistical calibration and prediction methods in quantitative analysis were reported (19,22–30). Digitized infrared spectra produced by FT spectrometers having a high signal-to-noise ratio and linearity over a wide dynamic range made the use of chemometrics feasible. In the case of Raman spectroscopy, the advent of FT-Raman spectroscopy and progress in NIR excited multichannel Raman spectroscopy enabled one to avoid the problems of fluorescence and photodegradation at the same time (31–33). Technological innovations in lasers, spectrometers, and detectors also made it possible to measure Raman spectra with high signal-to-noise ratios and good reproducibility in a very short time. Thus, we now have excellent techniques that can generate vast quantities of digitized infrared and Raman spectral data suitable for chemometric analysis.

The application of multivariate statistical methods to infrared and Raman data of biological materials stretch back to the early 1980s (19), when several research groups carried out infrared analysis of the components in blood by use of CLS (20) or multiple linear regression (MLR) (24). Later, **q**-matrix methods were used for similar problems (26). The determination of aspirin, acetaminophen, and caffeine in solution by infrared spectroscopy combined with CLS methods was attempted in 1987 (22). The CLS methods were also applied to estimate three protein conformations from the infrared spectra in the amide I and II regions of proteins (29).

Using the entire spectral range in the analysis has greatly increased the possibility of chemometrics in infrared and Raman spectroscopy of biological materials (19). Multivariate statistical methods have made possible even the estimation of physical and chemical properties of biological materials from infrared and Raman spectroscopy. From the beginning of the 1990s, chemometric analysis of Raman and infrared spectra of biological materials began to be widely tapped by a number of spectroscopists. Today, infrared and Raman spectra and chemometric techniques are used for the following purposes in biological sciences.

> *Calibration* (34–41): Determination of concentrations of components in biological materials by principal component regression (PCR) or partial least squares (PLS) regression is a representative example of calibration.
>
> *Discrimination of very similar biological materials* (42–48): For example, Shimoyama et al. (46) carried out discrimination among hard and soft ivories of African elephants and mammoth tusks based upon Raman principal component analysis (PCA). Ootake and Kokot (47,48) applied soft independent modeling of class analogy (SIMCA) combined with NIR, infrared, and FT-Raman spectroscopy to discriminate between glutinous and nonglutinous rice.
>
> *Classification of biological materials* (49–54): Cluster analysis is most often used for this purpose. Naumann et al. (49) developed a means based upon infrared spectroscopy and chemometrics for classifying intact microorganisms.
>
> *Prediction of physical and chemical properties of biological materials* (55–62): It is possible to predict physical and chemical properties of biological materials such as specific gravity of hard and soft ivories (46) and the strength of cotton fibers (54) from infrared and Raman spectral data by use of PCR or PLS regression.
>
> *Resolution enhancement of Raman and infrared spectra by factor analysis (FA)* (63–65): Raman imaging combined with a multivariate data-processing technique by Morris et al. (63–65) is a good example of this.

## 2   METHOD

Although there is a variety of experimental aspects for quantitative bio-infrared and Raman spectroscopy that must be considered, this chapter focuses on brief explanations for representative multivariate calibration techniques that are often used in the biological applications of infrared and Raman spectroscopy (13–17). We do not give any explanation about artificial neural networks (ANNs) because Chapter 10 deals with it.

## 2.1 Principal Component Analysis (PCA) and Factor Analysis (FA)

In spectral analysis it is often very important to represent data sets in more efficient ways than the original variables. Principle component analysis (PCA) is a powerful data-reduction technique that decomposes the data matrix $\mathbf{X}$ into a "structure part" and a "noise part." The purpose of PCA is to describe the main information in the variables $\mathbf{X}$ by a lower number of variables, the so-called principal components (PCs) of $\mathbf{X}$. In other words, PCA performs a transformation in which many original dimensions are transformed into another coordinate system, with fewer dimensions. The directions of the new coordinate axes, called PCs, factors, or $t$-variables, are selected to express the largest variations. A principal component is a linear combination of the original measurements and hence contains information from the entire spectrum. To represent the $n$-dimensional data structure in a smaller number of dimensions, usually two or three PCs permit one to observe groupings of objects, outlines, etc. that define the structure of a data set. The coordinates in the new coordinate systems related to the PCs are called *scores*, and the corresponding relationships between the original variables and the new PCs are called *loadings*. Principle component analysis has the following advantages. First, the resulting equation effectively uses all the data, in that each of the new variables involves all the original measurements. Second, complicated wavelength-selection procedures can be avoided. Many researchers use PCA to disentangle overlapping spectral data.

## 2.2 Principal Component Regression (PCR)

Principal component regression is "MLR (multiple linear regression)" where PCA scores, $\mathbf{T}$, is used instead of $\mathbf{X}$. One need not use a full PC decomposition in the regression step. Usually, first fewer components are employed, because the latter components generally correspond to noise. We compute the PC of $\mathbf{X}$ and use only a few of them in the regression equation. The PCA decomposition is performed with regard to the $\mathbf{Y}$ data structure. Thus, PCR does not guarantee that the separate PC decomposition of the $\mathbf{X}$ matrix necessarily gives exactly what we want—the structure that is correlated to the $y$ variable. A powerful alternative to PCR is partial least squares regression (PLS-R).

## 2.3 Partial Least Squares Regression (PLS-R)

Principal component regression and PLS-R are two of the most frequently used calibration methods in vibrational spectroscopy. Partial least squares employ the $y$ data structure in decomposing the $\mathbf{X}$ matrix as a guiding hand, and hence the outcome constitutes an optimal regression in the strict prediction validation sense. In other words, PLS uses the information in $\mathbf{Y}$ to find the $\mathbf{Y}$-relevant structure in $\mathbf{X}$. There are two versions in PLS, i.e., PLS 1 and PLS 2; the former models only one $\mathbf{Y}$ variable, while the latter models several $\mathbf{Y}$ variables simultaneously. The PLS approach can cope with the fully multivariate (both $\mathbf{X}$ and $\mathbf{Y}$ space) regression case. It has prediction ability with respect to fewer, and more interpretable, components compared to PCR. It is considered that PLS-R is especially suitable for calibration on a small number of samples with experimental noise. Figure 1 compares MLR, PCR, and PLS-R.

## 2.4 Cluster Analysis (CA)

Cluster analysis is an analysis for model groupings of data. It consists of various techniques classified into two major groups: Some CA techniques do not require assumptions made about the number of data groups or the group structure; others require such knowledge as input. Some kinds of criteria against which the grouping can be performed must be input to CA. The most popular display method for CA is dendogram.

## 2.5 Soft Independent Modeling of Class Analogy (SIMCA)

The SIMCA technique has been applied to solve many chemical pattern recognition problems. When one wants to see if one or more new samples belongs to an existing group of similar samples, SIMCA may be used. This technique aims at assigning a new object to the class to which it shows the largest similarity. In this approach the objects are allowed to have individualities and only the common properties of the classes are modeled.

The SIMCA technique has several advantages over conventional methods such as linear discriminant analysis (LDA). For example, SIMCA is not restricted to the cases where the number of objects is smaller than the number of variables. It is also applicable to cases where an object belongs to more than one class. Moreover, SIMCA can handle collinearities easily.

The SIMCA technique is based on separate PC models, so-called disjoint modeling, of the different classes. Disjoint class modeling is a training stage in the classification procedure. The subsequent classification stage employs the established class models to assess to which classes new objects belong.

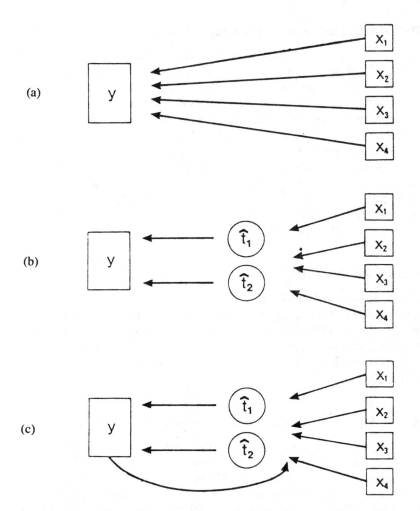

**Fig. 1** Comparison of regression methods: (a) MLR; (b) PCR; (c) PLS. One $y$ variable is to be determined from four $X$ variables. (Reproduced from Ref. 14 with permission. Copyright 1989, John Wiley & Sons.)

## 3  APPLICATIONS

### 3.1  Infrared Spectroscopy and Chemometrics Study of Biological Materials

Chemometrics has been widely used in combination with infrared spectroscopy for a variety of studies of biological materials from basic to applied research. In this chapter, the determination of secondary structure of proteins (55,56,59) and of concentrations of enzyme products (60), the discrimination between normal and abnormal cells (42–45), and the classification of microorganisms (27,28,49,50), human hair (53), and gallstones (51,52) are discussed. Other examples of the applications of chemometrics to bio-infrared spectroscopy can be found in Chapters 9 and 10 in this book.

### 3.1.1 Determination of Protein Secondary Structure Using Chemometric Analysis of Infrared Spectra and Vibrational Circular Dichroism

Infrared spectroscopy is a powerful way to investigate the secondary structure of proteins in a physiological environment (41,66). Since the polypeptide backbone of globular proteins is normally folded in more than one conformation, the amide bands of these proteins result from the superimposition of bands due to the different types of structure. Various methods have been proposed for the analysis of infrared amide bands (41,66). They can be classified into frequency-based methods and statistical methods. The former methods rely on peak assignments of either second-derivative or deconvolved spectra, which is not always an easy task for complex spectra such as those of proteins (41,66). In contrast, the statistical methods do not require any assumptions about the band assignments, the frequency limits of each band, or the bandshapes that are important in the cases of second-derivative spectra or curve-fitted deconvolved spectra.

As the chemometrics methods, CLS, PLS, and FA have been used. Here, we describe classical least squares (CLS) and PLS methods for the determination of the protein secondary structure developed by Dousseau and Pézolet (55). Dousseau and Pézolet (55) measured the infrared spectra of $H_2O$ solutions of 13 proteins of known crystal structure and corrected for the spectral contribution of water in the amide I and II regions. They first analyzed the amide I and II regions by use of the CLS method as a linear combination of bands due to only three classes of structure, i.e., $\alpha$-helix, $\beta$-sheet, and undefined structure. Parallel and antiparallel $\beta$-sheets were treated as one group, and turns were included in the undefined structure. As the first step, the pure-structure spectra were calculated from a set of 13 calibration proteins. The calculated spectra of each type of structure are shown in Fig. 2, where, for comparison, those of polylysine in the $\alpha$-helix, $\beta$-sheet, and random-coil conformations are also presented (55). Note that the calculated spectra are qualitatively in very good agreement with those of polylysine. The calculated pure-structure spectra were used to estimate the secondary-structure content of proteins. To do that, each protein was in turn eliminated from the calibration set, and the remaining proteins were employed to predict its secondary-structure content. The calculated secondary-structure contents were compared with the x-ray estimates, and the average standard deviations of the differences, or errors, between infrared and x-ray diffraction estimates and

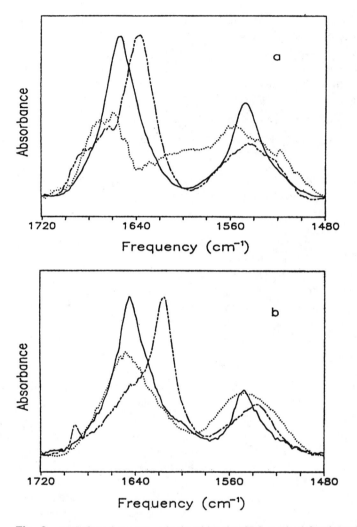

**Fig. 2** (a) Infrared spectra calculated by the CLS method for the α-helix (—), β-sheet (---), and undefined (...) conformations. (b) Infrared spectra of polylysine in the α-helix (—), β-sheet (---), and undefined (...) conformations. (Reproduced with permission from Ref. 55. Copyright 1990, American Chemical Society.)

the correlation coefficients between the two estimates were calculated (see Tables II and III in Ref. 55). The average absolute standard deviation for the prediction of the content of α-helix, β-sheet, and undefined structure is 11, 7, 6.6, and 6.7%, respectively.

Partial least squares was also used to improve the analysis of the infrared data (55). Various attempts were made to search for the best agreement between the secondary structure determined by x-ray crystallography and that predicted by infrared spectroscopy. For example, two types of α-helices, ordered and disordered helices, instead of one were introduced in the PLS calculation and both $H_2O$ and $D_2O$ solutions were compared. Figure 3 shows PLS correlation curves between infrared and x-ray estimates of the percentages of α-helix, β-sheet, and undefined conformation of the calibration proteins (55). The slopes of the correlation lines for the α-helix, β-sheet, and undefined structure were 0.96, 0.92, and 0.56, respectively. The low correlation coefficient for undefined structure was due partly to the narrow range of undefined-structure content of the calibration proteins. It was found that the best method to estimate the secondary-structure content of proteins in solution from their infrared spectra is to use the amide I and II regions ($H_2O$ solution) and to analyze the data with the PLS method, taking into account two kinds of α-helices, including only one type of β-sheet, and including turns in the undefined structure.

Lee et al. (56) applied the FA approach to infrared spectra of proteins for the determination of protein secondary structure. The infrared spectra of 18 proteins of known crystal structure were subjected to FA. The FA followed by MLR identified those eigenspectra that correlated with the variation in properties described by the calibration set. For the prediction of an unknown, the factor loadings required to reproduce its spectrum were substituted in the regression equation for each property to predict its secondary structural composition. The standard errors of prediction obtained by this method were 3.9% for α-helix, 8.3% for β-sheet, and 6.6% for turns (56).

Wi et al. (59) also used the FA approach for the prediction of protein secondary structure. In their case, Fourier self-deconvolution was performed on the amide I and II regions of infrared spectra of proteins to examine if the resultant increased bandshape variation would lead to improvements in the prediction with FA-based restricted multiple regression (RMR) methods. The deconvoluted spectra sets were subjected to PCA, which was followed by a selective RMR analysis of the PC loadings with regard to the fractional components of secondary structure. According to Wi et al. (59), the prediction quality varied, depending on the deconvolution parameters used.

The same group introduced a similar statistical method to interpret vibrational circular dichroism (VCD) spectra of proteins (57,58). In the approach of principal components (PCs) and FA, a set of VCD spectra of proteins whose

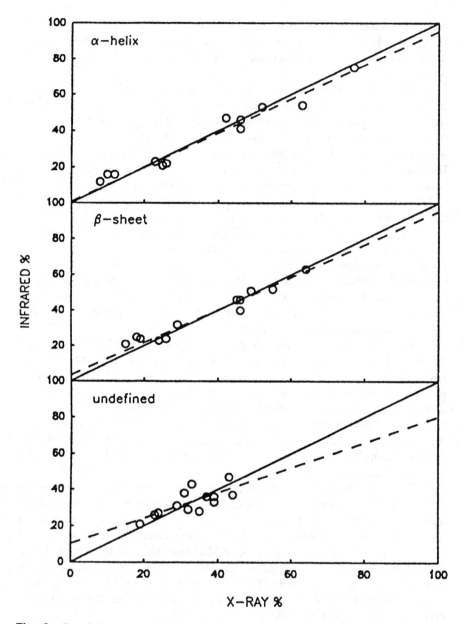

**Fig. 3** Correlation curves between infrared and x-ray estimates for the secondary-structure content. (Reproduced with permission from Ref. 55. Copyright 1990, American Chemical Society.)

crystal structures are known from x-ray studies were used as a training set. The obtained VCD spectra were reduced to a numerical form suitable for comparison with secondary-structure data from x-ray crystallography by the principal component method of factor analysis. Relationships between these parameters and secondary structural characteristics of proteins studied were then sought by using regression techniques. This method takes the guesswork out of empirical correlation and provides an impartial statistical approach to the correlation of spectra among themselves and to known molecular-structure motifs.

### 3.1.2  Diagnosis of Cervical Cancer by Infrared Microspectroscopy and Artificial Neural Networks (ANNs)

Papanicolaou (Pap) smears have long been employed for the initial screening of cervical cancer. Although this technique has been widely accepted, it is not always highly accurate, with a reported 20% false-negative rate. Because of the inconsistencies occurring from Pap smear misdiagnosis, a more reliable method for the diagnosis of cervical cancer has been desired.

Romeo et al. (42) demonstrated the usefulness of infrared microspectroscopy and ANN as an alternative automated means of screening for cervical cancer. It has been shown by a number of studies that infrared spectroscopy holds great promise in the diagnosis of cancer and in particular cervical cancer (43–45). It does not depend upon morphological observations but upon molecular differences, so it may lead to earlier detection of abnormalities.

To make the cancer diagnosis by infrared spectroscopy practicable, multivariate statistical analysis or ANN discrimination of the infrared spectra must be used. Principle component analysis can be successfully applied to classify the infrared spectra of cervical smears into abnormal and normal groups (45). However, ANN has some advantages over ordinary pattern recognition or modeling techniques in the data treatment for biomedical objects (42). Biomedical samples, in particular those exhibiting a disease state, lack data consistency because of variations in symptoms and degrees of severity of the given disease state. For these circumstances, ANN usually shows better performance than traditional multivariate statistical analysis. It seems that nonlinear statistical analysis of the data tolerates considerable amounts of imprecise or incomplete data.

Figure 4 shows representative infrared spectra of cervical cells diagnosed as normal and as dysplastic by biopsy (42). The spectrum of cultured HeLa cells is also presented in Fig. 4. The most marked changes between the spectra of normal and malignant cells are observed in the $1250$–$950$-$cm^{-1}$ region; bands at $1047$ and $1025$ $cm^{-1}$ due to C—O stretching modes of glycogen and those at $1244$ and $1082$ $cm^{-1}$ assigned to $PO_2$ antisymmetric and symmetric stretching modes, respectively, show significant intensity changes between them.

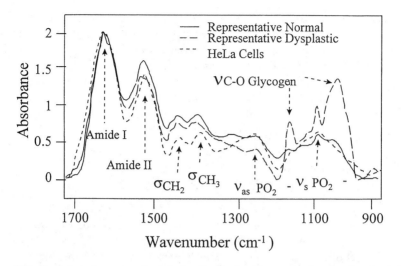

**Fig. 4** Infrared spectra of cervical cells diagnosed as normal and as dysplastic by biopsy and of cultured HeLa cells. (Reproduced from Ref. 42 with permission. Copyright 1998, C.M.B. Association.)

Infrared spectra of 88 normal and 32 abnormal (mild to severe dysplasia) cervical smear samples were first reduced to seven wavenumber values (1450, 1400, 1244, 1150, 1080, 1050, and 1026 cm$^{-1}$) by PCA. Figure 5 shows a plot of two PC scores from a data bank used to train the neural network (42). It can be seen from the plot that although most of the samples are separable, there are two obvious atypical spectral samples of the abnormal set that appear in the normal group. These were retained in the database.

For ANN, the infrared data were randomized to generate 11 data sets of training, validation, and testing sets, each containing 80, 20, and 20 samples, respectively. Training was performed systematically, starting with a simple architecture and gradually expanding the number of hidden modes and layers. Thirteen kinds of ANN architectures were developed that could differentiate between normal and abnormal cervical smears. Table 1 summarizes all the architectures prepared (42). It was very important to ascertain which data sets and architectures resulted in the best learning. For this purpose, 20 samples with known biopsy results, and previously unseen by the networks, were selected and assigned an output according to normality. This external data set of samples consisted of 10 normal and 10 abnormal samples. Comparison between the expected and actual outputs for each of the 20 samples in the external testing set was made to determine the network performance.

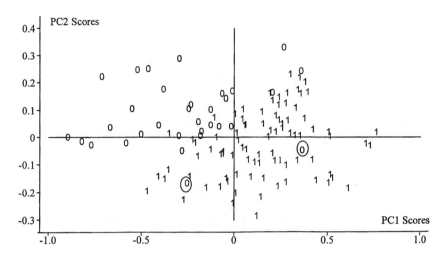

**Fig. 5** Plot of two principal component scores from a data bank used to train the neural network. (Reproduced from Ref. 42 with permission. Copyright 1998, C.M.B. Association.)

**Table 1** Architecture Types Investigated

| Architecture number | Architectural configuration[a] | Number of weights |
|---|---|---|
| 1 | 7:1:1 | 10 |
| 2 | 7:2:1 | 19 |
| 3 | 7:3:1 | 28 |
| 4 | 7:4:1 | 37 |
| 5 | 7:1:1:1 | 11 |
| 6 | 7:1:2:1 | 13 |
| 7 | 7:1:3:1 | 15 |
| 8 | 7:2:1:1 | 20 |
| 9 | 7:2:2:1 | 23 |
| 10 | 7:2:3:1 | 29 |
| 11 | 7:3:1:1 | 28 |
| 12 | 7:3:2:1 | 35 |
| 13 | 7:3:3:1 | 40 |

[a] Number of nodes in input layer:hidden layer:output layer.
*Source*: Ref. 42.

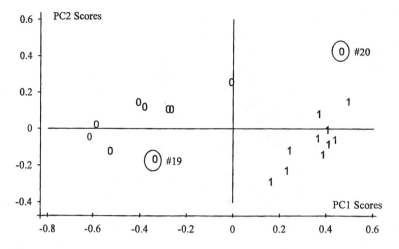

**Fig. 6** Plot of two principal component scores from the external testing set. (Reproduced from Ref. 42 with permission. Copyright 1998, C.M.B. Association.)

Figure 6 shows a plot of two PC scores from the external testing set obtained by architecture of 7:4:1 (42). The plot shows a separation between the abnormal and normal samples, with samples 19 and 20 (mild dysplasia) closer to the normal/abnormal divide. The results based upon this restricted data set indicated that ANN coupled to infrared microspectroscopy may provide an objective and automated screening means for the diagnosis of cervical cancer (42).

### 3.1.3 Assays of Enzyme Activity by Attenuated Total Reflection/Fourier Transform Infrared Spectroscopy and Partial Least Squares

Assays of the activity of major enzymes in blood are very important in laboratory medicine. Various colorimetric methods are usually employed to analyze the enzyme activities. Recently, Fujii and Miyahara (60) demonstrated the potential of attenuated total reflection (ATR)/FT-IR and the PLS method for enzyme-activity assays. The quantitative analysis of enzyme activity using infrared spectroscopy had been thought to be difficult because of insufficient sensitivity. To improve the sensitivity of ATR/FT-IR, they introduced a cryoenrichment method by which the band intensities of solutes in an aqueous solution are increased by 10–100 times compared to those measured at room temperature. The enzymes investigated are shown in Table 2 (60).

Figure 7 illustrates a flowchart of PLS calibration used in the study (60). The concentration change in a substrate or a product by enzyme reactions was

**Table 2** Composition of the Standard Solutions

| Objective enzyme | Component[a] | Concentration range (mg/dl) | Buffer |
|---|---|---|---|
| Lactate dehydrogenase | Lithium pyruvate | 0–20 | 50 mM sodium phosphate pH 8.0 |
| | Lithium lactate | 0–20 | |
| | NADH | 0–100 | |
| | NAD | 0–100 | |
| α-Amylase | Starch | 100–200 | 30 mM Tris-HCl pH 7.4 |
| | Maltose | 0–60 | |
| Creatine kinase | Creatine phosphate | 620–655 | 30 mM Tris-HCl pH 6.8 |
| | Creatine | 290–340 | 10 mM MgCl$_2$ |
| | ADP | 0–20 | |
| | ATP | 0–60 | |
| Alkaline phosphatase | pNPP | 470–500 | 30 mM Tris-HCl pH 8.8 |
| | pNP | 0–50 | 0.45 mM MgCl$_2$ |

[a] NADH: nicotinamide adenine dinucleotide, reduced form; NAD: nicotinamide adenine dinucleotide; ADP: adenosine diphosphate; ATP: adenosine triphosphate; pNPP: *p*-nitrophenyl phosphate; pNP: p-nitrophenol.
*Source*: Ref. 60.

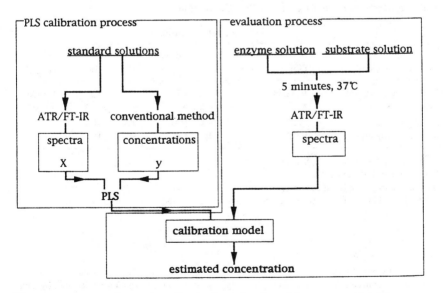

**Fig. 7** Flowchart of PLS calibration for the estimation of the concentration of an enzyme substrate. (Reproduced from Ref. 60 with permission. Copyright 1998, Society for Applied Spectroscopy.)

estimated by the cryoenrichment method for ATR/FT-IR combined with PLS, and the relationships between the measured concentration and enzyme activities were obtained.

Table 3 summarizes parameters employed for the PLS calibration and prediction error sum of squares (PRESS) obtained by the calibration (60). The correlation coefficients between the activity levels of lactate dehydrogenase, amylase, creatine kinase, and alkaline phosphatase and the concentration changes of the substrate on the product were 0.99, 0.93, 0.96, and 0.78, respectively. Therefore, except for alkaline phosphatase, good correlation coefficients were obtained. This method may be used for the simultaneous measurement of several enzymes if the optimum reaction parameters of the enzymes are similar.

### 3.1.4   Infrared Chemometric Study of Discrimination of Human Hair

The infrared microscope has become one of the indispensable tools for the characterization of biological materials. Kokot et al. (53) demonstrated the power of infrared microscopy combined with chemometrics in the discrimination of human hair. Infrared spectra of a set of six black, untreated, female hair samples (three Asian and three Caucasian) were measured by infrared microscopy. The samples

**Table 3**   Parameters and Results of PLS Calibration

|  | Parameters of calibration | | | | $\text{PRESS}^{1/2d}$ |
| Objective enzyme | Component[a] | Wavenumber range (cm$^{-1}$) | $N^b$ | $M^c$ | (mg/dl) |
| --- | --- | --- | --- | --- | --- |
| Lactate dehydrogenase | NADH | 1700–1300 | 50 | 104 | 1.40 |
| α-Amylase | Maltose | 1200–950 | 50 | 66 | 0.99 |
| Creatine kinase | ATP | 1400–1000 | 47 | 104 | 0.65 |
| Alkaline phosphatase | pNP | 1400–1000 | 50 | 104 | 0.29 |

[a] See Table 2 for explanation of abbreviations.
[b] $N$: the number of standard solutions.
[c] $M$: the number of spectral data points.
[d] PRESS: prediction error sum of squares, which was calculated from a cross-validation using the following equation:

$$\text{PRESS} = \frac{1}{M} \sum_{i=1}^{n} (y_i - \hat{y}_i)$$

where $y_i$ is the prepared concentration of the objective component in a standard solution $i$, and $\hat{y}_i$ is the estimated concentration of the objective component in a standard solution $i$ calculated by using cross-validation.
*Source*: Ref. 60.

were flattened by rolling, to reduce lensing effects, which distort the spectrum, fastened to a gold mirror. For each hair, 10 spectra were collected from different points on the sample.

Figure 8 displays infrared spectra of female hair in the 1700–700-cm$^{-1}$ region (53). Bands near 1650, 1550, and 1240 cm$^{-1}$ are assigned to amide I, amide II, and amide III modes of keratin, respectively. A reference data set was established of the spectra for two samples from each of the two racial groups; the remaining spectra were employed as validation data. The reference spectra

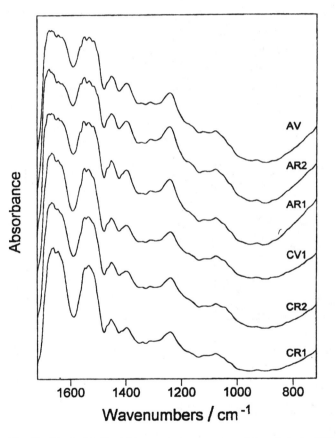

**Fig. 8** Typical infrared transmission spectra of female hair. CR1 and CR2, Caucasian subjects; AR1 and AR2, Asian subjects; CV and AV, validation data from additional Caucasian and Asian subjects, respectively. (Reproduced from Ref. 53 with permission. Copyright 1998, Royal Society of Chemistry.)

were subjected to PCA on a double-centered (column and row mean-scaled), variance-scaled matrix. Fuzzy clustering and SIMCA were used to identify atypical spectra, which were removed from the set.

Figure 9 illustrates a plot of PCA scores for the infrared spectra of female hair samples (53). PC1 explains 61.6% of the variance and PC2 explains a further 32.1%. Six clusters are revealed by the plot. PC1 discriminates among the spectra on the basis of the different races. The Asian black hair (CR1 and CR2) gives negative scores on PC1, whereas the Caucasian black hair (CR1 and CR2) yields positive scores. The two sets of validation spectra (AV and CV) fall within their respective groups. It seems possible to distinguish between black Asian hair and black Caucasian hair by infrared spectrometry. This may be of great significance in forensic science.

### 3.1.5 Identification of Microorganisms by Infrared Spectroscopy and Cluster Analysis

Naumann et al. (28,29,49) have developed a method using infrared spectroscopy chemometrics for identifying microorganisms rapidly (see Chapter 9). This

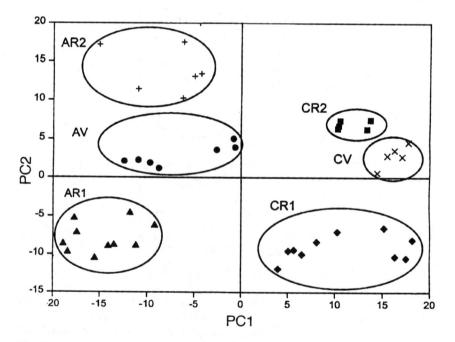

**Fig. 9** Plot of PCA scores for infrared transmission spectra of female hair samples. (Reproduced from Ref. 53 with permission. Copyright 1998, Royal Society of Chemistry.)

method has been modified for application to microorganisms important in the dairy industry (50).

Figure 10 shows infrared spectra of some *Clostridium* spp. (50). Note that the spectra are very similar to one another except for the 1300–1000-cm⁻¹ region. Therefore, a statistical method was attempted to extract the difference. Having a library containing all species of interest, the spectrum of an unknown microorganism can be compared with library spectra by use of cluster analysis. Specific spectral regions (1200–800 cm⁻¹ weighted 3 times, 3000–2800 cm⁻¹, and 1500–1200 cm⁻¹ each weighted once) were used for the cluster analysis. Pearson's product moment correlation coefficient was employed as the measure of similarity among the infrared spectra. Correlation coefficients were transformed to *d*-values, or spectral distances. The spectral distance for each pair of spectra thus computed was plotted in the form of a dendogram, as shown in Figure 11, where the vertical lines mark the spectral distances between spectra or between clusters of spectra (50). Some species, such as *Clostridium perfringens* and *C. tetanomorphum*, show little difference; the spectral distance is only 5–10. In contrast, other species, such as *C. butyricum* DSM 2477 and *C. sporogenes*, have large distances

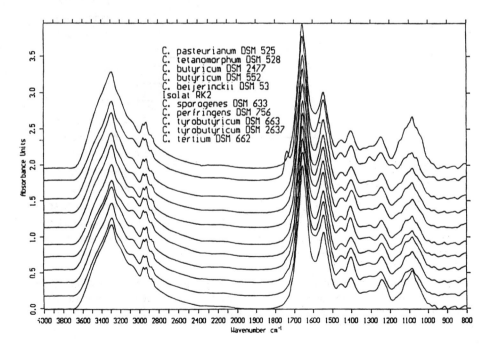

**Fig. 10**   Infrared spectra of Clostridium spp. (Reproduced from Ref. 50 with permission. Copyright 1995, American Association of Official Analytical Chemists.)

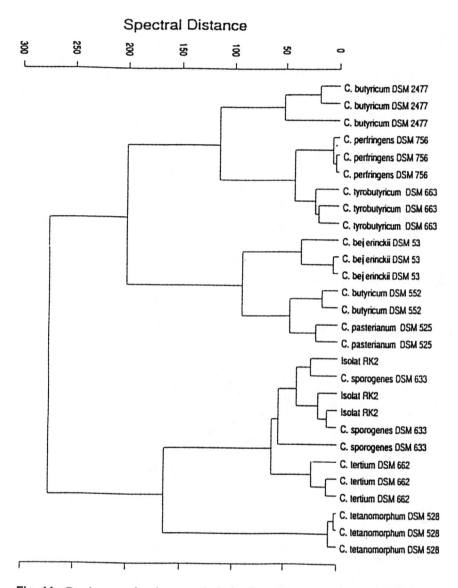

**Fig. 11** Dendogram of a cluster analysis for Clostridium spp. The vertical lines mark the spectral distances between spectra or between clusters of spectra. (Reproduced from Ref. 50. Copyright 1995, American Association of Official Analytical Chemists.)

of 50–60. In order to examine the power of identification, the result of one identification is shown in Fig. 11. The isolated spot RK2 was placed into the cluster of *C. sporogenes*. The result was in good agreement with the traditional method for the determination of the microorganisms.

### 3.1.6   Cluster Analysis of Infrared Spectra of Gallstones

Cluster analysis was employed to solve the analytical problems resulting from the number of compounds that can be present in human gallstones, some of them unidentified (51). It provided objective criteria (similarity of spectral profiles) to identify homogeneous groups of stones on the basis of their composition. For the CA of gallstones, Laloum et al. (51) carried out tests on simulated mixture spectra: 64 combination spectra of three major gallstones, i.e., cholesterol, calcium bilirubinate, and calcium carbonate, were simulated according to a ternary diagram. Figure 12 shows the ternary diagram of the coefficients for the 64 simulated mixtures (51). Each mixture is identified by a triplet indicating its position in the diagram. It was possible to show the effects of different normalization procedures, i.e., Euclidean norm, spectrum maximum, and area under spectrum set to 1. These spectra were subjected to CA, depending on different agglomerative links (single linkage, complete linkage, average linkage, and Ward's criterion). It was found that Ward's criterion best preserves the pattern of the spectral data, although the resultant trees give the same groups.

More than 100 human gallstones were classified based upon CA of their infrared spectra with Ward's criterion. Figure 13 shows a classification tree of infrared spectra (normalized to unit area) of 128 human gallstones clustered with Euclidean distance and Ward link (51). It can be seen from Fig. 13 that the 128 spectra are classified into two main groups, i.e., I and II, and each group is classified further into subdivisions. All in all, seven homogeneous groups were obtained. Typical spectra (mean spectrum and standard deviation spectrum) were calculated for each group and are presented in Fig. 14 (51). Figure 14(a), for example, corresponds to cholesterol and some traces of biliary pigments (peaks around 1600 $cm^{-1}$).

Laloum et al. (51) then made a comparison between morphological types and spectral groups. Table 4 summarizes the crossing between morphological and spectral classes. The identified spectral groups correspond well to morphological groups and confirm that three pathological processes are responsible for cholesterol stones, brown pigment stones, and black pigment stones.

Wentrup-Byrne et al. (52) carried out PLS analysis for infrared/photoacoustic spectroscopy (PAS) spectra of gallstones to model their cholesterol concentration. The use of the plot of their first and second PLS latent variable scores allowed them to make categorization of the stones.

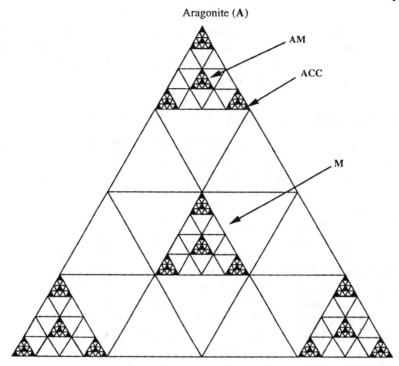

**Fig. 12** Ternary diagram of the coefficients for the 64 simulated mixtures. Each mixture is identified by a triplet indicating its position in the diagram. (A) Aragonite, (B) calcium bilirubinate, (C) cholesterol, (M) medium zone corresponding to balanced ternary mixtures of the three components. (Reproduced with permission from Ref. 51. Copyright 1998, Society for Applied Spectroscopy.)

## 3.2 Raman Spectroscopy and Chemometric Study of Biological Materials

Recently, chemometrics has been actively employed in applied bio-Raman spectroscopy. One can find interesting examples in nondestructive in situ analysis of biological materials (46), determination of components in medical samples (36,37), agricultural products (39), and foods (38), prediction of physical and chemical properties of biological materials (46,54), and quality evaluation of foodstuffs (61). A new stream in Raman chemometric applications is to use multivariate statistical techniques for simplifying complex spectral data and highlighting subtle interrelationships (63–65).

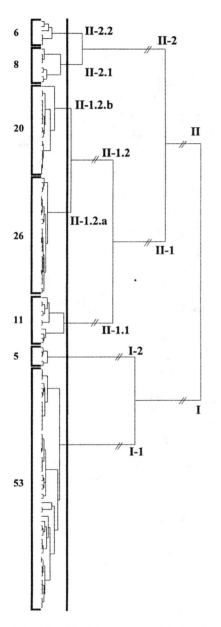

**Fig. 13** Classification tree of infrared spectra normalized to a unit area of 128 French and Vietnamese gallstones clustered with Euclidean distance and Ward link. (Reproduced with permission from Ref. 51. Copyright 1998, Society for Applied Spectroscopy.)

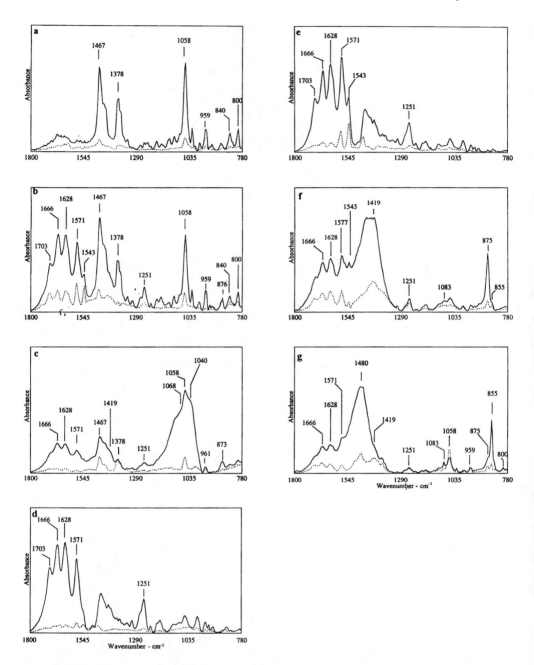

**Fig. 14** Characteristic spectra of the seven clusters. Mean (plain line) and standard derivation (dotted line) spectra of clusters: (a) I-1; (b) I-2; (c) II-1.1; (d) II-1.2.a; (e) II-1.2.6;
(f) II-2.1; and (g) II-2.2. (Reproduced with permission from Ref. 51. Copyright 1998,
Society for Applied Spectroscopy.)

**Table 4** Chi-square Table Crossing the Four Morphological and Seven Spectral Groups of Gallstones[a]

| Group | Mean composition | Pure C. | Mixed C. | Brown P. | Black P. | Total |
|---|---|---|---|---|---|---|
| I-1 | CHOL + traces (CB, BIL, CP) | 11 | 40 | 0 | 2 | **33** |
| I-2 | CARB, CHOL, CB, ACCP | 0 | 1 | 1 | 3 | **5** |
| II-1.1 | CB, CP, CHOL | 0 | 6 | 5 | 0 | **11** |
| II-1.2.a | CB | 0 | 1 | 16 | 9 | **26** |
| II-1.2.b | CB, CP | 0 | 0 | 20 | 0 | **20** |
| II-2.1 | CALC, CB, AP, ARA | 0 | 0 | 4 | 4 | **8** |
| II-2.2 | ARA, CB, CHOL, CALC | 0 | 1 | 1 | 4 | **6** |
| | **Total** | **11** | **49** | **47** | **22** | **129** |
| | Chi-square = 138.21    dof = 18 | | $P < 0.0001$ | | Cramer's $V = 0.60$ | |

[a] Note: CHOL = cholesterol; CB = calcium bilirubinate; BIL = bilirubin; CP = calcium palmitate; CARB = carbonate apatite; ACCP = amorphous carbonated calcium phosphate; CALC = calcite; ARA = aragonite.
*Source*: Ref. 51.

### 3.2.1 Nondestructive Discrimination of Biological Materials and Prediction of Their Specific Gravity by Near-Infrared Fourier-Transform Raman Spectroscopy and Chemometrics

Near-infrared FT Raman spectroscopy and chemometrics was employed for nondestructive discrimination of three kinds of biological materials (hard ivories, soft ivories, and mammoth tusks) and for prediction of the specific gravity of the ivories (46). Near-infrared (1064-nm) excited FT-Raman spectra were measured in situ for them, and PCA was carried out for the obtained spectra in the 1800–400-$cm^{-1}$ region. The two kinds of ivories were clearly discriminated from one another based upon a one-factor plot for the FT-Raman spectra. Principal component weight loadings showed that the discrimination relies upon the ratio of collagen and hydroxyapatite included in the two kinds of ivories. The discrimination among the hard and soft ivories and mammoth tusks was made by a three-factor plot for the spectra. Partial least squares regression allowed one to develop a calibration model that predicts the specific gravity of the hard and soft ivories.

Figures 15(a), (b), and (c) show the 1064-nm excited FT-Raman spectra of a hard ivory, a soft ivory, and a mammoth tusk, respectively (46). An FT-Raman spectrum of hydroxyapatite is shown in Fig. 15(d) for comparison purposes. Ivories consist largely of collagen and hydroxyapatite, so the Raman spectra of ivories overlap those of collagen and hydroxyapatite. An intense band at

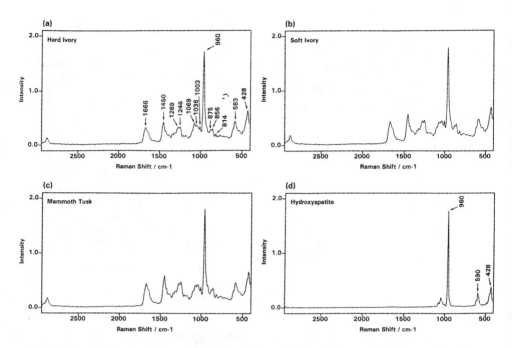

**Fig. 15** FT-Raman spectra of (a) a hard ivory, (b) a soft ivory, (c) a mammoth tusk, and (d) hydroxyapatite. (Reproduced from Ref. 46 with permission. Copyright 1997, Society for Applied Spectroscopy.)

960 cm$^{-1}$ and two medium bands at 583 and 428 cm$^{-1}$ are assigned to the symmetric stretching and bending modes of the PO$_4^{3-}$ group of hydroxyapatite, respectively, while bands at 1666, 1450, 1269, 1246, and 1003 cm$^{-1}$ are due to the amide I mode, the CH$_2$ bending mode, the amide III modes, and the breathing mode of phenylalanine of collagen, respectively.

The Raman spectra of the hard and soft ivories and mammoth tusk are so similar to one another that it is not easy to discriminate the three spectra with the naked eye. Therefore, Shimoyama et al. (46) carried out PCA for the FT-Raman spectra in the 1800–400-cm$^{-1}$ region of the hard and soft ivories (10 samples each). Figure 16(a) displays the overlapping of the 20 spectra of the hard and soft ivories (46). Of note is that the absolute intensities of the Raman bands in the 1800–400-cm$^{-1}$ region change considerably from one spectrum to another. Therefore, the spectra were subjected to the multiplicative scatter correction (MSC) treatment that is a powerful preprocessing method for removing additive (baseline shifts) and multiplicative differences in a spectrum (46). It

**(a)** **(b)**

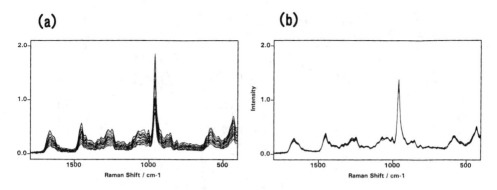

**Fig. 16** Overlapping of 20 FT-Raman spectra of the hard and soft ivories before (a) and after (b) the MSC treatments. (Reproduced from Ref. 46 with permission. Copyright 1997, Society for Applied Spectroscopy.)

corrects the spectra according to a simple linear univariate fit to a standard spectrum. In Fig. 16(b), the 20 FT-Raman spectra after the MSC treatments are presented.

Figure 17 shows the one-factor plot for a qualitative model of the hard and soft ivories based upon their 20 FT-Raman spectra after the MSC treatments (46).

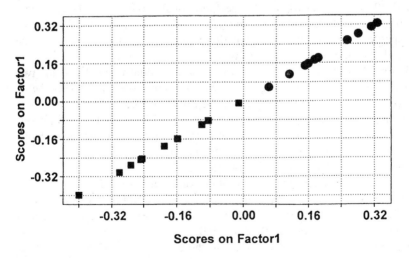

**Fig. 17** The one-factor plot for a quantitative model of the hard (■) and soft (●) ivories based upon their 20 FT-Raman spectra after the MSC treatments. (Reproduced from Ref. 46 with permission. Copyright 1997, Society for Applied Spectroscopy.)

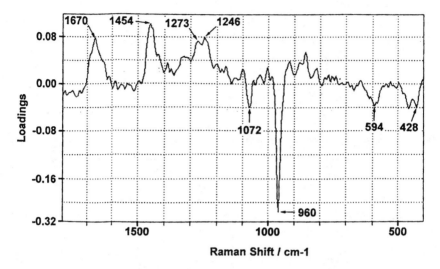

**Fig. 18**  A principal component weight loadings plot for factor1. (Reproduced from Ref. 46 with permission. Copyright 1997, Society for Applied Spectroscopy.)

The two kinds of hard and soft ivories are clearly discriminated. A PC weight loadings plot for factor 1 is shown in Fig. 18 (46). Upward peaks at 1670, 1454, 1273, and 1246 $cm^{-1}$ correspond to the vibrational modes of collagen, while downward peaks at 1072, 960, 594, and 428 $cm^{-1}$ are ascribed to the modes of hydroxyapatite. Probably, the discrimination was made by the difference in the ratio of collagen and hydroxyapatite involved in the ivories.

Figure 19(a) shows the two-factor plot for a qualitative model of the two kinds of ivories and tusk based upon the 25 FT-Raman spectra after the MSC treatments. The three kinds of samples can be discriminated by two factors. The corresponding three-factor plot is presented in Fig. 19(b) (46). There is no significant difference in the discrimination between the two- and three-factor plots, although the three-factor plot may be better visually.

It was also of interest to predict physical properties of biological materials by use of Raman-chemometric analysis. Shimoyama et al. (46) developed a PLS regression calibration model, based upon factor 1 shown in Fig. 18, that predicts the specific gravity of the hard and soft ivories. The correlation coefficient and standard error of calibration are 0.92 and 0.02, respectively. In the model, the plots for the hard and soft ivories appear in the upper right and bottom left regions, respectively. Thus, the results revealed that the hard and soft ivories can be discriminated by specific gravity.

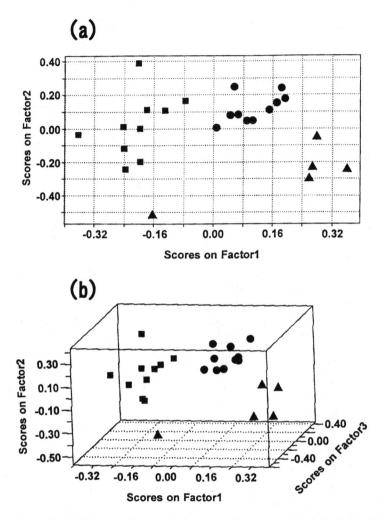

**Fig. 19** (a) The two-factor plot for a quantitative model of the hard (■) and soft (●) ivories and mammoth tusks (▲) based upon their 25 FT-Raman spectra after the MSC treatments. (b) The corresponding three-factor plot. (Reproduced from Ref. 46 with permission. Copyright 1997, Society for Applied Spectroscopy.)

### 3.2.2 Concentration Measurements of Aqueous Biological Analytes by Near-Infrared Raman Spectroscopy and a Partial Least Squares Analysis

Berger et al. (37) carried out rapid, noninvasive concentration measurements of glucose, lactic acid, and creatinine in saline solution by NIR Raman spectroscopy and a PLS analysis. The Raman spectra were measured remotely through optical fibers, and a root mean squared prediction error (RMSEP) of 1.3 mM or less for the analytes was achieved in 100 sec. Twenty-one samples with various dissolved analyte concentrations were prepared from 100 mM stock solutions of glucose, lactic acid, and creatinine in phosphate-buffered saline solution (pH 7.4).

Figure 20(a) shows a typical spectrum of an analyte mixture solution before background subtraction (37). The same spectrum after subtraction of the system background is shown in Fig. 20(b). Bands near 630, 710, and 930 cm$^{-1}$ are due to creatinine; those in the 1000–1200-cm$^{-1}$ region are ascribed mainly to glucose. An intense band near 850 cm$^{-1}$ contains contributions from both creatinine and lactic acid. The PLS cross-validation was carried out directly for the unsubtracted spectra with no loss of prediction accuracy. All three analytes required three loading vectors for good fits and yielded essentially optimal results with four. The PLS cross-validation results for the three analytes are presented in Fig. 21 (37). An RMSEP for glucose was 1.2 mM, which is the level of uncertainty desired for clinical measurements of blood glucose concentration. Lactic acid and creatinine were with an RMSEP of 1.3 and 1.2, respectively. These results indicated that NIR Raman spectroscopy coupled with PLS spectral analysis holds considerable promise in determining dissolved biological analytes such as glucose with clinical accuracy.

### 3.2.3 Aqueous Dissolved Gas Measurements by Near-Infrared Raman Spectroscopy Coupled with Partial Least Squares

The measurement of aqueous dissolved gas concentrations is important for understanding the mechanisms of many chemical and biochemical reactions and is of clinical interest. Raman spectroscopy has an inherent advantage as a technique for measuring concentrations of dissolved aqueous molecules because Raman scattering of water is weak. Berger et al. (36) reported the NIR Raman spectra of $CO_2$, $SO_2$, $N_2O$, dimethyl ether (DME), and $O_2$ dissolved in phosphate-buffered saline (PBS), together with spectra of the same species measured in the gaseous state. The transition from the gaseous to the aqueous dissolved state caused significant changes in Raman band frequencies, intensities, and widths. The reasons for the spectral changes were discussed. Quantitative extractions of dissolved $CO_2$ concentrations from spectra using the PLS technique were also presented.

Figures 22(a), (b), and (c) depict background-subtracted Raman spectra of the dissolved $CO_2$, DME, and $O_2$, respectively, and the corresponding gaseous-

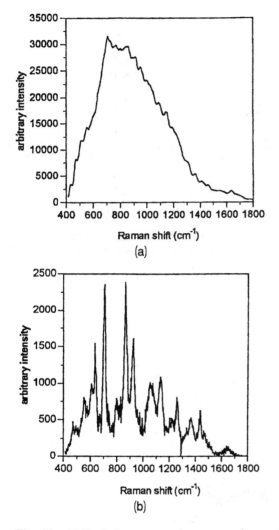

**Fig. 20** (a) Typical unprocessed spectrum of an analyte mixture solution. Spectra such as this were employed for the PLS analysis directly, without any wavelength correction or background subtraction. (b) Same spectral data as in (a) after background subtraction of a spectrum of pure solvent. The resulting spectrum shows a combination of Raman bands from the three dissolved analytes. (Reproduced from Ref. 37 with permission. Copyright 1996, Optical Society of America.)

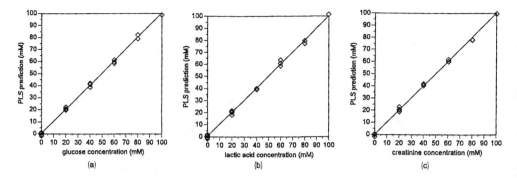

**Fig. 21** (a) PLS prediction of dissolved glucose concentrations from Raman spectra with six loading vectors. RMSEP is 1.2 mM. (b) PLS prediction of dissolved lactic acid concentrations from Raman spectra with six loading vectors. RMSEP is 1.3 mM. (c) PLS prediction of dissolved creatinine concentrations from Raman spectra with three loading vectors. RMSEP is 1.2 mM. (Reproduced from Ref. 37 with permission. Copyright 1996, Optical Society of America.)

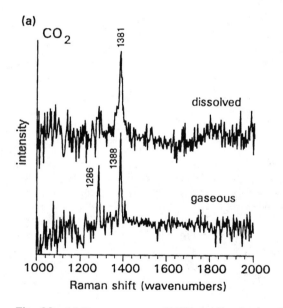

**Fig. 22** (a) Raman spectra of 760 torr dissolved and gaseous $CO_2$.

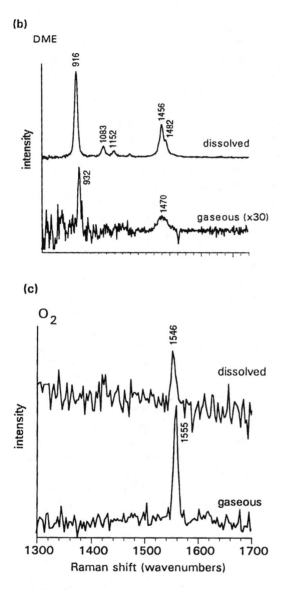

**Fig. 22** *Continued* (b) Raman spectra of 760 torr dissolved and gaseous DME. (c) Raman spectra of 760 torr dissolved and gaseous $O_2$. (Reproduced from Ref. 36 with permission. Copyright 1995, Society for Applied Spectroscopy.)

phase spectra (36). Of note in Fig. 22(a) is that the stretching band at 1388 cm$^{-1}$ in the spectrum of gaseous $CO_2$ shifts to 1381 cm$^{-1}$ in the dissolved state and that the intensity of the bending overtone at 1286 cm$^{-1}$ decreases relative to the 1381 cm$^{-1}$ band by about a factor of 2. It seems likely that the coupling between these two modes is weaker in the dissolved state, reducing the Fermi resonance. In the spectra of DME, the methyl band at 932 cm$^{-1}$ shows a downward shift by 16 cm$^{-1}$ in the dissolved state. For $O_2$, the stretching band at 1555 cm$^{-1}$ shifts to a lower frequency by about 10 cm$^{-1}$. In addition to these shifts, all the Raman bands from gases show an increased linewidth upon going from the gaseous state to the dissolved state. These shifts and broadenings can be explained by interactions with the surrounding water molecules. Since a dissolved molecule, in vibrating, must displace neighboring water molecules, the frequency of vibration should decrease; by the same token, since the distance to the nearest neighbors varies from place to place, this pattern should cause inhomogeneous broadening.

Dimethyl ether shows the most remarkable changes between its Raman spectra in the two states. "New" bands appear at 1083 and 1152 cm$^{-1}$ in the Raman spectrum of dissolved DME. The appearance of these two bands, which are Raman inactive but infrared active, can be explained by relaxed selection rules due to lowered symmetry in the dissolved state (36).

Another important observation in Fig. 22 is Raman scattering enhancements in the dissolved phase. When a molecule is surrounded by a dielectric medium such as water, its Raman scattering cross section increases because of local electric field enhancement. This effect is clearly seen in Fig. 22 (36).

For the determination of $CO_2$ concentration, data from the 22 samples were analyzed by PLS. A PLS cross-validation, in which the dissolved $CO_2$ concentration of one sample is predicted from its spectrum with the use of all of the others for calibration, was performed on the 22 samples, with each sample rotated out once for prediction. In Fig. 23 are plotted the concentrations predicted in the PLS cross-validation against the reference values measured by the blood gas analyzer (36). It was found that selecting the 1500–1000 cm$^{-1}$ region and a bin size of five pixels, corresponding to 15 cm$^{-1}$, gave the lowest prediction errors. This bin size corresponds roughly to both the resolution of the system and the typical natural linewidth of a Raman band in this regime.

The root mean squared uncertainty of prediction was 12 torr for the determination of $CO_2$. It was of note that even though the Raman signal from the dissolved $CO_2$ is relatively small and the background varies from sample to sample, accurate concentration prediction can be made. In this case the uncertainty in the prediction is inversely proportional to the signal-to-noise ratio, so it can be reduced by a longer integration time, higher power, increased CCD quantum efficiency, and more efficient collection geometry. The results indicated that remote,

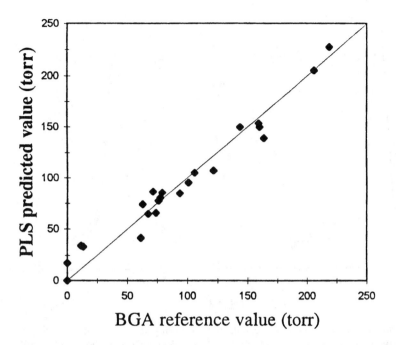

**Fig. 23** Prediction of dissolved $CO_2$ concentrations. Predicted values are plotted against the reference concentrations from the blood gas analyzer. (Reproduced from Ref. 36 with permission. Copyright 1995, Society for Applied Spectroscopy.)

real-time, quantitative dissolved gas detection may be achievable in the near future in industrial, biological, and medical settings.

### 3.2.4 Raman Chemometric Analysis of Foods

Raman spectroscopy is a promising technique for nondestructive, in situ analysis of foodstuffs and agricultural products. Li-Chan (61) applied PLS regression to Raman spectra of mixtures of canola oil and varying proportions of Greek virgin olive oil to predict the degree of adulteration of olive oil. Figure 24 shows Raman spectra of the mixtures in the $1700–1600$-$cm^{-1}$ and $1500–1400$-$cm^{-1}$ regions, where bands due to C=C stretching and $CH_2$ scissoring modes of the fatty acyl chains appear, respectively (61). It is noted that the addition of canola oil to olive oil changes the ratio of peak heights of the two Raman bands at 1660 and 1456 $cm^{-1}$ ($I_{1660}/I_{1445}$). They estimated the ratio of peak heights and that of peak areas of the two bands to predict the degree of adulteration. The results are compared in Table 5, together with results of PLS regression (61). A much better calibration

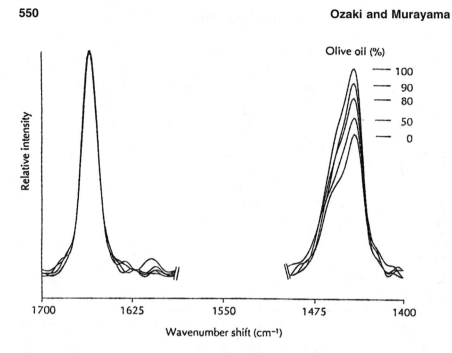

**Fig. 24** Raman spectra of mixtures of canola oil and varying proportions of Greek virgin olive oil, in the regions corresponding to C=C (1700–1600 cm⁻¹) and CH₂ (1500–1400 cm⁻¹) stretching modes of the fatty acyl chains. (Reproduced from Ref. 61 with permission. Copyright 1994, Elsevier.)

**Table 5** Results of PLS and Two Univariate Models for Predicting Degree of Adulteration of Olive Oil with Canola Oil

| Method | Cross-validation | | All samples | |
|---|---|---|---|---|
| | $r^2$ | SEP (%) | $r^2$ | SEP (%) |
| PLS | 0.96 | 7.0 | 0.99 | 6.5 |
| Ratio of peak heights | 0.85 | 20.2 | 0.95 | 16.2 |
| Ratio of peak areas | 0.75 | 23.1 | 0.92 | 19.3 |

Based on Raman spectra data in the 1700–1600- and 1500–1400-cm⁻¹ regions.

$r$: correlation coefficient; SEP (%): standard error of prediction.

*Source*: Ref. 61.

model could be obtained for PLS calibration. The results of cross-validation showed that the PLS model is more stable to sample variation than are the univariate models.

In this study, the addition of less than 10% canola oil could not be unequivocally detected. This is similar to the detection limit (>7.5% canola oil in olive oil) based on triacylglycerol profiles obtained by reversed-phase HPLC. However, Raman spectroscopy has the advantages of speed of analysis and lack of any sample preparation or dilution.

Ootake and Kokot (47,48) compared FT-NIR diffuse reflectance spectroscopy, PAS, and FT-Raman spectroscopy for the discrimination between glutinous and nonglutinous rice. When SIMCA was used to classify raw spectral data, FT-Raman results gave the best discrimination and PAS measurements followed. The FT-Raman spectra of some samples of nonglutinous rice showed strong fluorescence effects. When these samples were excluded from analysis, modeling and classification improved. For FT-Raman, the best result was obtained with the MSC pretreatment. Figure 25 compares Cooman's plots for discrimination of FT-Raman spectra of glutinous and nonglutinous rice between the two kinds of pretreatment, second derivative (upper) and MSC (lower) (48).

Archibald et al. (38) compared Raman with NIR spectroscopic methods of determination of total dietary fiber in cereal foods. They generated three Raman and NIR reflectance models for the spectroscopic determination of total dietary fiber of a wide variety of cereal foods by use of PLS regression. Various pretreatment methods were investigated for the two kinds of spectroscopy. They suggested that the Raman method is limited by its sampling technique and could be improved with more densely packed, larger-area specimens. The regression vectors of the Raman models seemed more easily interpreted than NIR models. Both methods appeared capable of achieving an acceptable level of error (38).

### 3.2.5 Nondestructive Determination of Wood Constituents by Fourier Transform Raman Spectroscopy and Chemometrics

The feasibility of FT-Raman spectroscopy coupled with chemometrics for nondestructive determination of various wood constituents was examined by use of five *Eucalyptus* species, including samples of various ages and colors, which are of importance as a plantation source (39). The percentage of wood constituents related to pulp properties (holocellulose, $\alpha$-cellulose, hemicellulose, lignin, extractives, alkali extractives, total extractives, and extractives-free (EF) wood constituents for holocellulose, $\alpha$-cellulose, hemicellulose, and lignin) were determined. Partial least squares regression for the second derivatives of the Raman spectroscopic data provided good correlations between wet chemical and Raman predicted values for all traits except EF hemicellulose, with standard error of

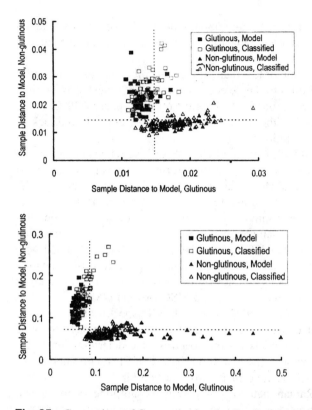

**Fig. 25** Comparison of Cooman's plots for discrimination of FT-Raman spectra of glutinous and nonglutinous rice between the two kinds of pretreatment, second derivative (upper) and MSC (lower). (Reproduced from Ref. 48 with permission. Copyright 1998, NIR
Publications.)

prediction (SEP) < 0.8 points in the calibration (for known samples) and SEP
< 3.4 points in the prediction (for unknown samples), respectively.

Figures 26(a) and (b) show an example of the Raman spectrum from *E.
camaldulensis* sample and its second-derivative form (39). In order to measure
the Raman spectra, wood meal samples (20 mesh pass) were packed into NMR
tubes. The spectra in the 1800–100-cm$^{-1}$ region were transformed to the second-
derivative form after the MSC. They were then subjected to the PLS regression
with the contents of wood constituents. A total of 63 samples for calibration and
30 for prediction were selected randomly.

As can be seen in Fig. 27 and Table 6, Ona et al. (39) obtained highly
significant correlation coefficients of over 0.99 between wet chemical and Raman

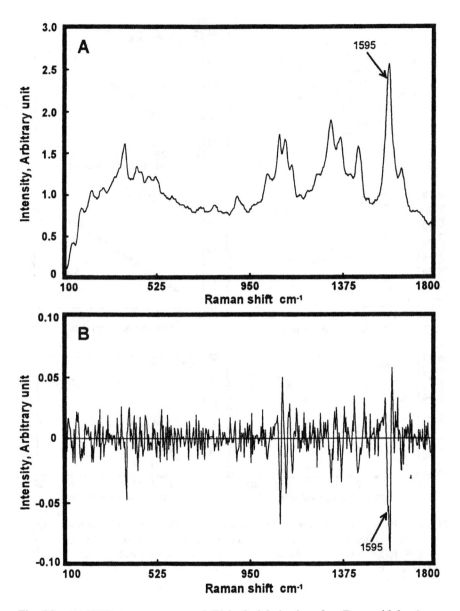

**Fig. 26** (A) FT-Raman spectrum and (B) its 2nd derivative of an *E. camaldulensis* sample. (Reproduced from Ref. 39 with permission. Copyright 1997, Marcel Dekker, Inc.)

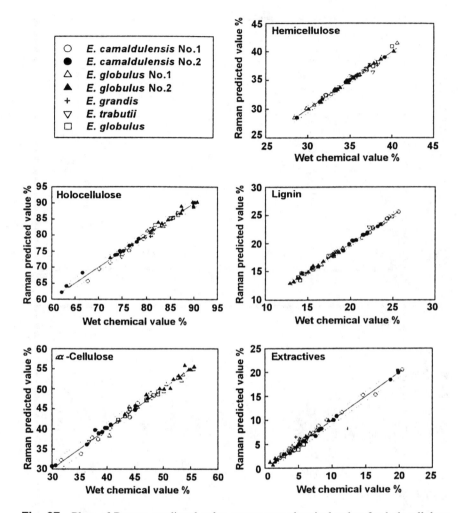

**Fig. 27** Plots of Raman predicted value versus wet chemical value for holocellulose, α-cellulose, hemicellulose, lignin, and extractives. The 95% confidence contours are represented by dashed lines. (Reproduced from Ref. 39 with permission. Copyright 1997, Marcel Dekker, Inc.)

**Table 6**   Results for Calibration

| Trait | r | SEP | Number of PCs used |
|---|---|---|---|
| Holocellulose | 0.994 | 0.760 | 5 |
| α-Cellulose | 0.993 | 0.762 | 5 |
| Hemicellulose | 0.994 | 0.309 | 6 |
| Lignin | 0.998 | 0.239 | 6 |
| Extractives | 0.994 | 0.495 | 5 |
| Alkali extractives | 0.994 | 0.334 | 6 |
| Total extractives | 0.997 | 0.344 | 6 |
| EF holocellulose | 0.996 | 0.384 | 6 |
| EF α-cellulose | 0.993 | 0.695 | 5 |
| EF hemicellulose | 0.994 | 0.462 | 6 |
| EF lignin | 0.994 | 0.448 | 5 |

EF: extractives-free; $r$: correlation coefficient; SEP: standard error of prediction; PC: principal component.
*Source*: Ref. 39.

predicted values for all traits. All the traits except cellulose and lignin had not been calibrated even by NIR, and hence the results obtained were quite valuable. Considering that wet chemical values had less than 2 points standard deviation by a small-scale method, the achieved level of calibration for each trait was highly significant, because the standard error of prediction (SEP) was less than 0.8 points (39).

Moreover, the samples employed for the calibration had important implications, since each calibration model was developed by use of four different *Eucalyptus* species, including samples taken of different ages from various parts within tree stems, and a sample of *E. grandis* chips of unknown age. In addition, it was not always easy to obtain good calibrations for the different species by Raman spectroscopy because of the baseline change in the spectra, due to the difference in florescence intensity of colored materials. In fact, some parts of the *E. camaldulensis* Nos. 1 and 2 and the *E. trabutii* samples were colored red due to the heartwood formation. The study overcame the difficulty by use of the second derivative and MSC (39).

### 3.2.6   Raman and Principal Component Analysis Study of Cotton Cellulose Fibers

Near-infrared excited FT-Raman spectroscopy and PCA study was carried out for seven types of raw naturally colored and white cotton cellulose fibers (54).

The plot of PCA scores showed that PC2 discriminates the spectra of the naturally colored cotton fibers and the white fibers. It was also possible to differentiate some of the different types of white cotton fibers on this PC, and the PC2 scores seemed to be related to the strength of the fibers.

Figure 28 shows FT-Raman spectra of seven cotton fibers in the 1800–800-cm$^{-1}$ region (54). The spectra in Fig. 28 are all similar to one another. Cotton

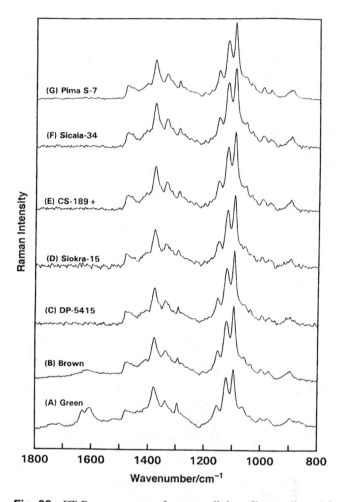

**Fig. 28** FT-Raman spectra of cotton cellulose fibers collected from New South Wales, Australia. Colored cotton fibers: (A) green and (B) brown. White cotton fibers: (C) DP-5415, (D) Siokra-15, (E) CS-189$^+$, (F) Sicala-34, and (G) Pima S-7. (Reproduced from Ref. 54 with permission. Copyright 1998, Royal Society of Chemistry.)

consists of linearly linked cellulose units, each unit containing two glucose molecules joined by an oxygen bridge. Thus, the spectra in Fig. 28 all reflect the common major chemical component, i.e., cellulose. The characteristic features in these spectra are bands due to the vibrations of the β-1,4-glycosidic ring linkages between the D-glucose units in cellulose; the bands at 1160, 1122, and 1096 cm$^{-1}$ are ascribed to the C—C ring asymmetric stretching, C—O—C glycoside link symmetric stretching, and C—O—C glycoside link asymmetric stretching modes, respectively. On the other hand, the most significant differences in Fig. 28 are the bands near 1720, 1625, and 1605 cm$^{-1}$, which appear only in the spectra of the green and brown cottons. Thus, spectroscopic differences between colored and white cottons are readily apparent. They may reflect changes in environmental conditions and the oxidative degradation of the cotton fibers during growth.

Principal component analysis was attempted for the data matrix of the FT-Raman spectra from seven types of natural cotton-cellulose fibers with 21 objects and 251 wavenumber variables. Two PCs were sufficient to account for 93.9% of the variance (PC1 = 50.6%, PC2 = 43.3%). Figure 29(a) shows a plot of full-scale representation of PC1 versus PC2 scores for the 21 FT-Raman spectra (54). It can be seen from Fig. 29(a) that the FT-Raman spectra fall into three distinct clusters, corresponding to the three colors: green, brown, and white. Both green and brown have negative scores on PC2, whereas white cottons have positive scores on this PC. An expanded feature of Fig. 29(a) is depicted in Fig. 29(b), which develops five clusters representing five white cotton fibers. PC1 separates Pima S-7 and Sicala-23 from CS-189$^+$, Siokra-15, and DP-5415; PC2 clearly differentiates the Pima spectra from the other groups, but there is no similar definite separation of the CS-189$^+$, Siokra-15, and DP-5415 groups. The spectra of these fiber types have similar scores to that of the Sicala cluster, indicating that PC2 distinguishes the samples on the basis of fiber type, i.e., Pima, Uplands (white), and Uplands (colored). A close examination of the Uplands (white) cluster reveals that there is a trend of positive scores on PC2 for the different fibers in the order DP-5415 < Siokra-15 < CS-189$^+$ < Sicala-23. Therefore, the combination of the effects of PC1 and PC2 provides a trend, which commences with the Pima S-7 cluster with high PC2 scores and is followed, in approximate order, by the clusters corresponding to Sicala-23, CS-189$^+$, Siokra-15, DP-5415, brown and green cotton fibers.

Figure 30 presents the relationship between the PC2 scores and the corresponding cotton strength for various cotton fibers (54). The PC2 scores increase with the increase in strength, suggesting that cotton fibers are separated according to their mechanical, physical, and structural properties, such as crystallinity, shape, fineness, and hydrogen bonding. The analysis of PC1 and PC2 loadings plots indicates that the difference between the colored and white cottons appears to be from the absorbed water, minor structural changes, and the relative differ-

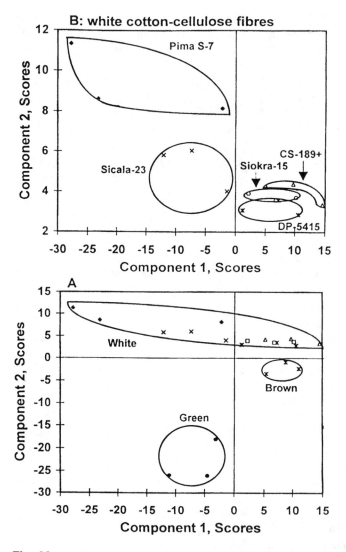

**Fig. 29** (a) Plot of full-scale representation of PC1 versus PC2 scores for the 21 FT-Raman spectra in the 1800–800-cm$^{-1}$ region of the different kinds of cotton fibers. (b) Expanded feature of (a). The curvilinear shapes denoting the cultures were drawn subjectively for clarity. (Reproduced from Ref. 54 with permission. Copyright 1998, Royal Society of Chemistry.)

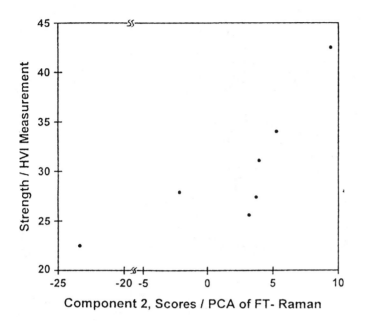

Component 2, Scores / PCA of FT- Raman

**Fig. 30**  Relationship between PC2 scores and strength property in various fibers. (Reproduced from Ref. 54 with permission. Copyright 1998, Royal Society of Chemistry.)

ence in crystallinity in the various cotton fibers. The foregoing study suggests that it may be possible to predict various physical properties of cotton fibers if sufficiently large calibration sets and appropriate validation spectra are available.

### 3.2.7  Raman Imaging Combined with Multivariate Data-Processing Techniques

Raman imaging such as hyperspectral Raman line imaging, combined with multivariate data-processing techniques, offers an excellent means of examining the chemistry and structure of various complicated chemical and biomedical systems (63–65). Here, hyperspectral Raman line imaging of an aluminosilicate glass reported by Jestel et al. (64) is described as an interesting example. The aluminosilicate glass is a model for glass formulations used as dental restorations. Hyperspectral Raman line imaging is well suited for the examination of materials with complicated spectra that might have small shifts of badly overlapping bands over a wide spectral window. In univariate Raman imaging, image contrast originates from the integrated intensity of particular vibrational bands. However, with multivariate analysis, an image (score image) is a map of the presence of an entire spectral signature (factor).

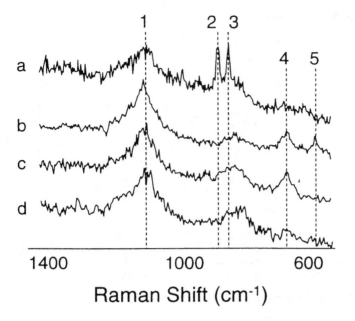

**Fig. 31**   Four raw spectra (a–d) from the over 30,000 spectra of a glass image data set. Labels 1–5 indicate changing spectral regions. (Reproduced from Ref. 64 with permission. Copyright 1998, Society for Applied Spectroscopy.)

The data set obtained for the hyperspectral Raman line imaging of the glass consisted of more than 30,000 spectra. Four spectra from them are shown in Fig. 31 (64). Band 1 changes in width and shape and slightly in frequency, while bands 2 and 3 are well defined only in spectrum a. Band 4 is present in spectra b–c and band 5 appears only in spectrum b. The spectra in Fig. 31 reveal that it is difficult to reach meaningful conclusions about the glass structure from simple band comparisons.

All the spectral data were subjected to FA, and nine factors were needed to reproduce 99.99% of the original data. Figure 32 depicts factors (A–I) recovered from a hyperspectral Raman line image of glass sample (64). The corresponding score images are presented in Fig. 33 (64). Factors G, H, and I represent different convolutions of background, offset, and noise. The other six factors represent Raman spectra of different bonding environments of the silicate tetrahedron. Factors A, B, and C, which show sharp Raman bands, are ascribed to a fully polymerized silica network, with a silicate tetrahedron with one nonbridging oxygen, and with an alumina-related inclusion or a silicate tetrahedron with two nonbridging oxygen. Factors D, E, and F, showing broad Raman bands, indicate

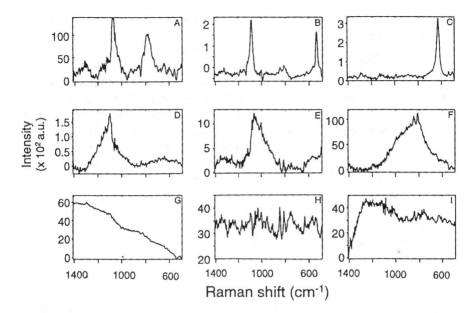

**Fig. 32** Factors (A–I) recovered from a hyperspectral Raman line image of glass sample. (Reproduced from Ref. 64 with permission. Copyright 1998, Society for Applied Spectroscopy.)

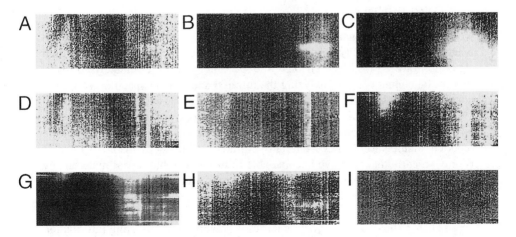

**Fig. 33** Score images (A–I) constructed from factors shown in Fig. 32. (Reproduced from Ref. 64 with permission. Copyright 1998, Society for Applied Spectroscopy.)

continua of slightly different bonding environments of silicate tetrahedra with 0–4 nonbridging oxygen.

Here, let us discuss only factor A. A band at 782 cm$^{-1}$ in factor A corresponds to the 800-cm$^{-1}$ band of vitreous silica, which has been attributed to several vibrations, including the symmetric motion of a silicon atom relative to the unmoving oxygen atoms of a fully bonded tetrahedral cage and the bending mode of the silica network. The low-frequency shift of this band suggests the decrease in the mole percent of silica. The 1064-cm$^{-1}$ band has been assigned to several related motions, including an antisymmetric Si—O—Si mode, the asymmetric stretching of two silica tetrahedra connected by a bridging oxygen, the symmetric stretching of a $(Si_2O_5)^{2-}$ unit, and general asymmetric Si—O stretching in the network. Both bands at 782 and 1064 cm$^{-1}$ are likely due to the same fully polymerized silica network.

Hyperspectral Raman line imaging, combined with multivariate statistical processing techniques, showed that the glass sample examined was not homogeneous and had a few regions with distinct heterogeneities. The score images revealed coexisting and intermingling glass networks with different degrees of polymerization.

Pezzuti and Morris (65) applied hyperspectral Raman imaging to investigate the distributions of chemical species within trabecular bone samples. They used multidimensional factor analysis to identify and map the phosphate salts present in the bone sections and to resolve the phosphate spectra from the overlapping bands of the mounting media universally employed in light microscopy of bone sections.

## 4 CONCLUSION

Nowadays, hardware and software of vibrational spectroscopy are well developed. The reproducibility of infrared and Raman spectra, combined with the high signal-to-noise ratios, has enabled the development of computerized dataprocessing techniques for analyzing the infrared and Raman spectra. We have a number of means for selecting useful information from a sea of spectral data. In 1911 Coblenz (67) had already suggested that biological materials can be analyzed by infrared spectroscopy. In the 1950s and '60s, a number of pioneers reported attempts to identify bacteria by infrared spectroscopy (49). However, it is only recently that we have become able to investigate biological materials by use of chemometrics and vibrational spectroscopy, because the combined method requires both sophisticated instrumentation and efficient computer hardware and software.

There is no doubt that chemometrics is powerful for extracting information from the infrared and Raman spectra. However, if chemometrics methods are applied to the spectra without an understanding or appreciation of their capabilities, assumptions, and limitations, these methods are quite capable of "magically" producing errors or meaningless results. Thus, it must be kept in mind that the traditional spectral analyses, such as band assignments, are always important even for chemometric analysis. Careful experimental design is also very important for the statistical data analyses.

In the future, chemometric analysis of infrared and Raman data of biological materials will become more and more popular. Applications of vibrational spectroscopy-multivariate statistical methods to real-life problems such as medical diagnosis and environmental analysis are very important and promising. Of course, NIR spectroscopy must also be involved in such studies. For the spectral analysis of real-world samples, ANN should be used more positively because, generally speaking, ANN is more suitable for those samples, such as medical samples, which often lack data consistency. Imaging may be a promising field for chemometrics and vibrational spectroscopy, as shown in Sec. 3.2.7.

Chemometric methods, such as multivariate analysis, should also be employed for spectral analysis of basic research. In order to gain more solid information from chemometric analysis, a comparison or combination of chemometrics and 2D correlation spectroscopy may be useful, because these methods are often complementary; the former highlights significant bands for the analysis, and the latter provides correlations among the bands.

## REFERENCES

1. PR Griffiths, JA de Haseth. Fourier Transform Infrared Spectroscopy. New York: Wiley, 1986.
2. JR Ferraro, K Krishnan, eds. Practical Fourier Transform Spectroscopy: Industrial and Laboratory Chemical Analysis. New York: Academic Press, 1990.
3. MW Mackenzie, ed. Advances in Applied Fourier Transform Infrared Spectroscopy. Chichester, UK: Wiley, 1988.
4. AA Christy, Y Ozaki, VG Gregoriou. Modern Infrared Spectroscopy: Principles and Applications. Amsterdam: Elsevier, in press.
5. JR Durig, ed. Vibrational Spectra and Structure. Vol. 1. Amsterdam: Elsevier, 1973.
6. RJH Clark, RE Hester, eds. Advances in Infrared and Raman Spectroscopy. Chichester, UK: Wiley, 1975.
7. HA Szymanski, ed. Raman Spectroscopy: Theory and Practice. Vols. 1, 2. New York: Plenum Press, 1967, 1970.
8. DA Long. Raman Spectroscopy. New York: McGraw-Hill, 1977.
9. JG Grasselli, BJ Bulkin, eds. Analytical Raman Spectroscopy. New York: Wiley, 1991.
10. JR Ferraro, K Nakamoto. Introductory Raman Spectroscopy. New York: Academic Press, 1994.
11. I Noda. Appl Spectrosc 44:550–561, 1990.
12. I Noda. Appl Spectrosc 47:1329–1336, 1993.

13. ER Malinowski. Factor Analysis in Chemistry. 2nd ed. New York: Wiley, 1989.
14. H Martens, T Neas. Multivariate Calibration. Chichester, UK: Wiley, 1989.
15. PL Massart, BGM Vandegiste, SN Deming, Y Michotte, L Kaufman. Chemometrics, a Text Book. Amsterdam: Elsevier, 1988.
16. E Morgan. Chemometrics: Experimental Design. Chichester, UK: Wiley, 1987.
17. K Esbensen, S Schonkopf, T Midtgaard. Multivariate Analysis in Practice. Trondheim, Norway: CAMO, 1994.
18. P Williams, K Norris, eds. Near-Infrared Technology in the Agricultural and Food Industries. 2nd ed. St. Paul, MN: American Association of Cereal Chemists, 1990.
19. DM Haaland. In: JR Ferraro, K Krishnan, eds. Practical Fourier Transform Infrared Spectroscopy. New York: Academic Press, 1990, pp 436–468.
20. HJ Kisner, CW Brownm, GJ Kavarnos. Anal Chem 54:1479–1485, 1982.
21. LL Tyson, YC Ling, CK Mann. Appl Spectrosc 38:663–668, 1984.
22. GL McClure. In: HA Willis, JH van der Maas, RGJ Miller, eds. Laboratory Methods in Vibrational Spectroscopy. 3rd ed. Chichester, UK: Wiley, 1987, pp 145–202.
23. DM Haaland, EV Thomas. Anal Chem 60:1193–1202, 1998.
24. HJ Kisner, CW Brownm, GJ Kavarnos. Anal Chem 55:1703–1709, 1983.
25. M Terrien, M Laufleur, M Pezolet. Proc Soc Photo-Opt Instrum Eng 553:173–176, 1985.
26. GL McClure, PB Roush, JF Williams, CA Lehman. ASTM Spec Tech Publ 934: 131–134, 1987.
27. MR Nyden, GP Forney, K Chittur. Appl Spectrosc 42:588–594, 1988.
28. D Naumann, V Fijala, H Labischinski, P Giesbrecht. J Mol Struct 174:165–170, 1988.
29. D Naumann, S Sällström-Baum, H Labischinski. In: A Bertoluzza, D Fagnano, P Monti, eds. Spectroscopy of Biological Molecules: State of the Art. Bologna: Società Editrice Esculapio, 1989, pp 475–479.
30. HM Heise, R Marbach, G Janatsch, JD Kruse-Jarres. Anal Chem 61:2009–2015, 1989.
31. IW Levin, EN Lewis. Anal Chem 62:1101A–1110A, 1990.
32. B Schrader. In: JR Ferraro, K Krishnan, eds. Practical Fourier Transform Infrared Spectroscopy. San Diego, CA: Academic Press, 1990, pp 168–202.
33. PJ Hendra, CH Johes, G Warnes. Fourier Transform Raman Spectroscopy: Instrumentation and Chemical Application. Chichester, UK: Ellis Horwood, 1991.
34. HM Heise. Horm Metab Reb 28:527–534, 1996.
35. HM Heise, A Bittner, T Koschinsky, FA Gries. Fresenius J Anal Chem 359:83–87, 1997.
36. AJ Berger, Y Wang, DM Sammeth, I Itzkan, K Kneipp, MS Feld. Appl Spectrosc 49:1164–1169, 1995.
37. AJ Berger, Y Wang, MS Feld. Appl Optics 35:209–212, 1996.
38. DD Archibald, SE Kays, DS Himmelsbach, FE Barton II. Appl Spectrosc 52:22–31, 1998.
39. T Ona, T Sonoda, K Ito, M Shibata, T Kato, Y Ootake. J Wood Chem Tech 17: 399–417, 1997.
40. HM Heise, A Bittner. Fresenius J Anal Chem 359:93–99, 1997.

41. HH Mantsch, D Chapman, eds. Infrared Spectroscopy of Biomolecules. New York: Wiley, 1996.
42. M Romeo, F Burden, M Quinn, B Wood, D McNaughton. Cell Mol Biol 44:179–187, 1998.
43. PTT Wong, RK Wong, MFK Fung. Appl Spectrosc 47:1058–1063, 1993.
44. H Mantsch, M Jackson. J Mol Struct 347:187–206, 1995.
45. BR Wood, MA Quinn, FR Burden, D McNaughton. Biospectroscopy 2:143–153, 1996.
46. M Shimoyama, H Maeda, H Sato, T Ninomiya, Y Ozaki. Appl Spectrosc 51:1154–1158, 1997.
47. Y Ootake, S Kokot. J NIR Spectrosc 6:241–250, 1998.
48. Y Ootake, S Kokot. J NIR Spectrosc 6:251–258, 1998.
49. D Naumann, CP Schultz, D Helm. In: HH Mantsch, D Chapman, eds. Infrared Spectroscopy of Biomolecules. New York: Wiley, 1996, pp 279–310.
50. A Fehrmann, M Franz, A Hoffmann, L Rudzik, E Wust. J AOAC Int 78:1537–1542, 1995.
51. E Laloum, NQ Dao, M Daudon. Appl Spectrosc 52:1210–1221, 1998.
52. E Wentrup-Byrne, L Rintoul, JM Gentner, JL Smith, PM Fredericks. Mikrochim Acta 14:615–616, 1997.
53. L Rintoul, H Panayiotou, S Kokot, G Geouge, G Cash, R Frost, T Bui, P Fredericks. Analyst 123:571–577, 1998.
54. Y Liu, S Kokot, TJ Sambi. Analyst 123:633–636, 1998.
55. F Doussear, M Pézolet. Biochemistry 29:8771–8779, 1990.
56. DC Lee, PI Haris, D Chapman, RC Mitchell. Biochemistry 29:9185–9193, 1990.
57. P Pancoska, SC Yasui, TA Keiderling. Biochemistry 30:5089–5103, 1991.
58. P Pancoska, TA Keiderling. Biochemistry 30:6885–6895, 1991.
59. S Wi, P Pancoska, TA Keiderling. Biospectroscopy 4:93–106, 1998.
60. T Fujii, Y Miyahara. Appl Spectrosc 52:128–132, 1998.
61. E Li-Chan. Trends Food Sci 5:3–11, 1994.
62. M Jackson, HH Mantsch. In: HH Mantsch, D Chapman, eds. Infrared Spectroscopy of Biomolecules. New York: Wiley, 1996, pp 311–340.
63. CA Hayden, MD Morris. Appl Spectrosc 50:708–714, 1996.
64. NL Jestel, JM Shaver, MD Morris. Appl Spectrosc 52:64–69, 1998.
65. JA Pezzuti, MD Morris. Abstracts of Pittcon '98, 1998, pp 486–487.
66. H Torii, M Tasumi. In: HH Mantsch, D Chapman, eds. Infrared Spectroscopy of Biomolecules. New York: Wiley, 1996, pp 1–18.
67. WW Coblenz. Bull Natl Bur Stand (US) 7:619–663, 1911.

# Subject Index

α-Amylase, 297
AATase (*see also* Aspartate), 484
Aa-tRNA (*see also* Aminoacyl), 506
Ab initio 41, 478, 493, 509
Absolute configuration, determination of, 24
Absorbance values, baseline-corrected, 101
Absorption, time-dependent, 177
Acanthopleura hirtosa, 447
α-Cellulose, 551
Acetobacter, 344
Acetylcholine, 415
Acetylcholinesterase (AChE), 161
n-Acetylglucosamine, 326
Acholeplasma laidlawii, 361
Acid:
  amino, 18, 27, 42, 325
  arachidic (ArAc), 163, 175
  fatty, 116, 442
    deuterated, 361
  γ-aminobutyric, 415
  glucoronic, 326

[Acid]
  glutamic, 325
  nucleic, 4, 6, 18, 45, 238, 259, 324, 328, 379
  poly-β-hydroxy fatty, 344, 374
  polyhydroxybutyric, 344, 374
  teichoic, 325
  teichuronic, 325
ADA (*see also* Adenosine deaminase), 497
Adaline, 383
Adaptivity, 381
Adenine, 334, 350
Adenosine deaminase (ADA), 497
Advantage:
  Fellgett or multiplex, 102, 194, 196
  Jacquinot or throughput, 102, 194, 196
  wavenumber accuracy, 196
A/G ratio, 287
Albumin, 67, 289
Aleurone cell, 122